ATMOSPHERIC SULFUR DEPOSITION
Environmental Impact
and
Health Effects

Edited by

David S. Shriner
Advanced Fossil Energy Program
Environmental Sciences Division
Oak Ridge National Laboratory
Oak Ridge, Tennessee

Chester R. Richmond
Associate Director
Biomedical and Environmental Sciences
Oak Ridge National Laboratory
Oak Ridge, Tennessee

Steven E. Lindberg
Earth Sciences Section
Environmental Sciences Division
Oak Ridge National Laboratory
Oak Ridge, Tennessee

Proceedings of the Second Life Sciences Symposium,
Potential Environmental and Health Consequences of
Atmospheric Sulfur Deposition
Gatlinburg, Tennessee
October 14–18, 1979

Sponsored by
Oak Ridge National Laboratory
Department of Energy
U.S. Environmental Protection Agency
Tennessee Valley Authority

ANN ARBOR SCIENCE
PUBLISHERS INC / THE BUTTERWORTH GROUP

Copyright © 1980 by Ann Arbor Science Publishers, Inc.
230 Collingwood, P. O. Box 1425, Ann Arbor, Michigan 48106

Library of Congress Catalog Number 80-68667
ISBN 0-250-40380-3

FOREWORD AND ACKNOWLEDGEMENTS

The objectives of the Second Life Sciences Symposium on "Potential Environmental and Health Consequences of Atmospheric Sulfur Deposition" were (1) to improve communication between scientists and decision-makers from a broad range of disciplines whose common focus is the atmospheric deposition of sulfur; (2) to present and discuss new research findings in the context of state of the art reviews of the subject area; and (3) through publication of the proceedings, to provide a timely compilation of research data and conclusions, areas of needed research, recent advances in problem areas, and experimental methodology and rationale.

We wish to express our particular thanks to a number of people. Special credit is due to the Session Chairmen and Planning Committee for their year-long participation in the planning and organization of the conference. The session chairmen also reviewed and assisted in editing the papers. The Conference Sessions and Session Chairmen were: Natural and Anthropogenic Sources—J. F. Meagher, Tennessee Valley Authority; Human Health Effects—P. J. Walsh and D. C. Parzyck, Oak Ridge National Laboratory; Emission Characteristics and Atmospheric Transformations—J. F. Meagher; Effluent Delivery and Air Mass/Landscape Interactions—S. E. Lindberg, Oak Ridge National Laboratory, and R. P. Hosker, National Oceanic and Atmospheric Administration; Process Level Effects of Sulfur Deposition—D. W. Johnson, Oak Ridge National Laboratory, and J. M. Skelly, Virginia Polytechnic Institute and State University; Ecosystem Level Effects of Sulfur Deposition—J. M. Kelly, Tennessee Valley Authority; Regional Scale Studies of Atmospheric Deposition Effects—J. N. Galloway, University of Virginia. We especially thank E. B. Cowling, North Carolina State University, for his skillful moderation of the keynote addresses.

We are very grateful to B. M. Roberts and B. S. Reesor, Oak Ridge National Laboratory, for their skillful and tireless management of conference details, and to L. Bradley, P. Ellis, S. Ward and P. Wright for their flawless attention to the mechanics of the symposium.

Requests for assistance with technical documents and related information during the symposium were provided through an information center staffed by L. W. Rickert, E. B. Lewis and J. Barker. The contributions of R. P. Metcalf and T. W. Robinson ensured the timeliness of the editorial production of the proceedings and is gratefully acknowledged.

The chapters in this proceedings have been organized differently, in some cases, from their order of presentation, in order to facilitate the flow of topics and ideas. We have attempted to optimize the utility of the proceedings through this organization to integrate the diverse disciplines and subjects addressed by the symposium. Hopefully, the mix of review papers and reports of new work have resulted in the best reflection of the current state of atmospheric sulfur deposition research.

Preparation of the proceedings has been an exciting and rewarding effort. We hope that this book can serve both those researchers and decision-makers to whom this topic is currently of vital concern.

David S. Shriner

Chester R. Richmond

Steven E. Lindberg

Oak Ridge National Laboratory
Proceedings Editors

INTRODUCTION

Chester R. Richmond*

This volume is the proceedings of the Oak Ridge National Laboratory's second annual Life Sciences Symposium, the subject of which is the Potential Environmental and Health Effects of Atmospheric Sulfur Deposition. This symposium is sponsored by the Assistant Secretary for Environment in the Department of Energy, the Environmental Protection Agency, and the Tennessee Valley Authority.

Last year marked the first symposium in this series. We gathered together people from industry, several levels of government, universities, and various research and development laboratories to discuss the potential health and environmental effects of synthetic fossil fuels. That topic has become extremely timely in view of the numerous actions recently initiated to make our nation less dependent on imported liquid fuels. These meetings are interdisciplinary in character and designed to view major problem areas from numerous perspectives. We want to avoid the usual meeting format where biologists talk to other biologists, chemists to chemists, or economists to economists. We want to bring the holistic approach to analyzing major problems in the life sciences that now confront our nation.

Our President recently delivered his second major Environmental Message—the first was in May 1977. President Carter called for a major increase in research on the acid rain problem. The proposed program will cover ten years, and the $10 million for the first year will double the current level of effort. President Carter called for the establishment of a ten-year comprehensive federal acid rain assessment program to be planned and managed by a standing Acid Rain Coordinating Committee. The assessment program will include applied and basic research on acid rain effects, trends monitoring, transport and fate of pollutants, and control measures. The committee will also establish links with industry to promote cooperative research wherever appropriate. This committee will also play a role in future research cooperation with Canada, Mexico, other nations, and international organizations. The committee will prepare a comprehensive ten-year plan for review by the end of the year. Many of us here today look forward to reviewing this plan.

I recently participated in the EPA Section II Environmental Evaluation hearings mandated by the Federal Nonnuclear Energy Research and Development Act (Public Law 93-577). Most participants agreed that environmental problems could be classified as:

---

*Associate Director, Biomedical and Environmental Sciences, Oak Ridge National Laboratory, Oak Ridge, Tennessee 37830.

those that are site-specific,

those that are regional or cumulative, and

those that are global in nature.

Sulfur deposition and its attendant effects are an example of a phenomenon that is regional and cumulative and, in one sense, becoming global in nature. In addition, there is lack of agreement and, in fact, some controversy concerning some of the effects of sulfur compounds on man. It is clear that much remains to be done. Perhaps this symposium will allow us to sharpen our focus on some of the problems still to be resolved.

This problem is truly international in scope, since some countries unwillingly import sulfur from others. I look forward to several exciting days during which we will examine this important topic of sulfur deposition. On behalf of the Oak Ridge National Laboratory, I welcome you to this symposium and to the Great Smoky Mountains.

CONTENTS

SECTION I
ENVIRONMENTAL VERSUS CONTROL COSTS

1. Conference Perspective and Challenge, *D. S. Shriner* 3
2. Environmental, Economic, and Energy Balancing in Fossil Energy Regulation, *David W. Tunderman* 5
3. Sulfur Dioxide Emission Controls—What Are the Costs? *Charles H. Goodman* 13
4. Environmental Versus Emission Control Costs—A Legislative Perspective, *Thomas H. Moss* 21

SECTION II
NATURAL AND ANTHROPOGENIC SOURCES

5. Overview, *James F. Meagher* 33
6. Estimates of Natural Sulfur Source Strengths, *D. F. Adams, S. O. Farwell, M. R. Pack and E. Robinson* 35
7. Direct Measurements of Emission Rates of Some Atmospheric Biogenic Sulfur Compounds and Their Possible Importance to the Stratospheric Aerosol Layer, *Viney P. Aneja* 47
8. Impact of Recent Measurements of OCS, CS₂, and SO₂ In Background Air on the Global Sulfur Cycle, *Alan R. Bandy and Peter J. Maroulis* 55

SECTION III
HUMAN HEALTH EFFECTS

9. Overview, *P. J. Walsh and D. C. Parzyck* 67
10. Laboratory Studies of Biological Effects of Sulfur Oxides *W. E. Dalbey* 69
11. Short-Term Respiratory Effects of Sulfur-Containing Pollutant Mixtures: Some Recent Findings from Controlled Clinical Studies *Jack D. Hackney, William S. Linn, Michael P. Jones, Ronald M. Bailey, Dennis R. Julin and Michael T. Kleinman* 77
12. Health Significance of Exposures to Sulfur Oxide Air Pollutants *Morton Lippmann* 85
13. The Six-City Study: A Progress Report, *B. G. Ferris, Jr., F. E. Speizer, Y. M. M. Bishop, J. D. Spengler and J. H. Ware* 99

14. An Overview of Epidemiologic Evidence on Effects of Atmospheric Sulfur, *Sati Mazumdar and Carol Redmond* 109

SECTION IV
ATMOSPHERIC TRANSFORMATION

15. Overview, *James F. Meagher* 121
16. Primary Sulfate Emissions From Stationary Industrial Sources *John S. Nader* .... 123 ✔
17. Atmospheric Oxidation of Sulfur Dioxides, *L. Newman* 131
18. The Fate of Sulfuric Acid Aerosol in the Atmosphere, *Cyrill Brosset* ... 145
19. The Dynamics of Secondary Sulfur Aerosols, *Peter H. McMurry* 153
20. Monte Carlo Simulation of Daily Sulfate Distribution in the Eastern United States: Comparison with SURE Data and Visibility Observations, *D. E. Patterson, R. B. Husar and C. Hakkarinen* 163

SECTION V
AIR MASS/LANDSCAPE INTERACTIONS

21. Overview, *Steven E. Lindberg and R. P. Hosker* 181
22. Dry Deposition of Sulfur Dioxide, *A. C. Chamberlain* 185
23. Turbulent Transfer Processes to a Surface and Interaction with Vegetation, *B. B. Hicks and M. L. Wesely* 199
24. Experimental Techniques for Dry-Deposition Measurements *J. G. Droppo* 209
25. Model Predictions and a Summary of Dry Deposition Velocity Data *George A. Sehmel* 223
26. Transfer and Deposition of Particles to Water Surfaces, *B. B. Hicks and R. M. Williams* 237
27. An Urban Influence on Deposition of Sulfate and Soluble Metal in Summer Rains, *Donald F. Gatz* 245
28. Overview of Wet Deposition and Scavenging, *M. Terry Dana* 263
29. A Review of Gaussian Diffusion-Deposition Models *Thomas W. Horst* 275

SECTION VI
PROCESS LEVEL EFFECTS

30. Overview, *Dale W. Johnson* 287
31. Sulfur Accumulation by Plants; The Role of Gaseous Sulfur in Crop Nutrition, *JC Noggle* 289
32. Effects of Simulated Acid Rain on Growth and Yield of Soybeans and Pinto Beans, *Lance S. Evans and Keith F. Lewin* .... 299 ✔
33. Vegetation: Effects of Sulfur Deposition by Dry-Fall Processes *S. V. Krupa, B. I. Cervone, J. L. Brechtwold and J. L. Wolf* 309
34. Sulfur Nutrition of Forests, *John Turner and Marcia J. Lambert* 321
35. Consequences of Sulfuric Acid Inputs to a Forest Soil *Cristopher S. Cronan* 335

36. Processes Limiting Fish Populations in Acidified Lakes
Carl L. Schofield... 345

37. Effects of Acidity on Primary Productivity in Lakes: Phytoplankton
George R. Hendrey 357

SECTION VII
ECOSYSTEM LEVEL EFFECTS

38. Overview, J. M. Kelly 375

39. Impact of Atmospheric Sulfur Deposition on Agroecosystems
U. S. Jones and E. L. Suarez 377

40. Impact of Atmospheric Sulfur Deposition on Forest Ecosystems
Gunnar Abrahamsen.. 397

41. Impact of Atmospheric Sulfur Deposition on Grassland Ecosystems
W. K. Lauenroth and J. E. Heasley 417

42. Impact of Sulfur Deposition on the Quality of Water from Forested
Watersheds, Gray S. Henderson, Wayne T. Swank and James W.
Hornbeck 431

43. Ecological Effects of Whole-Stream Acidification, Ronald J. Hall and
Gene E. Likens... 443

44. Ecological Effects of Experimental Whole-Lake Acidification
D. W. Schindler 453

SECTION VIII
REGIONAL SCALE STUDIES

45. Overview, James N. Galloway 465

46. National Atmospheric Deposition Program: Analysis of Data from the
First Year, John M. Miller 469

47. Atmosphere-Canopy Interactions of Sulfur in the Southeastern
United States, Geoffrey G. Parker, Steven E. Lindberg and
J. Michael Kelly 477

48. Sensitivity of Soil Regions to Long-Term Acid Precipitation
William D. McFee 495

49. Regional Pattern of Soil Sulfate Accumulation: Relevance to
Ecosystem Sulfur Budgets, Dale W. Johnson, J. W. Hornbeck,
J. M. Kelly, W. T. Swank and D. E. Todd 507

50. Geologic Factors Controlling the Sensitivity of Aquatic Ecosystems to
Acidic Precipitation, Stephen A. Norton 521

51. Implications of Regional-Scale Lake Acidification, D. W. Schindler ... 533

52. A Regional Ecological Assessment Approach to Atmospheric
Deposition: Effects on Soil Systems, Jeffrey M. Klopatek, W. Frank
Harris and Richard J. Olson 539

List of Participants 555

INDEX 561

Shriner **Richmond** **Lindberg**

David S. Shriner is an Environmental Scientist in the Advanced Fossil Energy Program, Environmental Sciences Division of Oak Ridge National Laboratory in Oak Ridge, Tennessee.

Dr. Shriner received a PhD in Plant Pathology from North Carolina State University, an MS in Plant Pathology from Pennsylvania State University and a BS in Forestry from the University of Idaho.

The editor's areas of research concentration in recent years include the effects of wet and dry deposition of atmospheric contaminants in terrestrial ecosystems, and the transport, fate and effects of sulfur in the terrestrial environment.

Author of more than 40 technical publications, Dr. Shriner is a member of the Biological Effects Advisory Board of the National Atmospheric Deposition Program, a member of the Air Pollution Control Association and Associate Editor of that organization's Effects Division Environmental Report. He is also a member of the American Phytopathological Society, the American Association for the Advancement of Science and Sigma Xi.

Chester R. Richmond is Associate Laboratory Director for Biomedical and Environmental Sciences at Oak Ridge National Laboratory, Oak Ridge, Tennessee.

He was previously Alternate Health Division Leader, Los Alamos Scientific Laboratory, Los Alamos, New Mexico, and earlier was with the Medical Research Branch, Division of Biology and Medicine, U.S. Atomic Energy Commission, Washington, DC. He earned his PhD and MS degrees in Biology-Physiology from the University of New Mexico and his BA from New Jersey State College, Montclair.

Dr. Richmond is a Professor, the University of Tennessee—Oak Ridge Graduate School of Biomedical Sciences at Oak Ridge. He has published a number of papers, is a Fellow of the American Association for the Advancement of Science, and a member of several other professional societies reflecting his interests in experimental biology, public health, radiation protection and risk assessment. He was honored by the U.S. Atomic Energy Commission in 1974 with the E. O. Lawrence Award, and in 1976 with the G. Failla Lecture and Award by the Radiation Research Society.

Steven E. Lindberg is a geochemist in the Earth Sciences Section, Environmental Sciences Division, Oak Ridge National Laboratory, Oak Ridge, Tennessee.

Dr. Lindberg received a BS in Chemistry from Duke University which included studies at the Duke Marine Laboratory. He earned his MS in Oceanography while on a research fellowship at Florida State University for work on sediment geochemistry, and his PhD from the same school for research on atmospheric deposition.

He has been Principal Investigator of the Environmental Sciences Division's research project on Atmospheric Deposition in the Forest Environment since 1976. He has published 30 technical papers and reports in the general field of trace element geochemistry, reflecting his interests in precipitation chemistry, dry deposition methodology, air pollutant/forest canopy interactions, power plant emissions and the environmental chemistry of mercury. He is currently a member of the Subcommittee on Data Interpretation of the National Atmospheric Deposition Program.

SECTION I

ENVIRONMENTAL VERSUS EMISSION CONTROL COSTS

Moderator: E. B. Cowling

CHAPTER 1

KEYNOTE ADDRESSES: ENVIRONMENTAL VERSUS EMISSION CONTROL COSTS

Moderator: **E. B. Cowling**
North Carolina State University

CONFERENCE PERSPECTIVE AND CHALLENGE

D. S. Shriner*

This Second Life Sciences Symposium, sponsored by the Oak Ridge National Laboratory, the Department of Energy, the Environmental Protection Agency, and the Tennessee Valley Authority, was organized to bring together as many as possible of the various disciplines and institutions concerned with the environmental and health consequences of atmospheric sulfur deposition. Major emphasis was placed on information exchange in a multidisciplinary setting for the purpose of defining the current state of knowledge. Special consideration was given to present and future levels of fossil energy utilization on regional and national scales.

It is our hope that, by providing the mechanism for free and informal exchange of perspective among scientists concerned with different aspects of the sulfur issue, each of us might become better equipped to deal with our own specific areas of research.

The keynote speakers to the conference were selected for the specific insight that they, as leaders in the arena of environmental policy, might share with the working scientists upon whose data future policy decisions will be made. The challenge of the keynote speakers to the scientists in attendance was clear: Policy decisions must have a basis with data which will permit clear evaluation of the benefits of emission controls in the light of the social cost, or risk, of uncontrolled emissions. As Thomas Moss states, problems such as atmospheric sulfur deposition "may have social implications so profound that only a scientific case of unprecedented credibility and completeness will be able to move the rest of society to undertake the necessary remedies." This challenge is symbolic of the heavy responsibility scientists must bear to do the things they do with "unprecedented credibility." Just as the keynote speakers were asked to provide perspective against which the technical papers could be assimilated, authors of technical papers were asked to present, in some cases, an

*Advanced Fossil Energy Program, Environmental Sciences Division, Oak Ridge National Laboratory, Oak Ridge, Tennessee 37830.

overview of the state of the art in their particular areas of specialization and, in other cases, new, original work which was not yet available to the scientific community. The results, plus topical summaries by the section editors, form the content of this volume.

As I have read the papers in this volume, I have felt sure that they will provide a stimulus for additional research for some time to come. There are significant new additions to our collective knowledge and, certainly, fuel for a continuing discussion and exchange of ideas.

CHAPTER 2

ENVIRONMENTAL, ECONOMIC, AND ENERGY BALANCING IN FOSSIL ENERGY REGULATION

David W. Tundermann*

INTRODUCTION

Thank you very much for inviting me to address this symposium on environmental and health effects of sulfur deposition. I am pleased to have this opportunity to discuss energy, economics, and the environment, an issue not always understood and in fact often misunderstood.

Somehow there has developed the myth that economic growth, wise energy use, and environmental protection are fundamentally at odds. The myth deserves to be debunked.

On the contrary, a healthy environment, a healthy economy, and wise energy use are all necessities. We will not be productive as a nation if we continue our profligately wasteful energy use patterns, high levels of inflation and unemployment, or habits of industrial waste disposal which threaten to poison air, water, or terrestrial resources. The balance we search for is between an environment that is sound enough to support a productive economy and an economy, including its energy consumption patterns, that makes good use of but does not consume its environment.

There is not an either/or relationship among sound economics, wise energy policy, and prudent environmental policy. They must go in tandem. Don't believe for a moment that when industries polluted the Cuyahoga River so thoroughly that it could catch fire, there was no cost to the people of Cleveland, even if the pollution was accomplished at next to no cost by the dumping industries. Don't think there is no cost to the people of Gary, Indiana, when United States Steel pollutes the air. But neither think there is no cost to Gary if U.S. Steel shuts down its plants there. The costs are there either way. We are talking about how they are distributed.

How do we decide how much sulfur dioxide or sulfate or other potentially acidic pollutant is too much? How do we measure those damages? Should we try to value the damages economically, and if so, how should we decide how much damage is too much damage?

These are fundamental issues. They have theoretical solutions but so far no practical ones.

Economic theory tells us we should reduce sulfur-related emissions until the incremental damage from reducing emissions another unit equals the incremental cost of paying for that reduction.

---

*Office of Policy and Planning, Environmental Protection Agency, Washington, DC 20460.

But the world is not so simple. We don't understand which pollutants cause how much damage. We don't understand which classes of sources contribute how much acidic pollution to which areas. We don't understand the health and environmental effects of this acid deposition. And we are hesitant to rely on limited and uncertain data on how much these damages are worth, and to whom.

Since this is a scientific conference, I won't presume to explain the complexities of sulfur dioxide and sulfate formation in combustion processes, sulfur dioxide to sulfate transformation, modeling problems in complex terrain, or methodological problems in the health studies that supported the original SO_2 standards. If you were financial analysts I might try to explain these phenomena. Instead, I'll take notes on these subjects for my next conference.

We environmental regulators are living in an environment in which we are being asked—or told—to be much more rigorous in our analyses and decision-making than earlier in this decade. And deservedly so. Government has historically been pretty sloppy analytically, from the shabby cost-benefit analyses used to support public works projects in the fifties and the sixties to the failures of economists and analysts—despite decades of theoretical discussion—to come up with very good measures or data for benefits of regulation. Analytic concepts such as incremental cost-benefit analysis and cost-effectiveness analysis are important ones, but they remain crude, frequently because of large shortcomings in scientific data and understanding.

Thus I hope that sessions like this, by advancing how much we know and how much we understand, will contribute to wise regulation in the future.

EPA had committed itself to rigorous regulatory analysis in response to pressures from the last administration to estimate the inflationary impact of EPA's regulations. Then came OMB's quality of life review, which forced the Agency even more to articulate the projected impacts of proposed regulations. Thus, by the time President Carter institutionalized regulatory analysis in Executive Order 12044, EPA was well along in developing and using the necessary analytic techniques.

REGULATION ANALYSIS AND DECISION-MAKING

Perhaps the most valuable contribution I can make to your discussions is to review the regulatory analysis and decision-making process for one of our most important regulations this year—the New Source Performance Standards (NSPS) for Utility Boilers. Coincidentally, the analysis for this regulation stands out as about the most rigorous EPA, or anyone else in the government, has done. It also happens to affect primarily the emissions of sulfur dioxide to the atmosphere. A nice coincidence for this conference.

The analytic structure for the NSPS was dictated in part by the Clean Air Act and in part by data limitations.

The Clean Air Act and its legislative history establish a presumption for full scrubbing of emissions from new power plants, regardless of the effect on ambient air quality. Section 111 of the Clean Air Act is designed more as an emissions-reduction provision than as a provision to achieve particular ambient air quality levels. The 1977 amendment reflected a strong push by an alliance of environmentalists and Eastern and Midwestern mining interests. The former were concerned about increasing emissions from Western power plants. The latter were worried about losing markets to Western low-sulfur coal. Together, they pushed through an amendment which requires EPA to set a percent reduction requirement as well as an emissions limitation for emissions from coal-fired plants, regardless of the quality of the coal used.

Section 111 thus reflects two congressional judgments, one environmental and one economic, that heavily influence how one approaches standard-setting. The environmental judgment is that we still do not understand relations between sulfur emissions and their effects well enough to play close to the vest in precisely relating emission levels to ambient standards from sources as big and numerous as coal-fired power plants. The second judgment is that the emissions standard for coal-fired power plants should not cause precipitous shifts in coal markets from west to east. The other major purposes for this section are set out succinctly in the report of the House Commerce Committee, but since they are not so germane as the first two, I won't recite them here.

In effect, the structure of Sect. 111 of the Clean Air Act made emission reductions the relevant regulatory benefit, more than other possible benefit measures such as improved air quality or reduced population exposure to sulfur dioxide.

Of course, even if Congress had written Sect. 111 differently, we would not have known how to analyze a national regulation in terms of its effect on air quality or population exposure in 243 air quality control regions. So Congress made our job easier, thankfully. Congress also made the Regulatory Analysis Review Group unhappy, for some there believed that EPA should have defined the benefits more precisely.

In addition to the impact of the standard on emissions, EPA and other groups, within and without the administration, brought other valid concerns to bear on the regulatory design of the NSPS. The effects of the revised standard on utility fuel choice, especially between coal and oil; on coal markets, especially between Western and Midwestern mines; and on utility investment and annualized costs were all valid concerns. The Regulatory Analysis Review Group sharpened EPA's general concern with cost-effectiveness and pushed hard for marginal analysis between alternative standards.

We structured our regulatory analysis by designing, through contractors, two elaborate linear programming models of the coal and electric utility industries. Given inputs such as alternative New Source Performance Standards, oil prices, delivered coal prices, electrical demand growth, and scrubber costs, the model would estimate 1995 coal and oil use by utilities, sulfur emissions, annualized utility costs, and coal production by region.

An important issue attending the use of a complex model was how much weight to place on its results. Changing an important assumption—such as projected electricity demand—would yield differences in results which overshadowed the differences among alternative standards. This might suggest that the model, despite its complexity, was too crude or simplistic to be relied on or that the differences among alternative standards were too small to be significant. Neither was true.

More important was how sensitive the differences in projected impacts of alternative standards were to changing assumptions in the model. We were less concerned with the sensitivity of absolute results than relative results. After all, EPA was trying to select a standard among competing alternatives, not trying to forecast the future with dependable accuracy.

The more important questions concerning use of the model were deciding whether to accept the behavioral assumptions it made about decision-making by utility executives, deciding whether the sensitivity of relative results to changes in key variables was low enough, and deciding how much weight to place on national and regional projections, as opposed to typical plant-specific ones. These are all judgment calls. The first two were decided at a senior analytic level. The Administrator effectively decided the third one.

After building these tools, the major tasks were defining alternative standards and managing a collaborative and open analysis.

EPA considered basically three alternative standards:

- a full-scrubbing standard, requiring the same percent reduction in sulfur emissions regardless of coal quality,

- a variable standard, which imposed a fairly high minimum percent reduction requirement—70%—which gets stricter as coal quality deteriorates,

- a partial-scrubbing or sliding-scale standard, with a very low minimum percent removal requirement—33%.

We also considered for a time a regional standard, with the percent removal requirement varying for geographic area. We rejected this entire approach for legal and technical reasons and also because few thought it credible to Westerners.

A host of very important additional design issues turned this list of three alternatives into a list of dozens:

- Alternative emissions limits could affect emissions and were thought to affect coal markets. Eastern mining companies believed that low emission ceilings would make high-sulfur coals unmarketable even with 85–90% scrubbing.

- Alternative percent removal requirements affected emissions, control costs, and utility fuel choice between oil and coal. DOE and the President's inflation advisors were thus concerned about this feature of the alternative standards.

- The emissions floor, or the crossover point between the most stringent percent removal requirement and any less stringent one applicable to the use of cleaner coals, would affect emissions and cost effectiveness significantly.

- The minimum percent removal requirement under the variable, partial, and regional options would significantly affect both emissions and costs.

In all, we examined over 14 alternative standards, plus sensitivity analyses on all major assumptions—all this in an effort to identify a set of alternatives which would give the Administrator the opportunity to balance economic, energy, and environmental considerations as efficiently as possible, i.e., without any inadvertent losses in possible emission reductions, control costs, or oil use that were avoidable.

POLICY MANAGEMENT

The managerial task was about as complex as the analytic one, maybe more so. Our goal was to be able to give Douglas Costle—and, as they involved themselves, other high-level administration officials—a common analysis. We wanted the administration's top energy, environment, and economic officials to be able to discuss policy alternatives without their staffs quibbling over projections or assumptions. All too often, the analytic squabbling that often accompanies a major policy debate obscures the key issues, exhausts and bores the decision makers, undermines the analytic effort, and, I believe, frequently leads to a second-rate or worse decision. Thus, achieving an analytic consensus was one of our major goals. And we openly and heavily involved staff from the Department of Energy, the Council of Economic Advisors, the Council on Wage and Price Stability, and the Department of the Interior and the Domestic Policy Staff as interested on every major analytical judgment call—choices of key assumptions and ranges in assumptions and key choices regarding model design. These analytic issues were also put out for public comment and hearing, and we held occasional informal meetings with all parties to go over important analytic judgment calls.

The most significant assumptions—delivered coal prices, including transportation rates, oil prices, and scrubber costs—were closely reviewed and agreed to by all the major participants. The agreed ranges in assumptions included scenarios specified by the utilities and the environmental public interest groups.

A fairly late development with highly significant results was the presentation by EPA's Office of Research and Development of performance and cost data on dry scrubbing. These data showed dry scrubbing to be about one-third cheaper than wet scrubbing at fairly high (e.g., 70%) removal efficiencies. The dry-scrubbing data are what enabled EPA to design a cost-effective alternative with a high minimum removal efficiency.

The projected impacts of the alternative standards appeared small in relation to total projected utility emissions and annualized costs. The emissions difference between the most stringent and the least stringent options was about 200,000 tons—less than 1% of projected 1995 power plant emissions. The annualized cost difference was about $1.7 billion—also less than 1% of projected 1995 costs. The present value of the difference was quite large, however—$17 billion.

On a plant-by-plant or regional basis, the alternative standards produced very different emissions and cost differences, especially in the West. Full scrubbing would yield one-seventh the emissions of partial control on an individual plant and one-third the emissions of variable control.

As you know, the Administrator chose a standard with a 70% minimum scrubbing requirement, a fairly low floor or "crossover point" where the scrubbing requirement would jump up to 90% and the same emissions ceiling as in the previous New Source Performance Standard. The emissions ceiling, 1.2 lb of sulfur per million BTU of heat input (monthly average), is the controlling factor on the highest-sulfur coals, about 5% sulfur. Little of this coal ever gets burned anyway. The 90% removal requirement controls emissions as coal quality improves down to about 2.5% sulfur. As coal quality improves to the point where 90% removal reduces emissions to 0.6 lb per million BTU (monthly average), the percent removal requirement gradually lessens, from the 90% level to no less than 70%. Thus, the 0.6-lb standard is the binding factor from about 2.5% sulfur to about 0.8% sulfur coal. Below that coal quality, the 70% minimum removal requirement determines the emissions.

What you have, in effect, are four closely dovetailed control devices, each designed to control emissions efficiently as coal quality changes. The standard is more cost effective than the more stringent one EPA originally proposed and more cost effective than less stringent ones advocated by the Department of Energy and the utility industry.

POLICY IMPLICATIONS

What are the implications of this process for other sulfur-controlling regulations? You can anticipate a similar concern with cost effectiveness in the review in any revision in the ambient standards for SO_2 and TSP, as well as in eventual regulation to protect visibility under Sec. 169A of the Clean Air Act. But in both cases there are statutory and analytic complexities that change the situation.

In the case of the ambient standards, there exists an active debate among the legal community regarding the extent to which EPA can explicitly consider costs in setting the ambient standards. Some argue that Congress intended the standard-setting to be free of economic considerations, reportedly based on health considerations alone. Others argue that this position is logically absurd, that there is no absolute health threshold, and that any standard reflects a balancing between risks and costs of reducing those risks. By this second view, EPA cannot rationally avoid making risk vs cost comparisons among alternative standards, and the task is to define the most reasonable combination.

With respect to visibility regulations, we know from the start that protection of public health is not an issue. Here we are protecting a value characterized in many different ways—an environmental value; an economic value, especially in the West, where tourism is so important; an aesthetic value; a spiritual value; or all of these. The ambiguity of defining what we are really protecting is going to make the analysis there more difficult and probably in the end more subjective. The best we can do is to try to identify the subjectivities, so that policy-decision makers can see where they are and make different choices if they choose.

I am confident that your work will help us enormously as we approach these coming regulatory tasks. I look forward to the day when we can agree on something as basic as what pollutants should be regulated and then proceed to determine what the controls are.

DISCUSSION

Orie Loucks, Institute of Ecology, Indianapolis: You made a quick passing reference to several alternative areas of emphasis, and the one I'm particularly interested in was local effects vs regional costs. I understood you to say that the administrator opted for an emphasis on consideration of local costs with less consideration for regional, large-area costs. Did I understand that correctly, or could you expand on that?

David Tundermann: No, I think he considered both, placing approximately similar weight on each. One couldn't ignore the national projections, because the cost and oil-consumption projections were the principal foci of the economic advisers in the administration and the Department of Energy, so they played an important role in the administrator's final decision. At the same time, the most dramatic differences in cost and in emissions were seen in the West, where the amendment was really intended to bite. Their fuel choice wasn't really an issue; everyone knew the Western utilities weren't going to burn oil and they weren't going to burn Illinois coal; they were going to burn Western coal, and it was going to be low-sulfur, so the real issue there was directly the cost vs emissions trade-offs. One could look at that either regionally, since the model had a regional forecasting capability, or locally by projecting on a typical plant basis the difference between cost and emissions. The latter was important because that was the way any given environmental group, utility, or permit issuer would see a case coming. Given the sensitivities in the West, particularly around the parks, to emissions from new plants and the opportunity cost you might bear by forgoing emission controls that might have been imposed, it was very useful to look at those plant-by-plant emissions.

Dave Odor, Public Service of Indiana: Last month there was a court case between Alabama Power and the Environmental Protection Agency in which portions of the Prevention of Significant Deterioration (PSD) characteristics of emission standards were remanded by the court. I wonder if you could address to this group how those remands affect the emission regulatory process that you perceive to happen in the future?

David Tundermann: The gentleman is referring to a decision by a federal district court called Alabama Power vs Costle in which that company challenged regulations EPA issued about a year and a half ago implementing those sections of the Clean Air Act amendments which specify the regulatory regime for areas designated as

Prevention of Significant Deterioration areas, that is, those areas where air quality is better than what the federal health standards require. In terms of a connection to the New Source Performance Standard, I think there is none. There are some petitions for reconsideration pending before the agency asking us to reevaluate several portions of the original decision. I won't presume to predict what the agency's responses to those petitions are, but I think they are not dependent on the court's decision in Alabama Power. Two of the more significant regulatory changes that the court mandated, however, are (1) a redefinition of the regulatory threshold, which determines the applicability of that entire regulatory regime to new sources. The EPA had interpreted the statute to say that you're subject to these Prevention of Significant Deterioration regulations if you emit more than 100 tons per year of uncontrolled particulate or SO_2. The utilities argue that to be the wrong benchmark, that the benchmark should be their emissions as controlled by whatever control techniques they had intended to design at the plant, in effect lowering the threshold and potentially knocking some plants out of the regulatory regime. On the other hand, EPA had had a small-plant cutoff which subjected small plants to a more informal, less rigorous review, which was a 50 ton per year threshold. The court said there was no basis in the statute for that and that the only basis for categorical exclusion of a new source from these requirements was the emission of deminute amounts of pollution, so the court in effect moved the regulatory reach of the PSD regime back, if you will, imposing the most stringent review requirements on fewer sources but imposing the less stringent review requirements on many more sources. The other significant thing that the court said is that EPA was statutorily wrong in regulating only sulfur dioxide and particulates through this regulatory regime. The statute includes the requirement that when a new source comes in for one of these permits, it has to apply best available control technology on its pollutants. EPA had said that the relevant pollutants were SO_2 and particulates, and the court said that was wrong, that the best available technology requirement applies to all regulated pollutants. A new industry coming in now therefore faces that control requirement, not just for sulfur and particulates, but for any other pollutant we regulate, which now includes lead and hydrocarbon, CO, and so forth.

CHAPTER 3

SULFUR DIOXIDE EMISSION CONTROLS— WHAT ARE THE COSTS?

Charles H. Goodman*

INTRODUCTION

Potential environmental and health effects of atmospheric sulfur deposition, as will be discussed in this symposium, have taken on increasing importance as our society wrestles with the problems of developing adequate energy supplies, preserving equitable environmental quality, and keeping the costs of environmental protection in balance with the benefits to be gained. Faced with regulations that will require the expenditure of over $100 billion of its ratepayer's dollars on air pollution control equipment and operations during the next 20 years, the electric utility industry has a very lively interest in assuring the establishment of factual data bases upon which the requirements for air pollution control strategies are established. This interest is reflected by the research programs of the Electric Power Research Institute (EPRI), by the regulatory efforts of the Utility Air Regulatory Group (UARG), and by independent work being carried on by individual utilities.

Each of us should be concerned with assuring that the finite resources (i.e., manpower, energy, capital, environmental) which are available to our society are used in the most productive and efficient manner. Squandering resources in attempts to reduce small environmental risks while leaving larger ones unattended creates in itself a major risk that other more serious societal needs will go unattended. Productivity of society goes down when valuable resources are used to produce less valuable resources. One basic effect of reduced productivity that we are all very much aware of is its contribution to higher inflation.

I do not want my presentation to be one where I would have you substitute my beliefs for scientific fact. In fact, it is my purpose to encourage you as members of the scientific community to develop a strong scientific data base and to assure that it is applied in an appropriate manner in order that our environmental control dollars are spent efficiently. A continuing key question with respect to sulfur emissions is whether the large-scale and expensive emission control systems now required for new electric generating units are indeed necessary. Possibly much of the capital and human resources currently directed toward sulfur emissions should be directed toward other more useful purposes.

*Research and Development Department, Southern Company Services, Inc., P.O. Box 2625, Birmingham, Ala. 35202.

In preparing for this presentation, I quickly came to the conclusion that no one knows with much certainty how much our current policy of environmental control for sulfur emissions is costing. There is no legislative mandate in the Clean Air Act which requires costs to be minimized or weighed against benefits. In a practical sense, control costs fall under no form of budget. However, in this presentation, I will present part of the results of an analysis to estimate the costs related to various SO₂ control strategies for new units. Although the case presented does not correspond exactly to the New Source Performance Standards (NSPS) which were recently established by EPA, the costs are of the proper order of magnitude, and the methodologies presented should aid you in understanding how the results of your scientific work might be used in future cost-benefit studies. However, these costs—which are clearly passed on to the consumer—cannot be used as a good measure of the total costs of control. For example, other facets of regulatory requirements also tighten control and restrict the siting of new power plants. All of these constraints severely restrict the formation of capital that generates economic growth and brings about improvement in productivity. A simple totaling up of control costs will not reflect this cost. In order to present the economic analysis, I will first need to briefly review the regulatory history related to sulfur oxide emissions.

REGULATORY HISTORY

Much federal legislation and regulation over the past 10 to 15 years has focused upon encouraging the application of technological and engineering solutions to the control of pollution. The Clean Air Act of 1964 and its subsequent amendments of 1967, 1970, and 1977 represent one part of these initiatives. The Clean Air Act amendments of 1970 introduced the concept that certain air pollutants should be controlled primarily by two mechanisms: (1) National Ambient Air Quality Standards (NAAQS), which were established for the explicit purpose of protecting public health and welfare, and (2) emission standards for new sources, or New Source Performance Standards (NSPS). In addition, the 1970 amendments required that each state develop a plan to provide for implementation, maintenance, and enforcement of the NAAQS. New electric generating units were required to comply with the NSPS, while existing units were required only to meet the conditions of the state plans. Since 1970, the trends in air pollution control regulations and legal interpretations of the 1970 amendments have had a profound impact on the electric utility industry from both a policy and a research standpoint. Foremost among these trends has been the development of NSPS based on the capabilities of certain technological control systems. The establishment of technology-based NSPS reflects a technology-forcing principle. This means that as more effective emission control technologies are demonstrated, the NSPS are revised to reflect the performance capabilities of the improved technologies. These revised NSPS, in effect, force new stationary sources to apply the improved technologies. The 1977 amendments have reaffirmed and strengthened this concept of technology-based NSPS.

As a result, the technology-forcing provisions of the 1977 amendments (and in some respects the 1970 amendments) require emission controls which are more restrictive and expensive than would otherwise be necessary in most instances to maintain air quality at levels which protect public health and welfare. This is especially true with respect to the control of sulfur dioxide emissions.

Key provisions of the 1977 amendments either develop or reinforce regulatory concepts related to:

NSPS

Prevention of significant deterioration (PSD)

Requirements for nonattainment area

Limits on modeling credits for stack heights

Establishment of short-term NO_2 NAAQS

Periodic review of the NAAQS

Periodic review of the NSPS

Coal conversion orders

Use of local coal

Visibility protection

Noncompliance penalties

Revisions of state implementation plans

Special requirements for the use of new and
innovative air pollution control technology

Due to the complexity of the issues addressed by these statutory provisions and the constraints necessitated by the topic of this meeting, I will not try to discuss each of these provisions. My comments concerning costs will be directed only toward the NSPS and PSD. However, the costs for each provision will be substantial. Unfortunately, very little work, if any, has been done to determine if the environmental benefits are commensurate with the costs, especially as related to sulfur emissions.

It is important to understand the PSD requirements in the 1977 amendments. The vast majority of the United States will be subject to these requirements, which will influence the control of sulfur dioxide emissions as well as other air pollutants. It is also important to realize that the NSPS is related to the emission limits required by PSD. In fact, this relationship is established by law. The NSPS is the minimum emission standard that may be required of sources which are subject to PSD. In other words, compliance with the NSPS does not prevent more stringent emission limits from being imposed on a source. PSD areas are those where the ambient air quality is currently better than the NAAQS. Under the 1977 amendments, PSD areas are designated either as class I, class II, or class III, depending upon the degree of future air quality deterioration that is to be allowed. Table 1 shows the air quality increments (increases over the background concentrations of certain air pollutants) which have been set for key classifications. The NAAQS are also shown for comparison. Initially all PSD areas are designated class II except all preexisting international parks, national wilderness areas and national memorial parks which exceed 5,000 acres, and national parks which exceed 6,000 acres. These are designated class I. Provisions also exist for changing the classification of certain areas, except that the class I areas described above cannot be redesignated.

These PSD provisions were apparently included in the 1977 amendments because of a concern for any, as yet unidentified, health and welfare effects caused by secondary pollutants. These effects include acid precipitation, visibility impairment, and sulfate formation which may result from the interaction of various atmospheric pollutants. As illustrated in Table 1, the PSD increments are much smaller than the current NAAQS. Of course, there are a few cases where the difference between the existing ambient concentration of a pollutant and the NAAQS for that pollutant is

Table 1. Comparison of National Ambient Air Quality Standards (NAAQS) and Prevention of Significant Deterioration (PSD) increments

In micrograms per cubic meter

Pollutant	Averaging time	NAAQS		PSD Increment[a]		
		Primary	Secondary	Class I[b]	Class II	Class III
Sulfur dioxide	Annual	80		2	20	40
	24 hr[c]	365		5	91	182
	3 hr[c]		1300	25	512	700
Particulate	Annual	75	60	5	19	37
	24 hr[c]	260	150	10	37	75
Nitrogen dioxide[d]	Annual	100		[e]		

[a] Allowable increase over background concentrations.

[b] The law contains provisions that allow the class I PSD increments to be exceeded in special cases after governmental review.

[c] Not to be exceeded more than once per year.

[d] A short-term ambient standard will be proposed for nitrogen dioxide in late 1980.

[e] The law requires EPA to establish PSD increments for all remaining criteria pollutants by mid-1979.

less than would otherwise be allowed by the PSD air quality increment. And in these instances the NAAQS cannot be exceeded regardless of the size of the PSD increment. Except for these cases, EPA has interpreted the PSD provisions as an additional ambient air quality standard which is more restrictive than the NAAQS.

In addition to creating increments which allow specific increases in ambient pollution concentrations, the PSD provisions require that new emission sources undergo an extensive review in order to receive approval for construction. This preconstruction review is required in order to assure that all applicable air quality requirements have been satisfied, to allow public review of the project, and to effect some control over future growth. During this review the possible effects of secondary pollutants, which may result from the proposed source, will be assessed, and air quality modeling, using mathematical techniques, will be used to assure that the allowable PSD increments are not exceeded. Thus, the emission limits required by the PSD provisions will reflect specific considerations of each individual power plant site.

The 1977 amendments require that the NSPS for fossil-fuel-fired power plants include not only an emission standard (as was required by the 1970 amendments) but also a percentage reduction in emissions that otherwise would occur from the burning of untreated fuel. The emission standard and the percentage reduction must reflect the best technological system of continuous emission reduction which has been adequately demonstrated. Furthermore, the 1977 amendments require that all new facilities which are subject to PSD requirements install the "best available control technology" (BACT) for reducing emissions of all criteria pollutants. The manner in which BACT will be determined is not presently clear, since it will be determined for each source individually, as part of the PSD preconstruction review. For the purposes of this paper, it is assumed that BACT will require the same level of control as the NSPS. Although this may be the case initially, BACT could become more restrictive than the NSPS in the future if sulfur dioxide control systems demonstrate removal capabilities better than those reflected by the revised NSPS. The revised NSPS for sulfur dioxide emissions from coal-fired generating stations are summarized in Table 2.

Table 2. Summary and comparison of current and prior NSPS—electric utility boilers

	Prior NSPS (1970 Air Act)	Current NSPS (1977 Air Act)
SO₂		
Ceiling	1.2 lb/MBtu	1.2 lb/MBtu
Percent reduction	None	70–90%
Averaging time	3-hr intermittent test	30-day rolling average
Continuous Monitoring for SO₂		
Data capture requirement	None (intermittent test— see above)	75% for inlet system and outlet systems individually

In summary, the combined effect of BACT, compliance with the PSD increments, and a more restrictive technology-based NSPS severely limits the opportunities for optimizing cost and pollution control levels when constructing new coal-fired electrical generation, while at the same time protecting the environment. These provisions act in conjunction, so that, in any specific case, the one requiring the maximum emission reduction will control.

SULFUR DIOXIDE NSPS—ITS COST EFFECTIVENESS

The Utility Air Regulatory Group (UARG) is an ad hoc group formed to provide the principal means of electric utility industry participation in EPA's rule makings to implement the provisions of the 1977 amendments to the Clean Air Act. UARG consists of the Edison Electric Institute, the National Rural Electric Cooperative Association, and 63 individual utility systems which collectively own a majority of the electrical generating capacity in the United States. As part of the evaluation of the various proposed versions of the NSPS, both UARG and EPA compared the nationwide impacts of their proposals through the use of economic models which estimate the amount and location of future electrical generation based on a predicted electrical demand growth rate. These models find the least-cost alternative for building and operating this additional capacity based upon coal mining and transportation costs and upon other constraints such as coal sulfur content and the cost of sulfur dioxide emission control. UARG's analysis was based upon the National Economic Research Associates, Inc. (NERA) Electricity Supply Optimization Model, and EPA's was based upon ICF, Incorporated's Coal and Electric Utilities Model. While there are minor differences between these models, the overall results were approximately the same. Therefore, only the key points of UARG's analysis (NERA, 1978) are discussed here.

Unfortunately, cost estimates which exactly compared the old regulations under the 1970 amendments with the recently promulgated NSPS were not available at the time this paper was prepared. The results I will present differ from the final NSPS in that a sliding scale of removal efficiencies ranging from 20 to 85% is used instead of 70 to 90%. In effect, the actual regulations are more expensive than the costs presented will indicate.

These analyses were performed using the base-case assumptions of (1) a 1990 oil price of $2.50 per million Btu (1977 dollars), (2) a growth rate in electricity demand of 5.3% per year, and (3) a range of flue gas desulfurization (FGD) module reliability varying from 70 to 90%, depending upon the sulfur content of the coal, and an availability penalty of 3% for FGD plants. Studies were performed on the sensitivity of the model to these and other variables. A detailed discussion of all the major assumptions is available in NERA (1978). Table 3 compares the annual nationwide impact in 1990 of a 20 to 85% removal NSPS with the sulfur dioxide emission controls (NSPS and state plans) established under the 1970 amendments. Under the 1970 amendments the annual cost in 1990 (expressed in 1977 dollars) would be $2.8 billion. The revised NSPS increases that by more than $1.7 billion, or 58.9%. Yet, the revised NSPS reduce emissions by only about 7.3%, whereas the 1970 amendments reduced emissions by over 40% when compared with an industry base without SO_2 controls. Thus the cost per ton of SO_2 removed increased by nearly a factor of 6.

One disturbing result of these regulations that was revealed by the analysis was that oil and gas consumption by utilities in 1990 increased by 70.3%. This is due to the fact that in order to minimize cost, the many generating units which make up the

Table 3. Comparison of the annual nationwide impact of the NSPS for sulfur dioxide emissions under the 1970 and 1977 Clean Air Act amendments

	1970 amendments	Relative impact of new NSPS compared with the old NSPS
Annual cost in 1990		
Billions of 1977 dollars	2.8	+ 1.7
Percent		+ 58.9
SO$_2$ emission in 1990		
10^6 tons	25.0^a	− 1.8
Percent		− 7.3
Cost per ton of SO$_2$ removed, dollars	158.0	912.0
FGD sludge production in 1990		
10^6 tons	31.5	+ 1.8
Percent		+ 5.6
Electric utility oil and gas consumption in 1990^b		
10^6 bbl equivalent	415.0	+ 292.0
Percent		+ 70.3
Present worth of additional cost between 1984 and 2020^b		
Billions of dollars	77.6	+ 29.5
Percent		+ 38.0

^{a}The annual emission of 25 × 10^6 tons of SO$_2$ shown here is a reduction of 17.9 × 10^6 tons over the industry base without any SO$_2$ controls.

^{b}Data presented are for SO$_2$ and particulate control; however, particulate control cost is estimated to be less than 3% for these calculations.

total capacity of an electric utility are operated so that the units which have the highest variable cost are operated the least. As a result of the 1977 amendments, the variable cost of operating new coal-fired generating units will increase relative to the cost associated with the old regulations. This increased variable cost will result in an increased use of existing oil- and gas-fired electrical capacity to minimize overall generating cost. As Table 3 indicates, the proposed NSPS will result in an increased annual consumption in 1990 of the equivalent of 292 million barrels of oil and natural gas compared with the old regulations. The result clearly runs counter to the objective of the national energy plan to reduce the consumption of oil and natural gas.

Finally, the real consequences of the revised NSPS are not completely evident unless the total cost is considered. Table 3 compares the present worth of the old regulations with the recently promulgated NSPS. As can be seen, the present worth (1977 dollars) of sulfur dioxide control costs in the electric utility industry between 1984 and 2020 which result from the 1970 amendments is $77.6 billion. The new NSPS, however, will require a 38% increase of $29.5 billion. These costs are not inconsequential, especially when it is realized that they represent only a portion of the expected costs (i.e., PSD not considered) for the control of one potential air pollutant from one U.S. industry.

CONCLUSIONS AND SUMMARY

The cost comparisons which have been presented here serve to highlight only one part of the sulfur dioxide control costs related to the 1970 and 1977 amendments. Whatever the ultimate cost of sulfur dioxide control, the costs given here understate them. In particular, the economic impact of the PSD regulations will be substantial and are not known. Whether these costs for sulfur dioxide emission controls are appropriate is simply not known, because EPA did not consider actual health and environmental benefits within the decision-making process.

If, in the future, society is to avoid unacceptable waste of its limited resources, we must develop methodologies which assure that major resources are not directed toward reducing small risks while larger societal needs go unattended. A key first step is to develop methodologies which differentiate between major and minor risks. In addition, economic comparisons similar to the one which has been described here will undoubtedly need innovative modifications in order to allow environmental variables to be considered. Nevertheless, it is a task which must be undertaken if we expect to maximize the quality of life in a world of scarcity.

For this symposium, I feel that the following questions should be addressed:

1. Do the currently available environmental and health effects data related to sulfur dioxide emissions indicate that the costs of control are justified? It may never be possible to completely answer this question, because the methodology for weighing risks associated with various environmental and societal problems has not been established; however, cost-benefit concerns should be addressed.

2. Are scientific investigations being performed and presented in a manner so that the conclusions will stand up to the highest professional standards of the peer-reviewed literature? If scientists cannot repeat and confirm or deny the conclusions of their peers, then credible risk and economic assessments are impossible. A more complete discussion of this issue was given by the late Cyril Comar (1979), where he discussed problems associated with what he called "bad science."

3. Finally, what form should risk assessments or cost-benefit analysis take when society is trying to place a value on an environmental effect? In order to avoid squandering resources, society must have assessment methodologies which assure that major resources are focused toward resolving major problems as opposed to minor ones.

REFERENCES

NERA. 1978. Comments on the Economic Impacts of EPA's September 19, 1978 Proposed Revision to New Sources Performance Standards for Electric Utility Steam Generating Units. Prepared for the Utility Air Regulatory Group by National Economic Research Associates, Inc., Dec. 15.

Comar, Cyril. 1979. *Has environmental regulation gone too far? A debate on the costs versus the benefits.* Chem. Eng. News, Apr. 23.

CHAPTER 4

ENVIRONMENTAL VERSUS EMISSION CONTROL COSTS— A LEGISLATIVE PERSPECTIVE

Thomas H. Moss*

I am grateful for the chance to be here today, for a reason broader than the obvious privilege of being on the podium of a significant conference. It is related to what I call the Winnie-the-Pooh syndrome, which I use to characterize my daily existence in working in the congressional setting. Those of you who have several small children, as I do, probably can guess what I mean. You recall that the first volume of the stories begins "bump, bump, bump" as Winnie-the-Pooh is taken down the stairs in the usual fashion by Christopher Robin—pulled by his leg, with his head bumping on each stair as they descend. Pooh says that he thinks there must be a better way to come down stairs than that, but the bumping in his head keeps him from being able to think what it is. It is the same with my own work. I often think that there must be a better way to deal with issues of science and public policy than the way I see myself and others doing it in the press of daily events, budget deadlines, and political cycles. However, the bumping in my head (the pressure of all those events, deadlines, and cycles) is my excuse, too, for not making much methodological progress. It is only the chance to get out of Washington and away from all that bumping, into the quiet of places like these beautiful mountains, which really gives me a chance to reflect. You can see why it means a great deal to me.

Getting to my designated title of a "Legislative Perspective on Environmental Versus Emission Control Costs," I can state as a basic observation that quantitative cost- or risk-benefit analysis has become a central feature of contemporary legislative thought. This development has been enthusiastically welcomed by a considerable faction in the technological community, as representing political acceptance of the methodology of science. I would like, however, to use my time to at least consider some of the not-so-obvious and perhaps not-so-desirable consequences, for both science and politics, of this trend. There may be profound hidden implications for how we do science, organize its findings, and even teach it. There may also be some pitfalls for the political community, with disillusionments and reexaminations in the offing.

Before I go too far in raising these implications of the dominant cost- and risk-benefit thinking, let me remind you, however, that this climate of thought has not

*Staff Director and Science Advisor to Congressman George E. Brown, Jr.

always existed. The 1970 Clean Air Act amendments were perhaps the best example of another era in which health protection was taken as an absolute, not to be compromised or balanced by any trade-offs with "practical" factors. To illustrate, I quote from an *Ecology Law Quarterly* analysis:

> Issues of technology and economics, at least with respect to the attainment of National Ambient Air Quality Standards, were removed from the executive and judicial branches altogether. Rather than allow the agencies or courts to balance these factors against the goal of restoring healthful air quality, the legislature clearly stated that the cost or technological problems of meeting the requirements of state Implementation Plans to attain these Standards were to be no defense to an enforcement action.

Further, in a widely quoted discussion, the Senate Committee's report made clear that commitment to this position:

> In the Committee discussions, considerable concern was expressed regarding the use of the concept of technical feasibility as the basis of ambient air standards. The Committee determined that (1) the health of people is more important than the question of whether the early achievement of ambient air quality standards protective of health is technically feasible; and (2) the growth of pollution load in many areas, even with the application of available technology, would still be deleterious to public health.

> Therefore, the Committee determined that existing sources of pollutants either should meet the standards of the law or be closed down.

Lastly, in a definitive 1976 decision in a Union Electric Company vs EPA case, challenging the legality of a state Implementation Plan, the Court of Appeals held that

> the language of §110(a)(A) provides no basis for the Administrator ever to reject a state implementation plan on the ground that it is economically or technologically infeasible.

The major amendments of 1977 presented quite a different mode of thinking. Literally dozens of studies and analyses were mandated to examine the cost- or risk-benefit trade-offs of proposed or existing regulations. Some of these were aimed at fine-tuning regulatory levels at the margin—that is, assessments of incremental benefits and incremental costs for incremental changes in the regulated levels of pollutants. All are heavily laced with those same considerations of economics and practicality which were so explicitly excluded in the 1970 language. The mandate to the newly established National Commission on Air quality, as just one example, required reports to Congress with emphasis on such points as

> (2) the economic, technology, and environmental consequences of achieving or not achieving the purposes of this Act and programs authorized by it;

> (3) the technological capability of achieving and the economic, energy, and environmental and health effects of achieving or not achieving required emission control levels for mobile sources of oxides of nitrogen in relation to and independent of regulation of emissions of oxides of nitrogen from stationary sources; . . .

> (6) the ability of (including financial resources, manpower, and statutory authority) Federal, State, and local institutions to implement the purpose of the Act; . . .

> (8) (A) the special problems of small businesses and government agencies in obtaining reductions of emissions from existing sources in order to offset increases in emissions from new sources for the purpose of this Act; . . . and

(G) the effects of such provisions on employment, energy, the economy (including State and local), the relationship of such policy to the protection of the public health and welfare as well as other national priorities such as economic growth and national defense, and its other social and environmental effects.

Reasons for this change in outlook are undoubtedly complex and not really the subject of today's discussion. They certainly include a greater consciousness of the cost of environmental control, exacerbated by the general stress on energy resources and the economy. Further, as the most acute of the environmental problems of the sixties appeared to come under control, there naturally began to be greater consciousness of the imperfection of the regulatory *solutions* than there was of the remaining environmental *problems*. Careful risk- and cost-benefit balancing thus seemed necessary, and the tools of science were assumed adequate to make it possible. That quantitative balancing also had political promise, of course, in that it seemed a way to resolve some very difficult political problems. Political judgments which could not avoid alienating at least some constituencies could be at least partly cloaked in scientific terms. For better or worse, a number of key regulatory issues, such as final standards for nitrogen oxide, particulate, and halocarbon emissions, and many lesser matters, were deferred for final judgment pending the result of very complicated cost- and risk-benefit analyses. These, to me, have seemed to be a set of political time bombs, going off in intervals from the 1977 amendments' passage, as each study is reported with its implications for increasing or decreasing the stringency of environmental regulation.

Where does this leave the science and technology community? I can see it as an enormous challenge and opportunity or, less positively, a terrible burden. Coupling the cost- and risk-benefit thinking directly to the legislative and policy decision process implies deadlines, formats, and completeness, which will not be easily achievable. The analyses usually asked for are not to be merely generally supportive of a standard-setting process but precise and detailed in picking the turning point where complicated sets of costs begin to outweigh similarly complex sets of benefits. The reemergence of the "pollution tax" notion in regulatory thinking is one striking example of the need for quantification of costs and benefits. Even when accepted philosophically, it still requires an unprecedented sophistication in valuing health and ecological factors in order to be equitable. The political decision makers are probably much less aware than they should be of the limitations of these kinds of data and analyses, again throwing a burden of very careful communication on the spokesmen for technical findings.

The responsibility of coupling directly to political decision-making will imply changes in the kind of science apparently demanded by society. *Discovery* of phenomena may seem less important than exploration of them under widely varying conditions. Interactions among them and perhaps endless exploration of parameterized dose-response curves could become the dominant preoccupation of very large numbers of scientific personnel. Organization of research programs to fulfill specific perceived data needs, and the organization and handling of the data obtained, could become as important a part of the scientific enterprise as doing the research itself. The well-known Environmental Protection Agency "CHESS"* study on health effects of sulfates perhaps epitomized the real needs. It was not the scientific frontiers which limited the usefulness of CHESS but the lack of ability to

---

*Community Health and Environmental Surveillance System.

organize and quality-control a large-scale decentralized data-gathering operation and to convince the lay public of its repeatability and definitive applicability to the relation of health effects on air pollutants.

Aside from changes in emphasis among scientific disciplines, the cost-benefit thinking will probably catalyze changes within them. It is difficult, for example, not to envision a great enhancement of the importance of epidemiological thinking in medicine. The traditional stress on individual diagnosis and care will have to give way to methodologies better able to detect and quantitate chronic, long-term health effects. In this and other fields the interactive nature of pollutant-related health and ecological threats may force a valuable new cross-disciplinary approach to many problems. I have found with my own graduate students at Columbia and New York University that it was often a student's determination to work on an inter-disciplinary environmental problem which forced isolated academic departments to reach out to one another and acknowledge the appropriate interconnections among their programs.

I see a special significance of the cost-benefit thinking for the problem of at-mospheric sulfur deposition, which is the focus of this conference. As many of you know, there is a growing perception that current levels of atmospheric SO_2 and its derivatives in most areas are not significant direct threats to human health—or that whatever threat there is is of such nonspecific nature that it will take an extraor-dinary and costly effort to detect. More generally and importantly, it is increasingly obvious that humans are not, in any case, very easy to handle in cost-benefit equa-tions. The complexity of determinants of their health, diversity, toughness, and resiliency, as well as their defiance of attempts to quantify their behavior and well-being, is well known to the health experts here at the conference. There is only limited agreement on the value of a life or on the cost of morbidity and mor-tality—ranging even to occasional notions that preserving life under some condi-tions may be more costly to society than sacrificing it to pollution effects. And adding to all these inherent uncertainties is the basic limitation that we cannot ex-periment on human beings, as we might in other situations where there were so many uncertainties in drawing cause-and-effect relations.

This all adds up to a sense that the agricultural, silvicultural, and ecological systems may be the most important and most easily quantified targets of many forms of pollution, but especially of atmospheric sulfur compounds. Crops, timber, and productive lakes have values which can be relatively easily quantified for cost-benefit purposes. Significantly too, effects in these areas show naturally the cost of *integrated* doses of pollutants—over time and space—so that endless discussions of the significance of standards based on hourly or daily averages can be avoided. Not only do these nonhuman effects lend themselves best to scientific analysis, they also may be most effective in political discussions. The public and politicians do not have to be reminded of the significance of agricultural productivity to our economy. Fishermen can be as determined as any societal group in conveying the importance of preserving lake productivity. These considerations all lead me to look forward with great excitement to hearing the conference's ecological findings on atmospheric sulfur effects, as they may become the catalyst for a new set of societal perceptions concerning the treatment of our valuable atmospheric resource.

What are the scientific and political dangers of the great emphasis on the cost-benefit approach? I am sure many of them are obvious by implication from what has already been said. I will just list a few; I imagine that other and better examples may come to your minds:

1. In some cases the demand for precise quantification may lead to oversimplified analysis, such as concentration on the effects of a single pollutant instead of on

synergistically interacting combinations. "Missing the forest for the trees" could become an all too typical result. The furor over EPA's reliance only on ozone levels as a basis for standard-setting for atmospheric oxidants, as opposed to some more complex notion of total oxidant concentration, may be a first example of this tendency. The current adversary climate in considering environmental regulation can only exacerbate this trend. The formality of "rules of evidence" may replace scientific instinct and intuition as a basis for argument, with the inevitable result of time and effort wasted on formality rather than substance and optimized opportunities for new insight.

2. The costs, in dollars and manpower, of gathering the needed data may be staggering. Even in the simplest cases, the variety of conditions and interacting factors affecting the response of natural systems is such that one can almost envision the entire trained technological manpower pool fully occupied just in checking out the details of generally well-understood phenomena. The solution may be unprecedented organization, on an international scale, to be sure that data are gathered as efficiently as possible—but the need for that kind of organization may be itself a serious negative factor.

3. Strict adherence to cost-benefit analysis as a policy determinant loses much of the valuable technology-forcing function of environmental regulations, a concept that was very much in the minds of the framers of the 1970 amendments. Trade-offs analyzed and compromised solely in terms of existing technology may leave little incentive for private-sector development of new and more efficient control methods.

4. It may be beyond the capabilities of responsible federal and nonfederal agencies to organize and work with one another in the complex data-gathering tasks which will be needed. The difficulty of the organizational challenge cannot be exaggerated. The ill-fated proposal to establish a Natural Resources and Environment Department brought forth the many parochial interests in environmental areas which will almost inevitably resist attempts at organization and unification. It may be for the best that political capital and organizational talent were not wasted on the simple formation of such a department, as opposed to the tasks which needed accomplishing, but the debate itself showed the many centrifugal forces which counter the best-motivated designs for coordinated activity. The current interagency Acid Rain Coordination Committee, under EPA and Department of Agriculture leadership, shows how we must proceed in the absence of more definitive concentration of responsibility. The emphasis on it in the President's environmental message is a hopeful sign of the administration's perception of the importance of getting a quantitative understanding of this crucial interdisciplinary and interdepartmental element of a number of our environmental cost-benefit equations. Nonetheless, we all realize that after several years of relative neglect of this area, and indeed of all of the ecological impacts of pollutants, EPA is going to have to play "catch-up ball" to reestablish viable programs. Perhaps one of the most hopeful signs in terms of our organizational prospects is the continued and strengthened existence of the National Atmospheric Deposition Program, with which many of you have been familiar. Here is a project filling a basic need for coordinated and standardized data gathering, which was originally organized and operated on an almost entirely ad-hoc basis, long before a perception took hold at national policy levels of the importance of this kind of effort.

5. The overall lack of emphasis on technological planning in the administration is the general ailment of which the difficulties in establishing a coordinated acid rain program have been symptoms. Neither the Office of Science and Technology Policy nor the Council on Environmental Quality has been able or willing to assume the broad coordinating role some envisioned for them in identifying and prioritizing

emerging problems and in marshaling forces to solve them. Unless such a planning operation can be established as a function of the President's Executive Office, or unless the outside scientific community can mount an incredibly effective and coherent self-definition of the necessary long-term coordinated programs to provide the data for intelligent cost-benefit thinking, organizational and continuity problems will undoubtedly continue to plague environmental cost-benefit research.

6. Perhaps the danger which weighs most on my mind concerning all this cost-and risk-benefit thinking is that to the health of science itself. Of course, I may be looking much too much at the dark side of things. Many would say that the mandate for scientific thinking is a blessing which will keep us all in business for years to come. However, I shudder at the thought of the time and effort it is going to take to organize our scientific findings to interface with the political community. Our happy anarchy and treasured scientific individuality may be difficult to preserve in the face of the need to make airtight arguments to balance in the complex cost-benefit equations. Some of the emerging problems, such as the acid deposition effects we will be discussing in the next few days, and certainly the effects of atmospheric carbon dioxide buildup may have social implications so profound that only a scientific case of unprecedented credibility and completeness will be able to move the rest of society to undertake the needed remedies. I only hope that the health and fun of science will not be a casualty of the heavy responsibility to build such a case.

I will conclude only by saying that it has been a privilege to appear before you, and it will be a greater privilege to hear any comments or questions you may have. I look forward to being part of the scientific exchange which I know we will have in the next few days.

DISCUSSION

Dr. Feder, University of Massachusetts: Your talk is a very disturbing one, also a very challenging one. I would like to enter a plea from the scientific community, if I might, to your committee and to EPA that if indeed this is the data you want and this is the data you're prepared to pay for, that you prepare to live with us for long-term periods rather than constantly changing short-term periods; further, that if you decide that some of us are worthy of support for the research we're doing, that you continue to support that research long enough to bear some fruit, and at the other end, if the fruit is borne and you eat it, that you're prepared for the consequences of what it tastes like.

T. H. Moss: And my point is that I don't know whether the political community is prepared for those consequences.

Dennis Parzyck, Oak Ridge National Laboratory: I was also extremely interested in the comments that you have made in your presentation. To a certain extent, though, I was very disturbed by the relative role that you assigned to ecological vs health effects information and input to this cost-benefit type of analysis. I think that in a certain area of work that I'm somewhat familiar with we now bear the legacy of being unable to communicate to society, to the people who represent us in Washington, the relative effect of risk on human health. This is, of course, in the nuclear power area. The legacy that we bear there is that this society in general and the people who must set the policy for this country have no full realization of what health effects really are for this technology. Admittedly there must be a characterization of impact, and I would offer that the major needs we have are ones of communication, of establishing the network whereby information developed by people doing basic research can be conveyed to the people who must make the policy decisions, based on these various cost-benefit analyses. It really does not suggest a change in the system so much as it does the addition of a specific means of conveying that information, of processing that data.

T. H. Moss: Well, let me just comment on that. I don't quite see why it disturbs you to go to the ecological emphasis for the health emphasis; can you clarify that for me a minute?

Dennis Parzyck: You had said that certain ecological-type impacts, the effects on crops or forest species, were more easily characterized. I would submit that the effect on human health can be characterized, but it represents a step in scientific development, a type of interactive collaboration, not yet achieved, but that we are making some tangible steps in that direction.

T. H. Moss: I don't disagree with you that they can be characterized, though I still would maintain my position that a lot of times, in the political arena, it's easier to talk about the corn crop than the most sensitive human population, for instance. But let me say on this communication thing, I agree with you tremendously. I didn't detail it, but that's what I was alluding to when I said that there was a chance that the scientific community in terms of formats of data, and so on, will have this in the CO_2 program. We have this wonderful anarchic system of publication where we want to make sure our graduate student is the first on the paper, so even if we could publish jointly with three different laboratories, we tend to publish individually so that the graduate student can get a good job, and so on. That's fine; that's the way we've always done it; I've done it in my laboratory at Columbia. From the point of view of society, the politician or the decision makers getting these pieces of data, however, they're so much more credible when we have organized joint publications—people monitoring acid sulfate in a series of networks over a number of different locations, for example, all publishing together—instead of trying to publish very quickly, spending two or three months trying to get their data in the same format. Agreed, it takes a lot of trouble sending it back and forth through the mail and getting everybody to sign off, but in the long run, if it is going to be in the decision-making process, I think we're going to have to give up some of this treasured anarchy and organize our scientific work to an unprecedented degree.

Moderator Cowling: Thank you, Tom. During the time that remains what I would hope we could do is engage in a general panel discussion, and I hope that those of you who ask questions will stand up and address them to any of the speakers individually or to the panel as a whole, so the session is now open to discussion.

John Skelly, Virginia Tech: All three of the speakers have talked about human health effects and the difficulty of getting those cost-benefit figures that are so important to the establishment of standards. I would ask the question—the cost of air conditioning major cities such as New York, Los Angeles, and Washington: one of the reasons we do not have any human health effects is that as we develop an air-pollution episode in these major cities, we tell people with various heart and lung disorders to stay inside, which eliminates the effect by avoiding it. In relation to plants and ecological systems, it is very difficult to move the soybean fields of eastern Virginia into an air-conditioned facility for the duration of exposure that might be encountered by people less than 25 miles away in Washington, D.C. I wonder why the cost of the air conditioning is not considered in cost-benefit studies?

David Tundermann: Maybe no one else has thought of it; that's certainly the first time I've heard that comment—a very thought-provoking question, John.

John Skelly: Just a final comment on that, the idea that we move into an episode and we therefore must air condition, creates more air pollution because of the demand for the air conditioning. We may be caught in a "Catch 22."

David Tundermann: If you simply shut down the air-conditioning systems of the city of New York you will increase the mortality rate. It's been demonstrated, and that's a very interesting connection to make.

Chuck Hakkarinen, Electric Power Research Institute: This question can go either to the EPA or the utility speaker. Can either of you give a simple percentage estimate for the electric customer as to how much his rates might increase in 1990 if the SO_2 regulations are implemented to the extent proposed now?

David Tundermann: The estimates were that current average monthly residential utility bills in 1978 dollars were around $27. If the standard were not changed, if the 1971 standard had continued in effect, the projection was that the typical average monthly bill in 1995, expressed in 1978 dollars, would be about $53, and by changing the standard to the one EPA adopted, that would go up by about a dollar, so the incremental percent change as a result of the standard was less than 2% in estimated monthly utility bills in 1995.

Tom Moss: Why does it go up at all in real dollars, just because people use more?

David Tundermann: No, because the cost of building power plants and of buying fuel is going up.

Owen Hoffman, Oak Ridge National Laboratory: I'm very happy to see that the main theme of all three speakers is the role of cost-benefit analyses in promulgating regulation to protect the environment. I would like to add that I don't think we should talk about a cost-benefit analysis for the evaluation of regulating one pollutant. I think our representative from industry mentioned that we should not spend great quantities regulating minor problems when we could spend that money rather on major problems. I think one of the problems that we have to face is that once we have this great network of information that we have obtained from our new generation of scientists, how do we compare the data? How do we compare health effects with ecological effects? How do we compare agricultural productivity with, let's say, the loss of visibility over the Grand Canyon? I think that this also extends to the impacts associated with coal combustion vs the impacts associated with hydroelectric or nuclear power production. All these aspects must be taken into account. My direct question would be placed to our representative from industry, Mr. Goodman, in that I would like him to define for me what a minor problem is.

Charles Goodman: I think that my personal perception of many issues related to sulfur oxides at the level of control we've now come to puts it minor compared to other concerns related to coal utilization, which is the thing I worry the most about. I think we have beat the problem to death. Obviously, there are people who disagree with me, but the problem is, I do not think we have the methodology for picking our priorities with respect to environmental issues, even as narrow as saying related to expanding coal combustion. We know that there, we were restricted to SO_2, which mocks the noncriteria pollutants, all those secondary pollutants that will have to be addressed under PSD, prevention of significant deterioration. I don't think there's any way right now where people have prioritized what's important and what's not. The scientific community hasn't been given, I think, the mandate to do that for themselves, so I surely can't. Sulfur oxides, though, to me, are lacking the data to indicate a real problem. This is very frustrating. The man brought up episodes a while ago, my answer should have been, I think there are very few air-pollution episodes with sulfur oxide as a major contributor; a lot of other pollutants, but not sulfur oxide. Maybe that's wrong.

Stephen Wilson, New York State Energy Research and Development Authority: I guess my question is aimed again at Mr. Goodman. Did I understand you to say that you assumed a growth in utility of 5.3% for your table?

Charles Goodman: For that particular set of numbers, the assumption that was agreed upon for those calculations with EPA, DOE, and others who were involved in doing model studies was 5.2.

Stephen Wilson: I guess there's a comment and sort of a question. The comment is that as some of you may know if you've read the press or seen the New York State Energy Master Plan, which does reflect a considerable amount of analysis with respect to growth and potential growth in utilities, specifically, electric utilities, we come up with 2.1% growth by 1994. I recognize that we're dealing with a part of the country in which there has been (1) a population exodus, (2) increasing taxes, and (3) possibly some saturation with respect to supplying energy, in other words, maturity of system. I guess I have two questions. One of them is, is 5.3%, in the face of that, a realistic figure at all, even on a national basis, and secondly, what would happen if you halved that 5.3%? Could you give us a brief example of what would happen?

Charles Goodman: To answer your first question, I think that EPRI recently had a national study of the utility industry that showed something like 2 to 2-½% growth nationally for energy as a whole and nationally for electricity, 3-½ to 4%. Of course that had some regional breakdowns for higher and lower depending on the part of the country you would be in. I don't know the answer to your second question; maybe the EPA representative could answer that from the ICF study, i.e., what it did to those numbers as you change the growth rate.

David Tundermann: The question you didn't ask was, So What? Perhaps that really was the second question. It really didn't change the relative relationships among the alternative standards. When you increase the assumed electrical growth rate, you increase the number of coal-fired power plants that you're projecting get built. That means that you increase dollar differences absolutely, you increase emission differences absolutely, from what they would have otherwise been, but across the range of standards that EPA considered, the relative differences among the alternative standards did not change as a function of changing the electrical demand assumption for the New Source Performance Standards. One might well envision from a scientific and a regulatory point of view the following circumstance: if electrical demand growth is 5.5% as opposed to 2.1%, one would expect that your absolute projection of emission loadings would go up, and that might cause you worry, particularly if you find increasing evidence of ecological effects that cause you great concern. Then you might think about other control devices or what not in order to keep that total loading figure down, but it did not affect this particular regulatory decision.

William Graustein, Yale University: I'd like to direct a comment and question to Dr. Moss. I can see your reservations about the effects of anarchy in the sciences, how it frequently muddies the water and makes it difficult for an outside observer to see what the truth is. But I think the truth may be as elusive to those in the scientific community as to those outside it. Is there a way that you see to reduce the negative effects of anarchy without stifling the positive effects of creative dissent?

T. H. Moss: Yes, I think so. First of all, I agree with you that anarchy is sometimes the best pathway, and that's what I meant by saying that all this pressure to be part of a cost-benefit equation may destroy the best of science. If we're not careful, we'll all become data collators instead of creative thinkers. But I think there is some way to get rid of some of the anarchy and accomplish some of the goal of providing good information without destroying science—that simple example of trying to publish certain kinds of things jointly. The climate I know best, and in that case if you look in science traditionally you see all sorts of miscellaneous papers about little characteristics of the antarctic ice sheet in a certain year and something about the ocean temperatures in a region of the North Pacific and something about the Siberian spring in another paper—completely incoherent sets or pieces of data, each of which individually has very limited implications. On the other hand, if the people involved took the trouble of thinking to themselves, Gee, I know so and so over here is working on the same thing, maybe I ought to call up and see if he has some results on this, maybe we can publish it together—it is an annoying thing to do, it makes a lot of phone calls, a lot of mail, and all that sort of thing, but it has enormous impact. There have been some key recent papers on the subject, for instance, "Is there a worldwide cooling trend going on?" published by five or six different laboratories all over the world. I dread to think of the problem of getting those papers together, but the papers had enormous impact because the people bothered to think of putting their data together with similar data of other laboratories. That's just a trivial example. I think there's a lot of other simple things that the scientific community can do without destroying itself, to make its data presentation more effective.

Moderator Cowling: I would like to thank our panelists for their presentations this morning and also those of you in the audience who participated in the discussion, either mentally or in your questions, and I would just like to make one closing comment. There was a time in the life of science when science was not supported by society, when it was supported in the way in which art is supported, by the voluntary contributions of interested wealthy people. The tradition developed during those early times when amateurs did science that made academic freedom a reality. Today science is a big business which is supported by society through taxation. If we accept the opportunity to continue to support science by society, I think we must understand the obligation inherent in that. We can go back to a situation in which science was done by amateurs, by a patronage system of wealthy individuals. Science progressed under that system, but very slowly, and we are, for the most part, servants of the public that pays the bills, and I am reminded of Sterling Hendrick's comment, "The opportunity to inquire into the nature of things is a privilege granted to a few by a permissive society," and I would submit to you that the extent of permissiveness of society and support of our pursuit to curiosity will be a direct relationship with their perception of the extent and magnitude of public benefit derived from that science.

SECTION II

NATURAL AND ANTHROPOGENIC SOURCES

Chairman: J. F. Meagher

CHAPTER 5

NATURAL AND ANTHROPOGENIC SOURCES

Chairman: **James F. Meagher**
Tennessee Valley Authority

OVERVIEW

James F. Meagher*

Virtually all atmospheric pollutants result from a composite of natural and anthropogenic inputs. The relative contribution of natural sources to ambient levels varies markedly depending upon the location of the measurement and the compound being measured. As the approach to air pollution control takes on more of a regional perspective, more emphasis needs to be placed on an assessment of natural source strengths. A recent estimate has placed man's contribution to the global atmospheric sulfur burden at approximately 60%. The bulk of the anthropogenic input is as sulfur dioxide resulting from fossil fuel combustion.

Natural inputs to the global sulfur burden result primarily from three source types—biogenic, volcanic, and marine. Biogenic sulfur is believed to be released as a result of bacteriological activity primarily in anaerobic environments. The emissions are composed of both organic and inorganic sulfides. Of lesser importance to the global sulfur burden is volcanic activity. Once again the emissions are composed primarily of reduced sulfur species. Sea salt aerosol entrainment contributes a variety of sulfur salts and dissolved gases to the atmosphere. Most of the aerosols are short-lived, however, due to their relatively large size. Sea salt aerosol is believed to be a less significant source than biogenic activity and contributes mainly in marine areas.

New data on emission rates for various naturally formed sulfur compounds were presented in this session by Adams and Aneja. Adams reported measurements of the biogenic sulfur gas flux at 21 sites, representing nine major soil orders, within the northeastern United States. The data obtained indicate that, contrary to what has been traditionally assumed, inland soils contribute significantly to the total sulfur flux of this area. The compounds H_2S, COS, and CS_2 were found to be the major constituents in the sulfur flux from most soils, water, and vegetation, with lesser amounts of dimethyl sulfide and dimethyl disulfide. Unidentified sulfur compounds in minor concentrations were found in several samples. The sulfur fluxes measured

---

*Air Quality Research Section, Tennessee Valley Authority, Muscle Shoals, Alabama 35660.

at the 21 sites were found to vary greatly as a function of soil order. Adams estimated that natural sources contribute less than 1% of the anthropogenic emissions in the SURE (Sulfur Regional Experiment) study area, which constitutes the northern two-thirds of the eastern United States.

Aneja's flux measurements were made on the coastal salt marshes of North Carolina. Sulfur fluxes were found to vary greatly as a function of soil condition and moisture. When the emission rates obtained from these data were applied on a global basis, it was estimated that biogenic sulfur from marshlands contributed approximately 1% of the global tropospheric sulfur burden. However, it was estimated that global marshlands could contribute up to 19% of the stratospheric sulfate. Both Adams and Aneja reported that the emission rates were strongly dependent on temperature, with higher rates at higher temperatures.

Bandy reported new background measurements for SO_2, COS, and CS_2. The SO_2 and COS data were obtained using aircraft and covered the region from 57° south latitude to 70° north latitude. The CS_2 data were taken at Wallops Island, Virginia. The average concentrations for SO_2, COS, and CS_2 were found to be 54, 518, and 30 parts per trillion, respectively, for the marine boundary layer in the absence of anthropogenic inputs. The SO_2 found over the Pacific Ocean appeared to be derived from natural sources and not from long-range transport of anthropogenic SO_2 from continental sources as has been previously suggested. An examination of the chemistry of the marine atmosphere resulted in the conclusion that CS_2 oxidation could account for no more than 20% of the measurable COS and 33% of the SO_2.

In order to better appraise and understand the contribution of natural sulfur sources to regional and global budgets, much additional information will be needed. Additional measurements of emission source strengths need to be made for each of the three major natural source classes listed above. In particular, the biogenic input needs to be better characterized. The biological fluxes reported indicate a strong dependence on source type and meteorology. Additional measurements are needed if these fluxes are to be adequately parameterized. Better source inventories for anthropogenic emissions are needed. More experimental data (both field and laboratory rate measurements) are needed to help quantify the fate of natural and anthropogenic sulfur compounds in the atmosphere.

CHAPTER 6

ESTIMATES OF NATURAL SULFUR SOURCE STRENGTHS

D. F. Adams*†

S. O. Farwell*

M. R. Pack‡

E. Robinson‡

ABSTRACT

Data are presented for the first systematic measurements of biogenic sulfur gas flux from nine major soil orders within the northeastern United States Sulfate Regional Experiment (SURE) area.

Sulfur gas enhancement of sulfur-free sweep air passing through dynamic emission flux chambers placed over selected sampling areas was determined by wall-coated open-tubular capillary-column cryogenic gas chromatography (WCOT/GC) using a sulfur-selective flame photometric detector (FPD).

Sulfur gas mixtures varied with soil order, ambient temperature, insolation, soil moisture, cultivation, and vegetative cover. Statistical analyses indicated strong temperature and soil order relationships for sulfur emissions from soils.

Fluxes ranged from 0.001 g to 1940 g of total sulfur per square meter per year. The calculated mean annual land sulfur flux, weighted by soil order, was 0.02 g of S per square meter per year for the SURE area, or 53,660 metric tons total. The estimated annual average sulfur flux increased from 65 metric tons per 6400 km² for the land grids in the northern tier to an average of 165 metric tons for the land grids in the southernmost SURE tier.

This systematic sampling of major soils provides a much broader data base for estimating biogenic sulfur flux than previously reported for three Atlantic coast intertidal sites. Thus, this study provides the first sulfur flux estimates for inland soils representing approximately 95% of the land of the northeastern United States.

BACKGROUND

Sulfur compounds, in both gaseous and aerosol phases, are ubiquitous in the earth's background atmosphere and are generally conceded to result from natural processes, including marine, volcanic, and biospheric. The biospheric source is the result of biological action in soils, water, or vegetation and thus includes land, water, and oceanic sources. Consequently, the biosphere is considered to be the major source of nonanthropogenic, background sulfur compounds in the troposphere. This conclusion has been reached, primarily by circumstantial reasoning, by many investigators of the atmospheric sulfur cycle who have postulated the magnitude(s) of the biospheric emissions.

---

*University of Idaho, Moscow, Idaho.

†Field studies accomplished while at Washington State University.

‡Washington State University, Pullman, Washington.

The first analytical study indicating that bacteriogenic sulfur might be an important contributor to atmospheric sulfur involved the sulfur isotope measurements by Grey and Jensen (1972). Subsequently, Lovelock et al. (1972), Berchard (1974), and Rasmussen (1974) suggested that organic sulfur compounds, in particular dimethyl sulfide (DMS), were important constituents of the biogenic emissions. Sandalls and Penkett (1977), using cryogenic trapping and packed-column gas chromatography (PC/GC), identified carbonyl sulfide (COS) and carbon disulfide (CS_2) in atmospheric samples near Harwell, England, and assumed that the source of CS_2 was seawater, as Lovelock had previously speculated.

The first published direct measurements of biogenic sulfur flux appear to be those of Aneja (1975) and Hill et al. (1978). In 1977, Aneja et al. (1979) explored the nature and magnitude of biogenic sulfur emissions from a salt marsh on Lockwood Folly Inlet near Long Beach, North Carolina. The measured flux ranged from 0.01 to 100 g of S per square meter per year. These data compare with similar measurements made at the same location by Adams et al. (1979) in 1977 and 1978, which showed the total biogenic flux to be in the range of 0.02 to 1940 g of S per square meter per year.

Until recently, detailed experimental appraisal of biogenic sulfur emissions has been severely limited by inadequate analytical sensitivity and the inability to speciate completely the many compounds present in the biogenic sulfur gas emissions. These analytical shortfalls no longer exist, and studies are under way to provide information on the nature and range of biogenic sulfur emissions for selected areas of the globe. It is essential that these measurements also be extended into the more remote areas and the tropics to provide a more reliable global biogenic sulfur flux pattern.

OBJECTIVE

This research was not intended to answer all of the outstanding questions about natural sources and strengths of sulfur gases. The objective was to examine both qualitatively and quantitatively the sulfur compound flux from different common soil orders to at least 6400 km^2 resolution during various seasonal weather conditions using an *in situ* field sample collection technique within an area defined by the Electric Power Research Institute Sulfate Regional Experiment (EPRI/SURE) in northeastern United States (Perhac, 1978) (Fig. 1). Within the SURE region, approximately 470 of the 600 80×80 km grids contain significant land masses; the remainder of these grids represent continental shelf and deep ocean. The first use of our field measurements of biogenic sulfur emissions was to be in EPRI's development of a SURE area simulation model to examine, for example, sulfate formation systems from all available data bases. Thus our major thrust was directed toward an extensive examination of the primary soil orders within the SURE area. Samples were collected over a five-day period at most selected sampling sites and for two or three five-day periods during the year at three sites selected to determine seasonal (temperature) variations in sulfur flux. The study herein reported represents the first systematic experimental effort to measure the range of biogenic sulfur flux from a wide variety of soil orders.

EXPERIMENTAL METHODS

Sampling Site Selection

The soils within the SURE area range from (a) coastal, saline marshes through (b) poorly drained inland organic soils to (c) dry mineral soils. The major soil orders and suborders within SURE were initially ranked in order of anticipated biogenic

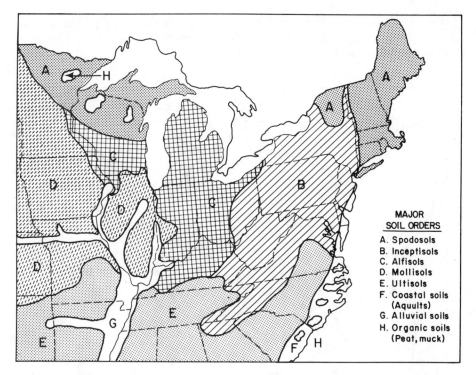

Fig. 1. The eight general soils categories in northeastern United States.

sulfur productivity by considering organic matter, aeration, soil moisture, vegetation regimes, salinity, stagnation, and possible pollutant levels of associated water. The resultant ranking (column 4, Table 1) was used in a "forced random" approach to select 21 sampling locales which directed our measurements more toward soils deemed to have the greatest biogenic productivity potential while minimizing (but

Table 1. Comparison between predicted and measured sulfur productivity

Sulfur productivity group by regression	Slope of regression with temperature	Soil order and suborder	Initially predicted productivity rank
I	0.1954	Marshes	1
II	0.1073	Histosol	2
		Mollisol	4
		Ultisol/spodisol Wet suborder	3
		Fresh water	
III	0.0561	Inceptisol/mollisol/ alfisol wet suborder	5
		Alfisol	6
		Inceptisol	7
		Spodosol	8
		Ultisol	9

not ignoring) soils of lower biogenic productivity potential, thereby providing more data on the range of sulfur fluxes for the nonuniform study area. The sampling sites and associated soil orders are shown in Figs. 1 and 2.

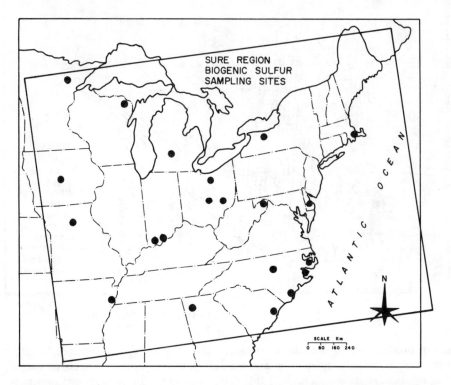

SURE REGION
BIOGENIC SULFUR
SAMPLING SITES

ATLANTIC OCEAN

N

SCALE Km
0 80 160 240

Fig. 2. WSU's biogenic sulfur flux sampling sites, SURE area, 1977 and 1978.

Sampling Procedure

Readily portable dynamic chambers similar to those used in other biogenic flux studies were used to enclose known surface areas. Sulfur-free enclosure sweep air was prepared by passing ambient air through a four-stage solid-phase reactor containing layers of Purafil, Drierite, activated charcoal, and soda lime. A measured flow of sweep air (usually 6 cfm) was passed into the enclosure via a distribution ring for 15–20 min prior to start of sampling and throughout each sampling period. A small aliquot of the vented sweep air (usually 30 ml/min) was drawn through a 15.2 cm specially deactivated Pyrex glass sample loop cooled to $-183°C$. The sampling rate and time were varied with the range of sulfur flux encountered. Sample volumes ranged from 15 to 720 ml. Both battery and 117-V ac portable field sampling packages were used as required (Adams et al., 1979).

Analytical Procedures

The sulfur compounds, cryogenically concentrated in the deactivated Pyrex sample loops, were quantified and speciated with a modified Hewlett-Packard model 5840 flame photometric gas chromatograph (FPD/GC) equipped with a cryogenic oven using a 30-m OV101 wall-coated open-tubular capillary column (WCOT). The detailed analytical procedure has been published (Farwell et al., 1979).

Known concentrations of H_2S, COS, methyl mercaptan (MeSH), DMS, CS_2, and dimethyl disulfide (DMDS) were obtained from low-loss permeation tubes maintained at $40 \pm 0.1°C$ in a constant-temperature air bath. Calibrations were conducted daily and weekly to provide least-squares calibration equations.

Statistical Analysis

The least-squares relationships between the measured sulfur fluxes (independent variable) and the dependent variables, including (a) ambient temperature, (b) soil order, (c) day and night, (d) vegetative cover, and (e) bare soil, were calculated using the Statistical Analysis System (SAS) package (Barr et al., 1976). The SAS regression program also determined the percent of the dependent variable associated with the independent variable relationship. An Amdahl model 470V06-11 computer was used.

RESULTS AND DISCUSSION

Air and Soil Temperatures vs Sulfur Flux

Temperature is a major factor controlling the rate of biogenic activity, with reactions being slow below 5°C and generally increasing two- to threefold for each 10°C increase up to about 50°C (Tauber, 1949). Although the sulfur flux would probably be more dependent upon soil temperature than the associated air temperature, only air temperatures are generally available in the published climatic data. Therefore, we used our 158 sulfur flux samples having concomitant soil and air temperatures to test the hypothesis that air temperatures could be used to estimate the seasonal sulfur flux from individual measurements.

Linear regression analysis showed that soil temperature vs sulfur flux and air temperature vs sulfur flux had correlation coefficients of $+0.685$ and $+0.647$, respectively. There was no statistically significant difference between the two regressions. The linear regression and correlation between the measured air temperatures and associated soil temperature were also calculated and confirmed that the field-measured sulfur fluxes at any given temperature could be adjusted for seasonal variation from the published air temperature data.

Soil Order vs Sulfur Flux

Regression curves for temperature effect on sulfur flux were computed by soil order for all bare soil samples, forcing a common intercept of 5°C, based on known temperature effects on biological activity. The soil flux for the soil order vs temperature relationships fell into three distinct slope regimes (Table 1). The resultant groupings of data were tested with Duncan's t test (1975) for multiple-range testing of group means having unequal numbers of replications. This test confirmed that the measured sulfur fluxes for the range of soil orders under study fell into three different slope regimes, thus simplifying extrapolation of our field flux measurements to the percentage of each soil order present in each SURE grid.

Estimation of Sulfur Flux by SURE Grids.

The mean seasonal and annual sulfur fluxes were then calculated for each of the SURE 80 × 80 km grids by using the three regression equations associated with the corresponding soil orders, weighted for the percent of each soil order present in each grid, and adjusted for the mean seasonal temperature for each grid. The calculated seasonal sulfur fluxes for each grid were summed to provide an estimate of the annual sulfur flux for each grid and for the entire SURE study area and the total land and ocean grids (Table 2).

Table 2. Summary of estimated total sulfur emissions for the SURE area

Soil/water group	Sulfur flux (metric tons per year)	SURE area (10^5 km²)
Soil I	11,329	1.5
Soil II	12,801	7.5
Soil III	29,530	21.2
Soil group subtotal	53,660	30.2
Oceanic	13,000	8.2
Estimated total	66,660	38.4

The calculated land surface annual sulfur fluxes for the grids within each of the 20 east-west SURE grid tiers were averaged and plotted against the mean annual temperature for each east-west grid tier. This gave a relationship with a slope of nearly 1 and a correlation of $r = +0.99$. The distribution of sulfur flux within the SURE area is presented graphically in Fig. 3 (Hakkarinen et al., 1979).

ANNUAL BIOGENIC SULFUR EMISSIONS FROM *SURE* GRIDS (TONS 6400 km^{-2})

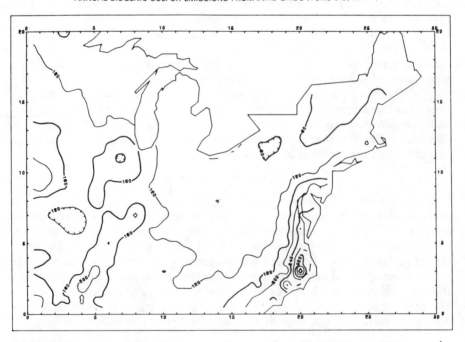

Fig. 3. Isopleths of estimated annual sulfur flux for the SURE area. Fluxes are in metric tons of S per grid (6400 km²).

Some Examples of Sulfur Emissions by Soil Order

Inland organic histosols

Soil flux sampling was conducted in two soils of Orleans and Genesee Counties, New York, between August 1 and 4, 1978: (*a*) the stagnant, swampy Iroquois

Wildlife Refuge and Oak Orchard Swamp and (b) the cultivated black histosol (Palms muck) of nearby Elba, New York. The data are presented in Table 3. Approximately 50% lower total sulfur flux was measured from the swampy wetlands than from the Palms muck, there was little difference in the H_2S flux, and there were appreciably higher DMS and CS_2 emissions from the cultivated muck than from the wetlands. Our data support Hitchcock's (1977) general hypothesis that a significant fraction of the summertime airborne sulfate in Monroe County, New York, is biogenic, originating south of the Monroe County high-volume sampling network. However, our flux measurements indicated that (a) H_2S may not be the major or only airborne sulfate precursor, (b) the cultivated histosol muck farmland may be a more significant source of biogenic sulfur than the nearby wetlands, and (c) DMS and CS_2 emissions are significantly higher from the Palms muck than from the relatively undisturbed wetlands.

Table 3. Comparison between biogenic sulfur flux from wetlands and histosol muck soil in northwestern New York, August 1-4, 1978

Soil	N^a	Sulfur flux (g S m^{-2} year^{-1})						
		H_2S	COS	DMS	CS_2	DMDS	U^b	Total S
Cultivated histosol	12	0.168	0.027	0.005	0.127	0.004	0.006	0.331
Cultivated histosol, potatoes	7	0.164	0.019	0.008	0.171	0.001	0.006	0.361
Stagnant swamp	2	0.166	0.005	0.004	0.006		0.006	0.182

[a]Number of measurements.
[b]Unknown sulfur compounds.

Tidal zones

Tidal effects on biogenic sulfur flux were observed in the Canary Marsh near Lewes, Delaware, from August 16 through 20, 1978. The central portion of Canary Marsh is relatively level, uniformly vegetated with marsh grass (*Spartina alterniflora*); is infrequently flooded except during extremely high tides and/or strong easterly winds; and is wet and boggy. The data in Table 4 show an increase of nearly two orders of magnitude in sulfur flux from the intertidal zone to the infrequently flooded marshy zone. The H_2S emissions are comparable, but the emissions of COS, DMS, CS_2, and DMDS increased significantly in the central marsh. The lower flux in the intertidal zone is probably related to twice-daily tidal flushing, whereas the central marsh is more anaerobic and thus more productive.

Relative contribution from coastal wetlands and inland soils

The percentages of each soil order for the SURE study area grids were summed within each of the three general soil groups noted previously (Table 2). Group I soils are wetlands, many of these being coastal, which comprise approximately 5% of the SURE land area and produce approximately 21% of the calculated sulfur flux.

Group II soils, which are inland soils with high organic matter content and/or exist under wet conditions, comprise approximately 25% of the SURE land area and produce nearly 24% of the calculated sulfur flux for the SURE land area. However, group II soils are widely distributed throughout the SURE grids (as contrasted with

Table 4. Tidal influence on biogenic sulfur emissions from a marsh—
Lewes, Delaware, August 16-20, 1978

Soil	Average sulfur flux (g S m^{-2} year^{-1})					
	H$_2$S	COS	DMS	CS$_2$	DMDS	Total S
Intertidal marsh	0.01	0.006	0.042	0.013	0.0004	0.016
Marsh, infrequently flooded	0.01	0.020	0.91	0.125	0.006	1.16

group I soils, which are frequently coastal). Even the more typical inland soils (group III) representing approximately 70% of the SURE land area contribute 55% of the measured sulfur flux. Therefore, approximately 95% of the soils in the SURE area which are inland and nonsaline contribute nearly 80% of the biogenic sulfur to the SURE study area. A total annual sulfur flux was calculated for the 130 oceanic SURE grids of approximately 13,000 metric tons using the published data for Atlantic Ocean sulfur flux (Maroulis and Bandy, 1977; Lovelock et al., 1972) (Table 2).

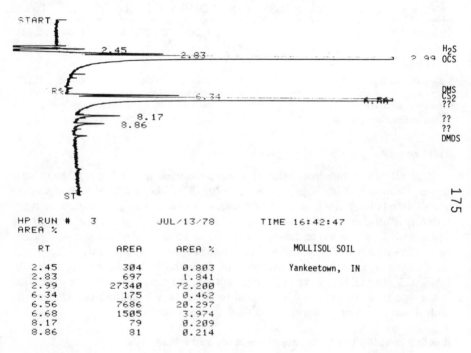

Fig. 4. Typical sulfur flux chromatogram showing known and unknown sulfur compounds.

Examination of selected WCOT chromatograms for DMS and CS$_2$

Some of the earlier chromatographic data (Lovelock et al., 1972; Rasmussen, 1974) indicated that DMS should be considered as an important biogenic product, together with H$_2$S, which was the initially suggested natural precursor of atmospheric SO$_2$ and sulfate aerosol. Subsequently, COS and CS$_2$ (Sandalls and

Penkett, 1977) were added to the reported atmospheric sulfur compounds. Through the use of WCOT/GC analysis in our field studies and GC single-ion-monitoring mass spectrometry (GC/SIM/MS), CS_2 has been positively identified (Holdren, 1978). The separation of CS_2 from DMS is greatly enhanced through the use of WCOT/GC. From our analysis of more than 600 soil flux samples from a wide range of soil orders within (and now south) of the SURE study area, we believe that most of the earlier PC/GC analyses reporting DMS could well have included substantial proportions of CS_2 as an unresolved component of the GC peak identified as DMS. Therefore, we conclude that CS_2 is a common, frequently occurring constituent in the natural emissions from soils, water, and vegetation. Such speciation is not routinely possible with PC/GC.

Unidentified sulfur gases from soils

Other sulfur compounds have also been observed in some of our soil flux samples. However, the GC retention times and the relative retention times (RR_t) based upon the CS_2 R_t of 1.000 did not compare with RR_t's for any of the 23 sulfur compounds published by Farwell et al. (1979). A typical chromatogram containing several unknown sulfur peaks from an Ohio River alluvial mollisol is shown in Fig. 4. These unidentified compounds, when present, generally represent less than 3% of the total sulfur present.

CONCLUSIONS

Aneja et al. (1979) measured a flux range of 0.01 to 100 g of S per square meter per year over two salt marshes. Adams et al. (1979) found an average land sulfur flux of 0.02 g $m^{-2}year^{-1}$ for the SURE study area. Jaeschke (1978) measured a surface flux equivalent to 0.04 g $m^{-2}year^{-1}$ over tide flats and swamps in Denmark. All of these data are in line with Granat's estimate (Granat et al., 1976) of global natural sulfur flux of 40 × 10^{12} g $year^{-1}$, or about 0.08 g m^{-2} $year^{-1}$ averaged over the entire earth (5 × 10^8 km^2), or 0.04 g of S per square meter per year for the land surface.

Hitchcock (1976) and others have assumed that the emission of H_2S (or any other volatile reduced sulfur compound) from inland areas is small. The data reported herein indicate that (a) H_2S is not necessarily the major constituent of the sulfur flux; (b) inland soils contribute significantly to the total sulfur flux of the SURE study area; (c) COS and CS_2 are also important constituents in the sulfur flux from most soils, water, and vegetation; (d) the sulfur flux is inhomogeneous over the expanse of a tidal marsh; (e) sulfur flux from inland swampy wetlands may be of lesser impact to the local sulfur budget than adjacent cultivated peat/muck soils; and (f) unidentified sulfur compounds in minor concentrations may exist from time to time in the flux from various soil orders.

It is obvious that there are many related subjects requiring further experimental clarification, including (a) the relative significance of various biogenic sources such as water surfaces, soils, and vegetation; (b) the identity and relative proportions of the sulfur compounds in the flux emissions under various natural conditions; (c) the fate of natural sulfur compounds once released to the atmosphere—sources, sinks, residence times, and reaction products, especially COS and CS_2; (d) the most appropriate experimental procedures to quantify sulfur compound emissions from the wide variety of natural sources; and (e) how anthropogenic sources compare with biogenic in determining the overall atmospheric sulfur budget.

We are presently continuing our 1979–1980 research south of the SURE area—in Georgia, Florida, Alabama, Louisiana, and Texas—to provide biogenic emission data from subtropical wetlands. It must be emphasized that our data represent only a small fraction of the global surface and do not yet provide fluxes for the most "sulfur-productive" areas of the globe, namely, the high-biomass subtropics and tropics.

ACKNOWLEDGMENTS

This work was supported by the Electric Power Research Institute under Contract 856-1. The assistance of W. L. Bamesberger in computer programming and data analysis is gratefully acknowledged. Reference to commercial products is for identification purposes only and does not constitute an endorsement of these products by EPRI, the University of Idaho, or Washington State University.

REFERENCES

Adams, D. F., S. O. Farwell, E. Robinson, and M. R. Pack. 1979. Assessment of Biogenic Sulfur Emissions in the SURE Area. Final report submitted to Electric Power Research Institute, Palo Alto, California.

Aneja, V. P. 1975. Characterization of Sources of Biogenic Atmospheric Sulfur Compounds. M.S. thesis, North Carolina State University, Raleigh, North Carolina.

Aneja, V. P., et al. 1979. Biogenic Sulfur Sources Strength Field Study. Report ESC-TR-79-01 submitted to U.S. Environmental Protection Agency, Research Triangle Park, North Carolina.

Barr, A. J., J. H. Goodnight, J. P. Sall, and J. T. Helwig. 1976. A User's Guide to SAS 76. SAS Institute, Raleigh, North Carolina.

Berchard, M. S. 1974. Emission of Volatile Organic Sulfides by Freshwater Algae. M.S. thesis, Washington State University, Pullman, Washington.

Duncan, D. B. 1975. t-tests and intervals for comparisons suggested by data. Biometrics 31:339–359.

Farwell, S. O., S. J. Gluck, W. L. Bamesberger, T. M. Schutte, and D. F. Adams. 1979. Determination of sulfur-containing gases by a deactivated cryogenic enrichment and capillary gas chromatography system. Anal. Chem. 51(6):609–615.

Granat, L., R. O. Hallberg, and H. Rodhe. 1976. The global sulphur cycle. Pp. 39–134 in Svensson, B. H., and R. Soderlund (eds.), SCOPE report 7, Ecol. Bull. (Stockholm).

Grey, D. C., and M. C. Jensen. 1972. Bacteriological sulfur in air pollution. Science 177:1099.

Hakkarinen, C., D. F. Adams, and S. O. Farwell. 1979. Measurement of Gaseous Sulfur Emissions from Soils in the Eastern United States. Paper presented at the Fourth International Conference of the Commission on Atmospheric Chemistry and Global Pollution, August 12–18, 1979, Boulder, Colorado.

Hill, F. B., V. P. Aneja, and R. M. Felder. 1978. A technique for measurement of biogenic sulfur emission fluxes. J. Environ. Sci. Health, Part A 13(3):199–225.

Hitchcock, D. R. 1976. Microbiological contributions to the atmospheric load of particulate sulfate. Pp. 351–367 in Nriagu, J. O. (ed.), Environmental Biochemistry, Ann Arbor Science.

Hitchcock, D. R. 1977. Biogenic Sulfur Sources and Air Quality in the United States. Final report to the National Science Foundation, NSF RANN Grant No. AEN-7514571.

Holdren, M. 1978. Unpublished information. Washington State University, Pullman, Washington.

Jaeschke, W. 1978. New methods for analysis of SO_2 and H_2S in remote areas and their applicability to the atmosphere. Atmos. Environ. 12:715–722.

Lovelock, J. E., and R. J. Maggs, and R. A. Rasmussen. 1972. Atmospheric dimethyl sulphide and the natural sulphur cyle. Nature 237:452.

Maroulis, P. J., and A. R. Bandy. 1977. Estimate of the contribution of biologically produced dimethyl sulfide to the global sulfur cycle. Science 196:647–648.

Perhac, R. M. 1978. Sulfate regional experiment in northeastern United States: the "SURE" program. Atmos. Environ. 12:641–647.

Rasmussen, R. A. 1974. Emission of biogenic hydrogen sulfide. Tellus 26:254–260.

Rodhe, H. 1978. Budgets and turnpover times of atmospheric sulfur compounds. Atmos. Environ. 12:671–680.

Sandalls, F. J., and S. A. Penkett. 1977. Measurement of carbonyl sulfide and carbon disulfide in the atmosphere. Atmos. Environ. 11:197–199.

Tauber, H. 1949. The Chemistry and Technology of Enzymes. John Wiley, New York: p. 4.

DISCUSSION

George Sehmel, Pacific Northwest Laboratory: When you sample for ten minutes do you completely exhaust the sulfur in the system? What happens if you sample for successive ten-minute intervals? Does sulfur continue to be released?

D. F. Adams: First of all, I should say that the chamber was flushed for a period of 15 to 20 minutes to allow an equilibrium to develop between the surface flux and the sulfur-free sweep air. The actual sampling periods varied from two minutes to four hours. In some cases, we left the chamber in position and took a second sample. There was no significant difference between these samples.

Malcom Ko, Atmospheric and Environmental Research, Inc.: Do you have enough data from your soil type correlations to estimate the biogenic input to global sulfur levels and how these sources compare with anthropogenic emissions?

D. F. Adams: The latest estimates by Granat et al.,[1] which are based on circumstantial reasoning, place biogenic sulfur emissions at 0.04 gram of sulfur per square meter per year, on a global basis. Our figures for the SURE area come out to 0.02 gram of sulfur per square meter per year, so we are in the same ball park. We haven't considered the subtropic area with its high biomass, nor have we considered the artic area with its relatively low biomass. These two ends of the spectrum may keep our global numbers somewhat in the same order as those which we are reporting for the SURE area.

[1]Granat, L., H. Rodhe, and R. O. Hallberg (1976) The Global Sulfur Cycle. In Svensson, B. H. and R. Soderlund (eds.) Nitrogen, Phosphorous and Sulphur–Global Cycles. SCOPE Report 7, Ecol. Bull. (Stockholm) 175:587–596.

CHAPTER 7

DIRECT MEASUREMENTS OF EMISSION RATES OF SOME ATMOSPHERIC BIOGENIC SULFUR COMPOUNDS AND THEIR POSSIBLE IMPORTANCE TO THE STRATOSPHERIC AEROSOL LAYER

Viney P. Aneja*

ABSTRACT

Atmospheric sulfur compounds of biogenic origin are thought to constitute a significant fraction of the atmospheric sulfur burden. A determination of fluxes of these compounds into the atmosphere is desirable in order to permit accurate assessment of the relative roles of anthropogenic and biogenic sources in contributing to such phenomena as the atmospheric sulfate burden and acidity in precipitation.

Direct measurements of sulfur emission rates were made in salt marshes on the coast of North Carolina during the summers of 1977 and 1978. An emission flux reactor (chamber) technique was used to determine the emission rates of sulfur compounds into the atmosphere. The sulfur gases were identified and their concentrations in the flux reactor measured with a gas chromatograph (GC) equipped with a flame photometric detector specific for sulfur (S). Hydrogen sulfide (H_2S, average flux $\sim$0.5 g of S per square meter per year), dimethyl sulfide [$(CH_3)_2S$, average flux $\sim$0.4 g of S per square meter per year], carbonyl sulfide (COS, average flux $\sim$0.03 g of S per square meter per year), and carbon disulfide (CS_2, average flux $\sim$0.3 g of S per square meter per year) were measured. In general, the emission rates of these gases varied over a wide range of concentrations under varying conditions of soils, soil moisture, temperature, and insolation. Based upon the measured fluxes, the emissions from marshes are important to the sulfate aerosol burden (19% or less) of the stratosphere but unimportant for the tropospheric sulfur burden.

INTRODUCTION

Natural processes and anthropogenic activity are the two sources of sulfur-containing compounds in the atmosphere. Natural sources are believed to constitute a large fraction ($\sim$0.1 to $\sim$0.7) of the atmospheric sulfur burden (Conway, 1943; Eriksson, 1963; Junge, 1963; Robinson et al., 1968; Kellogg et al., 1972; Friend, 1973; Granat et al., 1976). Only recently (Aneja, 1975; Aneja et al., 1979b, 1979c; Hill et al., 1978; Adams et al., 1978) has comprehensive experimental evidence been reported in which earth-atmosphere fluxes of biogenic sulfur compounds were measured. Moreover, physical and chemical properties thought to influence the release of these compounds were also measured. Identification and characterization

*Northrop Services, Inc., Environmental Sciences, Post Office Box 12313, Research Triangle Park, N.C. 27709.

of sources of atmospheric biogenic sulfur compounds are essential for the rational formulation of emission control policies designed to limit the atmospheric sulfate burden and for analysis of the origins of acid precipitation.

Biogenic sulfur is released from vegetation and is produced by the decomposition of organic matter and from bacterial sulfate reduction in the biosphere. Data on sulfur release by vegetation are scanty, although some higher plants are known to emit dimethyl sulfide (Aneja et al., 1979b) and carbon disulfide (Aneja et al., 1979c). Many fungi and bacteria release sulfur compounds (Starkey, 1964) during organic decomposition. The best-known microbial process for generating hydrogen sulfide (H_2S) is the reduction of sulfate (SO_4^{2-}) by bacteria of the genera *Desulfovibrio* and *Desulfotomaculum*. These anaerobes utilize sulfate and other inorganic sulfur compounds as specific hydrogen acceptors during the oxidation of organic energy sources (Clarke, 1953; Postgate, 1959). During this metabolic cycle, the original sulfur compounds are reduced to sulfides. The latter organisms are strictly anaerobic, while nonspecific reducers may be found in anaerobic or aerobic environments. Both processes of H_2S production require the presence of organic material. Hydrogen sulfide is produced in an aqueous medium such as a film of moisture in soil, interstitial water in sediments, or the bulk water of natural bodies of water.

Field investigations of biogenic sulfur compounds have involved concentration measurements of compounds of particular interest in the ambient air (Breeding et al., 1973; Natusch et al., 1972; Maroulis et al., 1977) and direct comprehensive measurements of emission fluxes of biogenic sulfur compounds (Aneja, 1975; Aneja et al. 1979a, 1979b, 1979c; Hill et al., 1978; Adams et al., 1978) along the east coast of the United States from tidal marshes and water surfaces. Inland nonaquatic environments are also of interest, especially if plants are emitting sulfur compounds.

This paper reports estimates of sulfur emission rates into the atmosphere based on sampling done at two salt marshes (Cox's Landing and Cedar Island) on the coast of North Carolina during the summers of 1977 and 1978, using an emission flux reactor. During the same period a similar comprehensive study was performed by Adams et al. (1978).

FLUX MEASUREMENT TECHNIQUE

An emission flux reactor (Aneja et al., 1979a) was employed for measuring earth-atmosphere fluxes of biogenic sulfur compounds (Fig. 1). This chamber technique has an important feature enabling high sensitivity for flux measurements without the necessity of measuring very low concentrations. In addition, gas residence times in the emission flux reactor are on the order of minutes; thus, chemical transformations occurring between emission and analysis are minimized. The sweep gas passed through the reactor was ambient air (aerated). The walls are made from 5-mil-thick (0.13 mm) FEP Teflon supported by an exterior aluminum frame. This design was chosen to afford negligible attenuation of ambient light and minimal wall reactions. A more complete design of the experimental field apparatus is shown in Fig. 2. Subsequently, the emission rate, n, was calculated from the following equation:

$$n = \frac{F \cdot \Delta C}{A},$$

where n is the emission rate of a particular compound, ΔC is the concentration increase in the gas leaving the reactor over its concentration in the gas entering the reactor, F is the steady volumetric flow rate of gas through the reactor, and A is the area of emitting surface covered by the reactor.

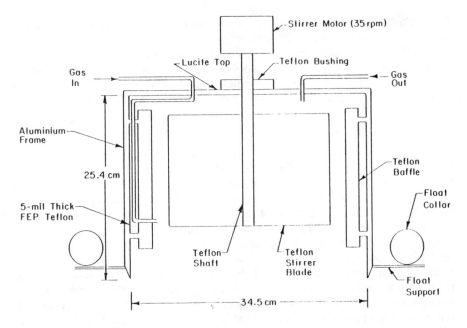

Fig. 1. Emission flux reactor. From Aneja et al. (1979a).

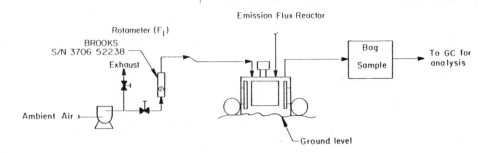

Fig. 2. Layout of field experiment using emission flux reactor with ambient air carrier gas. From Aneja et al (1979a).

The sulfur compounds were analyzed by a gas chromatograph equipped with a flame photometric detector. A 394-nm interference filter made the detector sensitive to sulfur compounds. An ~11-m FEP Teflon column (0.32 cm outside diameter) packed with 40–60 mesh Teflon coated with 5% polyphenyl ether (PPE) and 0.5% phosphoric acid (H_3PO_4) was used at 50°C to separate gaseous sulfur species (Stevens et al., 1971). The carrier gas was nitrogen with a flow rate of ~80 ml/min. The column and operating conditions gave good resolution of peaks for H_2S plus COS (retention time, 1.5 min), sulfur dioxide (SO_2 retention time, 3.1 min), methyl mercaptan (CH_3SH retention time, 5.0 min), dimethyl sulfide (DMS retention time, 12.2 min), and carbon disulfide (CS_2 retention time, 14.5 min). Overlapping DMS and CS_2 peaks were manually resolved into separate contributions. Hydrogen sulfide and COS were not separated by this column. The separation of these two species was obtained by incorporating a second (5.8 m) FEP Teflon column (0.32 mm outside diameter) packed with 50/80 mesh Porapak QS in parallel with the first.

The two columns were separately incorporated into the flow lines by a pneumatically controlled switching system. The carrier gas for the second column was nitrogen with a flow rate of ~40 ml/min. The retention time for COS under these conditions is ~8 min, allowing a good separation from H_2S (retention time ~5.5); however, this column retained CS_2. The gas chromatograph was calibrated for both columns in the laboratory and in the field using a dilution system and bag samples.

EMISSION FLUX MEASUREMENTS

Measurements were conducted for biogenic sulfur species at two sites on the east coast of the United States (Fig. 3); (1) Cedar Island Wild Life Refuge, North Carolina, and (2) Cox's Landing, Long Beach, North Carolina. At each location, sampling was done over marsh grass (*Spartina alterniflora* clipped to ~2.5 cm above ground level) and mud flats under aerated (ambient air) conditions. Time-integrated bag samples were collected every 30 min and analyzed within 30 min for various gaseous sulfur species. Experiments were performed during the daytime in the intertidal zone at low tide (absence of water column). During the experiments over *Spartina*, ambient temperatures ranged between 21° and 36°C, while sediment temperatures ranged between 25.5° and 32.5°C. For the mud-flat experiments, both ambient and sediment temperatures ranged from 29.5° to 32.5°C. The sulfate content at both locations was similar, ~2400 µg per milliliter of solution (Aneja et al., 1979b). The predominant sulfur species emitted over the *Spartina* zone was dimethyl sulfide, while the predominant sulfur species emitted over the mud-flat zone was hydrogen sulfide plus carbonyl sulfide (Aneja et al., 1979b).

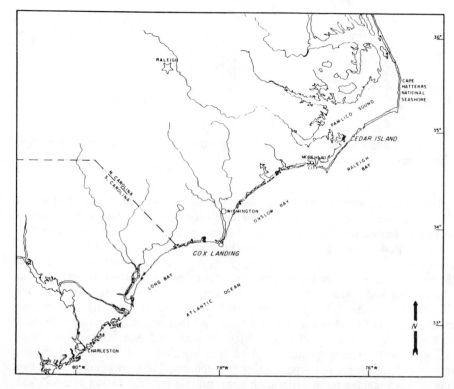

Fig. 3. Location of the sites. From Aneja et al. (1979b).

Climatic and environmental factors play an important role in natural emission phenomena. Aneja (1975), Hill et al. (1978), and Aneja et al. (1979b) have found that sulfur (H_2S and DMS) emission rates increase with increasing temperatures (Figs. 4 and 5). Aneja et al. (1979b) and Adams et al. (1978) observed the temperature correlation for sulfur emission rates from numerous sites.

In the present study, the measured fluxes of H_2S ($\sim$0.5 g of S per square meter per year) in saline marshes over mud flats are comparable with some earlier direct emission rate measurements (Aneja, 1975; Hill et al., 1978; Aneja et al., 1979b). Dimethyl sulfide, the predominant species over the *Spartina alterniflora* zone (Aneja et al., 1979b), has an emission rate of $\sim$0.4g of S per square meter per year, consistent with results reported earlier (Aneja et al., 1979b). However, the measured DMS emission rates over *Spartina* exceed the indirectly measured values (Maroulis et al., 1977). Carbonyl sulfide flux in saline marsh locales is $\sim$0.03 g of S per square meter per year, while CS_2 (predominant over the *Spartina* zone) has an emission rate of $\sim$0.30 g of S per square meter per year (Aneja et al., 1979c).

The fluxes reported in the present work ($\sim$0.5 g of S per square meter per year), based on diurnal and tidal changes from fresh and saline marshes, are compared with biogenic fluxes derived from theoretically calculated global sulfur budgets and anthropogenic emissions expressed as fluxes. Theoretical average terrestrial fluxes, calculated from recent global atmospheric sulfur budgets, range from 0.4 to 0.7 g of S per square meter per year, while marine fluxes, derived with identical methods, range from 0.1 to 0.5 g of S per square meter per year (Hill et al., 1978).

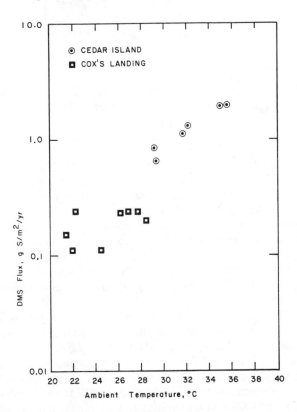

Fig. 4. Ambient temperature dependence of dimethyl sulfide flux over the *Spartina* zone at Cedar Island and Cox's Landing. From Aneja et al. (1979b).

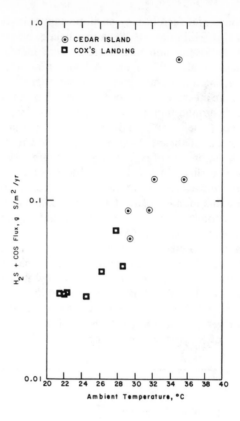

Fig. 5. Ambient temperature dependence of hydrogen sulfide plus carbonyl sulfide flux over the *Spartina* zone at Cedar Island and Cox's Landing. From Aneja et al. (1979b).

To estimate the percentage that biogenic sulfur (H_2S, DMS, CS_2, and COS) contributes to the global tropospheric sulfur cycle, we assumed that all marshes are completely made up of *Spartina alterniflora* and that they emit uniformly. The total marsh area was taken as 3.8×10^5 km^2 (Woodwell et al., 1973), and we assumed that 100×10^6 tons of gaseous sulfur are emitted per year from the biosphere by natural processes (Robinson and Robbins, 1968). Based on our H_2S, DMS, CS_2, and COS emission rates and the above assumptions, the contribution of biogenic sulfur (H_2S, DMS, CS_2, and COS) from marshes is a small portion (~1 %) of the global tropospheric natural sulfur cycle. Obviously, more work is needed to characterize the global process adequately.

Since only CS_2 and COS are able to diffuse through the troposphere and be converted to sulfate in the stratosphere, we also estimated the contributions of CS_2 and COS emissions to the stratospheric sulfate aerosol layer. According to Crutzen's (1976) estimate, a flux of $\sim1.8 \times 10^{-4}$ g of S per square meter per year into the stratosphere will account for the entire stratospheric sulfate aerosol layer. Applying Junge's (1974) one-dimensional eddy diffusion model to the troposphere and using the present CS_2 and COS average emission rate for global marshes and CS_2 and COS atmospheric lifetimes (Sandalls and Penkett, 1977) of 1 year and 20 years respectively, an upper limit to the contributions can be made by assuming zero CS_2 and

COS concentrations at 15 km height (average Junge aerosol layer $\sim$20 km). The upper-limit contributions are 8% for CS_2 and 11% for COS. Thus, although the observed emissions do not substantially contribute ($\sim$1% based on the global marsh area) to the natural tropospheric sulfur cycle, they may contribute up to $\sim$19% of the stratospheric sulfur aerosol layer (Aneja et al., 1979c).

ACKNOWLEDGMENTS

This work was supported by the Environmental Sciences Research Laboratory of the U.S. Environmental Protection Agency under contract No. 68-02-2566.

REFERENCES

Adams, D. F., S. O. Farwell, M. R. Pack, and E. Robinson. 1978. An initial emission inventory of biogenic sulfur flux from terrestrial surfaces. Paper 78-16 presented at the Annual Meeting of PNWIS/APCS, Portland, Oregon.

Adams, D. F., S. O. Farwell, M. R. Pack, and W. L. Bamesberger. 1979. Preliminary measurements of biogenic sulfur-containing gas emissions from soils. J. Air Pollut. Control Assoc. 29(4):380-383.

Aneja, V. P. 1975. Characterization of Sources of Biogenic Atmospheric Sulfur Compounds. M.S. thesis, Department of Chemical Engineering, North Carolina State University, Raleigh, North Carolina.

Aneja, V. P., E. W. Corse, L. T. Cupitt, J. C. King, J. H. Overton, Jr., R. E. Rader, M. H. Richards, H. J. Sher, and R. J. Whitkus. 1979a. Biogenic Sulfur Sources Strength Field Study. Northrop Services, Inc. Report No. ESC-TR-79-22, Research Triangle Park, North Carolina.

Aneja, V. P., J. H. Overton, Jr., L. T. Cupitt, J. L. Durham, and W. E. Wilson. 1979b. Direct measurements of emission rates of some atmospheric biogenic sulfur compounds. Tellus 31(2):174-178.

Aneja, V. P., J. H. Overton, Jr., L. T. Cupitt, J. L. Durham, and W. E. Wilson. 1979c. Carbon disulfide and carbonyl sulfide from biogenic sources and their contribution to the global sulfur cycle. Nature 282:493-496.

Breeding, R. J., J. B. Lodge, Jr., J. B. Pate, D. B. Sheeseley, M. B. Klonis, B. Fogle, J. A. Anderson, T. R. Englest, P. L. Haagenson, R. B. McBeth, A. L. Morris, R. Pogue, and A. F. Wortburg. 1973. Background trace gas concentration in the central United States. J. Geophys. Res. 78:7057-7064.

Clarke, P. H. 1953. Hydrogen sulfide production by bacteria. J. Gen. Microbiol. 8:397.

Conway, E. J. 1943. Mean geochemical data in relation to ocean evolution. Proc. R. Irish Acad., Sect. A 48:119-159.

Crutzen, P. J. 1976. The possible importance of CSO for the sulfate layer of the stratosphere. Geophys. Res. Lett. 3(2):73-76.

Eriksson, E. 1963. The yearly circulation of sulfur in nature. J. Geophys. Res. 68:4001-4008.

Friend, J. P. 1973. The global sulfur cycle. In Reasool, S. I. (ed.), Chemistry of the Lower Atmosphere. Plenum Press, New York. Pp. 177-201.

Granat, L., R. O. Hallberg, and H. Rodhe. 1976. The global sulfur cycle. In Svensson, B. H., and R. Soderlund (eds.), Nitrogen, Phosphorus, and Sulfur—Global Cycles. SCOPE Report 7. Ecol. Bull. (Stockholm) 22:39-134.

Hill, F. B., V. P. Aneja, and R. M. Felder. 1978. A technique for measurement of biogenic sulfur emission fluxes. Environ. Sci. Health 13(3):199-225.

Junge, C. E. 1963. Air Chemistry and Radioactivity. Academic Press, New York.

Junge, C. 1974. Sulfur budget of the stratospheric aerosol layer. Proceedings of the International Conference of Structure, Composition and General Circulation of the Upper and Lower Atmospheres and Possible Anthropogenic Perturbations, Melbourne, Australia.

Kellogg, W. W., R. D. Cadle, E. R. Allen, A. L. Lazrus, and E. A. Martell. 1972. The sulfur cycle. Science 175:587-596.

Maroulis, P. J., and A. R. Bandy. 1977. Estimate of the contribution of biologically produced dimethyl sulfide to the global sulfur cycle. Science 196:647-648.

Natusch, D. F. S., M. S. Klonis, M. D. Anelford, R. J. Teck, and J. P. Lodge, Jr. 1972. Sensitive method for the measurement of atmospheric hydrogen sulfide. Anal. Chem. 44:2067-2070.

Postgate, J. R. 1959. Sulfate reduction by Bacteria. Ann. Rev. Microbiol. 13:505.

Robinson, E., and R. C. Robbins. 1968. Sources, Abundance, and Fate of Gaseous Atmospheric Pollutants. SRI Project Report PR-6755, prepared for the American Petroleum Institute, New York.

Sandalls, F. J., and S. A. Penkett. 1977. Measurements of carbonyl sulfide and carbon disulfide in the atmosphere. Atmos. Environ. 11:197-199.

Starkey, R. L. 1964. Microbial transformations of some organic sulfur compounds. In Heukelekian, H., and N. C. Dondero (eds.), Principles and Applications in Aquatic Microbiology. John Wiley, New York. Pp. 405-429.

Stevens, R. K., J. D. Mulik, A. E. O'Keefee, and K. J. Krost. 1971. Gas chromatography of reacted sulfur gases in air at the parts-per-billion level. Anal. Chem. 43:827.

Woodwell, G. M., P. M. Rich, and C. A. S. Hall. 1973. Carbon and the Biosphere, eds. E. M. Woodwell and E. V. Pecan. AEC Symp. Ser. 30:221-240. U. S. Department of Commerce.

DISCUSSION

Malcolm Ko, Atmospheric and Environmental Research, Inc.: Some reaction rate measurements have been made recently which indicate that the lifetime of carbonyl sulfate is one year instead of the 20-year lifetime you use in your calculation. Would this shorter lifetime significantly change your estimate of the biogenic contribution to stratospheric sulfur?

Viney P. Aneja: It would change the estimate somewhat, but I don't think the answer would be significantly different.

S. Lindberg, ORNL: Were your annual biogenic emission estimates based on simple extrapolation of 30-minute measurements, or did you take into account possible diurnal and seasonal variations?

Viney P. Aneja: Diurnal and seasonal variations were taken into account.

CHAPTER 8

IMPACT OF RECENT MEASUREMENTS OF OCS, CS₂, AND SO₂ IN BACKGROUND AIR ON THE GLOBAL SULFUR CYCLE

Alan R. Bandy*

Peter J. Maroulis*

ABSTRACT

Background measurements of sulfur dioxide, carbon disulfide, and carbonyl sulfide are reported with emphasis on the marine atmosphere. The SO_2 and OCS data were collected during aircraft flights extending from 57°S to 70°N latitude. The CS_2 data were collected at Philadelphia, Pa., and Wallops Island, Va. Atmospheric oxidation of CS_2 was estimated to contribute no more than 20% of the OCS and no more than 33% of the SO_2 measured in the marine atmosphere.

INTRODUCTION

The effects of anthropogenic emissions of sulfur gases and their reaction products on human health and regional environment have received a great amount of attention in the literature and this symposium. Indirect effects of these materials on the global environment, especially weather and climate, however, have received considerably less attention. It is the purpose of this paper to point out what some of the global effects might be, to review some of the more important recent findings, and to suggest some areas where future research should be concentrated.

Some atmospheric particles, called condensation nuclei and formed in large part from sulfur gas precursors in background areas, are extremely effective in promoting the formation of fog and haze particles and raindrops. Presently the chemistry and physics of these condensation nuclei on a molecular level are in an elementary stage of development. However, through a series of reactions, most of which are not well characterized, sulfur gases are oxidized, forming molecular intermediates from which very hygroscopic sulfuric acid condensation nuclei are formed.

Regional reductions in visibility from haze and fog intensified by the condensation nuclei derived from anthropogenic materials have been linked to local decreases in visibility and concurrent reductions in the aesthetic qualities of the atmospheric environment. The potential effects of anthropogenic releases of condensation nuclei precursors on the radiation balance of the earth and thus the climate of the earth have not received nearly as much attention.

---

*Chemistry Department, Drexel University, Philadelphia, Pa. 19104.

Perturbations on the haze-, fog-, and rain-forming ability of the atmosphere are likely to be largest in regions of low condensation nuclei concentration. Natural levels of condensation nuclei exist in continental areas (more than 1000 cm^{-3}), and although the increment produced from anthropogenic precursors can be quite large, the fractional contribution, though significant, is not dominant.

In the marine atmosphere, however, the nominal condensation nuclei concentration is less than 100 cm^{-3}, an order of magnitude less than that found in continental air. Thus the haze-, fog-, and rain-forming ability of the marine atmosphere is certainly different and is probably much less than that of continental air. The chief effect is a dramatically larger mean particle size in marine clouds, rain, and fog.

The vulnerability of the condensation nuclei processes in the marine atmosphere is supported by studies which show that haze and fog are more intense in regions where marine air and continental air are mixed. Such effects would presumably be intensified by anthropogenic precursors of condensation nuclei.

A first step toward evaluating the potential effects of anthropogenic sulfur gas precursors on the haze-, fog-, and rain-forming characteristics of the marine atmosphere is a detailed understanding of the chemical processes in which the sulfur gases are converted to condensation nuclei. The remainder of this paper will deal with the data identifying the natural sulfur gases in the marine atmosphere, with a small section reviewing the current state of knowledge concerning the chemistry of these compounds in the marine atmosphere as it pertains to condensation nuclei formation.

SULFUR GASES IN THE MARINE ATMOSPHERE

At the present time, only SO_2 (Maroulis et al., 1980; Georgii, 1970; Georgii and Vitze, 1971; Jaeschke et al., 1976; Nguyen et al., 1974a, b), OCS (Torres et al., 1980; Maroulis et al., 1977) and H_2S (Goldberg et al., 1980; Slatt et al., 1978) have been definitively identified in the marine atmosphere; however, many other sulfur gases, such as CS_2 (Maroulis and Bandy, 1980), CH_3SH, $(CH_3)_2S$ (Maroulis and Bandy, 1977), and $(CH_3)_2SO$ have been suggested to be there. Adequate data are now available on SO_2 (Margoulis et al., 1980; Nguyen et al., 1974a, b) and OCS (Torres et al., 1980) so that the global distribution of these compounds is fairly well known.

A large fraction of the available background SO_2 and OCS data was obtained during a two-month period in the late spring and early summer of 1978 in the Global Atmospheric Measurement Experiment of Trace Aerosols and Gases Program, GAMETAG, sponsored by the National Science Foundation. The aircraft flights on which these data were taken extended from 57°S (south of Christchurch, New Zealand) to 70°N (north of Whitehorse, Canada). Emphasis here will be on the remote regions of the Pacific Ocean and northern Canada.

The SO_2 data taken in the GAMETAG program are shown as a plot of SO_2 concentration as a function of latitude in Fig. 1. Inspection of this figure reveals that the concentration of SO_2 over the continental areas of the Northern Hemisphere was larger than over the marine areas of the Pacific Ocean. As the distance from continental areas increased, the SO_2 concentration decreased to an average level of about 80 pptv (parts per trillion by volume) in the Pacific marine atmosphere, which is approximately a factor of 2 less than the mean SO_2 concentrations found in continental air. Clearly the SO_2 from the North American continent is not very important in Pacific Ocean air even in the Northern Hemisphere.

There appears to be a natural background of SO_2 in the marine atmosphere, which had been expected but which previously had been thought to be in large part SO_2 transported to these marine areas from continental anthropogenic sources.

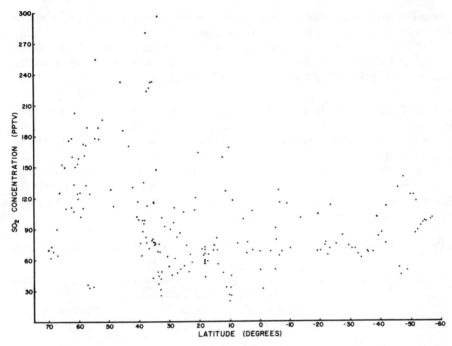

Fig. 1. Plot of all the individual GAMETAG SO₂ data as a function of latitude.

A statistical analysis of various subsets of the SO₂ data, shown in Table 1, reveals that the mean marine SO₂ concentrations of 54 pptv in the boundary layer and 85 pptv in the free troposphere are approximately one-half the mean continental SO₂ concentrations of 112 pptv in the boundary layer and 160 pptv in the free troposphere. The marine SO₂ could be the principal precursor of condensation nuclei in the marine atmosphere, particularly above the region where sea-salt particles are concentrated.

Table 1. Statistical analysis of sulfur dioxide grouped into various subsets

	Boundary-layer data			Free-troposphere data		
Category	Concentration, mean and standard deviation (pptv)	SEM* (pptv)	Number of samples	Concentration, mean and standard deviation (pptv)	SEM* (pptv)	Number of samples
All data	82 ± 63	10	42	113 ± 74	6	159
Northern Hemisphere	89 ± 69	12	32	122 ± 85	8	116
Southern Hemisphere	57 ± 18	6	10	90 ± 21	3	43
Continental	112 ± 79	18	20	160 ± 100	13	59
Marine	54 ± 19	4	22	85 ± 28	3	100

*Standard error of the mean.

The relatively large SO₂ levels in the remote regions of northern Canada are of particular interest. These high levels of SO₂, at times exceeding 200 pptv, are supported by either an unusually large natural source or the transport of anthropogenic SO₂ to these regions. Unusually high levels of methyl chloroform (Krasnec et al., 1978; L. Rasmussen and R. Rasmussen, 1978) and CO (Heidt and Krasnec, 1978; R. Rasmussen, 1978) in this region as well as the persistence at times of an inordinate amount of haze suggest that transport of anthropogenic material to this area may be the answer. Further work in this region is needed.

Carbonyl sulfide is the other major sulfur gas constituent of the marine background atmosphere. Measurements during the past three years have provided a clear picture of the global distribution of OCS (Torres et al., 1980; Maroulis et al., 1977).

Table 2 contains the statistical analysis of our 1978 aircraft OCS data grouped into various subsets. This table reveals that the OCS concentration was almost uniform throughout the regions sampled. There was no significant difference in OCS levels in the Northern Hemisphere and Southern Hemisphere. Also, the mean OCS levels in continental and marine areas were 524 and 511 pptv, respectively, which were statistically the same. The mean OCS concentration was 509 pptv in the free troposphere and 518 pptv in the boundary layer, indicating a homogeneous mixing of OCS throughout the troposphere.

Table 2. Statistical analysis of 1978 GAMETAG carbonyl sulfide data grouped into various subsets

Category	Concentration, mean and standard deviation (pptv)	Standard error of the mean (pptv)	Number of samples
Free troposphere	509 ± 63	4	240
Boundary layer	518 ± 74	8	92
Northern Hemisphere	523 ± 63	4	242
Southern Hemisphere	498 ± 71	7	104
Continental	524 ± 61	6	102
Marine	511 ± 62	4	244

There were four flights during 1978 in which the free-troposphere and boundary-layer OCS concentrations were significantly different. These data are shown in Table 3. For the three flights which took place over the Pacific Ocean, OCS levels were 51 pptv higher in the boundary layer than in the free troposphere. This was believed to be due to a reduction in the rate of reaction of OH with OCS due to a reduction in OH, which in turn was caused by a low O_3 concentration, a necessary precursor of OH in clean background air.

Table 3. Statistical comparison of OCS concentrations in the free troposphere and boundary layer

Flight[a]	Free troposphere Mean and standard deviation[b] (pptv)	95% confidence interval of the mean (pptv)	Boundary layer Mean and standard deviation[b] (pptv)	95% confidence interval of the mean (pptv)	T test of the equality of the means	Significance level[c]
JA to CI	507 ± 38(17)	488 to 527	555 ± 75(8)	492 to 618	−2.15	0.0474
CI to FI	529 ± 28(16)	514 to 544	596 ± 40(4)	533 to 659	−3.96	0.00128
CI to JA	537 ± 22(9)	520 to 553	575 ± 38(5)	527 to 622	−2.42	0.0405
GF to WH	583 ± 19(10)	570 to 597	538 ± 55(4)	451 to 625	2.39	0.0422

[a]JA: Johnston Atoll; FI: Fiji Island; WH: Whitehorse, Canada; CI: Canton Island; GF: Great Falls, Montana.

[b]Number of samples is included in parentheses.

[c]Two-tail test.

The fourth flight in which OCS levels in the boundary layer and free troposphere were different occurred over the continental areas of Canada. On this flight, OCS levels in the free troposphere were found to be 45 pptv higher than in the boundary layer. This part of our data has not been adequately explained, although the transport of OCS to this area from anthropogenic sources in northern Europe is likely. This is an intriguing aspect of the data which requires further investigation.

A plot of the OCS data obtained in the GAMETAG program as a function of latitude is shown in Fig. 2. Inspection of this figure reveals that the concentration of OCS is about five times higher than that of SO_2 in the marine atmosphere. Importantly, the percentage fluctuation in the OCS concentration is somewhat less than that of SO_2, indicating that, as expected, the OCS lifetime is somewhat longer than that of SO_2. A residence time longer than one year is suggested from these data, whereas a residence time less than a few days is predicted for SO_2.

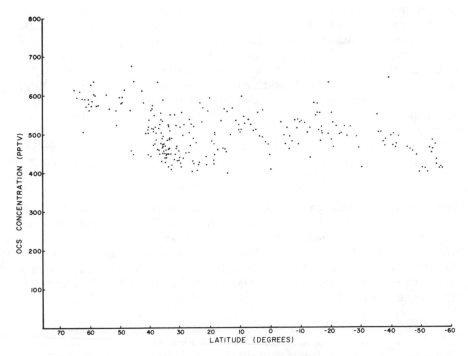

Fig. 2. Plot of all the individual GAMETAG OCS data as a function of latitude.

CHEMISTRY OF SO₂ AND OCS IN REMOTE MARINE ATMOSPHERES

Since these studies suggest that SO_2 concentrations in remote marine atmospheres appear likely to have a natural origin, a major question as to the source of the SO_2 becomes important. Except for volcanos, no surface source for SO_2 is known. Clearly the basic pH of seawater makes the sea surface an important sink, not a source, for atmospheric SO_2.

The reaction of OH and OCS probably does initiate a series of reaction in which SO_2 is formed as the final product:

$$OH + OCS \begin{array}{c} \nearrow \\ \searrow \end{array} \begin{array}{c} CO + HSO \\ CO_2 + HS \end{array} \tag{1}$$

$$HSO \xrightarrow{\frac{\text{many steps}}{?}} SO_2 \tag{2}$$

$$HS \xrightarrow{\frac{\text{many steps}}{?}} SO_2 \tag{3}$$

The SO_2 produced by these reactions is further converted to H_2SO_4 through a series of chemical reactions:

$$OH + SO_2 \xrightarrow{M} HSO_3 \xrightarrow{\frac{O_2}{?}} H_2SO_4 \tag{4}$$

$$SO_2 \rightarrow \text{water droplet or aerosol particle} \rightarrow H_2SO_4 \tag{5}$$

Little is known about the intermediate processes inferred in the above reaction scheme, although there has been an extensive amount of research done in this area.

There are some predictions that much of the SO_2 in the background atmosphere comes from processes (1)–(3) (Logan et al., 1979; Sze and Ko, 1979a, b; Turco et al., 1980). Arguments leading to these predictions, however, are based on a particular set of rate constants for the elementary reactions initiating (1)–(4) (Kurylo, 1978) and the neglect of (5). Except for (4), however, the rate data for the elementary reactions initiating the processes used in these predictions have not received widespread acceptance.

Other predictions suggest that much of the OCS and SO_2 comes from the reaction of OH and CS_2 (Logan et al., 1979; Sze and Ko, 1979a, b; Turco et al., 1980):

$$CS_2 + OH \rightarrow OCS + HS \tag{6}$$

To explain the relatively large OCS and SO_2 concentrations using current rate data for reactions (1)–(6) (Kurylo, 1978), a large background CS_2 level of almost 150 pptv is needed. From measurements made in air which we believe to have been over relatively unpolluted areas for several days, we estimated the background concentration of CS_2 to be 30 pptv. The CS_2 levels were 5–10 times higher in air that had been influenced by anthropogenic sources (Maroulis and Bandy, 1980). These data are summarized in Table 4.

Table 4. CS_2 measurements at Philadelphia, Pennsylvania, and Wallops Island, Virginia

Site[a]	Polluted source area			Unpolluted source area		
	Mean and standard deviation (pptv)	Standard error of the mean (pptv)	Number of samples	Mean and standard deviation (pptv)	Standard error of the mean (pptv)	Number of samples
Philadelphia, Pa.						
Channel 1	190 ± 114	16	52	37 ± 10	2	36
Channel 2	178 ± 115	16	52	34 ± 8	1	36
Wallops Island, Va.						
Channel 1	85 ± 17	3	26	39 ± 8	1	35
Channel 2	81 ± 17	3	30	40 ± 6	1	36
Both sites						
Channel 1	155 ± 106	12	78	38 ± 9	1	71
Channel 2	143 ± 103	11	82	37 ± 8	1	72

[a]Instrument measured CS_2 on two channels simultaneously.

To estimate the amount of OCS originating from CS_2, we followed the arguments used by Logan et al. (1979), Sze and Ko (1979a, b), and Turco et al. (1980) and solved the differential equations describing processes (1)–(6) in the steady-state approximation to yield the following expression:

$$\int_{atm} k_6[OH][CS_2] \, dV + S_2 = \int_{atm} k_1[OH][OCS] \, dV . \qquad (7)$$

Here S_2 is the non-CS_2 source strength for OCS, k_6 is the rate constant for reaction (6) $[k_6 = 1.9 \times 10^{-13} \, cm^3 \, mol^{-1} \, sec^{-1}$ (Kurylo, 1978)], and k_1 is the rate constant for (1) $[k_1 = 5.7 \times 10^{-14} \, cm^3 \, mol^{-1} \, sec^{-1}$ (Kurylo, 1978)]. An approximation to the amount of CS_2 needed to account for all atmospheric OCS can be found by setting S_2 equal to zero. This leads to the ratio of [OCS] and [CS_2] being equal to the ratio of the rate coefficients k_6 and k_1:

$$[OCS]/[CS_2] = k_6/k_1 . \qquad (8)$$

Using the atmospheric OCS concentration of 500 pptv (Maroulis et al., 1977; Torres et al., 1980; Sandalls and Penkett, 1977) yields a CS_2 concentration of 147 pptv needed to produce all the OCS. However, as Table 4 reveals, recent CS_2 data obtained from Philadelphia, Pennsylvania, and Wallops Island, Virginia, indicate that the background concentration of CS_2 which has passed over relatively unpolluted continental and marine areas is approximately 30 pptv (Maroulis and Bandy, 1980). Thus the background concentration of CS_2 can account for not more than 20% of the atmospheric OCS. More than 80% of the OCS originates from non-CS_2 sources.

Equations (1)–(6) were also used to derive a steady-state equation for SO_2:

$$\int_{atm} k_6[OH][CS_2] \, dV + \int_{atm} k_1[OH][OCS] \, dV + S_3 = \int_{atm} k_4[OH][SO_2] \, dV \qquad (9)$$

$$+ \int_{atm} (d[SO_2]_{HET}/dt) \, dV + S_4.$$

Here S_3 is a source term for SO_2, S_4 is a sink term for SO_2, and k_4 is the rate constant for (4) $[k_4 = 6.7 \times 10^{-13} \, cm^3 \, mol^{-1} \, sec^{-1}$ (Atkinson et al., 1976)]. To approximate the amount of SO_2 that comes from CS_2, we assume that terms S_3, S_4, and

$$\int (d[SO_2]_{HET}/dt) \, dV$$

are zero (this may or may not be valid but is current practice in the absence of other data). In addition, from the OCS calculations it is estimated that no more than 20% of the OCS comes from CS_2 and that more than 80% of the OCS originates from non-CS_2 sources. Importantly, one S atom of CS_2 is converted to SO_2 with OCS as the stable intermediate. The other S atom of CS_2 is converted directly to SO_2, presumably with no stable intermediate. Thus no more than 20% of SO_2 comes directly from CS_2, and no more than 33% of SO_2 indirectly and directly comes from CS_2.

REFERENCES

Atkinson, R., R. A. Perry, and J. N. Pitts, Jr. 1976. Rate constants for the reaction of OH radical with NO_2 (M = Ar and N_2) and SO_2 (M = Ar). J. Chem. Phys. 65:306–310.

Georgii, H. W. 1970. Contribution to the atmospheric sulfur budget. J. Geophys. Res. 75:2365–2371.

Georgii, H. W., and W. Vitze. 1971. Global and regional distribution of sulfur components in the atmosphere. Idojaras 75:294-299.

Goldberg, A. B., P. J. Maroulis, L. A. Wilner, and A. R. Bandy. 1980. Studies of H₂S emissions from a salt water marsh. Atmos. Environ., accepted for publication.

Heidt, L. W., and J. P. Krasnec. 1978. Tropospheric distributions of CO, CH₄, CO₂ and N₂O in the Northern and Southern Hemispheres. Eos 59:1077.

Jaeschke, W., R. Schmitt, and H. W. Georgii. 1976. Preliminary results of stratospheric SO₂ measurements. Geophys. Res. Lett. 3:517-519.

Krasnec, J. P., L. E. Heidt, R. A. Lueb, and W. H. Pollock. 1978. Interhemispheric distribution of CH₃CCl₂. Eos 59:1082.

Kurylo, M. J. 1978. Flash photolysis resonance fluorescence investigation of the reactions of OH radicals with OCS and CS₂. Chem. Phys. Lett. 58:238-242.

Logan, J. A., M. B. McElroy, S. C. Wofsy, and M. J. Prather. 1979. Oxidation of CS₂ and COS: sources for atmospheric SO₂. Nature 281:185-188.

Maroulis, P. J., and A. R. Bandy. 1977. Contribution of biologically produced dimethyl sulfide to the global sulfur cycle. Science 196:647-648.

Maroulis, P. J., and A. P. Bandy. 1980. Measurements of carbon disulfide in the eastern United States. Geophys. Res. Lett., submitted for publication.

Maroulis, P. J., A. L. Torres, and A. R. Bandy. 1977. Atmospheric concentrations of carbonyl sulfide in the southwestern and eastern United States. Geophy. Res. Lett. 4:510-512.

Maroulis, P. J., A. L. Torres, A. B. Goldberg, and A. R. Bandy. Atmospheric SO₂ measurements on Project GAMETAG. J. Geophys. Res., in press.

Nguyen Ba Cuong, B. Bonsang, and G. Lambert. 1974a. The atmospheric concentration of sulfur dioxide and sulfate aerosols over antarctic, subantarctic areas and oceans. Tellus 26:241-249.

Nguyen Ba Cuong, B. Bonsang, J. Pasquier, and G. Lambert. 1974b. Composantes marine et africaine des aerosols de sulfates dans l'hemisphere sud. J. Rech. Atmos. 8:831-844.

Rasmussen, L. E., and R. A. Rasmussen. 1978. Concentration distribution of methyl chloride in the atmosphere. Eos 59:1082.

Rasmussen, R. A. 1978. Trans Pacific CO and CH₄ measurements. Eos 59:1082.

Sandalls, F. J., and S. A. Penkett. 1977. Measurements of CS₂ and OCS in the atmosphere. Atmos. Environ. 11:197-199.

Slatt, B. J., D. F. S. Natusch, J. M. Prospero, and D. L. Savoie. 1978. Hydrogen sulfide in the atmosphere of the northern equatorial Atlantic Ocean and its relation to the global sulfur cycle. Atmos. Environ. 12:981-991.

Sze, N. D., and M. K. W. Ko. 1979a. CS₂ and COS in the stratospheric sulfur budget. Nature 280:308-310.

Sze, N. D., and M. K. W. Ko. 1979b. CS₂—a precursor for atmospheric COS. Nature 278:731-732.

Torres, A. L., P. J. Maroulis, A. B. Goldberg, and A. R. Bandy. 1980. Atmospheric OCS measurements on Project GAMETAG. J. Geophys. Res. 84, in press.

Turco, R. P., R. C. Whitten, O. B. Toon, J. B. Pollack, and P. Hamill. 1980. OCS, stratospheric aerosols and climate. Nature 283:283-286.

DISCUSSION

Questioner: Could you comment on the role of elemental sulfur in the atmospheric sulfur cycle?

Alan Bandy: Because it is so reactive, elemental sulfur would be present only as an intermediate with a very short half-life. Dr. William Zoller (University of Maryland) has made some atmospheric measurements and found only trace quantities of elemental sulfur.

SECTION III

HUMAN HEALTH EFFECTS

Co-Chairmen: P. J. Walsh and D. C. Parzyck

CHAPTER 9

HUMAN HEALTH EFFECTS

Chairmen: **P. J. Walsh and D. C. Parzyck**
Oak Ridge National Laboratory

OVERVIEW

P. J. Walsh*

D. C. Parzyck*

The characterization of health effects associated with exposure to sulfur-containing substances in the atmosphere requires a multifaceted, multidisciplinary approach. Among the skills required to carry out this analysis are those in the epidemiological, toxicological, and clinical areas. The papers within this session detailed research carried out in each of these three areas as a means of identifying the health implications of atmospheric levels of sulfur.

Mazumdar presented an overview of epidemiologic evidence on the health effects of atmospheric sulfur. In reviewing data for a large number of studies, Mazumdar emphasized the difficulty in being able to attribute an observed health effect to specific concentrations of any single pollutant. From the standpoint of short-term studies, confounding environmental factors such as weather and temperature must be considered in the evaluation of the health impact associated with acute exposure to high levels of sulfur-containing pollutants. In the case of long-term chronic exposures, the contribution of other pollutants and the effect of smoking must also be evaluated. Future studies must benefit from the inadequacies of past research so that confounding variables such as the presence of other pollutants (e.g., indoor as well as outdoor), personal habits (e.g., smoking), and environmental factors (e.g., weather and temperature) can be properly evaluated.

The results of a study designed to examine the chronic effects of air pollutants were presented by Speizer. In an attempt to determine the adequacy of present primary standards for sulfur dioxide and particulates, a study is being made of adults and children living in six United States cities of varying air quality. The study involves both the collection of air monitoring data and a health assessment of selected members of the population in the various cities. Monitoring data include measurement of nitrogen dioxide levels in homes using gas for heating and/or cook-

---

*Health and Safety Research Division, Oak Ridge National Laboratory, P.O. Box X, Oak Ridge, Tennessee 37830.

ing. Data collected to date do show that individuals living in the city with lowest ambient air quality do have higher disease rates. In addition, it has also been shown that children living in homes with gas stoves may have lower levels of pulmonary function. This last point serves to highlight again the need for additional research into the question of indoor air pollution and the relationship of indoor air pollutant sources to those present in the outdoor environment.

Lippman presented a review of laboratory studies designed to assess the human health significance of exposure to sulfur-containing pollutants in concert with other air pollutants and the effect upon individuals with specific respiratory problems. He concluded that, while various sulfur-containing pollutants have been shown to produce measurable effects at elevated concentrations, no clear evidence exists that this class of substances is producing any adverse effect on the United States population. Concern was focused primarily on sensitive members of the population such as those suffering from bronchitis or individuals subject to asthmatic attacks. Additional concern was directed toward the role that sulfur dioxide may play as a cofactor in the development of bronchial carcinoma or the possible effects on other lung diseases. Clearly, future research must be directed toward identifying the impacts of exposure to sulfur oxides among those individuals who are most sensitive.

The paper by Hackney and associates detailed some recent findings on clinical studies investigating the short-term effects of mixtures of sulfur-containing substances. Elevated levels of sulfur dioxide or sulfates have in the past been associated with respiratory morbidity in epidemiologic studies, while clinical exposures to ambient levels of the individual substances showed no significant effect. Conclusions drawn from this study suggest that individuals exposed to relatively low levels of sulfur dioxide, ozone, and sulfuric acid showed some significant decrease in lung function. Exposures to sulfur dioxide and nitrogen dioxide produced some increase in specific respiratory symptoms among asthmatics. While these findings were not proposed to be conclusive, they did demonstrate the fact that mixtures of pollutants do produce discernible health effects at levels where the individual substances do not produce adverse impact. Studies of this type highlight the need to identify pollutant mixtures that simulate actual environmental situations and to test these mixtures in a clinical setting to determine the potential for hazardous interactions.

Laboratory studies of the biological effects of sulfur oxides were reviewed by Dalbey. Acute exposure of laboratory animals to elevated levels of sulfur oxides produced bronchoconstriction and a resultant increase in pulmonary resistance. Sulfate aerosols were more potent than sulfur dioxide in producing this effect in animals. Chronic exposures to high levels of sulfur dioxide produced experimental bronchitis in several species of laboratory animals. While long-term exposure to levels of sulfur dioxide approximating ambient produced little or no change in respiratory function, significant alterations in pulmonary function and morphology were seen after chronic exposure to sulfuric acid aerosols. Dalbey also referenced the fact that animal studies have been performed to examine the influence of sulfur oxides on susceptibility to infection with both positive and negative results.

All of the participants in the session acknowledged the inadequacy of current information on the health effects associated with exposure to sulfur-containing substances. Particular emphasis was placed on the need to characterize confounding factors in the indoor as well as the outdoor environment. These confounding factors include personal habits such as smoking, which complicate attempts at exposure assessment, and simultaneous exposures to other, sometimes more harmful, substances.

CHAPTER 10

LABORATORY STUDIES OF BIOLOGICAL EFFECTS OF SULFUR OXIDES*

W. E. Dalbey†

ABSTRACT

Selected results from controlled exposures of laboratory animals to airborne sulfur oxides are briefly summarized. The main effect observed during acute exposures was reflex bronchoconstriction and a resultant increase in pulmonary resistance. The increase in resistance due to sulfur dioxide (SO_2) was potentiated by simultaneous exposure to aerosols under conditions that would increase the transfer of sulfur oxides into the respiratory tract and promote transformation to a higher oxidation state, especially one that is acid. Sulfate aerosols, particularly sulfuric acid aerosols, were more potent than SO_2 in causing bronchoconstriction. Chronic exposure to high concentrations (400–650 ppm) of SO_2 resulted in experimental bronchitis in several species. Long-term exposure to more realistic concentrations of SO_2 produced little or no change in respiratory function or morphology. Significant alterations in both pulmonary function and morphology have been reported after chronic exposure to sulfuric acid aerosols. Recent data indicate that changes in the lung may progress after cessation of such exposures.

INTRODUCTION

Several animal studies with sulfur oxides have been reported during past years. Greenwald (1954) provided a comprehensive survey of early work. More recent reviews (Rall, 1974; Ferris, 1978; Weir, 1979) have discussed the relation of experimental results with laboratory animals to human epidemiological findings. The authors reached varying conclusions about the utility of experimentation with animals in the establishment of standards for human exposure. However, the value of animal exposure was apparent in emphasizing the variables in exposure conditions that are important in the toxicity of sulfur oxides and in establishing the greater toxicity of particulate sulfates relative to sulfur dioxide (SO_2). I will summarize several of these animal experiments, covering areas not addressed by previous speakers. We will first discuss the response of laboratory animals to acute exposures to SO_2 or particulate sulfates and then briefly review the results of a series of chronic exposures to SO_2 or sulfuric acid aerosols.

---

*Research sponsored jointly by the Environmental Protection Agency under Interagency Agreement 79-D-X0533 and the Office of Health and Environmental Research, U.S. Department of Energy, under contract W-7405-eng-26 with the Union Carbide Corporation.

†Biology Division, Oak Ridge National Laboratory, Oak Ridge, TN 37830.

ACUTE EXPOSURES

A highly soluble gas, SO_2 dissolves rapidly into the lining of the respiratory tract during inhalation. Speizer and Frank (1966) were among those who demonstrated that only a small portion of the original SO_2 remains in the inhaled gas after it passes through the nose. The researchers exposed humans to SO_2 and observed that a very small fraction of the gas inhaled through the nose was found in the pharynx. This experiment was prompted by the observation that people who inhaled SO_2 through their noses had fewer symptoms of pulmonary distress or pulmonary irritation than those who inhaled through their mouths.

One of the primary responses to acute exposures to SO_2 is bronchoconstriction and a resultant increase in resistance to air flow in the respiratory tract. Bronchoconstriction resulting from acute SO_2 exposures has been observed in several species: cat, guinea pig, rabbit, rat, dog, and man (Nadel et al., 1965). The onset of the increased resistance is very rapid, on the order of minutes. The bronchoconstriction is, however, readily reversible after exposure (Amdur and Underhill, 1968). Nadel et al. (1965) have demonstrated that the increased resistance can be blocked by administering atropine or by cooling the cervical vagosympathetic nerves. This bronchoconstriction is therefore probably a nervous reflex initiated by sensory neurons in the respiratory epithelium and mediated through the vagosympathetic nerves. The reflex constriction of the airways is not specific to SO_2 but has been observed with other irritants (Nadel, 1977). During exposure to 5 ppm SO_2, virtually all persons exhibit some degree of increased pulmonary resistance. A few sensitive individuals respond to concentrations of 1 ppm. However, it has been emphasized (Amdur, 1969) that the bronchoconstriction is probably not a stress in the respiratory system of normal humans exposed to normal ambient SO_2 concentrations. Therefore, SO_2 must be considered only as a mild upper respiratory tract irritant to normal individuals.

Sulfate-induced bronchoconstriction has been extensively investigated in guinea pigs by Amdur and others. These investigators have made several significant observations, which will be summarized below. The guinea pig was chosen for this work because it is reportedly highly sensitive to SO_2. The method for measuring resistance involved anesthetizing the animal and implanting a saline-filled catheter for measuring intrapleural pressure around the lungs. Respiratory flow, tidal volume, and intrapleural pressure were simultaneously recorded in the animals after they recovered from the anesthesia. Resistance was expressed as the change in intrapleural pressure at midinhalation and midexhalation divided by the change in respiratory flow between the same points in the respiratory cycle. Resistance measurements were commonly taken during the latter part of a 1-hr exposure.

In the guinea pig model it has been observed that exposures to SO_2 with concentrations as low as a few ppm rapidly produced the bronchoconstriction described above. This increase in pulmonary resistance was found to be potentiated by simultaneous exposure to aerosols of certain salts (sodium chloride, potassium chloride, or ammonium thiocyanate). None of these aerosols increased pulmonary resistance by themselves. However, it was observed that the potentiation of the SO_2-induced increase in resistance was proportional to the solubility of SO_2 in a water solution of the individual salts (Amdur and Underhill, 1968). These data were consistent with the concept that the potentiation is a result of the absorption of SO_2 into the liquid aerosol of these salts and the increase in the transfer into the lung in some form.

McJilton et al. (1976) reported similar observations under more controlled conditions. They observed that a sodium chloride aerosol potentiated SO_2-induced bronchoconstriction in guinea pigs only when the relative humidity during exposure was high enough for water to condense on the sodium chloride aerosol to form liquid droplets (i.e., over 80% relative humidity). When the relative humidity was too low for the liquid droplets to form (40%), there was no such potentiation. The authors also observed a decrease in the pH of the liquid particles at high relative humidities in the presence of SO_2. In addition, bisulfite ion was present in the aerosol, presumably a result of the formation of sulfurous acid. McJilton and his co-workers suggested that the toxicity of SO_2 was likely to be increased by circumstances that favored either penetration of SO_2 into the upper airways (such as mouth breathing), absorption of SO_2 by liquid aerosols (lowered ambient temperature), or transformation of SO_2 to a higher oxidation state, especially one that is acid.

Amdur and her co-workers further demonstrated the importance of the conversion of SO_2 to an acid. They exposed guinea pigs to combinations of SO_2 and aerosols of metal salts (ferrous, vanadium, or manganese chloride) that would catalyze the conversion of SO_2 to sulfuric acid (Amdur and Underhill, 1968). All three aerosols did potentiate the increase in resistance during SO_2 exposures, at much lower aerosol concentrations than required with NaCl. Aerosols of ferrous sulfate or manganous sulfate alone were not irritating. Therefore, the irritant potential was not related to the sulfate ion as such (Amdur, 1971). Amdur and Underhill (1968) also reported a lack of potentiation by aerosols of ferric oxide, manganese dioxide, fly ash, and carbon particles, although some of these particles might potentiate the pulmonary response to SO_2 if conditions were such that they served as nuclei for fog droplets.

In reviewing what was known about the acute toxicity of sulfates, Amdur (1969) emphasized that the nature of particulate matter is critical to its potential enhancement of the response to acute exposures to SO_2. She made the following points: (1) Submicron particles are most likely to potentiate. (2) The degree of potentiation decreases with decreasing concentration of aerosol. (3) The potentiation is mediated through the formation of an irritant aerosol and not by simple transfer of SO_2 into the lungs. (4) Relative to the time required after exposure to SO_2 alone, the return of pulmonary resistance values to control levels is delayed after exposure to SO_2 in combination with aerosol.

It is apparent that the acute toxicity of some sulfate aerosols is greater than that of SO_2 alone. For example, 115 mg/m³ of SO_2 are required to produce a 50% increase in flow resistance in guinea pigs while only 1.9 mg/m³ of H_2SO_4 particles (0.8 μm) yield the same effect. However, it is also apparent that not all sulfates elicit bronchoconstriction. The relative potency of various sulfur oxide particles inhaled by guinea pigs has been investigated by Amdur and others. A full description of relative irritancy must be made with caution because the irritant potential of various sulfate aerosols differs with particle size. Irritancy increases markedly with decreasing particle size (Amdur and Corn, 1963). The mechanism for this size dependence might involve an increased penetration of the aerosol into the lung due to the small particle size or an increased number of particles for a given concentration. The slope of the dose-response curves increased with the smaller sized particles; thus, small changes in concentration produced larger increments in the increase of pulmonary resistance. These observations emphasize the great importance of submicron particles in acute toxicity of sulfate aerosols.

In a comparison of the relative irritancy of different sulfate aerosols, one must therefore take account of particle size in addition to aerosol concentration. By making such analyses, Amdur (1969) reported that sulfuric acid is the most irritating of all sulfate aerosols tested, including the products of its neutralization with ammonia. Most recently, Amdur has demonstrated that aerosols of sulfuric acid produced significant bronchoconstriction in guinea pigs at concentrations as low as 0.1 mg/m³ with a particle size of 0.3 μm (Amdur et al., 1978).

Two additional factors should be kept in mind while interpreting data from animal exposures to sulfuric acid. The first is ambient levels of ammonia. Larson et al. (1977) reported that even endogenous ammonia in the respiratory tract may neutralize some of the inhaled sulfuric acid. Secondly, the particle size of sulfuric acid is dependent on the molarity of the acid and the relative humidity. Particles take up water from (or lose it to) the surrounding air depending on the relative humidity and their acid concentration (Cavender et al., 1977). Since the relative humidity in the respiratory tract is nearly 100%, one must consider the equilibrated particle size at high humidity rather than at normal atmospheric conditions when predicting deposition in the respiratory tract.

CHRONIC EXPOSURES TO SULFUR OXIDES

From the acute studies cited above, it appears that, relative to sulfate aerosols, SO_2 is a comparatively nontoxic upper respiratory tract irritant. This concept was supported by results of chronic animal exposures, several of which are summarized in Table 1. Goldring (1970) exposed hamsters for several weeks to 650 ppm of SO_2 and observed morphologic evidence of stimulation of mucus cell secretion. Reid (1963) reported morphologic changes in the airways of rats exposed to 400 ppm of SO_2 for six weeks; these mimicked human bronchitis. Subsequent biochemical analysis of secretions from affected airways further supported the idea that exposed rats had a form of experimental bronchitis (Clark et al., 1979). Similar observations have been reported in dogs (Chakrin and Saunders, 1974).

Longer exposure of animals to more realistic concentrations of SO_2 produced little or no change in respiratory morphology or function (see Table 1).

Chronic exposures of animals to sulfuric acid are also summarized in Table 1. Here, significant alterations in both pulmonary function and morphology were observed in the guinea pig, dog, monkey, and rat. Although the sulfuric acid concentrations used in these experiments might be considered high, the observations of significant physiological and morphological impairment of the lung indicate that chronic sulfuric acid exposures may be more deleterious than SO_2 exposures and may in fact represent a potentially significant factor in human exposures to ambient air.

In the series of experiments with dogs cited in Table 1 which involved simultaneous exposure to SO_2 and sulfuric acid, pulmonary lesions apparently progressed after cessation of the exposure. Because these data were from the longest chronic exposure in the table, the observation of a progressive morphological and physiological change in the lung is of particular significance.

SUSCEPTIBILITY TO INFECTION

As discussed earlier, sulfur oxides have been implicated in the susceptibility of humans to infection. Animal studies have been performed in this important area, but I will not attempt to summarize the observations here. One can find reports of both positive (Giddens et al., 1972; Fairchild et al., 1972; Ukai, 1977) and negative (Goldring et al., 1967; Battigelli et al., 1969; Rylander, 1969) effects. It is interesting to note that viral infections were enhanced in mice inoculated intranasally and exposed to SO_2 at concentrations as low as 0.03–0.1 ppm over several days (Ukai, 1977).

Table 1. Summary of chronic exposures of animals to sulfur dioxide or sulfuric acid

Reference	SO$_2$ (ppm)	H$_2$SO$_4$ (mg/m^3)	Species	Duration	Observations
1) Goldring et al., 1970	650		Hamster	76 days	Stimulation of mucus cell secretion
2) Reid, 1963	400		Rat	6 weeks	Experimental bronchitis
3) Chakrin and Saunders, 1979	550		Dog	5 months	Experimental bronchitis
4) Weiss and Weiss, 1976	40		Mice	6–7 days	Increased lung compliance
5) Hirsch et al., 1975	1		Dog	1 year	No change—pulmonary function
6) Alarie et al., 1972 and 1975	0.1–5		Monkey	1.5 years	No change—pulmonary function and morphology
7) Alarie et al., 1970	0.1–5		Guinea pig	1 year	No change—pulmonary function and morphology
8) Lewis et al., 1969	5		Dog	7.5 months	Decreased lung compliance and increased lung resistance
9) Lewis et al., 1973	5		Dog	1.7 years	Altered N$_2$ washouts; no altered morphology
10) Cavender et al., 1978		10	Rat	6 months	No change in morphology
		10	Guinea pig	6 months	Minimal proliferation of alveolar macrophages and mild loss of cilia in trachea
11) Thomas et al., 1958		4	Guinea pig	140 days	No change in lung morphology
12) Lewis et al., 1969		0.9	Dog	7.5 months	Decreased diffusing capacity
13) Lewis et al., 1973		0.9	Dog	1.7 years	Decreased diffusing capacity, residual volume, total lung volume, increased expiratory resistance, no altered morphology
14) Alarie et al., 1973		0.08–1	Guinea pig	1 year	No altered pulmonary function or morphology
		0.08–1	Monkey	1.5 years	1) Increased respiratory rate and altered N$_2$ washout 2) Bronchial hyperplasia
15) Bushtueva, 1964		1.8	Rat	65 days	Interstitial pneumonitis and thickened alveolar septa
16) Vaughan et al., 1969	0.5	0.1	Dog	18 months	No change in pulmonary function
17) Gillespie et al., 1976	0.5	0.1	Dog	5.7 years + 2 years recovery	Increased total lung capacity
18) Hyde et al., 1978	0.5	0.1	Dog	5.7 years + 3 years recovery	1) Enlarged airspace in proximal acinar region 2) Hyperplasia of nonciliated bronchiolar cells

REFERENCES

Alarie, Y., W. M. Busey, A. A. Krumm, and C. E. Ulrich. 1973. Long-term continuous exposure to sulfuric acid mist in cynomolgus monkeys and guinea pigs. Arch. Environ. Health 27:16–24.

Alarie, Y., C. E. Ulrich, W. M. Busey, A. A. Krumm, and H. N. MacFarland. 1972. Long-term continuous exposure to sulfur dioxide in cynomolgus monkeys. Arch. Environ. Health 24:115–128.

Alarie, Y., C. E. Ulrich, W. M. Busey, H. E. Swann, and H. N. MacFarland. 1970. Long-term continuous exposure of guinea pigs to sulfur dioxide. Arch. Environ. Health 21:769–777.

Amdur, M. O. 1969. Toxicologic appraisal of particulate matter, oxides of sulfur, and sulfuric acid. J. Air Pollut. Control Assoc. 19:638–644.

Amdur, M. O. 1971. Aerosols formed by oxidation of sulfur dioxide. Review of their toxicology. Arch. Environ. Health 23:459–468.

Amdur, M. O., and M. Corn. 1963. The irritant potency of zinc ammonium sulfate of different particle sizes. Am. Ind. Hyg. Assoc. J. 24:326–333.

Amdur, M. O., M. Dubriel, and D. A. Creasia. 1978. Respiratory response of guinea pigs to low levels of sulfuric acid. Environ. Res. 15:418–423.

Amdur, M. O., and D. Underhill. 1968. The effect of various aerosols on the response of guinea pigs to sulfur dioxide. Arch. Environ. Health 16:460–468.

Battigelli, M. C., H. M. Cole, D. A. Fraser, and R. A. Mah. 1969. Long-term effects of sulfur dioxide and graphite dust on rats. Arch. Environ. Health 18: 602–608.

Bushtueva, K. A. 1964. The effect of absorption of oxides of sulfur. Gig. Sanit. 10:7–12.

Cavender, F. L., B. Singh, and B. Y. Cockrell. 1978. Effects in rats and guinea pigs of six month exposures to sulfuric acid mist, ozone, and their combination. J. Toxicol. Environ. Health 4:845–852.

Cavender, F. L., J. L. Williams, W. H. Steinhagen, and D. Woods. 1977. Thermodynamics and toxicity of sulfuric acid mists. J. Toxicol. Environ. Health 2:1147–1159.

Chakrin, L. W., and L. Z. Saunders. 1974. Experimental chronic bronchitis. Pathology in the dog. Lab. Invest. 30:145–154.

Clark, J. N., W. E. Dalbey, and K. B. Stephenson. 1979. Effect of sulfur dioxide on the morphology and mucin biosynthesis by the rat trachea. J. Environ. Pathol. Toxicol., in press.

Fairchild, G. A., J. Roan, and J. McCarroll. 1972. Atmospheric pollutants and the pathogenesis of viral respiratory infection. Arch. Environ. Health 25:174–182.

Ferris, B. G. 1978. Health effects of exposure to low levels of regulated air pollutants. A critical review. J. Air Pollut. Control Assoc. 28:482–497.

Giddens, W. E., and G. A. Fairchild. 1972. Effects of sulfur dioxide on the nasal muscosa of mice. Arch. Environ. Health 25:166–173.

Gillespie, J. R., J. D. Berry, and J. F. Stara. 1976. Pulmonary function changes in the period following termination of air pollution exposure in beagles. Am. Rev. Respir. Dis. 113(4):92.

Goldring, I. P., P. Cooper, I. M. Ratner, and L. Greenburg. 1967. Pulmonary effects of sulfur dioxide exposure in the Syrian hamster. I. Combined with viral respiratory diseases. Arch. Environ. Health 15:167–176.

Goldring, I. P., L. Greenburg, S. S. Park, and I. M. Ratner. 1970. Pulmonary effects of sulfur dioxide exposure in the Syrian hamster. II. Combined with emphysema. Arch. Environ. Health 21:32.

Greenwald, I. 1954. Effects of inhalation of low concentrations of sulfur dioxide upon man and other mammals. Arch. Ind. Hyg. Occup. Med. 10:455–475.

Hirsch, J. A., E. W. Swenson, and A. Wanner. 1975. Tracheal mucous transport in beagles after long-term exposure to 1 ppm sulfur dioxide. Arch. Environ. Health 30:249–253.

Hyde, D., J. Orthoefer, D. Dungworth, W. Tyler, R. Carter, and H. Lum. 1978. Morphometric and morphologic evaluation of pulmonary lesions in beagle dogs chronically exposed to high ambient levels of air pollutants. Lab. Invest. 38: 455–469.

Larson, T. V., D. S. Covert, R. Frank, and R. J. Charlson. 1977. Ammonia in the human airways: neutralization of inspired acid sulfate aerosols. Science 197: 161–163.

Lewis, T. R., K. I. Campbell, and T. R. Vaughan. 1969. Effects on canine pulmonary function via induced NO_2 impairment, particulate interaction, and subsequent SO_4. Arch. Environ. Health 18:596–601.

Lewis, T. R., W. J. Moorman, W. F. Ludmann, and K. I. Campbell. 1973. Toxicity of long-term exposure to oxides of sulfur. Arch. Environ. Health 26:16–21.

McJilton, C. E., R. Frank, and R. J. Charlson. 1976. Influence of relative humidity on functional effects of an inhaled SO_2-aerosol mixture. Am. Rev. Respir. Dis. 113:163–169.

Nadel, J. A. 1977. Autonomic control of airway smooth muscle and airway secretions. Am. Rev. Respir. Dis. 115:117–126.

Nadel, J. A., H. Salem, B. Tamplin, and Y. Tokiwa. 1965. Mechanism of bronchoconstriction during inhalation of sulfur dioxide. J. Appl. Physiol. 20:164–167.

Rall, D. P. 1974. Review of the health effects of sulfur oxides. Environ. Health Perspect. 8:97–121.

Reid, L. 1963. An experimental study of hypersecretion of mucus in the bronchial tree. Brit. J. Exp. Pathol. 44:437–445.

Rylander, R. 1969. Alternations of lung defense mechanisms against airborne bacteria. Arch. Environ. Health 18:551–555.

Speizer, F. W., and N. R. Frank. 1966. The uptake and release of SO_2 by the human nose. Arch. Environ. Health 12:725–728.

Thomas, M. D., R. H. Hendricks. F. D. Gunn, and J. Critchlow. 1958. Prolonged exposure of guinea pigs to sulfuric acid aerosol. Arch. Ind. Health 17:70–80.

Ukai, K. 1977. Effect of SO_2 on the pathogenesis of viral upper respiratory infection in mice. Proc. Soc. Exp. Biol. Med. 154:591–596.

Vaughan, T. R., L. F. Gennelle, and T. R. Lewis. 1969. Long-term exposure to low levels of air pollutants. Effects on pulmonary function in the beagle. Arch. Environ. Health 19:45–50.

Weir, F. W. 1979. Toxicology of the sulfur oxides. J. Occup. Med. 21:281–284.

Weiss, K. D., and H. S. Weiss. 1976. Increased lung compliance in mice exposed to sulfur dioxide. Res. Commun. Chem. Pathol. Pharmacol. 13:133–136.

CHAPTER 11

SHORT-TERM RESPIRATORY EFFECTS OF SULFUR-CONTAINING POLLUTANT MIXTURES: SOME RECENT FINDINGS FROM CONTROLLED CLINICAL STUDIES

Jack D. Hackney*
William S. Linn*
Michael P. Jones*
Ronald M. Bailey*
Dennis R. Julin*
Michael T. Kleinman*

ABSTRACT

Elevated ambient levels of sulfur dioxide (SO_2) or sulfate have been associated with increased respiratory morbidity in a number of epidemiologic studies. Conversely, controlled laboratory exposures of volunteers to SO_2 or individual sulfate compounds generally have shown no significant respiratory changes at concentrations similar to ambient, suggesting that interactions between sulfur oxides and other airborne substances may be important in producing adverse health effects of ambient exposure (if any actually occur). A number of recent controlled exposure studies have thus addressed pollutant mixtures. The SO_2-ozone (O_3) mixture has been investigated repeatedly; the more recent results suggest that SO_2 does not enhance the respiratory responses to O_3 if relatively little sulfate aerosol is present. We studied 19 healthy volunteers exposed to a mixture containing SO_2, O_3, and sulfuric acid (H_2SO_4) aerosol under conditions simulating "worst-case" ambient exposures. Modest but significant decrements in forced expiratory performance occurred, but it is not yet possible to determine the extent to which H_2SO_4 and/or SO_2 contributed to these effects. A similar study of mixed SO_2 and nitrogen dioxide (NO_2) showed little effect on lung function; however, respiratory symptoms showed a small significant increase with exposure.

INTRODUCTION

Of all air pollutants, sulfur compounds have received the most sustained attention as possible health hazards. They were strongly implicated in historical air pollution disasters and still are often reported to show associations with increased morbidity, increased mortality, or reduced pulmonary function in epidemiologic studies, even though ambient concentrations have been greatly reduced by modern emission controls. As is well known, interpretation of epidemiologic findings is complicated by the inability to monitor and control many environmental and demographic

---

*Environmental Health Service, Rancho Los Amigos Hospital Campus, University of Southern California School of Medicine, Downey, California.

variables potentially capable of influencing the health measures of interest. Laboratory investigations, in which volunteers are deliberately exposed to specific pollutants while their physiological and clinical responses are monitored, have been employed frequently in recent years in the quest for a firmer understanding of the health consequences of sulfur oxides (SO_X). Within their ethical and practical limits, controlled laboratory studies of humans can provide a more rigorous test of short-term pollutant effects than one can determine from epidemiological studies, although uncertainties in characterizing the exposure atmosphere and possible interfering variables are by no means eliminated. Controlled studies of individual pollutants, including sulfur dioxide (SO_2) and sulfuric acid (H_2SO_4) aerosol and its ammonia neutralization products, have generally failed to show any substantial respiratory effects of exposure, in contrast to epidemiologic findings. One possible explanation for the apparent discrepancy between laboratory and epidemiologic results is that SO_X affects the respiratory system by acting additively or synergistically with other pollutants. This possibility has stimulated considerable interest in controlled exposure studies employing mixtures of pollutants.

Past laboratory and epidemiologic studies of SO_X have been reviewed in detail elsewhere (NAS, 1978, USDHEW, 1970). This article is therefore limited to a brief and selective review of recent controlled clinical studies of SO_X-containing mixtures or sulfate aerosols and a preliminary report on two recent mixed-exposure studies in our laboratory.

OZONE–SULFUR DIOXIDE STUDIES

Ozone (O_3) is clearly among the most toxic of common air pollutants, on the basis of findings from animal toxicology, clinical studies, and epidemiology (NAS, 1977). Although O_3 pollution episodes are commonly associated with hot, sunny weather and SO_X episodes with cold, foggy weather, elevated concentrations of both types of pollutant can occur simultaneously under appropriate conditions. Hazucha and Bates (1975) first reported results of human exposures to mixed O_3 and SO_2, each at a nominal concentration of 0.37 ppm, conducted at McGill University in Montreal. Subjects were exposed for 2 hr in a chamber, in which they alternately rested and exercised lightly on a stationary bicycle for 15-min periods. Lung function was measured before, during, and after exposure. While only a small number of subjects were studied and individual responses varied substantially, the over all results indicated a marked synergism between O_3 and SO_2: the mean loss in lung function test performance was considerably larger in subjects exposed to the mixture than in others exposed to O_3 alone at the same nominal concentration. Subjects exposed to SO_2 alone or to laboratory air with no added pollutants showed little or no change in function.

A series of studies was undertaken in our laboratory at Rancho Los Amigos Hospital near Los Angeles to follow up Hazucha and Bates' findings (Bell et al., 1977). Initially, a repeat of Hazucha and Bates' experiment was conducted in the Rancho environmental control chamber with Los Angeles residents. This was followed by a cooperative study in which Montreal subjects were restudied in the Rancho chamber and Rancho monitoring equipment was used to test the Montreal chamber atmosphere retrospectively under conditions simulating the original Hazucha and Bates study. Subjects were generally less reactive in the Rancho chamber than they had been in Montreal. Little if any synergism between O_3 and SO_2 was evident in the Rancho chamber. A possible important difference between the two exposure atmospheres was observed: large numbers of fine particles, probably H_2SO_4 along with neutralized species, were formed wherever O_3 and SO_2 were mixed. In the Rancho chamber, which had elaborate air purification equipment and considerable dilution of the gases before mixing, the total mass of these particles remained small—on the order of 1 μg/m^3. In the Montreal chamber, where less gas dilution and purification of background air were available, the particulate concentration was perhaps 100 times as great. While other explanations could not be entirely ruled out, sulfate aerosol seemed a likely source of apparent O_3-SO_2 synergism in the original Montreal study. (This experience illustrates the technical problems affecting every controlled exposure study to a greater or lesser extent: the exposure atmosphere likely differs from what the investigator intended, since the purification of the background air supply, the pollutant generation, and the atmospheric monitoring can never be perfect.) Bedi et al. (1979) recently reported another study of mixed O_3 and SO_2 with little aerosol present; their results confirm those of the Rancho study in that little or no synergistic response was observable.

SULFATE AEROSOL STUDIES

There have been few controlled exposure studies with particulate pollutants until recently, because particles are much more difficult to generate and monitor than gases. Sackner et al. (1977a, b) have exposed volunteers to H_2SO_4, as well as to sodium, ammonium, zinc, and zinc ammonium sulfates. The exposures were for brief periods (10 min) at concentrations much higher than can be expected to occur in ambient air (1000 μg/m^3). No cardiopulmonary changes or symptoms were found. A recent series of studies in our laboratory (Avol et al., 1979) used the approach of attempting to simulate "worst-case" ambient exposure conditions, as has been done in studies with gaseous pollutants. The exposures lasted 2 hr and included intermittent exercise and heat stress. The aerosols studied were ammonium sulfate [$(NH_4)_2SO_4$], nominally at 100 μg/m^3; ammonium bisulfate (NH_4HSO_4), nominally at 85 μg/m^3; and H_2SO_4, nominally at 75 μg/m^3. The aerosols were polydisperse, simulating the size distribution of sulfur-containing particles expected in ambient air. Even with the sophisticated air purification capability available, neutralization of the acidic aerosols by background ammonia and/or ammonia generated by the subjects themselves while in the exposure chamber proved to be a problem. Thus part of the NH_4HSO_4 exposure aerosol was actually $(NH_4)_2SO_4$, and part of the H_2SO_4 aerosol was actually NH_4HSO_4 (excess acid aerosol was added in the H_2SO_4 studies to compensate for the amount neutralized). None of these exposure studies showed any clear evidence of respiratory changes attributable to the aerosols. Thus the currently available laboratory evidence does not indicate that common sulfate species in ambient air, in and of themselves, produce short-term health effects.

OZONE–SULFUR DIOXIDE–SULFURIC ACID STUDY

The aforementioned results appear consistent with the hypothesis that H_2SO_4, SO_2, and O_3 together are more toxic than any of them alone or than O_3 and SO_2 without H_2SO_4. As one step in testing this possibility, we recently exposed 19 generally healthy volunteer Los Angeles residents to a mixture containing O_3 at 0.37 ppm, SO_2 at 0.37 ppm, and polydisperse H_2SO_4 aerosol nominally at 100 μg/m^3. Monitoring data indicated that neutralization of the H_2SO_4 aerosol tended to increase with the passage of time, so that late in the exposure period the aerosol probably consisted largely of NH_4HSO_4. Exposures lasted 2 hr and included intermittent light exercise and heat stress, as in previous gas and aerosol studies. Each subject underwent a control study consisting of a sham exposure (only purified air in the chamber) 24 hr before his actual exposure. Lung function was evaluated before exposure with the subject breathing clean air and again at the end of the exposure period with the subject breathing the exposure atmosphere. Symptoms were evaluated via a standardized questionnaire (giving emphasis to respiratory symptoms) before exposure, near the end of exposure, and later in the day. Lung function data were evaluated by repeated-measures analysis of variance. Symptom reports were converted to semiquantitative scores based on the number and severity of reported symptoms. Individual scores recorded on the actual exposure day were adjusted by subtracting the corresponding control-day scores, then ranked for the three scoring periods (preexposure, end-of-exposure, later-in-day). Ranks were analyzed by the Friedman rank-sums test. An effect of exposure on lung function was expected to produce a significant decrement in one or more function indices during the pollutant exposure, relative to the control study. An effect of pollutant exposure on symptoms would be reflected in higher adjusted symptom scores (higher ranks) during exposure and/or later in the day, with lower scores for the preexposure period. If the acid aerosol were not an important contributor to exposure effects, only minor symptoms or function changes should have resulted, judging from previous studies of O_3 with or without SO_2 in healthy Los Angeles residents. If the aerosol markedly enhanced the irritancy of the O_3-SO_2 mixture, however, some subjects could be expected to show marked respiratory symptoms and function decrement with exposure, as observed in Hazucha and Bates's original study.

Selected lung function results are shown in Table 1. These include mean preexposure values and mean changes during the exposure periods for forced vital capacity (FVC), the volume of a maximal forced expiration; forced expired volume in 1 sec (FEV$_1$), the maximum volume which can be expired in the first second of a

Table 1. Mean lung function results for 19 volunteers exposed to 0.37 ppm O_3 plus 0.37 ppm SO_2 plus 100 $\mu g/m^3$ H_2SO_4

| Test[a] | Exposure | | |
	Sham	Pollutant	p Value[b]
FVC (ml)			
Preexposure	4629	4641	0.623
Change with exposure	+15	-130	0.007
FEV$_1$ (ml)			
Preexposure	3802	3810	0.596
Change with exposure	+8	-139	0.001
FEF 50% (ml/sec)			
Preexposure	4677	4695	0.804
Change with exposure	+70	-279	0.001
FEF 75% (ml/sec)			
Preexposure	1939	1956	0.735
Change with exposure	+69	-90	0.005

[a]For explanation of abbreviations, see text.

[b]From analysis of variance. p Value less than 0.050 indicates significant difference between sham and pollutant exposures.

forced expiration; and instantaneous forced expiratory flow rates when 50% and 75% of the FVC has been expired (FEF 50%, FEF 75% respectively). These indices have previously shown sensitivity to effects of O_3 or O_3-SO_2 exposure. All four measures showed modest but significant decrements during the pollutant exposure relative to the sham/control study, while all preexposure values were similar on both study days.

The totals of symptom ranks were expected to average 36 if there were no effect of exposure. The actual totals were 29.5 before exposure, 39.0 at the end of exposure, and 39.5 later in the day. Thus there was an increase in reported symptoms during pollutant exposure relative to sham, which persisted later in the day after completion of the study. The differences approached statistical significance.

These results do not support the possibility that H_2SO_4 aerosol markedly enhances the respiratory toxicity of O_3-SO_2 mixtures. The mean loss in FEV_1 (the most commonly used index of response severity) in the present group with O_3-SO_2-aerosol exposure was 3.6%—almost identical to the 3.7% loss in the previous Los Angeles group exposed to O_3 and SO_2 without aerosol (Bell et al., 1977). However, the previous subjects, though in good general health, all had respiratory allergies and were selected with the expectation that they would be atypically reactive. The present group was not selected for high reactivity, and only 3 of its 19 members gave a history of respiratory allergy; therefore this group might have been expected to show a smaller percentage change in FEV_1 had they been exposed to O_3 and SO_2 without aerosol. The results are thus consistent with the hypothesis that H_2SO_4 aerosol enhances the respiratory response to O_3 and SO_2 to a mild degree. On the other hand, given the inherent variability of lung function measurements, subjects, and experimental situations, the results do not exclude the possibility that the H_2SO_4 aerosol had no effect on the subjects' responses. If responses to atmospheres differing only modestly in irritancy are to be compared, it will be necessary to match subjects and exposure conditions with great care and to employ subject groups large enough so that small differences in response can be detected statistically.

NITROGEN DIOXIDE–SULFUR DIOXIDE STUDY

Nitrogen oxides frequently coexist with sulfur oxides in ambient pollution, particularly in areas where coal or fuel oil is burned. We recently studied the respiratory effects of this combination under simulated "worst-case" ambient conditions, using the experimental protocol described previously for the O_3-SO_2-H_2SO_4 study. A group of 24 healthy volunteer Los Angeles residents was exposed to 0.5 ppm NO_2 plus 0.5 ppm SO_2. No significant effects would be expected from either gas alone at this concentration. Another group of 19 asthmatic volunteers was studied subsequently. They were exposed to 0.5 ppm NO_2 plus 0.3 ppm SO_2 because of a suggestion that 0.5 ppm SO_2 might be hazardous to individuals with severely compromised lung function (Lawther, 1978). Neither group showed statistically significant lung function changes of the kind seen in the exposures to O_3 plus SO_2 plus H_2SO_4; however, both the normals and the asthmatics showed significant increases in symptom scores attributable to exposure (Table 2). In the normals, total reported symptoms increased during the exposure period and increased further later in the day. The asthmatics' symptoms showed little change during the exposure period but increased significantly later in the day. In both groups, the increases were small and were scattered among the various symptom categories. Further investigation will be required to provide a better understanding of these responses. Because a more complete control experiment (separate sham exposure study on two successive days) could not be done, we cannot exclude the possibility that the symptom increases

Table 2. Total symptom scores for 24 normals exposed
to 0.5 ppm NO_2 plus 0.5 ppm SO_2 and for 19
asthmatics exposed to 0.5 ppm NO_2 plus 0.3
ppm SO_2[a]

	Normals	Asthmatics[b]
Expected rank sum[c]	48.0	36.0
Preexposure rank sum	37.5	30.0
End-of-exposure rank sum	50.5	29.5
Later-in-day rank sum	56.0	48.5
p Value[d]	0.012	0.001

[a]Symptoms evaluated are cough, sputum, dyspnea,
chest tightness, wheezing, substernal irritation, throat
irritation, nasal congestion, headache, and fatigue.
Preexposure, end-of-exposure, and later-in-day scores
for pollutant exposure day are adjusted by subtracting
corresponding control-day scores, then ranked (higher
ranks correspond to more symptoms). Sums of ranks
are compared statistically among the three time periods.
[b]One subject's data incomplete; excluded from analysis.
[c]Value which would be obtained if there were no variation in reported total symptoms.
[d]From Friedman rank-sums test. p Value less than 0.050
indicates significant variation.

were a result of undergoing the exposure protocol on two successive days, rather
than a response to pollutant exposure. However, a similar study of an apparently in-
nocuous pollutant, ammonium nitrate aerosol, showed no significant function or
symptom changes over two days (Hackney et al., unpublished results), suggesting
that the present symptom responses are probably not related to the exposure protocol per se.

CONCLUSIONS

The serious investigation of sulfur-containing pollutants by controlled human
studies is still in a relatively early stage, and the number of firm, independently
verified findings available is small. To derive any wide-ranging conclusions from
this data base would be unwise, but some trends are discernible. Studies of individual sulfur oxide pollutants, including H_2SO_4, have nearly always produced
negative results at concentrations simulating ambient exposures and even at
somewhat higher concentrations. When one begins to examine pollutant mixtures,
the picture quickly becomes complicated, given the essentially infinite number of
possible combinations, the difficulty of knowing how to formulate mixtures to
represent ambient conditions realistically, and the need to evaluate possible chemical
and biological interactions among pollutant substances. The issue of most concern is
the possibility of interactions to produce adverse health effects substantially greater
than would be expected from any individual component of a pollutant mixture. No

such hazardous interaction has been demonstrated conclusively as yet, but the possibility of at least slight enhancement of respiratory irritancy in mixtures seems to be supported by our current studies as well as by some previous findings elsewhere.

ACKNOWLEDGMENTS

The current studies described in this article are supported by project CAPM-31-79 of the Coordinating Research Council and by project RP 1225-1 of the Electric Power Research Institute.

REFERENCES

Avol, E. L., M. P. Jones, R. M. Bailey, N. M. Chang, M. T. Kleinman, W. S. Linn, K. A. Bell, and J. D. Hackney. 1979. Controlled exposures of human volunteers to sulfate aerosols: health effects and aerosol characterization. Am. Rev. Respir. Dis. 120:319–327.

Bedi, J. F., L. J. Folinsbee, and S. M. Horvath. 1979. Effects of ozone, sulfur dioxide, heat and humidity on the pulmonary function of young male non-smokers (abstract). Am. Rev. Respir. Dis. 119(4, part 2):200.

Bell, K. A., W. S. Linn, M. Hazucha, J. D. Hackney, and D. V. Bates. 1977. Respiratory effects of exposure to ozone plus sulfur dioxide in Southern Californians and Eastern Canadians. Am. Ind. Hyg. Assoc. J. 38(12):696–706.

Hazucha, M., and D. V. Bates. 1975. Combined effect of ozone and sulfur dioxide on pulmonary function in man. Nature 257:50–51.

Lawther, P. J. 1978. Looking backward and forward. Bull. N.Y. Acad. Med. 54(11):1199–1208.

NAS, Committee on Medical and Biologic Effects of Environmental Pollutants. 1977. Ozone and Other Photochemical Oxidants. National Academy of Sciences, Washington, D.C.

NAS, Committee on Medical and Biologic Effects of Environmental Pollutants. 1978. Sulfur oxides. National Academy of Sciences, Washington, D.C.

Sackner, M. A., D. Ford, R. Fernandez, E. D. Michaelson, R. M. Schreck, and A. Wanner. 1977a. Effect of sulfate aerosols on cardiopulmonary function of normal humans (abstract). Am. Rev. Respir. Dis. 115(4, part 2):240.

Sackner, M. A., M. Reinhardt, and D. Ford. 1977b. Effect of sulfuric acid mist on pulmonary function in animals and man (abstract). Am. Rev. Respir. Dis. 115(4, part 2):240.

U.S. Department of Health, Education, and Welfare. 1970. Air Quality Criteria for Sulfur Oxides. National Air Pollution Control Administration, Publication No. AP-50, Washington, D.C. Available from U.S. Government Printing Office.

DISCUSSION

Dave Odor, Public Service of Indiana: Did you determine for those subjects who were asthmatics whether they were on any medication, such as bronchodilators or steroid-dependent medication?

J. D. Hackney: You can philosophize about the best way to study people who are on medication regimes. You might want to take them off of all medication, but that's clearly disruptive and stressful to them and may be a problem ethically in some instances. On the other hand, there are those who believe that it's better to let subjects stay on long-lasting medication but to restrict the use of any aerosol bronchodilator or any short-acting bronchodilator during the study. In all of our work with asthmatics we have chosen the latter path, and in later publications we will documant the types of medication regimes for the subjects. Some were on no medication, some were on varying degrees of long-acting medication, some were on steroids.

Gail Bingham, Lawrence Livermore Labs: I'm a plant scientist and have a question in general on procedure here. In our studies, one of the things that we've been working on is the response of ponderosa pine in the San Bernadino forests. We find that some individuals of that species are very sensitive and others seem to go on and grow and photosynthesize at the same approximate rate as they would anywhere else. I would assume that humans are the same way in that some humans respond much more severely than others do. Since we can't thin them as we do pine trees and since they are the people responding first, why not set levels based on the more sensitive individuals rather than the population as a whole?

J. D. Hackney: There's no question some individuals are more sensitive than others. It's unclear just which group is or will be most sensitive; much more data is needed. Ethically, it's better to study normals first and then to add groups that may be more sensitive. With regard to which group ought to be used to set standards, I don't have a ready answer for that. Proposed legislation has pointed to the most sensitive, but it's not clearly defined what is meant by most sensitive or exactly what proportion of the population that is. Studies ought to address the issue of which groups are more sensitive and probably ought to try to bias whatever sampling is done to include at least the ones thought to be more sensitive, and we try to do that. For example, we study healthy people first and then try to bias everything toward including people who might be more sensitive in the final group that we study.

CHAPTER 12

HEALTH SIGNIFICANCE OF EXPOSURES TO SULFUR OXIDE AIR POLLUTANTS

Morton Lippmann*

ABSTRACT

The ambient air in the eastern United States generally contains SO_2 and H_2SO_4 and its ammonium salts at levels well above continental background levels, but also well below levels having demonstrated adverse effects on human health. However, the ambient concentrations are sufficiently high to generate concern about some specific health effects which have been produced in clinical and laboratory studies at concentrations within an order of magnitude of ambient levels. The most significant of these are the studies which demonstrate that H_2SO_4 in the ambient air could (1) contribute to the pathogenesis of chronic bronchitis and (2) cause airway constriction sufficient to affect individuals with reactive airways. Additional concern is generated by animal studies which indicate that (1) SO_2 can act as a cocarcinogen when inhaled with benzo(a)pyrene, (2) SO_2 can inhibit alveolar clearance of mineral dust particles, and (3) H_2SO_4 in combined exposures with ozone can enhance mortality in animals infected with bacterial aerosols. However, the concentrations required to produce these effects were very much higher than anticipated ambient levels.

INTRODUCTION

The significance to human health of inhalation exposures to the various sulfur oxide contaminants found in the ambient air has not been definitely established. Among the many specific sulfur oxide compounds generally present in significant concentrations are sulfur dioxide (SO_2), sulfuric acid (H_2SO_4), ammonium bisulfate (NH_4HSO_4), and ammonium sulfate [$(NH_4)_2SO_4$]. Even if their individual effects were better understood, there would still remain a need to evaluate their influence on health in representative combinations, since they are almost always present as mixtures in the ambient air. Furthermore, the effects of other common pollutant species, such as nitrogen dioxide (NO_2), nitric acid (HNO_3), and ozone (O_3), which may also influence the extent of damage to the respiratory epithelium or its adaptation to the exposure stress, either alone or in combination with the sulfur oxide species, must also be considered. The reasons for our woefully inadequate knowledge base lie only to a minor extent in the utility or reliability of our past work. The more significant limitations lie in the complexities of the environmental mixtures

---

*Institue of Environmental Medicine, New York University, New York, N.Y. 10016.

and their biological effects and in the inadequacies of our investigative resources.

Epidemiological techniques provide only a blunt probe even in the best of circumstances, such as when the agents causally related to the end effects of interest are established. Effective dose-response studies in human populations require a characterization of the effective doses. Such characterizations are hardly possible when we cannot be sure what compounds should be analyzed or over what averaging times. All we can be sure of is that we are probably not currently measuring concentrations of much relevance. We have significant data bases only for SO_2 and total suspended particulate (TSP), with some more limited data on water-soluble sulfates. SO_2, at recent and current ambient levels, is certainly not likely to be a significant stressor on human health. TSP presumably contains the sulfur oxide aerosol species of interest but is generally dominated by the presence of other constituents. Water-soluble sulfate also may be expected to contain either the compounds of interest or their conversion products formed on the collection substrate during the sample collection, storage, or analysis, but it tends to be dominated by ammonium sulfate, one of the least toxic compounds in the mixture.

Clinical studies are, of course, limited to studies of reversible physiological responses and can only provide insight into chronic respiratory disease pathogenesis when combined with high-level or chronic animal exposure studies performed with similar exposure protocols. A brief review of the human studies of respiratory mechanical function responses and of mucociliary clearance function responses will be presented.

Our laboratory studies data base has also been of limited value. Studies directed at the characterization of pathological effects of the sulfur oxides have, for the most part, been conducted at concentrations far higher than those that can be found in the ambient air. Those studies focused on physiological responses have been concerned primarily with only one type of response, namely, changes in respiratory mechanical function. As a result, we know a great deal about airway constriction in the guinea pig for a large number of compounds and about the influence of particle size on such responses. We also know that all other animals, including healthy humans, are much less responsive. On the other hand, it appears to be prudent to utilize the guinea pig as a model for asthmatic humans. The only other physiological function for which significant *in vivo* laboratory responses have been produced at concentrations representative of ambient air is mucociliary particle clearance. A summary of the limited amount of available data on these responses, and of their implications to the pathogenesis of chronic bronchitis, will also be presented.

The influence of sulfur oxide pollutants as cofactors in carcinogenesis, emphysema, lung fibrosis, and respiratory tract infectivity deserves further investigation. Laskin et al. (1976) demonstrated that combined SO_2 plus benzo(a)pyrene (BaP) exposures produced a major increase in the incidence of squamous carcinoma in rats, compared to BaP alone. Ferin and Leach (1973) demonstrated that SO_2 exposures could significantly alter the alveolar clearance of insoluble mineral dust particles. Gardner et al. (1977) demonstrated that combined exposure to ozone and sulfuric acid produced enhanced mortality in mice infected with *Streptococcus pyogenes,* whereas exposure to either pollutant alone did not. The results of these studies and their implications to human health will also be reviewed.

EFFECTS ON MUCOCILIARY CLEARANCE

In vivo exposures to SO_2 at concentrations that have been measured in the ambient air are not likely to affect mucociliary clearance. At higher levels, more typical of some occupational exposures, effects have been observed. Wolff et al. (1975) exposed nine nonsmokers to 5 ppm (13 mg/m³) of SO_2 for 3 hr after a ^{99m}Tc-tagged albumin aerosol exposure. The tracheobronchial mucociliary clearance of the tagged aerosol was essentially the same as in control tests, except for a transient acceleration at 1 hr after the start of the SO_2 exposure. In further tests by Wolff et al. (1977), it was shown that exercise accelerates bronchial clearance and 5 ppm of SO_2 during exercise significantly speeds clearance beyond that produced by the exercise alone.

High concentrations of SO_2 can slow bronchial clearance. Thirty-minute exposures to SO_2 via nasal catheters produced delayed bronchial clearance and severe coughing and mucus discharge via the nose in donkeys when the concentration exceeded 300 ppm (Spiegelman et al., 1968). Mean residence times following exposures of 53 to 300 ppm were not significantly different from control test levels. The one test performed at a lower concentration (27 ppm) produced an acceleration in bronchial clearance, which would be consistent with the clearance accelerations seen by Wolff et al. (1975, 1977) with 5-ppm exposures.

Fairchild et al. (1975) showed that 4-hr exposures to high concentrations of H_2SO_4 (15 mg/m³ of 3.2-μm droplets) reduced the rate of ciliary clearance of a tagged streptococcal aerosol from the lungs and noses of mice. At concentrations of 1.5 mg/m³ of 0.6-μm droplets, there were no significant effects.

Schlesinger, Lippmann, and Albert (1978) demonstrated that 1-hr exposures to 0.5-μm H_2SO_4 mist at concentrations in the range of 200 to 1000 μg/m³ produced transient slowings of bronchial mucociliary particle clearance in three of four donkeys tested. In addition, two of the four animals developed persistently slowed clearance after about six exposures. Similar exposures had no effects on regional particle deposition or respiratory mechanics, and corresponding exposures to $(NH_4)_2SO_4$ had no measurable effects. In subsequent tests, the two animals showing only transient responses and two previously unexposed animals were given daily 1-hr exposures, 5 days/week, to H_2SO_4 at 100 μg/m³. Within the first few weeks of exposure, all four animals developed erratic clearance rates, i.e., rates which, on specific test days, were either significantly slower than or significantly faster than those in their preexposure period. However, the degree and the direction of change in rate differed to some extent in the different animals. The two previously unexposed animals developed persistently slowed bronchial clearance during the second three months of exposure and during four months of follow-up clearance measurements, while the two previously exposed animals adapted to the exposures in the sense that their clearance times consistently fell within the normal range after the first few weeks of exposure (Schlesinger et al., 1979).

The sustained, progressive slowing of clearance observed in two initially healthy and previously unexposed animals is a significant observation, since any persistent alteration of normal mucociliary clearance can have important implications.

In a study of the possible effects of transient H_2SO_4 inhalation exposures on human pulmonary physiology (Lippmann et al., 1980), we exposed ten healthy nonsmokers to 0.5-μm H_2SO_4 aerosol for 1 hr at concentrations of 0 (distilled water control), 100, 300, and 1000 μg/m³. Nasal breathing was selected for the exposures in this study in order to provide a realistic simulation of normal human exposure conditions and to minimize *in situ* neutralization of the H_2SO_4 to $(NH_4)_2SO_4$ by endogenous NH_3 (Larson et al., 1977). A ^{99m}Tc-tagged Fe_2O_3 aerosol was inhaled $\sim$10 min before each H_2SO_4 exposure. Lung retention of the deposited radioactivity was monitored by external scintillation detectors. A tracheal probe containing six rectangular collimated scintillation detectors was used to determine the tracheal mucus transport rates (TMTRs) of boli of activity. Mucociliary bronchial clearance was monitored throughout the day following each acid exposure, by which time at least 90% of the tagged aerosol was cleared in most individuals. Each subject returned the following day in order to determine the 24-hr retention. Previous studies have shown that mucociliary clearance is essentially complete within this time, with the fraction of material remaining having initially been deposited in nonciliated pulmonary spaces (e.g., alveoli and respiratory bronchioles) (Albert et al., 1973).

No significant changes in respiratory mechanics or group mean TMTRs were observed following H_2SO_4 exposure at any level, but bronchial mucociliary clearance was markedly altered. Following the exposure at 100 $\mu g/m^3$, the mean bronchial clearance half-time ($TB_{1/2}$) was decreased from 80 to 50 min, indicating a significant increase ($p < 0.02$) in the rate of particle clearance. Following exposure to 1000 $\mu g/m^3$, the mean $TB_{1/2}$ was 118 min, which was significantly greater than the control mean ($p < 0.03$). At both 300 and 1000 $\mu g/m^3$, the $TB_{1/2}$ values were much more variable than those in the sham and 100-$\mu g/m^3$ exposures, and the responses at 300 $\mu g/m^3$ included equal numbers showing substantially faster and substantially slower $TB_{1/2}$ times than in the control tests. There was a correlation ($r^2 = 0.81$) between the control values of TMTR and $TB_{1/2}$ when paired by subject, but this relationship was disrupted following the H_2SO_4 exposures, with r^2 values being 0.11, 0.08, and 0.21 at 100, 300, and 1000 $\mu g/m^3$ respectively. Mucociliary transport in the airways distal to the trachea was therefore affected more by the H_2SO_4 exposures than transport in the trachea.

One other study on the effects of short-term inhalation exposures to H_2SO_4 in humans has been performed, and it produced some different but not necessarily inconsistent results. Newhouse et al. (1978) exposed exercising nonsmokers to 1000 $\mu g/m^3$ of H_2SO_4 and reported that the exposure produced an acceleration of bronchial clearance. At this exposure level, we observed a depression of clearance in most cases. This discrepancy may be partially accounted for by differences in the method of exposure and level of activity. In the Newhouse et al. study, the subjects were mouth breathing in an exposure chamber. Ammonia released by the chamber occupants could have resulted in partial neutralization of the H_2SO_4 to the more inert $(NH_4)_2SO_4$ (Larson et al., 1977, Schlesinger et al., 1978).

The absence of a consistent effect on tracheal transport rates is consistent with the results of other studies. Sackner et al. (1978) failed to find significant changes in tracheal mucus velocity following short-term exposures to 14 mg/m^3 (0.12 μm) in sheep. Similarly, we saw no effect on tracheal transport for H_2SO_4 exposures up to 1.4 mg/m^3 (0.5 μm) in donkeys. On the other hand, Wolff et al. (1979) reported a depression in tracheal transport rate in anesthesized dogs exposed to 1.0 mg/m^3 (0.9 μm) and a mixed response of both increased and decreased velocity following exposures to 0.5 mg/m^3. The latter results are quite similar to those observed in the bronchi of individual humans in our study, although we recorded no significant change in the mean tracheal mucociliary transport rates.

Clearly, the results of our human and donkey studies indicate that H_2SO_4 aerosol affects mucociliary clearance in the distal conductive airways. Mucociliary clearance is dependent upon both the physicochemical properties of the mucus and the coordinated beating of the underlying cilia. Mucus is excreted into the airway lumen in an alkaline form which is then acidified by CO_2 (Holma et al., 1977). In vitro studies have shown that mucus is a sol in high-pH solutions, while at lower pH it becomes viscous (Breuninger, 1964). The H^+ supplied by the H_2SO_4 may stiffen the mucus and increase the efficiency of removal. This is consistent with the increase in bronchial clearance rate observed in humans following exposure to 100 $\mu g/m^3$. In vitro studies have shown that in vivo exposures at $\sim$1000 $\mu g/m^3$ can cause a depression of ciliary beating frequencey (Grose et al., 1978) which may lead to a depression in overall bronchial clearance.

The effects of ambient H_2SO_4 exposures on mucociliary clearance may be influenced by the action of atmospheric cocontaminants. Last and Cross (1978) exposed rats for 3 and 14 days (23.5 hr/day) to $\sim$0.5-μm H_2SO_4 at 1.1 mg/m^3 and to O_3 at 0.5 ppm and produced significant increases ($\sim$30%) in the secretion of mucus glycoproteins into the trachea, while exposures to each alone at these concentrations did not. When the H_2SO_4 concentration was lowered to 11 μg/m^3, the three-day exposure to it plus 0.5 ppm O_3 still produced a 15% increase in secretion.

The dose-response trends observed for H_2SO_4 effects on bronchial clearance parallel those of previous studies conducted in our laboratory with cigarette smoke, a known causal factor in the complex etiology of chronic bronchitis. Bronchial clearance is markedly accelerated in both humans and donkeys exposed to the fresh smoke from several cigarettes (Albert et al., 1974, 1975). However, the fresh smoke from ten or more cigarettes produces substantial retardations of bronchial clearance in donkeys (Albert et al., 1974). As indicated previously, recent donkey studies in this laboratory have shown that repeated exposures to H_2SO_4 at 100 μg/m^3 are able to produce a marked and persistent depression in bronchial clearance comparable to that produced by chronic cigarette smoking, even for those animals which demonstrated no transient effects following single exposures at 100 μg/m^3 of H_2SO_4.

Although direct evidence of an association between chronic low-level exposures to H_2SO_4 and chronic bronchitis is lacking, the similarity in response between H_2SO_4 and cigarette-smoke exposures suggests that such an association is possible. At concentrations of H_2SO_4 much higher than those used here, there was an increase in the number of attacks of bronchitis in exposed workers, though the number of persons with respiratory disease was not increased (Williams, 1970), and persistent bronchitic symptoms developed in volunteer subjects following repeated acid exposures (Sim and Pattle, 1957).

Based on the results of this study, those of our coordinate animal studies, and those reported for human exposures by Williams and by Sim and Pattle, it appears that chronic H_2SO_4 exposures at concentrations of $\sim$100 μg/m^3 could produce persistent changes in mucociliary clearance in previously healthy individuals and exacerbate preexisting respiratory disease.

EFFECTS ON RESPIRATORY MECHANICS

Attempts to produce significant changes in airway resistance or conductance in healthy humans or animals with sulfur oxide aerosols at concentrations of up to 1 mg/m^3 have been largely unsuccessful, except in the guinea pig. Amdur and coworkers (1952a, 1978a, 1978b), in an extensive series of studies with random-bred guinea pigs, found significant increases in resistance and decreases in compliance following exposures to 0.1 to 1.0 mg/m^3. At these levels, 0.3-μm H_2SO_4 was more irritating than 1.0 μm aerosol. Slight increases in pulmonary resistance were reported for monkeys exposed to H_2SO_4 by Greenberg et al. (1978) but only at concentrations equal to or greater than approximately 3 mg/m^3.

Chronic H_2SO_4 exposures have also resulted in alterations in ventilatory mechanics, which were not readily reversible after the cessation of exposure. In one study by Alarie et al. (1973), continuous exposures to 2.4 and 4.8 mg/m^3 produced mild to moderate changes in respiratory rate and the distribution of ventilation in cynomolgus monkeys, first observable at 12 to 49 weeks after initiation of exposure. Exposure to 2.2-μm H_2SO_4 at 0.4 mg/m^3 produced slight microscopic changes in the lungs and increases in respiratory rate, while exposure to 0.5-μm H_2SO_4 at 0.5 mg/m^3 produced no significant effects. Hartley-strain guinea pigs also showed no response following 52-week exposures of 0.08 or 0.10 mg/m^3, 0.8 and 2.8 μm respectively. Lewis et al. (1969, 1973) observed other respiratory effects, including increased expiratory resistance and decreased CO_2 diffusing capacity and lung volumes, in beagle dogs exposed to 0.9 mg/m^3 for 32 weeks. In both studies, the authors found it difficult to correlate physiological changes with pathological changes due to a lack of dose-response trends. Lewkoski et al. (1978) found the pulmonary resistance and respiratory rate to increase in adult rats exposed to 4 mg/m^3 for 50 days, but little or no effect occurred in juvenile rats exposed to 2 mg/m^3. Similarly, short-term exposures failed to produce changes in respiratory resistance or compliance in anesthetized dogs exposed to 4 mg/m^3 ($\leqslant 0.1$ μm) for 4 hr (Sackner et al., 1978), in unanesthetized donkeys exposed to 1.4 mg/m^3 (0.5 μm) for 1 hr (Schlesinger et al., 1978), or in Hartley-strain guinea pigs exposed to 1.2 to 14.6 mg/m^3 (1 μm) for 1 hr (Fairchild et al., 1978, Silbaugh et al., 1979).

These apparent discrepancies may be accounted for by interspecies, and perhaps even intraspecies strain, differences. Such variability seems quite plausible, explainable by such factors as regional shifts in aerosol deposition, varied in situ ammonia concentration in expired air, adaptation, and varied intrinsic host susceptibility. Direct clinical experimentation is therefore quite important.

Investigations of possible alterations in the mechanics of ventilation in humans have been studied by several investigators. Amdur et al. (1952b) observed increases in respiratory rate and decreases in tidal volume (rapid, shallow breathing) in healthy persons breathing 1-μm sulfuric acid through a face mask at concentrations ranging from 0.4 to 5.0 mg/m^3. These changes occurred within 2 min after the start and lasted throughout a 5- to 15-min exposure. Subjects were unable to detect by odor, taste, or irritation exposures of 1 mg/m^3 or less. Above this level, subjects described a sensation of breathing dusty air, which was detectable by all persons exposed to 3 mg/m^3. Coughing was noted upon deep inhalation of 5 mg/m^3.

Exposures to 1.0 μm (MMD) sulfuric acid in much higher concentrations (up to 39 mg/m^3 for 10 to 60 min) by Sim and Pattle (1957) caused exposed persons to develop transient cough, rales and bronchoconstriction. Pulmonary resistance was increased more than 20% in most subjects and in some cases was as much as 150% above control values. Cardiovascular functions—blood pressure and pulse rate—remained unchanged. The two experimenters noted subjective respiratory effects, symptomatic of bronchitis, resulting from personal exposures to sulfuric acid or sulfur dioxide. During the ten-month period in which these experiments were conducted, one author developed a moderately severe but persistent bronchitis, which was immediately exacerbated into an uncomfortable period of coughing and wheezing when exposed to either agent. The other author noted chest symptoms lasting throughout the exposure day and a persistent wheezing lasting for four days following a 30-min exposure to 39 mg/m^3 of H_2SO_4.

Most recent clinical studies at concentrations equal to or less than 1.0 mg/m^3 have found no significant changes in parameters of respiratory mechanics. Sackner et al. (1978) reported no changes in several tests of cardiopulmonary function in healthy and asthmatic persons exposed to 10, 100, or 1000 μg/m^3 for 10 min. The mean particle size of the aerosol was quite small (~ 0.1 μm). No immediate or delayed effects upon ventilatory mechanics or bronchial asthma were noted in subjects followed for a few weeks after exposure.

Avol et al. (1979) also found minor changes in respiratory mechanics or reported symptoms in six healthy and six asthmatic persons exposed to 0.3-μm H_2SO_4 for 2 hr. Ammonia levels in the exposure chamber, presumably resulting from the expired air of the test subjects, were high enough to neutralize a portion of the H_2SO_4. The H_2SO_4 was subsequently increased so that the resulting exposure atmosphere contained 75 $\mu g/m^3$, as well as 40 $\mu g/m^3$ of ammonium salts. In the asthmatic group, two of the six subjects showed slight increases in pulmonary resistance following exposure, but no statistically significant changes in resistance or other test of ventilatory mechanics were recorded for the group as a whole.

In investigations conducted with ten healthy exercising nonsmokers, Newhouse et al. (1978) found no change in $FEV_{1.0}$ and only a slight (1.4%, $p \leqslant 0.09$) depression of MMES following a 2-hr exposure to 1000 $\mu g/m^3$ (0.5 μm) H_2SO_4.

In our own studies at New York University involving ten nonsmoking normal volunteers, we found no demonstrable alterations in measurements of pulmonary mechanics following exposures to 0.5-μm H_2SO_4 at concentrations of 1000 $\mu g/m^3$ and less.

The only study reporting statistically significant alterations in respiratory mechanical function following inhalation exposures to humans at H_2SO_4 concentrations of 1000 $\mu g/m^3$ is that of Morrow et al. (1980), who reported that small ($<$ 10%), but statistically significant, changes in specific airway conductance (SG_{aw}) were produced by 16-min H_2SO_4 exposures at 1 mg/m^3 in normal volunteer subjects. Much greater changes were produced when the exposures were followed by a carbachol challenge aerosol, which, when given alone, produced a 10% reduction in SG_{aw}. The carbachol challenge following the H_2SO_4 exposure increased the average reduction in SG_{aw} to 27%. Using the same protocols with NH_4HSO_4, $NaHSO_4$, and $(NH_4)_2SO_4$, the reductions in SG_{aw} without carbachol challenge were not statistically significant. However, with the carbachol challenge, the reductions in SG_{aw} were ~18%, 15%, and 14%, respectively, and were statistically significant. These results suggest that the effect was closely related to the concentration of hydrogen ions delivered to the airways by the inhaled sulfate aerosol.

OTHER HEALTH EFFECTS

While the data on other health effects, such as carcinogenesis and emphysema, are meager, these health problems are significant in terms of their potential for causing serious damage.

The role of the sulfur oxide pollutants in carcinogenesis has been addressed by Laskin and co-workers at New York University. In a limited number of experimental studies in rodents they obtained positive results with combined sulfur dioxide–benzo(a)pyrene (BaP) inhalation exposures (Laskin et al., 1976). On the other hand, inhalation exposures to sulfuric acid following administration of BaP by intratracheal instillation did not result in an increased cancer yield (Sellakumar, 1979).

The mechanism for the action of SO_2 as a cofactor in benzo(a)pyrene carcinogenesis is unknown. As an irritant on the larger conductive airways it could be altering the structure, physiological function, or metabolism of the airways in a manner which favors the expression of initiated cells. The SO_2 could also affect the dosimetry for the carcinogen. As a bronchoconstrictor, it would reduce the cross section for airflow within the larger bronchial airways, greatly increasing the surface density of the carcinogenic aerosol deposition on the bronchial bifurcations. The absence of a similar effect in the H_2SO_4 study could have been due to the absence of significant amounts of BaP deposition on the bronchial bifurcations.

Ferin and Leach (1973) demonstrated that SO_2 can reduce the clearance of titanium dioxide (TiO_2) particles from the alveoli. Exposures of 1 ppm of SO_2 for 7 hr/day for 25 days increased the alveolar retention of TiO_2 at 25 days from 61 to 92%, a significant decrease in clearance ($p < 0.01$). Virtually nothing is known about the effects of sulfur oxide aerosols on the clearance of insoluble particles from the alveolar regions of the lung parenchyma. Interference with clearance from the deep lung would have serious implications to the pathogenesis of emphysema and fibrosis, and experimental inhalation studies with respirable sulfur oxide aerosols need to be performed.

INVESTIGATION OF COMBINED EFFECTS OF SULFUR OXIDES AND OXIDANTS

Summertime haze episodes in the eastern half of the United States are characterized by high concentrations of ozone and a very high percentage of sulfuric acid within the aerosol. These observations have interesting implications for our concepts of potential health effects within human populations and of the subpopulations who should be considered to be at special risk. In a physical sense, the high proportion of acid within the aerosol is probably related to the efficiency with which ozone oxidizes SO_2. Toxicologically, it is significant because of the several demonstrations that ozone enhances the effects of SO_2 and H_2SO_4. For H_2SO_4, these include the increase in mucus glycoprotein secretions reported by Last and Cross (1978) and the increase in mortality in infected mice reported by Gardner et al. (1977). It also appears that combined exposure to O_3 and SO_2 may be synergistic with respect to mechanical functional decrements and athletic performance. On this basis it may be prudent to devote some of our air pollution epidemiological resources to rural populations, because the haze episodes characteristically produce the highest concentrations of sulfuric acid and ozone in "clean" rural areas. Because the populations in these areas are less affected by point-source pollutants and urban dirt, they may conceivably show greater transient responses in physiological functions when exposed to a summer haze episode.

HUMAN HEALTH IMPLICATIONS OF EXPOSURES TO AIRBORNE SULFUR OXIDES

Various sulfur oxide air pollutants have been shown to produce measurable effects on human health when the exposures take place at elevated concentrations. However, because the concentrations required to produce effects relevant to human disease are, in most cases, higher by orders of magnitude than those currently found in ambient air, there is no clear evidence that the United States population is currently suffering any adverse health effects from the inhalation of this class of compounds. On the other hand, our current state of knowledge on mechanism of action and dose-response is so poor at present that we clearly have no basis for complacency either.

On the basis of the studies reviewed in this presentation, the following areas of concern, and need for further investigation, are most acute:

1. The effects of sulfuric acid on the pathogenesis of chronic bronchitis. Concern is warranted because of the close similarity of responses with respect to the production of mucociliary clearance anomalies by sulfuric acid and cigarette smoke in experimental inhalation studies in humans and animals by Lippmann and co-workers and the known role of cigarette smoke in the pathogenesis of bronchitis. These studies suggest that periodic exposures to sulfuric acid at concentrations of less than 100 $\mu g/m^3$ may be sufficient to initiate the development of bronchitis.

2. The effects of sulfuric acid and ammonium bisulfate on the induction of asthmatic attacks in sensitive individuals. Concern is warranted on the basis of the studies of Morrow et al., which show that major changes in specific airway conductance are inducible by a carbachol aerosol challenge in healthy normals following exposures to sulfuric acid at 1 mg/m^3 for only 16 min.

The studies of Ferin and Leach (1973), which demonstrate that SO_2 can affect the clearance of insoluble particles from the alveolar region, the studies of Laskin et al. (1976), which showed that SO_2 can act as a cofactor in the development of bronchial carcinoma, and the studies of Gardner et al. (1977), which show that combined exposures to H_2SO_4 and O_3 can increase mortality in mice infected with streptococci, clearly raise significant concerns about the possible effects of the sulfur oxides on other lung diseases. However, the levels of SO_2 exposure required to produce the effects were many orders of magnitude higher than ambient levels. These results are considered to be of lesser immediate concern than those cited in the preceding paragraphs.

ACKNOWLEDGMENTS

The experimental work on mucociliary clearance was performed in collaboration with George Leikauf, Richard B. Schlesinger, Donovan B. Yeates, and Roy E. Albert. This work was supported by the U.S. EPA under Contract No. 68-02-1726. The entire review was prepared as part of a Center Program supported by the National Institute of Environmental Health Sciences—Grant No. ES 00260.

REFERENCES

Alarie, Y., W. M. Busey, A. A. Krumm, and C. E. Ulrich. 1973. Long-term continuous exposure to sulfuric acid mist in cynomolgus monkeys and guinea pigs. Arch. Environ. Health 27:16–24.

Albert, R. E., M. Lippmann, H. T. Peterson, Jr., J. Berger, K. Sanborn, and D. Bohning. 1973. Bronchial deposition and clearance of aerosols. Arch. Intern. Med. 131:115–127.

Albert, R. E., J. Berger, K. Sanborn, and M. Lippmann. 1974. Effects of cigarette smoke components on bronchial clearance in the donkey. Arch. Environ. Health 29:96–101.

Albert, R. E., H. T. Peterson, Jr., D. E. Bohning, and M. Lippmann. 1975. Short-term effects of cigarette smoking on bronchial clearance in humans. Arch. Environ. Health 30:361–367.

Amdur, M. O., R. Z. Schulz, and P. Drinker. 1952a. Toxicity of H_2SO_4 acid mist to guinea pig. Arch. Ind. Hyg. Occup. Med. 5:318–319.

Amdur, M. O., L. Silverman, and P. Drinker. 1952b. Inhalation of sulfuric acid mist by human subjects. Arch. Ind. Hyg. Occup. Med. 6:305–313.

Amdur, M. O., M. Dubriel, and D. A. Cresia. 1978a. Respiratory response of guinea pig to low levels of sulfuric acid. Environ. Res. 15:418–423.

Amdur, M. O., J. Bayles, V. Ugro, and D. W. Underhill. 1978b. Comparative irritant potency of sulfate salts. Environ. Res. 16:1–8.

Avol, E. L., M. P. Jones, R. M. Bailey, N.-M. N. Chang, M. T. Kleinman, W. S. Linn, K. A. Bell, and J. D. Hackney. 1979. Controlled exposures of human volunteers to sulfate aerosols. Health effects and aerosol characterization. Am. Rev. Respir. Dis. 120:319–327.

Breuninger, H. 1964. Uber das physikalisch-chemische Verhalten des Nasenschleims. Arch. Ohren- Nasen- Kehlkopfheilkd 184:133–138.

Fairchild, G. A., P. Kane, B. Adams, and D. Coffin. 1975. Sulfuric acid and streptococci clearance from respiratory tracts of mice. Arch. Environ. Health 30:538–545.

Fairchild, G. A., S. Stulz, and D. L. Coffin. 1978. Sulfuric acid effect on the deposition of radioactive aerosol in the respiratory tract of guinea pigs. Am. Ind. Hyg. Assoc. J. 36:584–594.

Ferin, J., and L. J. Leach. 1973. The effect on lung clearance of TiO_2 particles in rats. Am. Ind. Hyg. Assoc. J. 34:260-263.

Gardner, D. E., F. J. Miller, J. W. Illing, and J. M. Kirtz. 1977. Increased infectivity with exposure to ozone and sulfuric acid. Toxicol. Lett. 1:59-64.

Greenberg, H. L., R. M. Bailey, K. A. Bell, E. L. Avol, and J. D. Hackney. 1978. Health effects of selected sulfate aerosol in squirrel monkeys. Presented at the Annual Meeting of the American Industrial Hygiene Association, Los Angeles, May 1978.

Grose, E. C., F. J. Miller, and D. E. Gardner. 1978. The effects of ozone and sulfuric acid on ciliary activity of Syrian hamster (abstract). Pharmacologist 20:211.

Holma, B., J. Lindegren, and J. M. Andersen. 1977. pH effects on ciliomotility and morphology of respiratory mucosa. Arch. Environ. Health 32:216-226.

Larson, T. V., D. S. Covert, R. Frank, and R. J. Charlson. 1977. Ammonia in the human airways: neutralization of inspired acid sulfate aerosols. Science 197:161-163.

Laskin, S., M. Kuschner, A. Sellakumar, and G. V. Katz. 1976. Combined carcinogen-irritant animal inhalation studies. Pp. 190-213 in Aharonson, E. F., A. Ben-David, and M. A. Klingberg (eds.), Air Pollution and the Lung. Wiley, Jerusalem. Pp. 313.

Last, J. A., and C. E. Cross. 1978. A new model for health effects of air pollutants: evidence for synergistic effects of mixtures of ozone and sulfuric acid aerosols on rat lungs. J. Lab. Clin. Med. 91:328-339.

Lewkoski, J. P., L. Hastings, A. Vinegar, J. Leng, and G. P. Cooper. 1978. Inhalation of sulfate particulates. I: Effects on growth, pulmonary function and locomotor activity (abstract). Toxicol. Appl. Pharm. 45:246.

Lewis, T. R., K. I. Campbell, and T. R. Vaughan, Jr. 1969. Effects on canine pulmonary function: via induced NO_X impairment, particulate interaction and subsequent SO_X. Arch. Environ. Health 18:596-601.

Lewis, T. R., W. J. Moorman, W. F. Ludmann, and K. I. Campbell. 1973. Toxicity of long-term exposure to oxides of sulfur. Arch. Environ. Health 26:16-21.

Lippmann, M., R. E. Albert, D. B. Yeates, K. Wales, and G. Leikauf. 1980. Effect of sulfuric acid mist on mucociliary bronchial clearance in healthy non-smokers. In Proc. GAF 7—Aerosols in Science, Medicine, and Technology, Dusseldorf 3-5 Oct. 1979.

Morrow, P. E., M. J. Utell, F. R. Gibb, and R. W. Hyde. 1980. Studies of pollutant aerosol simulants in normal and susceptible human subjects. In Proc. GAF 7—Aerosols in Science, Medicine, and Technology, Dusseldorf 3-5 Oct. 1979.

Newhouse, M. T., M. Dolovich, G. Obminski, and R. K. Wolff. 1978. Effect of TLV levels of SO_2 and H_2SO_4 on bronchial clearance in exercising man. Arch. Environ. Health 33:24-31.

Sackner, M. A., D. Ford, R. Fernandez, J. Cipley, D. Peroz, M. Kwoka, M. Reinhardt, E. O. Michaelson, R. Schreck, and A. Wanner. 1978. Effects of sulfuric acid aerosol on cardiopulmonary functions in dogs, sheep and humans. Am. Rev. Respir. Dis. 118:497-510.

Schlesinger, R. B., M. Lippmann, and R. E. Albert. 1978. Effects of short-term exposures to sulfuric acid and ammonium sulfate aerosols upon bronchial airways function in donkeys. Am. Ind. Hyg. Assoc. J. 39:275-286.

Schlesinger, R. B., M. Halpern, R. E. Albert, and M. Lippmann. 1979. Effect of chronic inhalation of sulfuric acid mist upon mucociliary clearance from the lungs of donkeys. J. Environ. Path. Toxicol. 2:1351-1367.

Sellakumar, A. 1979. Personal communication.

Silbaugh, S. A., R. K. Wolff, and J. L. Mauderly. 1979. Acute pulmonary function effects of sulfuric acid inhalation in Hartley guinea pigs (abstract). Fed. Proc. 38:1234.

Sim, V. M., and R. E. Pattle. 1957. Effect of possible smog irritants on human subjects. J. Am. Med. Assoc. 165:1908-1913.

Spiegelman, J. R., G. D. Hanson, A. Lazarus, R. Bennett, M. Lippmann, and R. E. Albert. 1968. Effect of sulfur dioxide exposure on bronchial clearance in the donkey. Arch. Environ. Health 17:321-326.

Williams, M. K. 1970. Sickness absence and ventilatory capacity of workers exposed to sulfuric acid mist. Brit. J. Ind. Med. 27:61-66.

Wolff, R. K., M. Dolovich, Rossman, and M. T. Newhouse. 1975. Sulfur dioxide and tracheobronchial clearance in man. Arch. Environ. Health 30:521-527.

Wolff, R. K., M. Dolovich, G. Obminski, and M. T. Newhouse. 1977. Effect of sulfur dioxide on tracheobronchial clearance at rest and during exercise. In Inhaled Particles IV, pp. 321-330, ed. by W. H. Walton. Pergamon Press, Oxford.

Wolff, R. K., B. A. Muggenburg, and S. A. Silbaugh. 1979. Effects of sulfuric acid mist on tracheal mucous clearance in awake beagle dogs. (abstract). Am. Rev. Respir. Dis. 119:242.

DISCUSSION

Questioner: We know that in the air we are expiring there is a lot of ammonia, something like 1 ppm. Have you ever measured whether your input of sulfuric acid in animals or human beings exceeded the production of ammonia? What I mean is, if you just inhale a certain level, there may be ammonia enough to neutralize the acid, but if you exceed a certain level, there may not be enough ammonia.

M. Lippmann: That was a point in the written paper that I didn't mention because of time. We are well aware of the fact that there is ammonia excreted in the air from humans and animals. This was pointed out by Larsen and Frank, of the University of Washington, Seattle. The levels aren't quite as high as you suggest. Stoichiometrically, ammonia could neutralize up to about 150 μg of sulfuric acid per cubic meter if air is inhaled and exhaled through the nose. In our human studies we did exposure by a nasal mask to minimize the problem that exposure through a mouthpiece presents a greater opportunity for in situ neutralization. This problem arises because there is more ammonia in the oral cavity than in the nasal airways. The recent work at the University of Washington suggests that while ammonia is available for neutralization, in fact it doesn't. Only a fraction of the ammonia does in fact neutralize the aerosol during its flight before deposition. Clearly, if the aerosol is neutralized before deposition, I would expect it to have no effect, since we've tested ammonium sulfate and have seen no effect. Perhaps part of the reason is that the ammonia is picked up on exhalation, mostly at the end, and not so much on inhalation. There may be a basis for chemical thresholds, but ammonia excretion varies a great deal from animal to animal and human to human, and possible chemical thresholds would also vary.

Leonard Newman, Brookhaven: I noticed in a German work that they tested ammonium bisulfate. Did you test ammonium bisulfate, and what were the results?

M. Lippmann: Experiments on ammonium bisulfate are in progress. Initially we did only the ammonium sulfate and the sulfuric acid.

Questioner: If I understand what you're saying, when we breathe SO_2, most of it is removed in the nose, and very little gets down into the lungs. If you look at the physics of it, is that saying that the Brownian diffusion coefficient of sulfur dioxide is very high? Is it possible that sulfur dioxide is attached to submicron particles and these particles are getting down into the lung and causing damage?

M. Lippmann: Clearly that is the generally accepted and probably correct explanation. Any gas diffuses rapidly to the wall, and SO_2 happens to be soluble. The reason NO_2 is different is that it's less soluble and doesn't get washed out.

Questioner: If it's really this particle size that is causing the damage, can't we generate enough monodispersed particles and still get a response as a function of particle size?

M. Lippmann: The generation of monodisperse aerosols in the submicron range is difficult and can only be done with a mobility separator. Below 1 micron, particle size itself no longer plays much role in regional deposition. Mixing between tidal air and reserve air accounts for penetration and deposition of particles below 1 micron. There's a broad minimum in deposition between 1/10 and 1 micron where the size practically makes no difference. Larger particle size makes a tremendous difference, but again the intrinsic motion of the particles has very little to do with deposition for particles below 1 micron until you get down to less than a tenth micron, where Brownian diffusion takes over and deposition starts to increase again.

Moderator: Would the speakers comment on the general impression given by this session that SO_2 is of little concern except as a source of other pollutants?

M. Lippmann: I agree about SO_2. However, I don't agree about sulfuric acid. It seems to me that sulfuric acid and possibly ammonium bisulfate, which is also a strong aerosol acid, can certainly play a great role in the induction of bronchitis. Bronchitis is the only human disease which could be influenced by concentrations in this range. The only other system that shows any response is the guinea pig smooth airway muscle at 100 μg per cubic meter, so I think we can't neglect sulfuric acid. I think we can neglect SO_2 and ammonium sulfate in terms of human health effects. As far as I know there is not likely to be any aerosol acid other than sulfuric acid or ammonium bisulfate in the ambient air. If we had nitric acid in the ambient air, which we probably do, it would be present as a vapor, and be a highly soluble vapor at that, and behave more like SO_2 than like sulfuric acid.

Moderator: How about NO_2?

M. Lippmann: NO_2 might conceivably behave more like sulfuric acid because it's a less soluble gas and affects the deeper part of the lung more. I don't know that anybody has ever looked at the effect of NO_2 on the small conductive airways in terms of mucociliary clearance, and this is probably something we ought to do.

I think the number of studies and the quality of some are clearly inadequate to dismiss sulfuric acid or some of the other pollutants. Most mixtures haven't been studied in very much detail, and even the ones that have been studied have been studied in very limited numbers of subjects, under limitations on the characterization of the subjects, and it's not clear that the most sensitive subjects have been studied.

Dr. Speizer: I would like to give a little historical epidemiological perspective on this. Sulfur dioxide in the 1952 London episode was measured in Kew Gardens on the third day of the episode to be 1.34 ppm. I'm not sure that the measurement was accurate or representative. That episode was associated with 4,000 deaths. I'm not sure that anyone ever said that the extra deaths were due to 1-1½ ppm of SO_2; what they were measuring was dirty air. In most of the epidemiology since that time, SO_2 has been an indicator of dirty air, and efforts to clean out the SO_2 have been a little bit misdirected in the sense that what you may be doing is just taking away an indicator of dirty air. I don't think anyone has tried to say that the SO_2 itself was the agent that was the cause of these excess episodes.

M. Lippmann: I certainly agree, but perhaps to amplify that comment further, what we did was attack the messenger, and not only was it irrelevant but it was counterproductive because the Federal Government got the message that it was the SO_2, subsequently, the National Ambient Air Quality Standard was directed at reducing SO_2 at ground level. Reduction was accomplished by building much taller stacks, so that in most of the East, an order of magnitude reduction in sulfur dioxide concentrations was achieved, but we have not seen any significant decrease in that other index, sulfate. Sulfate is also a poor index, but it is presumably more related to sulfuric acid than most other things. What we've done is reduce the scrubbing action of the vegetation on the SO_2 near the source and put the SO_2 higher in the atmosphere, where it can better and more efficiently be oxidized and converted to sulfuric acid. If there is any health problem associated with the sulfur oxide, that problem has been exacerbated by the wrong application of the misguided control strategy adopted.

CHAPTER 13

THE SIX-CITY STUDY: A PROGRESS REPORT*

B. G. Ferris, Jr.†
F. E. Speizer‡
Y. M. M. Bishop§
J. D. Spengler‖
J. H. Ware#

ABSTRACT

The Six-City Study was designed to test the adequacy of the present primary standards for SO_2 and particulates and whether they should be relaxed or made more stringent. The study was originally designed to examine the chronic effects of air pollutants. To accomplish this we selected two relatively clean cities, two with levels slightly below the primary standards, and two dirty cities that had levels often above the standards. The cities are: clean—Topeka, Kansas, and Portage, Wisconsin; moderate—Watertown, Massachusetts, and Kingston-Harriman, Tennessee; and dirty—Steubenville, Ohio, and the southern tip of St. Louis, Missouri.

METHODS

The study in each of these sites can be separated into an assessment of the effect on health and air monitoring. Table 1 gives a summary of the methods involved in assessing the effect on health and Table 2 our air monitoring procedures. The health effects can be further separated into effects on adults and children. Random samples of the adult population have been selected from various census listings of the communities. About 1500 to 1800 persons 25 to 74 years of age were selected initially from each community and will be followed prospectively every three years. Health

*Supported in part by grants from the National Institute of Environmental Health Sciences (ES 0002, ES 01108), Electric Power Research Institute Contract No. RP 1001, and EPA Contract No. EP 68-02-3201.

†Department of Physiology, Harvard School of Public Health.

‡Channing Laboratory, Department of Medicine, Harvard Medical School, and Peter Bent Brigham Hospital.

§Department of Biostatistics, Harvard School of Public Health. Present position: Acting Deputy Assistant Administrator, Energy Data Operations, Energy Information Administration, Washington, D.C.

‖Department of Environmental Health Sciences, Harvard School of Public Health.

#Department of Biostatistics, Harvard School of Public Health.

Table 1. Summary of methods, Six-City Study: health effects
See text for explanation of abbreviations

| Adults (1500) | Questionnaire
Pulmonary function (FVC, FEV_1) | Q 3 yr |
| Children (2000) | Questionnaire
Pulmonary function (FVC, FEV_1,
$FEV_{0.75}$, MMEF) | Q 1 yr |

Table 2. Summary of methods, Six-City Study: air monitoring
See text for explanation of abbreviations

Central sites	Continuous: SO_2, NO_2, O_3 24-hr: TSP, MRP[a] SO_2, NO_2
Satellite sites (10–12)	24-hr: SO_2, NO_2, MRP[a]
Personal	24-hr: MRP[a] 5-day: NO_2

[a]MRP analyses—SO_4^{2-}, NO_3^-, XRF, NA.

assessment is based on a standard questionnaire on respiratory symptoms and past respiratory disease history, occupational history, tobacco usage, home characteristics, and residential history. Also, pulmonary function is measured by means of spirometry on a Collins portable survey spirometer. In general, five trials are made, and the mean of three best forced vital capacities (FVC) and forced expiratory volumes in the first second ($FEV_{1.0}$) are used. To have consistency, a maximum of eight trials is allowed, and the FVC is measured at 6 sec. Thus it is really a FEV_6. The rationale behind this and further discussion has been published elsewhere (Ferris et al., 1978). All values are corrected to BTPS (37°C, saturated, and at ambient pressure).

For the children, all the children in the first and second grades were eligible. Parochial schools were included where indicated. Because there were so many children in Topeka a random sample of slightly more than half the schools was used. These children are seen annually, and if the number was not about 2000, the new first-year class was added in subsequent years. Health effects were assessed by a standard questionnaire on respiratory symptoms and past history; home characteristics; and parental smoking habits, occupation, and educational attainment. Pulmonary function is measured in the school. The means of the three best values for FVC and $FEV_{1.0}$ are used. These are also corrected to BTPS. These tracings will also be analyzed for $FEV_{0.75}$ and midmaximal expiratory flow (MMEF).

These studies are done by two field teams. One surveys the adult population and the children in two communities. The other surveys the children in four communities. Repeat surveys of a community are done at the same time of year.

Air monitoring, as summarized in Table 2, also can be separated into categories, depending upon site. At all cities there is a central station. In some there are two. These central stations have continuous monitors for SO_2, NO_2, and ozone and 24-hr samplings for total suspended particulates (TSP), mass respirable particulates (MRP), and SO_2, and NO_2 by bubbler techniques. These are operated to coincide with EPA's sampling program and with our satellite station sampling. The central stations are also operating dichotomous samplers on a 24-hr basis: they sample two size distributions, 15–2.5 μm and less than 2.5 μm. The satellite sites are selected to yield information on the spatial distribution of the pollutants. At the satellite stations, indoor and outdoor sampling is done for 24-hr MRP, SO_2, and NO_2. Site selection also requires that a household be available for indoor monitoring in an activity room—not a kitchen, bathroom, or bedroom. In order to see how an individual's activity pattern might modify a person's exposure, personal samplers have been used in three of the cities. Because of technical problems, these have largely been limited to 24-hr MRP sampling. More recently we have used diffusion tubes (Palmes et al., 1976) for NO_2 that have to be exposed for five days to obtain an adequate sample.

In addition to weighing, the MRP samples are analyzed for sulfate (SO_4^{2-}) and nitrate (NO_3^-). The dichotomous samples are also analyzed by x-ray fluorescence (XRF). Some of the MRP samples have been subjected to neutron activation (NA). These procedures and possibly others in the future will be used in order to have a better characterization of the chemical composition of the various particles to help identify sources and thus the contribution of outdoor levels to indoor levels as well as indoor sources of particulates. This emphasizes our belief that equal mass does not necessarily imply equal effects.

RESULTS

Air Monitoring

The historical levels for the pollutants are shown in Fig. 1. It can be seen that Topeka and Portage have low levels, Kingston-Harriman and Watertown are closer to the standards, and Steubenville and St. Louis are higher than the standards. There is also a gradient across the cities for suspended sulfates, for which no Federal standard presently exists.

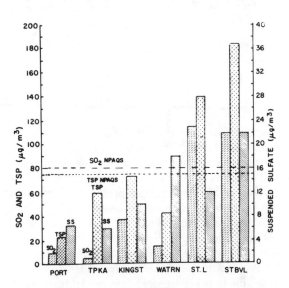

Fig. 1. Historical levels of SO_2, TSP, and suspended sulfate (SS) in the six cities. National air quality standards for SO_2 and TSP are indicated by dashed lines.

Figure 2 contains the results of our monitoring during the study period and is similar to the results obtained historically, except that the Watertown values seem to be lower. In our monitoring we have been collecting mass respirable samples. These are collected on a membrane filter after a cyclone that cuts off at 10 μm. The mass median diameter collected is 3.5 μm. These results indicate that for MRP only Steubenville stands out as different from the others. The other five are not markedly different from each other. Most of the suspended sulfate is found in the MRP. We do not have enough data as yet to report results of x-ray fluorescence or neutron activation (NA) analyses.

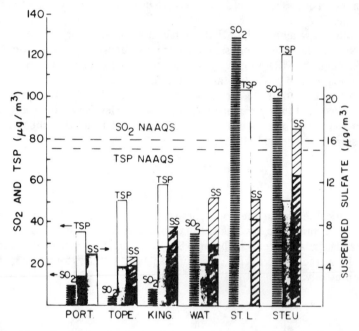

Fig. 2. Air pollution levels obtained during study period for SO_2, TSP, MRP, and SS. The darker fraction of SS represents the amount that appeared in the MRP samples.

Adults

These results represent the first cross-sectional data. Various respiratory symptoms have been tabulated for each city. White populations only have been used. The data have also been adjusted for age and smoking. Preliminary analyses have indicated that socioeconomic status will also have to be taken into consideration.

Table 3 presents the prevalence of phlegm production in the various cities by sex. Two categories are included, one in which the symptom was stated to have been present for less than three years, the other for three years or more. The more chronic form is probably more important. No clear pattern emerges, although Steubenville and St. Louis seem to have higher rates than the other cities. On the other hand, Topeka males and Portage females have relatively high rates.

If we look at wheezing in the chest most days or nights, particularly if it has been present for three years or more, Steubenville males and females have the highest rates (Table 4). If these are plotted as in Fig. 3 against the levels of pollution from

Table 3. Phlegm production by city, Six-City Study
Rate per 1000 adjusted by age and smoking, white population only

	Present for less than 3 years	Present for 3 or more years
Males		
Portage	62	165
Topeka	54	188
Watertown	82	161
Kingston-Harriman	69	176
St. Louis	102	235
Steubenville	92	201
Females		
Portage	46	82
Topeka	55	73
Watertown	83	79
Kingston-Harriman	57	103
St. Louis	70	119
Steubenville	57	91

Table 4. Wheeze rate per 1000, Six-City Study
Adjusted for age and smoking, white population only

	Occasionally apart from colds	Present for less than 3 years	Present for 3 or more years
Males			
Portage	110	18	68
Topeka	113	16	62
Watertown	159	22	68
Kingston-Harriman	109	36	72
St. Louis	84	21	71
Steubenville	94	29	93
Females			
Portage	100	9	37
Topeka	127	8	30
Watertown	132	14	26
Kingston-Harriman	130	8	51
St. Louis	72	17	41
Steubenville	69	16	64

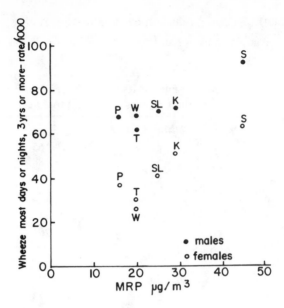

Fig. 3. Prevalence of wheeze most days or nights for three or more years in males and females by level of mass-respirable particulates (MRP) in each city. City indicated by letters. Rates are per 1000 and are age- and smoking-standardized for each sex.

Fig. 2, there does appear to be a trend across the cities, although the results could be interpreted that five of the cities are quite similar and one stands out as different. In Fig. 3 the prevalences are plotted against mass respirable particulates. Similar trends are apparent for each of the other pollutants in Fig. 2, but there is more scatter. These data have not been standardized for socioeconomic status (SES). Final conclusions must await this standardization and whether the observation can be replicated, as we have noted that SES does seem to be a factor in these respiratory symptom rates.

We do not see such a relationship for phlegm production, probably because this is so highly correlated with cigarette smoking; so when we standardize for cigarette smoking the differences in phlegm production disappear. Also a past history of pneumonia, chronic bronchitis, asthma, or emphysema did not seem to have any trend across the cities in association with the levels of pollution.

Children

The history of a variety of respiratory diseases has been examined in the children and related to or standardized for a number of home characteristics such as type of cooking fuel, air conditioning, parental smoking, heating fuel, parental educational level, number of children at home, and whether the child has his/her own bedroom.

The history of having had bronchitis in white children six to eight years of age for males and females is presented in Table 5 by each city. All rates have been age-standardized and then standardized for a selected home characteristic. Parental educational level was used to estimate socioeconomic status and represents the summation of schooling for both parents. It is apparent that Steubenville consistently has the highest rates, followed by Kingston-Harriman. Note that rates for females are slightly lower than for males but not as markedly as in adults. Also an examina-

Table 5. Bronchitis in children ages 6–8 years
Rate per 1000, age-standardized

	Age only	Cooking fuel	Parental educational level	Parental smoking
		White males		
Portage	154	155	139	146
Topeka	130	123	122	131
Watertown	133	134	133	140
Kingston-Harriman	220	(162)*	228	219
St. Louis	136	161	135	134
Steubenville	232	224	229	232
		White females		
Portage	109	106	118	112
Topeka	125	121	119	125
Watertown	109	102	103	106
Kingston-Harriman	166	167	171	161
St. Louis	116	122	110	115
Steubenville	187	190	191	185

*Estimate unreliable because of small numbers of households with gas stoves.

has the highest rates, followed by Kingston-Harriman. Note that rates for females are slightly lower than for males but not as markedly as in adults.

Gas Cooking and the Effect on NO₂ Levels Indoors

As a result of our indoor monitoring we have noted an increased level of NO_2 in homes using gas as the cooking fuel. These levels can be higher than those outdoors. Figure 4 compares the indoor and outdoor levels in gas and electric homes for the

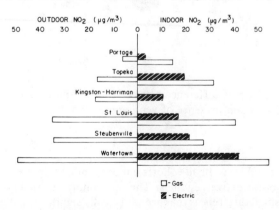

Fig. 4. Comparison of indoor and outdoor levels of NO_2 in homes using gas or electricity as cooking fuel by city.

various cities. We had no gas homes in Kingston-Harriman. In all other cities, homes using gas for cooking had higher levels than electric homes. The overall levels tend to be controlled by the outdoor levels.

Figure 5 summarizes some of the NO_2 data related to the gas and electric stoves. Here the average difference between indoor and outdoor levels of NO_2 are presented by nature of the home cooking device in selected homes. The data are further broken down by whether there was venting in the kitchen. Note that when there is venting the indoor levels in homes that use gas for cooking are much lower and tend to be similar to those in the electric homes.

It must be emphasized that these are 24-hr levels and the short-term peak values have not been recorded. We and others have shown that NO_2 peaks of 10 to 15 min duration can be as high as 1100 $\mu g/m^3$, or almost 0.6 ppm. It is probable that these peak values are more important than the average values. To test the effect of these levels of NO_2 associated with the nature of the home cooking device, we compared levels of pulmonary function in children five to nine years of age by whether they lived in households with gas or electric stoves. The children were evaluated in 12 separate cohorts, which represent for each cohort the first time the child was seen. The residuals from predicted levels for FVC in children six to nine years of age by

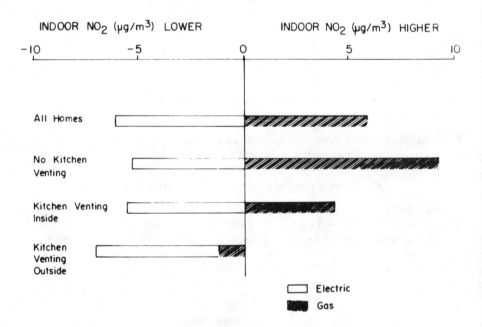

Fig. 5. Comparison of difference between indoor and outdoor levels of NO_2 by gas vs electrical fuel and presence of venting over stove.

cohort are shown in Fig. 6. In all cohorts but one, children in homes with gas stoves have slightly more negative residuals. The one cohort out of line is in St. Louis. These are very small differences that are statistically significant. We are not yet in a position to define their medical significance. We shall need to follow these children prospectively to see whether these observations persist.

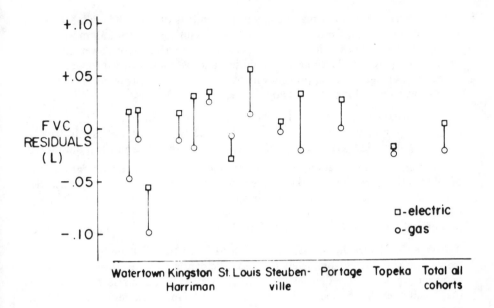

Fig. 6 Residuals from predicted of FVC in children 6 to 9 by year of age or cohort by city by gas vs electric cooking fuel.

SUMMARY

Thus both adults and children appear to have higher disease rates in Steubenville than in the other cities upon analysis of the initial cross-sectional data.

The use of gas as a cooking fuel can produce high short-term peaks of NO_2 indoors. This results in higher 24-hr values. With venting over the stove, these levels tend to fall toward levels seen in homes with electric stoves. It does appear that children who live in households with gas stoves may have lower levels of pulmonary function. This observation needs to be replicated, and the medical significance of this small difference needs to be assessed with further follow-up studies.

REFERENCES

Ferris, B. G., Jr., F. E. Speizer, Y. Bishop, G. Prang, and J. Weener. 1978. Spirometry for an epidemiologic study: deriving optimum statistics for each subject. Bull. Eur. Physiopathol. Respir. 14:145–166.

Palmes, E. D., A. F. Gunnison, J. DiMattio, and C. Tomczyk. 1976. Personal sampler for nitrogen dioxide. Am. Ind. Hyg. Assoc. J. 37:570–577.

DISCUSSION

Owen Hoffman, Oak Ridge: To what extent will your studies attempt to identify other confounding factors that may play an important role in the results that you have obtained so far? The first things that come to mind would be the presence of different allergens in the communities that you are studying—pollens, etc.—that would directly bring about the respiratory effects that you are looking for.

Dr. Speizer: We are not making any measurements of the pollen contents in the various communities. Certainly this could be a significant factor and must be considered. The asthma rates in the first rounds, in the first 12 cohorts, were highest in Topeka compared to the other communities, and we think this is probably due to allergens, but again the approach will be to identify the asthmatics within the analysis and try to look at trends over time.

Owen Hoffman: Of course, just the occurrence of normal bronchitis can be an allergic response, and I would imagine that one would not necessarily have to measure pollens but just get the information from the local medical authorities.

Dr. Speizer: Well, local medical authorities are not the best sources for this kind of information. We have asked both the adults and children about allergic histories, but their reporting is unreliable. It depends very much on access and ability to get medical care; in fact, it is probably a reflection of social class more than it is a reflection of real allergic history. It's a difficult issue to get at in the community studies.

Jeanne Calle, Oak Ridge National Laboratory: In terms of the health questionnaires, are you relying completely on the participant's recall over the one-year and two-year periods, or are they instructed in any way to fill out diaries? I would think that recall over that period of time might be a little unreliable, especially for acute respiratory diseases.

Dr. Speizer: Yes, I think their recall probably is unreliable, but we're not asking them to keep diaries in this particular part of the study.

Jeanne Calle: Do they receive any instruction in terms of recording health history over this period of time?

Dr. Speizer: No, but the objective measurements provide information on lung function that we hope to rely upon.

Chuck Hakkarinen, EPRI: Do you also record the type of heating and the air conditioning systems in the homes, and do you find any correlation with having gas-fired cooking with a gas-fired or oil-fired heating system? Does an electric home with an electric stove also have an electric heater?

Dr. Speizer: They tend to go together, but it varies from one community to another. In the Northeast you tend to have oil-fired heating systems and possibly either electric or gas stoves. In St. Louis it tends to be gas; in Kingston-Harriman, electric. Economic factors have an important influence on the type of heating and air conditioning.

CHAPTER 14

AN OVERVIEW OF EPIDEMIOLOGIC EVIDENCE ON EFFECTS OF ATMOSPHERIC SULFUR

Sati Mazumdar*
Carol Redmond*

ABSTRACT

The present review of the epidemiologic studies on effects of atmospheric sulfur shows the findings, problems, and limitations of existing studies in delineating adverse health effects of sulfur compounds from other confounding factors, especially cigarette smoking, in the investigation of *long-term effects*. The need for subjecting all studies, old and new, to critical evaluation as to their methodology and validity is stressed. In investigating the *short-term effects,* the surrogating behavior of SO_2 has been pointed out, and the need for proper adjustment for the confounding health effects of weather variables has been discussed.

INTRODUCTION

Evidence for an association of air pollution with adverse effects on human health is obtained from three sources: animal experiments, experimental human exposures, and epidemiologic studies of exposed human populations. Only through epidemiologic studies can one evaluate effects of real-life exposures in various population subgroups; however, in these studies it is difficult to attribute the observed health effects to specific concentrations of any one pollutant.

Sulfur oxides, especially SO_2, were among the first air pollutants for which human health effects were considered to be reasonably well documented. This fact stemmed from the early air pollution episodes in the Meuse Valley, Belgium; Donora, Pennsylvania; and particularly in London, England (Roholm, 1936; Logan, 1949; Schrenk et al., 1949). In all these cases, the particulates and sulfur oxides sources were from coal combustion.

The documentation on sulfur oxides prepared by the Committee on Sulfur Oxides (1978) for the Environmental Protection Agency (EPA) provides an update of the air pollution aspects of the sulfur oxides—their emission, transformation, and transport; their effects on aquatic ecosystems and vegetation; and toxicological, clinical, and epidemiologic studies of their effects on humans. This manual, together with a review paper by Ferris (Ferris, 1978), provides an excellent and exhaustive

*Graduate School of Public Health, University of Pittsburgh, 130 DeSoto Street, Pittsburgh, Penn. 15261.

bibliography of this subject matter. In the present paper we will discuss selected epidemiological studies, focusing on recent information and main issues about the health effects of atmospheric sulfur together with particulates, because, in general, they come from a common source and are highly correlated.

STUDIES CONCERNING SHORT-TERM EFFECTS

There is no question that in the above-mentioned episodes one sees evidence of health effects. But subsequent analysis with the Donora follow-up data (Ciocco et al., 1961) fails to establish or demonstrate a consistent and substantial relationship between environmental variables and health experience. The authors in that study pointed out that, although the Donora data do point to some effects on the cardiorespiratory system, they do not provide the means for relating particular pollutants to specific problems.

The episodes in London occurred after World War II; the most severe one lasted for four days in December 1952. The authorities instituted smoke-abatement measures, and the most severe of subsequent episodes occurred in 1962.

The London episode data were studied by different people over a period of time, which resulted in several publications. These studies include both mortality and morbidity statistics. The investigations by Martin and his co-workers (Martin and Bradley, 1960; Martin, 1964) revealed a general relationship between mortality, morbidity, and the concentrations of smoke and sulfur dioxide throughout the winter months but failed to distinguish among the effects of the various common air pollutants. In these studies, no attempt was made to calculate the partial correlation coefficients between individual pollutants and health measures. The use of these values (converted from smoke units) in establishing any dose-response relationship or intercepts of zero excess mortality or morbidity is questionable (Committee on Sulfur Oxides, 1978).

In evaluating the British studies, particular attention should be given to their method of measuring particulates, which is different from the hi-vol method used in the United States. British workers have remained attached to the assessment of smoke purely in terms of its blackness, whereas in the United States the emphasis has been mainly on the measurement of total suspended particulates by gravimetric methods. Any conversion from one set of figures to the other is very rough and subject to criticism (Critical review session, 1978).

During the 1960s there were two groups actively studying air pollution effects in New York City, one at Albert Einstein College of Medicine (Greenberg) and another at Cornell University Medical Center (McCarroll). Both groups studied episodes which were believed to be acute episodes of air pollution. The essential point made by the Cornell group (Cassell et al., 1968) is that, while one can find periods of apparent excess mortality during which pollution was relatively high, one also finds them when pollution is low, and only a systematic long-term approach will allow one to unravel the specific mechanism by which the various environmental influences produce their effect on health.

The Einstein group consisting of Greenberg and Schimmel has been engaged for more than a decade in making studies of the possible adverse health effects of air pollution using New York City data and regression analysis. In one of the earlier papers (Schimmel and Greenberg, 1972), using 1963-1968 as a data base, the authors introduced a new technique which eliminates the problem of considering many factors in addition to air pollution that influence longer-term, slower trends.

The superiority of a method wherein annual and seasonal trends are eliminated by Fourier analysis from the variables before being introduced into a regression model was not recognized, and results using a different method giving larger estimates were selected erroneously as providing the best estimates. The methodology was improved in the second study (Schimmel and Murawski, 1976), and using a larger data base (1963–1972), the authors reported that the SO_2 variable tended to behave as an indicator or proxy; i.e., as levels went down, regression coefficients went up. When SO_2 and smoke shade were included in a joint regression, coefficients for SO_2 were usually not statistically significant, although they tended to be weakly positive. Further, when regressions were carried out on particulate measures alone, the combined effect associated with SO_2 and smoke shade usually could be accounted for by the particulate measure alone. Schimmel extended his study to the data base for 1963–1976 and refined his methodology, including more in-depth corrections for weather influences and day-of-week variability, and the results are reported (Schimmel, 1978). Particular importance was attached to comparing results for 1963–1969 with those for 1970–1976, because the SO_2, levels in the first seven years were about five times as high as in the second seven years. Schimmel found no significant statistical association between sulfur oxide and mortality levels and only a weak and inconsistent association with particulate levels.

Buechley and his co-workers conducted a study in the New York–New Jersey metropolitan region to investigate the effect of SO_2. The first-phase analysis of deaths for the New York metropolitan region, with daily SO_2 measurements in Manhattan at the Davis Laboratory, showed that the level of SO_2 pollution contributes significantly to prediction of deaths. Mortality was found to be 2% greater than expected on 260 days with SO_2 levels above 500 $\mu g/m^3$ (Buechley et al., 1973). The study was extended by Buechley using a 1967–1972 data base. The results, although not published, have been reported on several occasions (Buechley, 1977a; Buechley, 1977b; Buechley, 1978). The investigators reported that the predicting equation continues to significantly predict slightly higher mortality on those days with higher SO_2 values in 1971–1972, even though the absolute amounts of SO_2 are one-tenth of what they were in 1967–1968. One explanation of this behavior is that SO_2 was acting as a surrogate for some other substances that had not changed as much.

The New York study by Schimmel developed a new methodology for the isolation and quantification of acute effects using techniques of time-series analysis. Despite this improved methodology, the study has been criticized for use of data from a single monitoring station. It has also been suggested that mortality may not be sufficiently sensitive for an epidemiological study of daily low-level effects.

With an awareness of these criticisms and of the potential contribution of replicating an epidemiological study in a different geographical area, a study has been designed in the Pittsburgh area to investigate mortality and morbidity as the adverse health responses to ambient pollution levels using air quality data from several monitoring stations. The first-phase findings have been reported on several occasions (Schimmel, 1978; Mazumdar et al., 1979a, b). Results indicate a possible association between heart disease mortality and morbidity with particulate levels only.

Lave and Seskin made extensive analyses to relate minimum levels of sulfate and levels of suspended particulates to mortality. They used data from standard metropolitan areas (SMSAs) in the United States, as well as British data (Lave and Seskin, 1970; Lave and Seskin, 1973). Their decade-long research has resulted in a book by the same authors (Lave and Seskin, 1977). Their first set of studies consider in detail the relationship between air pollution and mortality in the United States in 1960 (with replications for 1961 and 1969), using cross-sectional data on standard metropolitan statistical areas (SMSAs). The work includes examination of age-sex-race-adjusted total mortality rates as well as age-sex-race specific mortality rates and 15 disease-specific mortality rates, inclusion of other explanatory variables such as home heating characteristics, etc. A partial replication of the 1960 work was performed, several time-series analyses were done, and finally the authors investigated the implication of their findings by enumerating the benefits and costs associated with strategies of air pollution abatement. They concluded that the levels of certain pollutants in the air (as they prevailed in U.S. cities during the period 1960–1969) caused increases in mortality and presumably in morbidity as well and claimed that their analyses demonstrated that the benefits associated with a rather substantial abatement of air pollution from stationary sources are greater than the costs of abatement. Despite the large volume of work and effort put forth by these authors, the scientific community has difficulty in accepting their results due to numerous technical deficiencies, including demographic problems, absence of representative data base, treatment or lack thereof of age and smoking information, and drawing causal inferences from regression analysis (Ferris, 1978; Committee on Sulfur Oxides, 1978; Landau, 1978; Tukey, 1976). Considerable reanalysis of these data has been suggested before such results could be used to establish a standard.

In 1974, the first community health and environmental surveillance system (CHESS) report, "Health Consequences of Sulfur Oxides," was published (U.S. Environmental Protection Agency, 1974). Under this program a variety of epidemiological studies were carried out, including descriptive studies of the prevalence rates of acute and chronic respiratory symptoms and of morbidity patterns and panel studies for "attack rates" and "visits to hospital emergency rooms." The CHESS studies have been extensively criticized (see, for example, U.S. Environmental Protection Agency, 1976). We do not intend to enter into a detailed discussion of the CHESS results and the criticisms, but we can note that in the CHESS studies, for the first time, efforts were made to take account of the sulfate portion of the sulfur products.

The effect of air pollution on persons with chronic respiratory disease is particularly important to study from a public health viewpoint because these persons are believed to include a segment of the population susceptible to air pollution exposures. Many investigators have studied the temporal patterns of symptom status in these persons in relationship to concentrations of air pollution and weather factors.

An extensive series of studies on the effects of air pollution on bronchitic patients attending chest clinics was conducted in Britain between 1955 and 1970 and summarized in a paper by Lawther, Waller, and Henderson (Lawther et al., 1970).

The major findings of the study by Lawther and his co-workers are that (1) patients are most sensitive to changes in pollution at the beginning of each winter and (2) the minimum pollution leading to any significant response is about 500 $\mu g/m^3$ of SO_2 together with about 250 $\mu g/m^3$ of smoke (each representing the average over 24 hr, at a group of sites in inner London). There is no evidence that either of these pollutants would, by itself, produce the same response. The type of pollution found in London in later years of the study period, with much less smoke and fewer days of high pollution of any kind, has led to a reduced response among the patients (Lawther et al., 1970).

The study by Cohen in the United States is in some ways analogous to Lawther's study (Cohen et al., 1972). A group of asthmatics reported daily symptoms for over seven months while air pollution and weather parameters were noted. The analyses showed independent effects of temperature and of air pollution on attack rate but could not specify which pollutants were most important. It is important to note that after temperature and any one of the five measured pollutants (total suspended particulates, sulfur dioxide, soiling index, suspended sulfates, and suspended nitrates) have been taken into account, none of the other four variables explained a significant amount of attack rate variation. It is important to note here that these types of studies suffer from the lack of knowledge of the "attack rate" patterns of a similar group of patients living under a different pollution situation but the same weather conditions.

A comparatively recent air pollution episode in Pittsburgh, Pennsylvania (November 1975) prompted several investigators to assess its health impact. The effect of the episode upon pulmonary function in school children was studied by Stebbings and his co-workers (Stebbings et al., 1976). The 24-hr peak values were about 500–700 $\mu g/m^3$ for TSP and 0.130 ppm for SO_2. The investigators were not able to measure pulmonary function until a day or so after the pollution peak. They did not find any impairment of lung function in their initial analysis. They also noted that the variability of the results over the weekend was much greater than on the school days.

Stebbings has a recent paper reporting on a reanalysis of his pulmonary function test results collected during and after the Pittsburgh air pollution episode of November 1975. The findings are limited by the small number of subjects with strong postepisode upward trends in forced vital capability (FVC) and by lack of validation by replication of the study design, but they do suggest that episode levels of suspended particulates induce lung damage and that this may occur only in a small susceptible subgroup (Stebbings et al., 1979).

STUDIES CONCERNING LONG-TERM EFFECTS

There are a number of studies on the effects of chronic exposure to sulfur oxides and particulates, and most of the studies need cautious critical review due to lack of sufficient pollution measurements and possible effects of confounding factors. Among these studies, those done in Berlin, New Hampshire, by Dr. Ferris and his colleagues have been of particular interest (Ferris et al., 1973; Ferris et al., 1976). A random sample of the adult population of Berlin, New Hampshire, was surveyed in 1961 by means of a standard questionnaire and simple tests of pulmonary functions. Those who were still living in Berlin in 1967 were resurveyed using the same techniques. Air pollution was slightly less in 1967 than in 1961, and the authors noted that this decrease could have accounted for the decrease in prevalence of disease and the improvement in pulmonary function noted in the population. The 1967 sample was resurveyed in 1973. Measurements of the levels of air pollution showed a decrease in the number of suspended particulates and an increase in the concentration of sulfation. No differences in respiratory symptoms, prevalence of chronic nonspecific respiratory disease, or pulmonary function were detected. In the first of Ferris's studies, particulate measurements are limited to two months only, and SO_2 was assessed only by lead peroxide candle. In the second study, the authors noted a wide discrepancy in the SO_2 values measured by different methods, which may have an important role in the interpretation of the results. Another problem of this kind of study is that the type of cigarette smoke changed, as use of a low-tar filter cigarette may result in a lessening in respiratory symptoms and/or disease.

The dominating relationship between smoking habits and the prevalence of chronic nonspecific respiratory diseases and their dynamics has been revealed in the results of a Polish study (Sawicki et al., 1977). The investigators there failed to find a consistent relationship with various levels of air pollution and different morbidity measures and commented that the results indicated that the differences in air pollution levels in the contrasting areas were too small to produce convincing evidence of the relationship between these differences and prevalence and dynamics of respiratory symptoms.

We do not intend to enter into a detailed discussion of Fletcher's study in London (Fletcher et al., 1976), which is a prospective study of bronchitis in relation to smoking effects, because air pollution played a very minor role in the analysis presented.

In studies of adults designed to assess the long-term effect of air pollution, one finds a dominating effect of smoking. This effect can be avoided to some extent in studies using children. In this context the series of studies by Lunn and his co-workers should be mentioned. In the first study (Lunn et al., 1967), data were presented on 819 infant school children living in areas of Sheffield with widely ranging air pollution levels. Upper respiratory tract illnesses showed an association with area which generally reflected the pollution levels. Illness was less common in areas with lower SO_2 levels. No effort was made to separate the effect of SO_2 from particulates, and, as before, we did not attempt to convert the British measurements to comparable figures used in the United States. In the later follow-up study (Lunn et al., 1970), there has been a drop in air pollution, and advantage was taken of the vastly differing pollution levels throughout the city to choose areas with large contrasts. The changes in the morbidity noted are consistent with the overall improvement in pollution levels in the city, but from this analysis it is not possible to say whether SO_2 or smoke contributed more toward this improvement.

Among the most recent ongoing studies we can mention the "Six-Cities Study," which is a prospective epidemiologic study of the impact of environmental air quality on the respiratory health status in both children and adults. Although the goal of this study is to measure changes in health effects over time, it is also of interest to compare the health effects among cities using the available cross-sectional data on lung function and disease rates.

No evidence relating sulfur compounds in the atmosphere and lung cancer is found in the literature. Winkelstein and his co-workers showed a positive relation between the particulate component of air pollution and stomach cancer, but his study suffers from a lack of knowledge of smoking history and occupational exposure (Winkelstein and Kantor, 1967). In this context, we should mention the results from the most recent study of Pike (Pike et al., 1978). The investigators prepared a case-control study of smoking, occupation, and residential history in south-central Los Angeles County to determine if long-term residence in this area of "high" air pollution was associated with the excess of male lung cancer observed there. No association was found between long-term residence and risk to lung cancer. This study again confirmed the importance of cigarette smoking in the rate of lung cancer.

DISCUSSION

At present, many investigations attempt to define the relationship between environmental pollutants and health response, and such research is necessary for determining safe or permissible levels of airborne contaminants. When many determinants are present, not only is inference difficult, but spurious association is likely to occur. This is due to the fact that the correct determinants are rarely identified fully and accurately and the hypothetical determinants that are examined serve only as surrogates for the correct ones.

For this reason, associations drawn from regression analysis should not be considered as demonstrations of causality. Rather, they are useful only in the context of a larger body of information.

These difficulties are exacerbated when a study has as its focus hypothetical determinants (air pollution) which account for a minor portion of a health measure. The presence of certain systematic associations between air pollution levels and strong forces affecting mortality and morbidity will cause positive associations between air pollution and adverse health measures. The association will persist regardless of the absolute pollution level.

Because area studies are especially vulnerable to covariate influences, it is important to control properly the strong factors affecting the health measures being examined. In investigating the long-term effects the present review shows us the limitation of the epidemiologic studies in separating the effect of sulfur compounds from the effect of other pollutants and the effect of smoking. Any future research should give careful attention to these problems.

Time-series analysis of a specific population is also subject to systematic error when the adverse health being measured is influenced more significantly by covariates than by the air pollution. In investigating the short-term effects, our conclusion is that any valid result concerning these pollutant-health relationships can be obtained only if proper adjustment is made for the confounding health effects of temperature and other weather variables in the analysis. At present we (together with Schimmel and Bloomfield) are engaged in a study which will carry out extensive examination of the highly complex weather-health response and weather-pollutant relationships, using multisite data, for the purpose of providing insight and methods for weather adjustments. In particular, the investigation will attempt to determine whether ambient pollution levels influence the weather as well as being influenced by the weather. The influence of weather on mortality and morbidity, a subject which has been studied to a limited extent thus far, will constitute a secondary but equally important objective of the study.

REFERENCES

Buechley, R. W., W. B. Riggan, V. Hasselblad, and J. B. VanBruggen. 1973. SO_2 levels and perturbations in mortality—a study in the New York–New Jersey metropolis. Arch. Env. Health 27:134.

Buechley, R. W. 1977a. SO_2 levels, 1967–1972 and perturbations in mortality: a further study in the New York–New Jersey metropolis. Report, Contract No. 01-ES-52101 NIEHS, Research Triangle Park, N.C.

Buechley, R. W. 1977b. SO_2 levels, 1967–1972 and perturbation in mortality. Presented at the Conference of International Epidemiological Association, San Juan, Puerto Rico, Sept. 18–23.

Buechley, R. W. 1978. Discussion of paper by Herbert Schimmel. Proceedings Symposium on Environmental Effects of Sulfur Oxides and Related Particulates, New York Academy of Medicine.

Cassel, E. J., et al. 1968. Reconsideration of mortality as a useful index of the relationship of environmental factors to health. Am. J. Public Health 58(9):1653.

Ciocco, A., and D. J. Thompson. 1961. A follow-up of Donora ten years after methodology and findings. Am J. Public Health 51(1):155.

Cohen, A. A., et al. 1972. Asthma and air pollution and a coal-fueled power plant. Am J. Public Health 62:1181.

Committee on Sulfur Oxides, Board on Toxicology and Environmental Hazards, Assembly of Life Sciences, National Research Council. 1978. Sulfur Oxides. National Academy of Sciences, Washington, D.C.

Critical review session on the health effects of exposure to low levels of regulated air pollutants, discussion papers. 1978. J. Air Pollut. Control Assoc. 28:883.

Ferris, B. G., et al. 1973. Chronic nonspecific respiratory disease in Berlin, New Hampshire, 1961-1967. A follow-up study. Am. Rev. Respir. Dis. 107:110.

Ferris, B. G., et al. 1976. Chronic nonspecific respiratory disease in Berlin, New Hampshire, 1967-1973. A further follow-up study. Am. Rev. Respir. Dis. 113: 475.

Ferris, B. G. 1978. Health effects of exposure to low levels of regulated air pollutants. A critical review. J. Air Pollut. Control Assoc. 28:482.

Fletcher, C. M., et al. 1976. The Natural History of Chronic Bronchitis and Emphysema. Oxford: Oxford University Press: p. 272.

Landau, E. 1978. Review of *Air Pollution and Human Health*. Nation's Health, March.

Lave, L. B., and E. P. Seskin. 1970. Air pollution and human health. Science 169: 723.

Lave, L. B., and E. P. Seskin. 1973. An analysis of the association between U.S. mortality and air pollution. J. Am. Stat. Assoc. 63, No. 342.

Lave, L. B., and E. P. Seskin. 1977. Air Pollution and Human Health. Baltimore: Johns Hopkins University Press.

Lawther, P. J., et al. 1970. Air pollution and exacerbation of bronchitis. Thorax 25:525.

Logan, W. P. D. 1949. Fog and mortality. Lancet 256:78.

Lunn, J. E., J. Knowelden, and A. J. Handyside. 1967. Patterns of respiratory illness in Sheffield infant school children. Brit. J. Prev. Soc. Med. 21:7.

Lunn, J. E., J. Knowelden, and J. W. Roe. 1970. Patterns of respiratory illness in Sheffield junior school children. A follow-up study. Brit. J. Prev. Soc. Med. 24:223.

Martin, A. E., and W. H. Bradley. 1960. Mortality fog and atmospheric pollution. Mon. Bull. Minist. Health, Public Health Lab. Serv. 19:56.

Martin, A. E. 1964. Mortality and morbidity statistics and air pollution. Proc. Roy. Soc. Med. 57:969.

Mazumdar, S., and N. Sussman. 1979a. Evidence of possible acute health effects of ambient air pollution: results based on Pittsburgh area daily mortality and morbidity. Presented in Park City Environmental Health Conference at Park City, Utah, April 4-7.

Mazumdar, S., and N. Sussman. 1979b. Allegheny County air pollution, mortality, morbidity. Final Report Part I. Submitted to NIEHS (Grant No. ESO-1129).

Pike, M. C., et al. 1978. Occupation: "explanation" of an apparent air pollution related localized excess of lung cancer in Los Angeles County. Proceedings of a SIMS Conference, Alta, Utah, June 26-30. Sponsored by SIAM Institute of Mathematics and Society.

Roholm, K. 1936. On the cause of the fog catastrophe in the valley of the Meuse, December, 1930. Hospitalstidende 79.

Sawicki, F., and P. S. Lawrence. 1977. Chronic non-specific respiratory disease in the city of Cracow. Report of a 5-year follow-up study among adult inhabitants. National Institute of Hygiene, Warsaw.

Schimmel, H., and L. Greenberg. 1972. A study of the relation of pollution to mortality. J. Air Pollut. Control Assoc. 22:607.

Schimmel, H., and T. J. Murawski. 1976. The relation of air pollution to mortality. J. Occup. Med. 18:316.

Schimmel, H. 1978. Evidence for possible health effects of ambient air pollution from time series analysis: methodological questions and some new results based on New York City daily mortality, 1963-1976. Proceedings: Symposium of Environmental Effects of Sulfur Oxides and Related Particulates. New York Academy of Medicine.

Schrenk, H. H., et al. 1949. Air pollution in Donora, Pennsylvania. Public Health Bull. No. 306.

Stebbings, J. H., D. G. Fogleman, et al. 1976. Effect of the Pittsburgh episode upon pulmonary function in school children. J. Air Pollut. Control Assoc. 26(6):547.

Stebbings, J. H., and D. G. Fogleman. 1979. Identifying a susceptible subgroup: effects of the Pittsburgh air pollution episode upon school children. Am. J. Epidemiol. 110(1):27.

Tukey, J. W. 1976. Discussion of paper by Lave and Seskin, page 37. Proceedings of the 4th Symposium of Statistics and Environment, March 3–5, Washington, D.C.

U.S. Environmental Protection Agency. 1974. Health Consequences of Sulfur Oxides: A Report from CHESS, 1970–1971, Office of Research and Development, U.S. Environmental Protection Agency, Washington, D.C. Available from: U.S. Government Printing Office.

U.S. Environmental Protection Agency. 1976. The Environmental Protection Agency's Research Program with Primary Emphasis on the Community Health and Environmental Surveillance System (CHESS): An Investigative Report. A report prepared for the Committee on Science and Technology, U.S. House of Representatives, Ninety-Fourth Congress, Second Session, Serial SS, Washington, D.C. Available from: U.S. Government Printing Office.

Winkelstein, W., and S. Kantor. 1967. Stomach cancer. Arch. Environ. Health 14:544.

SECTION IV

EMISSION CHARACTERISTICS AND ATMOSPHERIC TRANSFORMATIONS

Chairman: J. F. Meagher

CHAPTER 15

EMISSION CHARACTERISTICS AND ATMOSPHERIC TRANSFORMATIONS

Chairman: **J. F. Meagher**
Tennessee Valley Authority

OVERVIEW

James F. Meagher*

The atmosphere contains both primary and secondary pollutants. A substance is said to be a primary pollutant if it exists in the same chemical form as when emitted. A primary pollutant, whether natural or anthropogenic, may combine with other atmospheric constituents, sometimes with the aid of sunlight, and is thus transformed to a secondary pollutant. The majority of anthropogenic sulfur is released into the atmosphere as sulfur dioxide. Once in the atmosphere the sulfur dioxide can be oxidized to various sulfate aerosols. These aerosols are more toxic than their precursors and, because of their size (less than 1 μm), they can be transported long distances and penetrate deep into the respiratory tract. The development of an effective control strategy for a secondary pollutant, such as sulfate aerosols, requires an understanding of the complete atmospheric transformation processes.

The emphasis of the papers presented in this session was on sulfate aerosols, existing in the atmosphere as both primary and secondary pollutants. In reviewing the relative contribution of various anthropogenic sources of sulfur oxides, Nader pointed out that the various forms of fuel combustion account for 78% of all man-made emissions, with electricity alone accounting for 63% of the total. The sulfate emission factors reported for coal- and oil-fired power plants were quite different. Sulfate aerosols accounted for 1% to 2% of the total sulfur oxide emissions, expressed as sulfur, in coal-fired power plants, while in oil-fired power plants, sulfates accounted for 5% to 10% of the sulfur oxide emissions. Since conventional limestone scrubbers are not effective for removing primary sulfate from flue gas, Nader suggested that primary sulfate emissions may increase if utilities turn to high-sulfur coals in combination with limestone scrubbing to meet emission standards.

The bulk of the available information on the rate of conversion of sulfur dioxide to sulfate in the atmosphere has been obtained using instrumented aircraft to probe urban, power plant, and smelter plumes. In reviewing the results of these studies,

---

*Air Quality Research Section, Tennessee Valley Authority, Muscle Shoals, Alabama 35660.

Newman concluded that the average conversion rate in "clean air" is less than 1% per hour, whereas in polluted urban air the average rate may exceed 2% per hour. Recent data indicated that the rate exhibited a strong diurnal dependence, with near zero conversion at night and midday rates of approximately 3% per hour. Newman discussed the various homogeneous and heterogeneous mechanisms which have been postulated for SO_2 oxidation. Currently, insufficient data are available to allow the selection of any one of these mechanisms with some degree of certainty. Newman presented the first measurements of nitrate formation in power plant plumes, which indicated that nitrate was formed two or three times as fast as sulfate and that the rate showed a similar diurnal dependence. Newman called for additional research to more thoroughly characterize the spatial distribution of sulfate in plumes, to perform measurements of the rates of neutralization of sulfate and nitrate aerosols, and to perform measurements of the sulfite transformation products.

The physical and chemical characteristics of the sulfate aerosols, both primary and secondary, were discussed by Brosset and McMurry. Brosset presented a phase diagram for the ammonium sulfate–sulfuric acid–water system augmented by tie lines for the ammonia equilibrium concentration over the liquid phase. If the temperature, relative humidity, and ammonia levels are known, the ultimate (equilibrium) mix of sulfate aerosols [H_2SO_4, NH_4HSO_4, $(NH_4)_3H(SO_4)_2$, and $(NH_4)_2SO_4$] can be determined using this diagram. The agreement between predicted and measured aerosol compositions was fair and could probably be improved if the phase diagram was available for a wider range of temperatures.

McMurry discussed how the time rate of change of the particle size distribution could be used to obtain information about the mechanism or mechanisms which are responsible for aerosol production. Theoretical growth laws were presented which distinguish among gas-phase, surface, and droplet-phase reactions. Although the present formulation of the model restricts its application to aerosol growth in an initially particle-free system, McMurry hopes that with further development it can be extended to more realistic situations.

A Monte Carlo long-range transport model was presented by Patterson. This model was used to predict 24-hr sulfate levels and to estimate visibility impairment over the eastern United States during August 1977. The extensive data base established during the SURE (Sulfate Regional Experiment) was used for model validation. This model was able to locate the position of a large polluted air mass within 300 to 500 km for 24 of the 31 days during the period modeled. Patterson found a high correlation between the sulfate levels and the degree of light extinction. He further suggested that the existing United States visibility data base could serve as a semiquantitative surrogate for sulfate concentrations in the atmosphere.

CHAPTER 16

PRIMARY SULFATE EMISSIONS FROM STATIONARY INDUSTRIAL SOURCES

John S. Nader*

ABSTRACT

The genesis of sulfur emissions impacting on ambient sulfate levels includes both natural and anthropogenic sources. Sulfate salts and sulfuric acid emitted as such directly from anthropogenic sources are called primary sulfates to distinguish them from secondary sulfates resulting from atmospheric transformations. Studies of sulfate emissions from various combustion sources are reviewed. Current experimental data on physical and chemical properties, emission factors, operating parameters, and fuel composition affecting the emissions are briefly summarized. Oil-fired combustion sources were found to have sulfate emissions that ranged from 5% to 10% of the SO_x emissions, about a factor of 5 times the historical values. The sulfate emissions are calculated for a New Jersey area which utilizes large amounts of oil in its combustion sources. Oil combustion alone accounted for 97% of the sulfate emissions in the combustion source category.

INTRODUCTION

The genesis of ambient sulfate levels is attributed to both natural and anthropogenic sources. In the area of anthropogenic activity which is within man's control, emphasis has been on control of sulfur dioxide as the predominant sulfur pollutant in the atmosphere and as the precursor of the sulfates that contribute significantly to the ambient sulfate levels (Commission on Natural Resources, 1975). These sulfates, resulting from the transformation of sulfur dioxide in atmospheric reactions, are called "secondary" sulfates. This distinguishes these transformation products from the sulfuric acid and sulfate salts that are emitted as such directly from sources and are called "primary" sulfates (Nader, 1978).

Total sulfur oxide emissions have been calculated for various source categories on the basis of sulfur content of the fuel used in combustion sources (U.S. EPA , 1978a), the sulfur in the industrial process, and the emission factors associated with the source (U.S. EPA, 1975). Total sulfur oxide emissions (SO_x) have been compiled for various source categories in a U.S. EPA emissions report (1978b). All the criteria pollutant emissions were tabulated for the entire United States, for individual states, and for air quality control regions (AQCRs). Table 1 summarizes for the year 1975

---

*Recently retired as Chief of Stationary Source Emissions Research Branch, Environmental Sciences Research Laboratory, U.S. Environmental Protection Agency, Research Triangle Park, N.C. Present address: 2336 New Bern Avenue, Raleigh, NC 27610.

Table 1. Summary of sulfur oxide emission inventories for 1975 for the United States and selected sites

Source category	United States Emission (metric tons per year)	Percent	New Jersey Emission (metric tons per year)	Percent	AQCR 043[a] Emission (metric tons per year)	Percent	AQCR 119[b] Emission (metric tons per year)	Percent
Fuel combustion	23,073,280	78	222,098	74	123,915	74	97,455	93
Industrial process	5,966,511	20	44,208	15	31,752	19	1,174	1
Solid waste disposal	55,367	0	2,651	1	1,345	1	740	1
Transportation	647,496	2	28,648	10	10,314	6	4,895	5
Miscellaneous	2	0	0	0	0	0	0	0
Total	29,742,656	100	297,605	100	167,326	100	104,264	100

[a]The New Jersey portion of the New Jersey–New York–Connecticut region.
[b]Metropolitan Boston.
Source: U.S. Environmental Protection Agency. 1978. 1975 National Emissions Data Report. National emissions data system of the aerometric and emissions reporting system. EPA-450/2-78-020.

the sulfur oxide emissions for the main source categories for the United States, for the state of New Jersey, for the New Jersey portion of AQCR 043 that encompasses portions of New Jersey, New York, and Connecticut, and for AQCR 119, the Boston metropolitan area. The New Jersey area was selected to represent part of the Northeastern area that has been under study for ambient levels of sulfur oxides and sulfates. The Boston area is just outside the study area.

Nationwide more than 78% of all source SO_x emissions result from fuel combustion in stationary sources. Further source breakdown shows that electric generation accounts for 81% of the SO_x emissions from fuel combustion in stationary sources. Both New Jersey and AQCR 043 show patterns of SO_x emissions for the main source categories similar to the national pattern. In the Boston region the industrial processes contribute 5% of that in the New Jersey–New York–Connecticut region. This demonstrates that the relative contribution of various source categories to SO_x emissions can vary significantly from region to region.

The emissions report (U.S. EPA, 1975) does not present data on the sulfate portion of the SO_x emissions. Emission factors are given for sulfate emissions from combustion sources in EPA's emission factors handbook (1975). Recent emission factors obtained experimentally for oil-fired combustion sources were found to be significantly greater than those in the emission factors handbook.

Since combustion sources are the predominant sources of primary sulfate emissions, this paper focuses its discussion on these sources and the relative impact of coal and oil as the significant sulfur-bearing fuels. Current experimental data are presented, characterizing primary sulfate emissions from various combustion sources and the impact of these emissions on ambient air quality. Many questions are raised by these early and preliminary efforts to identify the role that primary sulfate emissions play in their contribution to ambient sulfate levels. These questions may help to guide future research needs.

DATA CHARACTERIZING PRIMARY SULFATE EMISSIONS

The Environmental Sciences Research Laboratory of EPA has been carrying out intensive research studies of primary sulfate emissions from various fossil-fuel-fired combustion sources. Results have shown that primary sulfate emissions from oil-fired combustion sources are 5 to 10 times those from sources burning coal of a similar sulfur content. The sulfates also have a larger fraction of free sulfuric acid (Homolya and Cheney, 1978). Data have been obtained that characterize these primary sulfate emissions in terms of their chemical and physical properties, free sulfuric acid and particulate sulfate content, composition, size distribution, and light-scattering properties; the effects of fuel composition, combustion operating parameters, pollution control equipment, and design of combustion equipment have been studied by a number of investigators (Nader, 1978).

Physical and Chemical Properties

No EPA standard method has been established for the measurement of sulfate particulate matter and sulfuric acid in combustion source emissions. Volume 1 of the Workshop Proceedings (Nader, 1978) identifies and discusses the methods used to sample and to analyze total water-soluble sulfates (TWSS), particulate sulfur, SO_2, and gas-phase sulfuric acid (H_2SO_4). The sulfuric acid methods measure only the free acid in the gas phase and do not measure the acid adsorbed on particulate matter. At present there is no single measurement capable of determining the total sulfuric acid (free and adsorbed) that occurs in the emission gas stream. The procedure is to make concurrent measurements (TWSS, adsorbed H_2SO_4, and free

H_2SO_4) and to estimate the amount of total acid present in the gas stream. Preliminary estimates indicate that about 67% of TWSS is acid in coal-fired emissions and 70% of TWSS is acid in oil-fired emissions (Homolya, 1979).

Bimodal size distributions for sulfur have been found in coal-fired combustion emissions with electrostatic precipitators (Bennett, 1979). The larger peak was in the submicron size range below 0.3 μm, and a second peak appeared at about 2 μm in the size range often associated with the mean size of the total particulate emissions. Both monomodal and bimodal size distributions have been reported for sulfur emissions from oil-fired combustion sources (Knapp et al., 1976). The larger peak appeared in the submicron range at about 0.5 μm and the second peak was at 5 μm. Natusch (1976) pointed out that sulfur increases in specific concentration with decreasing particle size; the sulfur associated with coal fly ash is present in a layer of the order of 50 Å thick at the particle surfaces.

Parameters Affecting Emission Rate

Characterization work done at oil-fired power plants on the effect of operating parameters, control devices, and fuel composition on primary sulfate emissions (Nader, 1978) can be briefly summarized by the following observations.

Oil-fired sources emitted a significantly larger amount of sulfates (salts and acid) than coal-fired sources of a similar sulfur content. The vanadium content, the amount of excess boiler oxygen, and the sulfur content of the fuel had important impact on the amount of sulfate emitted. An increase in any of these resulted in a measured increase in sulfate emissions. Excess oxygen increased the sulfuric acid component of the sulfate. A buildup of deposits on the boiler tubes of oil-fired power plants has been shown to increase both the sulfate and acid production (Homolya, 1978).

Coal-fired sources with electrostatic precipitators (ESPs) have shown that these devices can reduce sulfate emissions in excess of 50%, primarily, the metal sulfates component. With a few exceptions, oil-fired sources are not equipped with any emission controls. Corrosion inhibitors used in these sources have been shown to reduce sulfate emissions. A wet scrubber flue gas desulfurization (FGD) control system on a 5% sulfur coal-fired power plant was evaluated for its efficiency of removal of SO_2 and sulfates (Homolya, 1979). The average SO_2 scrubbing efficiency was 76%, and the total sulfate efficiency averaged about 29%. The sulfate emissions measured in the scrubber exit gas consisted of about 85% acid as a fine aerosol.

Emission Factors

Field studies on coal-fired boilers without SO_2 controls but with ESP particulate controls substantiate that the emissions of sulfates are in the range of 1% to 2% of the sulfur in the fuel as presented in the emission factors handbook (U.S. EPA, 1975). Comparative measurements of elemental sulfur were conducted on emissions from power plants with different particulate control systems (Radian Corp., 1975). The percentages of the elemental sulfur in the coal that were discharged in the stack gas stream were 62, 88, and 98 for wet scrubber, hotside ESP, and cyclone collectors, respectively. Data on the FGD system using a wet limestone scrubber (Homolya, 1979) suggest a sulfate emission factor of about 1% of the SO_x emissions for coal-burning sources using this type of control system.

Experimental data for sulfates from oil-fired sources indicated a range from 5% to 10% of the SO_x emissions, depending upon the source category, the amount of excess oxygen typical of the operation, and the vanadium in the fuel. This range of

respectively. This is a reversal of roles when compared with the emissions nation-wide. Industry emitted 36% of the total sulfur oxides. All of this 36% was the result of oil combustion by industry.

Table 4 zeroes in on the New Jersey (nine counties) portion of the air quality control region (AQCR 043) that encompasses portions also of New York and Connecticut. Unfortunately, the fuel usage source did not include data on the basis of AQCRs. Nonetheless, one can reasonably assume that the state data (Table 3) adequately represent the fuel usage in this region within the state. Since this portion involves the more populated area, very likely the large consumption of oil is reinforced. This appears to be manifested indirectly by the SO_x emissions, 15% and 85% of the total SO_x attributed to coal and oil, respectively. Although the use of coal is small, electric generation remains the primary consumer.

The sulfate emissions values in Table 4 were developed with the emission factors discussed earlier as percentages of the SO_x emissions. The results are as might be expected. Because of the heavy consumption of oil, the sulfate (SO_4^{2-}) emissions for oil combustion are 97% of the total. The industrial category emits 51% of the sulfates, with oil as its sole source of fuel. Interestingly enough, the commercial/institutional category emits twice as much as electric generation as a result of oil-fired operations.

No attempt was made at projecting the impact of high-sulfur coal combustion using FGD controls. It is inappropriate at this time for several reasons. The data used in this paper emphasize an area where coal utilization is very low. Secondly, sulfate data on emissions from various FGD systems are not available to properly evaluate what may be happening. This is obviously an area that needs considerable research effort. Other areas of needed research include sulfate emission data on industrial and residential source categories and a number of sources in the main category of industrial processes. These include smelters, pulp and paper, cement, and other process sources.

Table 4. Summary of SO_x and sulfate emission inventories for stationary fuel combustion sources for 1975 in the New Jersey portion of AQCR 043 (New Jersey, New York, and Connecticut)

Source category	SO_x emissions[a] (10^3 metric tons per year)			Sulfate emissions[b] (metric tons per year)		
	Coal	Oil	Total	Coal	Oil	Total
External combustion	18	100	118	234	7,600	7,834
Residential	1	5	6	13	500	513
Electric generation	17	21	38	221	1,050	1,271
Industrial	0	54	54	0	4,050	4,050
Commercial/ institutional	0	20	20	0	2,000	2,000
Internal combustion	0	2	2	0	100	100
Electric generation	0	2	2	0	100	100
Industrial	0	0	0	0	0	0
Total	18	102	120	234	7,700	7,934

[a]Source: U.S. Environmental Protection Agency. 1978. NEDS Report. National emission data system of the aerometric and emission reporting system. EPA-450/2-78-020.

[b]Calculated values based on experimental and estimated emission factors identified in the text.

REFERENCES

Bennett, R. L., 1979. Communication.

Commission on Natural Resources, National Academy of Sciences, National Academy of Engineering, and National Research Council. 1975. The relationship of sulfur oxide emissions to sulfur dioxide and sulfate air quality. Pp. 233–275 a report for the Committee on Public Works, U.S. Senate, March 1975.

Homolya, J. B., and J. L. Cheney. 1978. An assessment of sulfuric acid and sulfate emissions from the combustion of fossil fuels. Pp. 3–12 in Nader, J. S. (Chm.), Workshop Proceedings on Primary Sulfate Emissions from Combustion Sources, Vol. 2, Characterization. EPA-600/9-78-020b, U.S. EPA, Research Triangle Park, N.C.

Homolya, J. B., and C. R. Fortune. 1978. The measurement of the sulfuric acid and sulfate content of particulate matter resulting from the combustion of coal and oil. Atmos. Environ. 12:2511–2514.

Homolya, J. B., and J. L. Cheney. 1979. A study of primary sulfate emissions from a coal-fired boiler with FGD. J. Air Pollut. Control Assoc. 29(9):1000–1004.

Homolya, J. B. 1979. Communication.

Knapp, K. T., W. D. Conner, and R. L. Bennett. 1976. Physical characterization of particulate emissions from oil-fired power plants. In Theodore, L., et al. (eds.), Energy and the Environment Proceedings of the Fourth National Conference, AICHE, Dayton, Ohio. Pp. 495–500.

Nader, J. S. (Chm.). 1978. Workshop Proceedings on Primary Sulfate Emissions from Combustion Sources. Vol. 1, Measurement Technology, EPA-600/9-78-020a, 280 pp.; Vol. 2, Characterization, EPA-600/9-78-020b, 278 pp.

Natusch, D. F. S. 1976. Characterization of fly ash from coal combustion. In Nader, J. S. (Chm.), Workshop Proceedings on Primary Sulfate Emissions from Combustion Sources. Vol. 2, Characterization, EPA-600/9-78-020b. Pp. 149–163.

Radian Corp. 1975. Coal-fired power plant trace element study. Vol. I. A three station comparison. NTIS PB-257-293.

U.S. EPA. 1975. Compilation of air pollutant emissions factors. 2nd ed. Public Health Service Publication 999-AP-42.

U.S. EPA. 1978a. 1975 National emissions data system (NEDS) fuel use report. EPA-450/2-78-018.

U.S. EPA. 1978b. 1975 NEDS Report. National emissions data system of aerometric and emissions reporting system. EPA-450/2-78-020.

DISCUSSION

R. D. S. Stevens, United Technology and Science, Inc., Downsview, Ontario: You mentioned that sulfate emissions could increase as a result of increased use of flue gas desulfurization. Do you mean that there will be an absolute increase, or will sulfate only increase relative to sulfur dioxide?

John Nader: Since present scrubber systems remove only a small fraction of the combustion-generated sulfate, there will certainly be an increase in the sulfate emission relative to that for sulfur dioxide. The availability of flue gas desulfurization (FGD) systems tends to make the cheaper, high-sulfur fuels more attractive. A power plant which presently burns 1 to 2 percent sulfur coal to meet emission standards could switch to 5 to 6 percent sulfur coal and meet a more stringent standard through the use of an FGD system. This would result in a net increase in sulfate emissions. Thus, in the attempt to control sulfur dioxide, we might, in fact, increase sulfate emissions while meeting the emission standard.

CHAPTER 17

ATMOSPHERIC OXIDATION OF SULFUR DIOXIDE*

L. Newman†

ABSTRACT

An overview is presented of significant historical, recent, and new power plant and smelter plume studies which have been directed at understanding the atmospheric oxidation of sulfur dioxide. It can be concluded that the average rate of oxidation of sulfur dioxide in plumes entering into clean air is generally less than 1% per hour but that in polluted urban air the rate can be at least twice as fast. In addition, there is a diurnal variation in the rate, being near zero at night and approximately 3% per hour during midday. Although there is a tendency to select homogeneous over heterogeneous as the dominant pathway, there is no basis for a definitive choice, and most likely both mechanisms are at times operative. The suggestion is made that important and definitive new studies can be performed with technologies just becoming available.

INTRODUCTION

An assessment of the impact of pollutants on the environment can be made only in part through an understanding of the details of the emissions and transport, but in addition it is necessary to understand the nature and the extent of the transformation from one chemical form to another. It has become exceedingly important to understand the mechanisms of transformation in the case of sulfur since the aerosol sulfate product has been identified as a more serious health hazard than the originally emitted gaseous sulfur dioxide. Corrective and effective control legislation can only be made with a knowledge of the mechanisms of sulfur pollutant transformation; consequently, in recent years much attention has been devoted to obtaining such information.

Many laboratory studies have been conducted to determine the rate constants of potential homogeneous gas-phase reactions involving the oxidation of sulfur dioxide with suspected atmospheric constituents. A detailed model based on measured rate constants (Calvert et al., 1978) has been developed which predicts oxidation rates of tenths to a few percent per hour depending on solar intensity and the mix of pollutants. The kinetics of the association reaction of sulfur dioxide with the hydroxyl radical (Castleman and Tang, 1976) has been identified as particularly important.

---

*This work was performed under the auspices of the United States Department of Energy under Contract No. EY-76-C-02-0016.

†Environmental Chemistry Division, Department of Energy and Environment, Brookhaven National Laboratory, Upton, NY 11973.

The role of heterogeneous processes for sulfur dioxide to sulfate conversion involving solid and liquid aerosols has been reviewed (Environmental Protection Agency, 1975; and Hidy and Burton, 1975; Beilke and Gravenhorst, 1978; Hegg and Hobbs, 1978). In a recently completed laboratory investigation, Judeikes et al. (1978) studied the rates of removal of gaseous sulfur dioxide over solids commonly found in urban aerosols. Atmospheric projections of the results from this study suggest that freshly emitted aerosols can be quite effective in converting gaseous sulfur dioxide to particulate sulfate. It is concluded that reactions taking place in adsorbed water films are of primary importance.

Studies of the transformation of sulfur dioxide in the ambient air are necessary in order to determine the effective and relative importance of homogeneous and heterogeneous reactions. Such studies are particularly difficult to perform in an unequivocal manner because of the dilute concentrations and complex nature of the reactants and products, not to mention the complexities imposed due to variations in meteorology. However, effective investigations can be performed by studying the transformation occurring when power plant plumes mix with ambient air. Such plumes contain high concentrations of sulfur dioxide, making it in principle possible to follow them for long periods of time. The rates and mechanisms of oxidation of sulfur dioxide can be determined by performing these studies under a variety of meteorological and pollutant conditions. Such studies serve as a "laboratory" to obtain information on the nature of the oxidation in diluted ambient air (Newman, 1977), provided that the mechanism is transferable to ambient atmospheres as described below. Furthermore, transformation studies in power plant plumes are important in their own right since they are by far the major anthropogenic source of sulfur dioxide. Smelter plumes also offer a unique opportunity for study since they are large sources of sulfur dioxide with a different mix of coemitted particulates and are frequently located in remote areas, which permits them to be followed for great distances. Much attention has been directed to such plume studies, and it is the purpose of this paper to present an overview of the highlights of recent findings obtained by many investigators and the results of some new studies performed by Brookhaven National Laboratory. Conclusions will be drawn on the significance of these investigations in providing a basis for understanding the mechanisms through which sulfur dioxide is oxidized to sulfate in ambient air.

HISTORICAL POWER PLANT PLUME STUDIES

An effective means to study transformations in power plant plumes is to employ aircraft to obtain samples from the plumes as they are progressively diluted by air. Early attempts frequently resulted in observing apparent high oxidation rates, whereby, as examples, Gartrell et al. (1964), Dennis et al. (1969), and Stephens and McCaldin (1971) all reported oxidation rates at times as great as 50% per hour. Since aged air masses are not devoid of sulfur dioxide but generally contain a sulfur dioxide to sulfate mol ratio as large as 5 (Newman, 1979), it is obvious that oxidation rates as great as one cannot generally pertain, be sustained, or possibly ever occur. These early plume studies were probably flawed, due mainly to the limitations in the technologies available for collection and measurement of the reactants and products.

The development of the filter pack by Forrest and Newman (1973) presented a new and powerful tool to be used for measuring ambient oxidation rates. A tandem arrangement of a glass-fiber filter for aerosol sulfate followed by a base-impregnated paper filter for gaseous sulfur dioxide gave investigators the opportunity to collect the species of interest simultaneously and in a high-volume mode. Utilizing this collection technique, whereby both the concentration ratio (SO_4^{2-}/SO_2) and the isotopic ratio $(^{34}S/^{32}S)$ of the collected sulfur dioxide were measured, investigators at Brookhaven National Laboratory and now elsewhere have been able to determine the oxidation rate as a function of downwind distance from the source.

In a study of a coal-fired power plant plume, Newman et al. (1975b) observed that the total extent of sulfur dioxide oxidation seldom exceeded 5% even when the plume was followed for distances of 50 km. The oxidation rate thus observed was certainly much lower than heretofore claimed. In a companion study of an oil-fired power plant plume, Newman et al. (1975a) found that the extent of oxidation could be as much as five times as great as in the coal study. However, most importantly, they observed that the oxidation rate was high near the source and decreased with distance. They concluded that the data could be interpreted as a second-order mechanism arising from coemitted reactants. A heterogeneous oxidation mechanism was invoked to explain the observations in both these studies. Originally it was suggested that the reaction was quenched due to catalyst depletion. However, in a subsequent analysis (Schwartz and Newman, 1978) it was determined that dilution itself could cause quenching of a second-order reaction and that no inference concerning catalyst depletion could be drawn from the data. The interpretation of the oil-fired data has come under criticism, particularly in terms of the utilization of the isotope ratio data to uniquely arrive at mechanistic decisions (Wilson et al., 1976). Nevertheless, what most assuredly has been established is that the oxidation rate of sulfur dioxide in power plant plumes is significantly slower than had been previously reported.

RECENT POWER PLANT PLUME STUDIES

The Brookhaven group did a follow-up study of four coal-fired power plant plumes (Forrest and Newman, 1977a). Generally the extent of oxidation seldom exceeded 5%, with essentially all of the observed oxidation occurring within the first few kilometers after emission. No distinct correlation was found between the extent of sulfur dioxide conversion and any measured parameter, including distance, travel time, temperature, relative humidity, time of day, or atmospheric stability. A hallmark of these investigations is the presentation of all the completed runs, avoiding the "good day" syndrome, which can give unique but not necessarily representative data. An important observation in this data set is a confirmation of the slow oxidation rate but also a recognition that the measurement reproducibility at these low rates was not much better than a factor of 2. By reproducibility here is meant the agreement in ratio of particulate to total sulfur on successive measurements. The scatter in such measurements is due to variation in primary sulfate emitted or to inherent variability in atmospheric oxidation rate, or both. This wide variability well exceeds the analytical error and sets a limit on the precision in the rate of oxidation that can be measured in a given plume study. Despite this large variability, nonetheless, from a large number of studies, meaningful values of the average rate of sulfur dioxide oxidation may be obtained, and the dependence of this rate upon such variables as temperature, relative humidity, and distance from stack may be examined.

Husar et al. (1976) reported on results of a coal-fired plume study whereby they showed an initial induction period with little oxidation followed by a rapid (greater than 5% per hour) oxidation rate at 2 hr downwind. This observation was used strongly to support the dominance of the photochemical mechanism of oxidation. Mass flows of sulfur were used to arrive at the conclusions, and the results are somewhat controversial, since it is problematical as to whether such measurements can be made with sufficient accuracy. Indeed, let it be said that only one such "good day" has been reported upon in the literature.

A more extensive follow-up study was performed at the same coal-fired power plant (Husar et al., 1978; Gillani et al., 1978). This time the workers generally relied on calculating the ratio of sulfur particulate to total sulfur from simultaneous measurements of particulate and gaseous sulfur. This data set certainly establishes a new landmark insofar as the authors clearly show, from extensive measurements, that the oxidation rate is negligible (less than 1% per hour) in the early morning and early evening hours and that the rate increases with solar radiation (to a maximum of about 3% per hour). The authors established another first in being able to follow a plume for more than 10 hr and to a distance of 300 km. In essence this study is in agreement with the low oxidation rates observed by Forrest and Newman (1977a) but also points out the shortcoming of the Forrest and Newman work insofar as they investigated only the atmospherically stable early morning and evening conditions. In addition to establishing a very fine data set, Husar et al. very carefully state their conclusions, and they are worth quoting: "The above conversion data show that the rate controlling parameters of sulfur dioxide oxidation are associated with time of day. The informed reader could be tempted to interpret the daytime conversion as a clue that photochemistry is the dominant mechanism. However, the extent of our current data analysis is insufficient to exclude in-cloud conversion, mixing with background air or other mechanisms as being rate controlling."

A major problem in studying power plant plumes during daytime convective mixing conditions is that the plume soon becomes indistinguishable from background air. This can in part be overcome by going to an area where the background concentrations are very low, as was done in a study by Lusis et al. (1978) of a refinery coke plant. They generally corroborate the above results by finding the oxidation rate to be less than 0.5% per hour during February and early morning June flights. However, during daytime flights in June they found increased oxidation rates approaching 3% per hour. They conclude that "these elevated rates are possibly due to homogeneous gas-phase reactions involving sulfur dioxide and various free radical species."

Another study of a coal-fired power plant by Meagher et al. (1978) confirms the observations of Forrest and Newman (1977a) insofar as they observe that most of the atmospheric oxidation occurs in the immediate vicinity of the power plant and that the total extent of oxidation was less than 4%, with an average being only 1%. Of some interest is that they found the sulfate to be associated with submicrometer fly ash particles and conclude that this "may have resulted from heterogeneous reactions."

A most recent and extensive study of two coal-fired power plants (Hobbs et al., 1979) makes use of gas-to-particle conversion rates to determine the extent of sulfur dioxide oxidation. At one location the results of a few tenths of 1% per hour are similar to that reported by Forrest and Newman (1977a), but at the other, a few percent per hour is observed. However, the authors emphasize that the reported magnitude of conversion is quite sensitive to the assumption that the excess aerosol volume produced is sulfuric acid. The presence of nitrates, organics, and/or neutralization of the acid by ammonia would reduce the calculated rate. Contrary to the observations of Gillani et al. (1978), no correlation with ultraviolet intensity was observed, but, in agreement with the suggestion of Forrest and Newman (1977b), the plume with the higher particulate loading (factor of 5) has a higher oxidation rate.

However, by utilizing an extensive model of the evolution of the particulates, Eltgroth and Hobbs (1979) conclude that gas-to-particle conversion due to heterogeneous reactions on soot particles is negligible, but even homogeneous rates in their model predict lower rates of sulfate formation than those derived from their measurements.

SMELTER PLUME STUDIES

Lusis and Wiebe (1976), in a study of the plume of a large nickel smelter, usually observed less than 10% of the sulfur transformed to sulfate. The oxidation rates were generally less than 3% per hour, with an average value of 1%. A slight decrease in the oxidation rate with plume age was observed, and Forrest and Newman (1977b) interpreted this effect as arising from a second-order heterogeneous mechanism with a rate of reaction of 0.2 ppm^{-1} hr^{-1}. This value is one-fifth of that found for the second-order mechanism at an oil-fired plume (Newman et al., 1975a). Of significance to a heterogeneous mechanism is that the emitted particulate rate of 0.04 g per liter of emitted sulfur dioxide at the smelter operation was lower but possibly only coincidentally exactly equal to one-fifth of the oil-fired operation.

Low oxidation rates were again found in a follow-on study (Chan et al., 1979) at the same smelter plume, but this time they were typically observed to be less than 0.4% per hour. There is a clear indication that the oxidation rate decreases as the plume age increases, and this is suggestive of a reaction mechanism that is second (or higher) order in plume constituents and that is being quenched by dilution or by depletion of catalyst activity (Schwartz and Newman, 1978). Chan et al. (1979) conclude that heterogeneous processes may be involved in the oxidation and also suggest that this mechanism is limited to the initial stages after emission. The authors go on to suggest that photochemical oxidation can also occur, since oxidation rates in the summer are higher than in the winter. However, the measured differences might not be statistically significant. Utilizing a technique developed by Leahy et al. (1975), they were able to determine that the particulate sulfur formed and remained predominantly as sulfuric acid. They go on to suggest that the reason for observing the overall slow oxidation rate is that the high acidity quenches heterogeneous reactions, and they argue further that in the clean atmosphere of these investigations (free of NO$_x$ and hydrocarbons), the concentrations of active free radicals are so low that they cannot support extensive photochemical oxidation.

A study of a copper smelter plume (Eatough, 1979) employing a novel technique of measurement presents some new and different results. The author determined that arsenic coemitted with the sulfur dioxide can serve as a conservative tracer of the plume. Consequently, increasing ratios of sulfate to arsenic with downwind distance can be interpreted as a measure of the extent of oxidation. Quite precise data were obtained in this fashion, and the author showed that the reaction mechanism must be first and not second order. The observed oxidation rate in this copper smelter plume is about 2% per hour, which is somewhat higher than in the nickel smelter plume discussed above. This higher oxidation rate and different mechanism might arise from the differences in the particulate emissions of the two plumes or possibly from the observed rapid neutralization of sulfuric acid from a background metal oxide and/or carbonate particulate. Although previous investigators studying smelter (Chan et al., 1979) or power plant (Forest and Newman, 1977a) plumes were not able to observe a relationship between oxidation rate and temperature, the results of this investigation show a clear positive temperature dependence.

Roberts and Williams (1979) performed a study of plumes arising from the comixture of lead and copper smelter operations. These authors assumed that lead could be used as a conservative tracer, and from ratios of sulfate to lead they were able to measure the extent of oxidation and follow the effluents for as long as 12 hr after emission. They were particularly aided by the low background concentrations associated with the remote area of the study. A clear relationship was established between the extent of oxidation and ultraviolet insolation, and the authors concluded that the sulfate is formed via a gas-phase homogeneous oxidation of sulfur dioxide. The rate of oxidation for this winter study is 0.25% per hour, averaged over a 24-hr period. Roberts and Williams show from kinetic considerations (Castleman and Tang, 1976) and modeled atmospheric concentrations of reactive species that the reaction rate of sulfur dioxide with the hydroxyl radical is sufficient to produce the observed results.

PLUME FRINGE REACTIVITY

Increased chemical reactivity on the fringes of a power plant plume was first reported by Davis and Klauber (1975). Late in the transport of the stack gas plume they observed an ozone bulge on the plume fringe which they attributed to photochemical reactivity involving sulfur dioxide, in a very speculative series of reactions. That photochemical reactions are involved is unquestioned, but the ozone bulge more likely arises from conventional reactions involving nitrogen oxide to nitrogen dioxide conversion chains of typical smog (Calvert et al., 1978). Ozone bulges are now a commonplace observation (e.g., Gillani et al., 1978; Lusis et al., 1978; Miller et al., 1978) and are usually associated with high ambient photochemical reactivity. Additional evidence for plume fringe activity is supplied by Wilson et al. (1976), who observed Aitken nuclei peaks on the plume fringe. This observation has not been generally made, but when it occurs it is most assuredly due to photochemical activity. However, it is not necessarily directly associated with sulfur dioxide oxidation.

Eltgroth and Hobbs (1979) developed a model, supported by data, that indicates that homogeneous gas-to-particle conversion of sulfur dioxide to sulfuric acid is greatest at the edges of a plume, where the concentrations of hydroxyl free radicals from the ambient are greatest. White (1979) has shown from b_{scat} measurements that significant aerosol production can occur on plume fringes and results presumably from sulfate as an oxidation product of sulfur dioxide. The claim is made by White that the aerosol production was diffusion-limited and was inhibited in the core of the plume by a shortage of essential reactants furnished by the background. In this regard, Gillani and Wilson (1979) make the connection between the correspondence in formation of ozone and aerosols in power plant plumes as arising from the photochemical activity due to free radicals. It should be pointed out that, although these measured aerosol profiles are most likely related to sulfate, such sulfate profiles have not been made. In this regard, Forrest and Newman (1977a) looked for enhanced production of sulfate in plume fringes by comparing data obtained at downwind distances from cross-plume flights to those obtained by circling within the plume and also by comparing flights at the top, center, and bottom of the plume. Higher plume fringe reactivity would have resulted in an observation of greater sulfate production in the top, bottom, and cross-plume samples; however, within the reproducibility of the measurements (factor of 2), no differences could be observed.

NEW BNL PLUME STUDIES

A study at an oil-fired power plant (Forest et al., 1979a) was directed at measuring the rate of formation of sulfate and nitrate, with the subsequent incorporation of ammonium into these products. The extent of oxidation of sulfur dioxide to sulfate for reaction times up to 100 min and distances of 50 km ranged between 1 and 3%. About one-half of this amount could be accounted for by primary emissions arising from the combustion process. A generalized trend toward increasing ammonium content of plume particulates with increasing distance was observed by utilizing total sulfur concentrations as a conservative tracer. Based on changes in equivalence ratios it can be concluded in some cases that ammonium is neutralizing sulfate, but no such conclusion may be made regarding nitrate. Although the extent of sulfur dioxide oxidation was observed to be a factor of 2 greater, comparing summer with winter measuring periods, the rate could not be correlated with reaction time, humidity, temperature, stability class, or cloud coverage. Possibly the very low rate of ambient oxidation might be related to the low levels of particulate emissions (about 0.05 g per liter of SO_2), since these loadings are similar to the values found at well-controlled coal-fired plants (Newman et al., 1975b); and Forrest and Newman, 1977a), whose plumes also experience slow oxidation rates.

An extensive new study, consisting of some 60 flights, of the original oil-fired power plant plume is just being completed, and preliminary data analysis is available (Dietz and Garber, 1979). Only a few of these experiments duplicate the original observation of an initial fast reaction being quenched due to the nature of the properties of a second-order mechanism (Schwartz and Newman, 1978). It is felt that this might be due to the power plant operating conditions, which have changed over the years and now give rise to a different mix of emissions. As an example, the particulate load has been markedly reduced in the plant's conversion from cyclone to electrostatic precipitators. Indeed, even the nature of the particulates could have changed with changes in the oil. In the recent observations, sulfate rarely accounted for more than 5% of the total sulfur in the plumes. The oxidation rate is generally 1% per hour or less, which is now similar to that observed in essentially all the studies discussed in this paper. Preliminary indications are that there is increased reactivity during daylight convective mixing conditions. Strong acid determination and sulfuric acid extractions indicate that free sulfuric acid is a common component of the plume aerosol. Attempts will be made to assess nitrate production and ammonium incorporation.

Another study of a coal-fired plant has recently been completed (Forrest et al., 1979b). Conversion of sulfur dioxide to sulfate is observed to range generally between 0.1 and 0.8% per hour during night and early morning hours. Late morning and afternoon rates range between 1 and 4% per hour, with one measurement at 7%. Interestingly, particulate emissions (0.13 g per liter of SO_2) averaged a factor of 2 greater than that measured at the other coal-fired plants that have been studied. Particulate sulfur to total sulfur was found to increase directly with total solar radiation. This is in corroboration of the observations of Gillani et al. (1978) and Husar et al. (1978). Particulate ammonium normalized against total sulfur generally increased with plume age, but incorporation rates varied widely, probably symptomatic of wide variabilities in ambient ammonia concentrations. Nevertheless, ratios of ammonium to sulfate generally indicated extensive neutralization of sulfuric acid during transport. Total nitrate (sum of gaseous and particulate) measurements were made and related to total sulfur as a conservative tracer. Estimates could be made of the formation rate of nitrate. Nighttime and early-morning rates of nitric oxide oxidation to nitrate were generally between 0.1 and 3% per hour and late-morning and afternoon rates between 3 and 12% per hour.

CONCLUSIONS AND PROSPECTS

Based on the plume studies presented in this paper it can be asserted that the diurnal average oxidation rate of sulfur dioxide to sulfate is quite low, probably less than 1% per hour. Apparently, little or no oxidation occurs from early evening through to early morning. The rate sometimes does get as high as 3% per hour during midday conditions. An unequivocal selection cannot be made between whether homogeneous or heterogeneous mechanisms dominate the oxidation, and probably both are contributing.

The frequent observation of an initially high oxidation rate which is subsequently quenched makes it tempting to suggest that the early stages of oxidation in a plume are dominated by the heterogeneous mechanism. However, this could be disputed based on the modeling activities of Calvert et al. (1978) and Levine (1979), who suggest that there can be an early burst in photochemical activity due to the high concentrations and mix of emissions. The correspondence of oxidation to solar intensity makes it tempting to suggest that oxidation generally takes place through the homogeneous mechanism. However, the observations could be consistent with a heterogeneous mechanism, since increased solar intensity causes increased mixing, which can bring in ambient particulates to serve as surfaces for heterogeneous reactions. Rigorous homogeneous chemical models (i.e., models based on laboratory-measured rate constants) can be constructed to predict oxidation rates consistent with measurements. Due to the complexities of the reactions, such rigorous models cannot be constructed in the case of the heterogeneous mechanism, although one can conclude that the reactivity is high enough to also satisfy the measurements (Beilke and Gravenhorst, 1978; Hegg and Hobbs, 1978). There has been a tendency to give greater credence to the homogeneous reaction path, possibly because it can better be described and tested through mathematically rigorous expressions.

Originally Newman et al. (1975a) and Forrest and Newman (1978) observed that an oil-fired plume could be more reactive than coal-fired plumes, but this observation has not been generally borne out. However, in this regard, oil-fired power plant operating conditions can be such that there are higher primary sulfate emissions (Dietz and Garber, 1978; Nader, 1979) and consequently possibly higher ambient reactivity.

Aerosol and ozone profiles of plumes showing plume fringe reactivity suggest the dominance of the photochemical mechanism. However, a study on aerosol dynamics (McMurry, 1979) presents evidence for the importance of the heterogeneous mechanism of aerosol formation. In this regard, there is a need to make measurements of sulfate profiles in plumes, and fortunately the technology to do so is just now becoming available (D'Ottavio et al., 1979).

If the mechanism giving rise to sulfur dioxide oxidation in plumes is first-order in plume constituents, then the rate of oxidation observed in plumes is directly transferable to general ambient air into which the plume is mixing. In unpolluted areas the rates appear to remain essentially as derived from these plume studies. Transformation in urban areas appears to be significantly (factor of 2) higher (Husar et al., 1976; MacCracken, 1979), and this is probably due to the higher levels of pollutants providing higher concentrations of either active free radical or catalytic agents.

It has been observed that sulfate concentrations in urban and nonurban areas have not decreased concomitantly with decreases in sulfur dioxide (Altshuller, 1973). In this regard, Judeikes et al. (1978) suggest that heterogeneous reactions could give rise to this observation from the fact that such reactions are limited by atmospheric particle burdens rather than sulfur dioxide concentrations. However, an alternative suggestion might be that there has not actually been a decrease in the amount of sulfur emitted but rather a change in the distribution and then the regional average sulfate concentrations might not diminish concomitantly with sulfur dioxide concentrations (Schwartz, 1979) since the residence time of sulfate is substantially greater than that of sulfur dioxide.

More information is still needed before a definitive description can be made of the atmospheric oxidation of sulfur dioxide. More plume studies need to be made with real-time instrumentation to obtain details of plume profiles. Studies on the rate of formation of nitric acid are just beginning, and they are particularly important as a test of photochemical models which can predict an order-of-magnitude faster rate of formation of nitrate than sulfate (Davis et al., 1979). Definitive studies are needed on the rate of neutralization of sulfuric and nitric acids, since the chemical form is of particular importance insofar as health effects are concerned. An evaluation has to be made of the extent and rapidity of the existence of phase equilibria of the gaseous components of nitric acid and ammonia with the aerosol components containing sulfates and nitrates. In this regard, Tang (1979) has calculated the pertinent vapor pressures above mixed salts containing H^+, NH_4^+, SO_4^{2-}, and NO_3^-, and Tanner (1979) has designed a detailed experiment utilizing new capabilities to measure nitrates (Stevens, 1979) and ammonia (Tanner et al., 1980). Finally, oxidation products are not the only concern, and more information is needed on sulfite transformation products, such as have been identified by Eatough et al. (1978). It just might be that these are the compounds that are most significant in terms of adverse health effects.

ACKNOWLEDGMENTS

This work was done under Contract No. EY-76-C-02-0016 with the United States Department of Energy. In addition, some financial assistance was received from the Environmental Protection Agency, Electric Power Research Institute, Empire State Electric Energy Research Corporation, and the Long Island Lighting Company.

REFERENCES

Altshuller, A. P. 1973. Atmospheric sulfur dioxide and sulfate. Distribution of concentration at urban and nonurban sites in the United States. Environ. Sci. Technol. 7:709–712.

Beilke, S., and G. Gravenhorst. 1978. Heterogeneous SO_2 oxidation in the aqueous phase. Atmos. Environ. 12:231–239.

Calvert, J. G., Fu Su, J. W. Bottenheim, and O. P. Strausz. 1978. Mechanism of the homogeneous oxidation of sulfur dioxide in the troposphere. Atmos. Environ. 12:197–226.

Castleman, A. W., Jr., and I. N. Tang. 1976. Kinetics of the association reaction of SO_2 with the hydroxyl radical. J. Photochem. 6:349–354.

Chan, W. H., R. J. Vet, M. A. Lusis, J. E. Hunt, and R. D. S. Stevens. 1979. Airborne $SO_2 \rightarrow SO_4^=$ oxidation studies of the INCO 381 m chimney plume. Atmos. Environ., in press.

Davis, D. D., and G. Klauber. 1975. Atmospheric gas phase oxidation mechanisms for the molecule SO_2. Int. J. Chem. Kinet. Symp. 1:543–556.

Davis, D. D., W. Heaps, D. Philen, and J. McGee. 1979. Boundary layer measurements of the OH radical in the vicinity of an isolated power plant plume: SO_2 and NO_2 chemical conversion times. Atmos. Environ. 13:1197–1203.

Dennis, R., C. E. Billings, F. A. Record, P. Warneck, and M. L. Arin. 1969.

Measurements of sulfur dioxide losses from stack plumes. APCA paper No. 69-156, 62nd Meeting of the Air Pollution Control Association, New York, N.Y.

Dietz, R. N., and R. W. Garber. 1978. Power plant flue gas and plume sampling studies. Progress report No. 2, BNL 25420. December.

Dietz, R. N., and R. W. Garber. 1979. Power plant flue gas and plume sampling studies. Progress report No. 3, BNL (in press) October.

D'Ottavio, T., R. L. Tanner, R. Garber, and L. Newman. 1979. Determination of ambient sulfur using a continuous flame photometric detection system II. The measurement of low level sulfur concentrations under varying atmospheric conditions. Atmos. Environ., submitted.

Eatough, D. J., T. Major, J. Ryder, M. Hill, N. F. Mangelson, N. L. Eatough, L. D. Hansen, R. G. Meisenheimer, and J. W. Fischer. 1978. The formation and stability of sulfite species in aerosols. Atmos. Environ. 12:263-271.

Eatough, D. J. 1979. Transformation in industrial plumes. Included in the progress report "The Multistate Atmospheric Power Production Pollution Study— MAP3S," DOE/EV-0040, pp. 209-214. Available from NTIS Department of Commerce, Springfield, Va.

Eltgroth, M. W., and P. V. Hobbs. 1979. Evolution of particles in the plumes of coal fired power plants II. A numerical model and comparisons with field measurements. Atmos. Environ. 12:953-975.

Environmental Protection Agency. 1975. Position paper on regulation of atmospheric sulfates. PB-245 760. Available from NTIS, Department of Commerce, Springfield, Va.

Forrest, J., and L. Newman. 1973. Sampling and analysis of atmospheric sulfur compounds for isotope studies. Atmos. Environ. 7:561-573.

Forrest, J., and L. Newman. 1977a. Further studies on the oxidation of sulfur dioxide in coal-fired power plant plumes. Atmos. Environ. 11:465-474.

Forrest, J., and L. Newman. 1977b. Oxidation of sulfur dioxide in the Sudbury smelter plume. Atmos. Environ. 11:517-520.

Forrest, J., and L. Newman. 1978. Oxidation of sulfur dioxide in power plant plumes. AIChE J. 74:48-53.

Forrest, J., R. Garber, and L. Newman. 1979a. Formation of sulfate, ammonium and nitrate in an oil-fired power plant plume. Atmos. Environ., in press.

Forrest, J., R. Garber, and L. Newman. 1979b. Conversion rate in the Cumberland plume based on filter pack data. Manuscript in preparation.

Gartrell, F. E., F. W. Thomas, and S. B. Carpenter. 1964. Atmospheric oxidation of sulfur dioxide in coal-burning power plant plumes. Am. Ind. Hyg. Assoc. J. 24:113-120.

Gillani, N. V., R. B. Husar, J. D. Husar, D. E. Patterson, and W. E. Wilson, Jr. 1978. Project MISTT: kinetics of particulate sulfur formation in a power plant plume out to 300 km. Atmos. Environ. 12:589-598.

Gillani, N. V., and W. E. Wilson. 1979. Formation and transport of ozone and aerosols in power plant plumes. N.Y. Academy of Sciences Conference on Aerosols, New York, N.Y., Jan. 9-12. Ann. N.Y. Acad. Sci., in press.

Hegg, D. A., and P. V. Hobbs. 1978. Oxidation of sulfur dioxide in aqueous systems with particular reference to the atmosphere. Atmos. Environ. 12:241-253.

Hidy, G. M., and C. S. Burton. 1975. Atmospheric aerosol formation by chemical reactions. Int. J. Chem. Kinet. Symp. 1:509-541.

Hobbs, P. V., D. A. Hegg, M. W. Eltgroth, and L. F. Radke. 1979. Evolution of particles in the plumes of coal-fired power plants—I. Deductions from field measurements. Atmos. Environ. 13:935-951.

Husar, R. B., N. V. Gillani, and J. D. Husar. 1976. Particulate sulfur formation in power plant, urban and regional plumes. Sympsoium on Aerosol Science and Technology, 82nd National Meeting of AIChE, Atlantic City, N.J., Aug. 30–Sept. 1.

Husar, R. B., D. E. Patterson, J. D. Husar, N. V. Gillani, and W. E. Wilson, Jr. 1978. Sulfur budget of a power plant plume. Atmos. Environ. 12:549–568.

Judeikis, H. S., T. B. Stewart, and A. G. Wren. 1978. Laboratory studies of heterogeneous reactions of SO_2. Atmos. Environ. 12:1633–1641.

Leahy, D., R. Siegel, P. Klotz, and L. Newman. 1975. The separation and characterization of sulfate aerosol. Atmos. Environ. 9:219–229.

Levine, S. Z. 1979. Private communication.

Lusis, M. A., and H. A. Wiebe. 1976. The rate of oxidation of sulfur dioxide in the plume of a nickel smelter stack. Atmos. Environ. 10:793–798.

Lusis, M. A., K. G. Anlauf, L. A. Barrie, and H. A. Wiebe. 1978. Plume chemistry studies at a northern Alberta power plant. Atmos. Environ. 12:2429–2437.

MacCracken, M. C., Coordinator. 1979. The Multistate Atmospheric Power Production Pollution Study—MAP3S, DOE/EV-0040, Pollutant Transformation, pp. 194–243. Available from NTIS, Department of Commerce, Springfield. Va.

McMurry, P. 1979. Sulfate aerosol dynamics. This proceedings, following paper.

Meagher, J. F., L. Stockburger, E. M. Bailey, and O. Huff. 1978. The oxidation of sulfur dioxide to sulfate aerosols in the plume of a coal-fired power plant. Atmos. Environ. 12:2197–2203.

Miller, D. F., A. J. Alkezweeny, J. M., Hales and R. N. Lee. 1978. Ozone formation related to power plant emissions. Science 202:1186–1190.

Nader, J. C. 1979. Primary sulfur emission from stationary industrial sources. This proceedings, preceding paper.

Newman, L., J. Forrest, and B. Manowitz. 1975a. The application of an isotopic ratio technique to a study of the atmospheric oxidation of sulfur dioxide in the plume from a oil-fired power plant. Atmos. Environ. 9:959–968.

Newman, L., J. Forrest, and B. Manowitz. 1975b. The application of an isotopic ratio technique to a study of the atmospheric oxidation of sulfur dioxide in the plume from a coal-fired power plant. Atmos. Environ. 9:969–977.

Newman, L. 1977. Plume characteristics. Presented at the American Nuclear Society meeting on Aerial Techniques for Environmental Monitoring, Las Vegas, Nevada.

Newman, L. 1979. General considerations on how rainwater must obtain sulfate, nitrate, and acid. American Chemical Society meeting, Honolulu, Hawaii, April 1–6.

Roberts, D. B., and D. J. Williams. 1979. The kinetics of oxidation of sulfur dioxide within the plume from a sulphide smelter in a remote region. Atmos. Environ. 13:1485–1500.

Schwartz, S. E., and L. Newman. 1978. Processes limiting oxidation of sulfur dioxide in stack plumes. Environ. Sci. Technol. 12:67–73.

Schwartz, S. E. 1979. Private communication.

Stephens, N. T., and R. O. McCaldin. 1971. Attenuation of power station plumes as determined by instrumented aircraft. Environ. Sci. Technol. 5:615–621.

Stevens, R. K., editor. 1979. Current methods to measure atmospheric nitric acid and nitrate artifacts. EPA-600/2-79-051, March. Available from NTIS, Department of Commerce, Springfield, Va.

Tang, I. N. 1979. On the equilibrium partial pressures of nitric acid and ammonia in the atmosphere. Atmos. Environ., in press.

Tanner, R. L., R. Abbas, and J. Lepore. 1980. Determination of trace ammonia using fluorescent derivitization I. Continuous analysis of ppb levels of aqueous ammonia. Anal. Chem., submitted. Chim. Acta

Tanner, R. L. 1979. Private communication.

White, W. H. 1979. On the crosswind distribution of secondary aerosol in a power plant plume. Unpublished manuscript.

Wilson, W. E., R. J. Charlson, R. B. Husar, K. T. Whitby, and D. Blumenthal. 1976. Sulfates in the Atmosphere. APCA paper No. 76-30-06, 69th Annual Meeting of the Air Pollution Control Association, Portland, Oregon, June 27–July 1.

DISCUSSION

Questioner: You mentioned that in some of your measurements you found the rate of sulfur dioxide oxidation to be second order. How did you come to that conclusion?

L. Newman: It was based primarily on the observation that the rate of oxidation was high immediately after emission and slower as the flue gases became more dilute. This is obvious from the smelter and oil-fired power plant studies. This appeared to also be the case in the coal-fired power plant studies since we saw a rapid conversion of approximately 1 percent of the emitted sulfur dioxide very near the stack. Relatively little change was observed at further distances downwind.

Questioner: Do you know of any studies or are you involved in any studies to measure particle deposition from these plumes, and how do you account for deposition in your rate calculations?

L. Newman: I think that this has been a somewhat neglected area. We have presented data in the literature to support the apparent loss of particles from the plume, which could account for an amount equal to the one to two percent that is formed by oxidation during the initial stages. However, it is hard to construct a mechanism by which this deposition could occur for the sulfate particle sizes that we would expect in the plume; thus, people tend to want to believe that relatively little deposition can occur in the time scales over which power plant plumes can be followed.

Charles Hakkarinen, Electric Power Research Institute: Based on your own work of measuring primary sulfate emissions and your review of atmospheric oxidation of sulfur dioxide, could you estimate what percentage of the ambient sulfate comes from direct emission of sulfate?

L. Newman: Our measurements of primary sulfate are a little bit at odds with the data coming out of the EPA laboratories, especially in regard to the amount of sulfate that is produced by an oil-fired power plant. We generally observe that the primary sulfate accounts for one percent or less of the sulfur that is being emitted. Oil-fired plants can produce higher concentrations of sulfate if they do not minimize their excess oxygen levels, but most oil-fired power plants keep their oxygen levels low. Otherwise, they have serious corrosion problems. So, as a working number, I will take one percent of the emitted sulfur as sulfate for power plants. I will use an atmospheric oxidation rate of one percent per hour for sulfur dioxide. In fact, I believe that half a percent per hour is more reasonable and maybe even as low as a quarter of a percent per hour for an annual average. If you combine these values with reasonable numbers for the sulfur dioxide and sulfate deposition velocities, I would calculate that ∼20% ± 10% of the sulfur dioxide that is emitted gets oxidized to sulfate before it leaves the continental United States. Thus, five percent of the steady-state sulfate in the atmosphere is due to primary emissions.

CHAPTER 18

THE FATE OF SULPHURIC ACID AEROSOL IN THE ATMOSPHERE

Cyrill Brosset*

ABSTRACT

The phase diagram for $(NH_4)_2SO_4-H_2SO_4-H_2O$ presented by Tang et al. (1978) was complemented with tie lines for the ammonia equilibrium concentration over the liquid phase (Lee and Brosset, 1979).

Such a complete diagram provides information essentially on the phase composition attained by a sulphuric acid droplet in equilibrium with the atmosphere. This composition can also be calculated if the relative humidity (r.h.) and the concentration of ammonia (pNH_3) of the atmosphere are known.

Whether such calculations reflect real conditions or not depends on the equilibrium adjustment, i.e., if it is sufficiently fast relative to the variation in temperature (t), r.h., and pNH_3 of the atmosphere.

It was shown that the graphic correlation between r.h. and log pNH_3 provides direct information on the phases present and their possible transformations. Such information, obtained through measurement of r.h. and pNH_3, was in good agreement with particle composition established through chemical analysis.

INTRODUCTION

A large number of investigations have shown that the ion composition of the water-soluble part of fine airborne particles consists predominantly of the ions NH_4^+, H^+, HSO_4^-, and SO_4^{2-}.

The reason for this predominance is probably that the fine particles often appear initially in the form of sulphuric acid droplets. These then react in the air with water vapour and ammonia.

It has previously been stated (Brosset, 1978) that the concentration ratio $(cNH_4^+) \cdot (cH^+)^{-1}$ (more correct, activity ratio) in the water solution constituting the whole or a part of airborne particles is dependent on the partial pressure of ammonia (pNH_3) in the air. Thus, the ammonia concentration of the air will also affect the composition of the solid phases which under certain conditions may crystallise from the solution in question. However, this latter process is also dependent on the relative humidity (r.h.) of the air.

In other words, a complete description of the phase composition attained by a sulphuric acid droplet in a state of equilibrium with air requires access to a relevant

---

*Swedish Water and Air Pollution Research Institute (IVL), P.O. Box 5207, S-402 24 Gothenburg, Sweden.

part of the phase diagram for the system $H_2SO_4-H_2O-NH_3$. To be able to draw up such a phase diagram, solubilities of solid phases and the equilibrium pressures of water and ammonia over the liquid phase at relevant compositions must be known.

At the present time, the following data are available:

The composition of the liquid phase in equilibrium at 25°C with relevant solid phases was determined by d'Ans (1913).

About one year ago, the water vapour pressure was measured at 30°C in a system consisting of the components $(NH_4)_2SO_4$, H_2SO_4, and H_2O (Tang et al., 1978).

Just recently, the ammonia pressure was calculated for some compositions of the liquid phase of the system at 25°C (Lee and Brosset, 1979).

As a result, it has now become possible to construct, at least approximately, an important part of the phase diagram for the system $(NH_4)_2SO_4-H_2SO_4-H_2O$ (i.e., the acid side of the system $H_2SO_4-H_2O-NH_3$). By means of this diagram, certain conclusions concerning the fate of sulphuric acid aerosol in the atmosphere could be drawn.

PHASE DIAGRAM FOR THE SYSTEM $(NH_4)_2SO_4-H_2SO_4-H_2O$ AND ITS INTERPRETATION

The Diagram

The phase diagram concerned is shown in Fig. 1a. It is largely identical with that presented earlier by Tang et al. (1978). However, tie lines for the ammonia pressure have been added.

The following should be noted as regards the diagram.

The solubility curve L1-L2-L3- is based on the 60-year-old data obtained by d'Ans and applies at 25°C. Renewed measurements with more modern methods should perhaps be made.

As pointed out above, the tie lines for r.h. were determined by Tang et al. at 30°C. However, according to them, r.h. does not alter very much within a narrow temperature range. Their tie lines should therefore, with good approximation, apply at 25°C as well.

Tie lines for pNH$_3$ at 25°C have been obtained through interpolation based on calculated values for some 30 points within the area concerned.

Some of the mean activity coefficients required for the calculation had to be estimated due to lack of basic data. The values arrived at must therefore be regarded as probably good approximations, at least in that part of the diagram where the ionic strength in the liquid phase is not too high (Lee and Brosset, 1979).

Interpretation of the Diagram

To facilitate understanding in the discussion to follow, the designations used in the diagram and the meaning of its different parts will be explained.

The line S1-S2-S3-

Point S1 represents the pure phase $(NH_4)_2SO_4$, point S2 the phase $(NH_4)_3H(SO_4)_2$, and point S3 the phase NH_4HSO_4. Thus, the line parts S1-S2 and S2-S3 correspond to mixtures of phases S1 and S2 and of S2 and S3, respectively.

The curve L1-L2-L3-

This is the solubility curve determined by d'Ans. The part L1-L2 represents solutions in equilibrium with S1. In the same way, the part L2-L3 represents solutions in

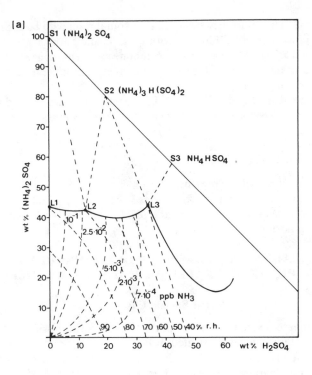

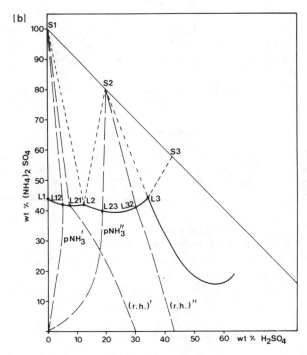

Fig. 1. Phase diagrams for the system $(NH_4)_2SO_4–H_2SO_4–H_2O$ at 25°C. (*a*) Diagram including tie lines for ammonia pressure; (*b*) diagram describing four special cases. See text.

equilibrium with $S2$. The solution at point $L2$ is in equilibrium with both $S1$ and $S2$; the solution at point $L3$ is in equilibrium with both $S2$ and $S3$.

At points $L2$ and $L3$ there are, in this tricomponent system, consequently, four phases, which means one degree of freedom. At chosen constant temperature, both r.h. and pNH_3 are thus given.

At the curve parts denoted $L1$–$L2$ and $L2$–$L3$ there are three phases and thus two degrees of freedom. Consequently, at chosen constant temperature, there is a further choice of, e.g., either r.h. or pNH_3.

The area below the solubility curve

In this area, no solid phases are present, which means that the number of phases is 2, and, thus, the number of degrees of freedom is 3. At chosen constant temperature there is a further choice of, e.g., both r.h. and pNH_3. A chosen such pair of values determines the composition of the solution.

THE PHASE DIAGRAM AS A BASIS FOR DETERMINING THE FATE OF A SULPHURIC ACID AEROSOL IN THE ATMOSPHERE

Relevant Special Cases

The use of the phase diagram for determining the composition which a sulphuric acid aerosol may attain in air is illustrated in Fig. 1b.

In this figure, the phase diagram from Fig. 1a has been limited to four special cases, which are combinations between two tie lines for pNH_3 [$(pNH_3)' \cong 0.1$ ppb and $(pNH_3)'' \cong 5 \cdot 10^{-3}$ ppb] as well as two tie lines for r.h. [(r.h.)$' \cong 75\%$ and (r.h.)$'' \cong 50\%$]. As will be seen from Fig. 1b, all points of intersection between these tie lines and the solubility curve ($L12$, $L21$, $L23$, and $L32$) are located to the left of $L3$. The reason for this choice is, inter alia, that the area in the phase diagram to the right of $L3$ is less well investigated.

In the atmospheres corresponding to the four combinations mentioned, a sulphuric acid droplet may be assumed to react in the following manner.

Case 1: The Combination (r.h.)$'$, $(pNH_3)''$

This is the simplest of the four cases. The curves intersect in the liquid range. The coordinates of the intersection point give the equilibrium composition of the liquid phase, which in the present case will be about 27 wt % $(NH_4)_2SO_4$, 18 wt % H_2SO_4, and thus 55 wt % H_2O.

The course followed by the sulphuric acid droplet to reach this point depends on whether water or ammonia uptake is the faster. Since the water vapour concentration in the gas phase here is much higher than the ammonia concentration, it is possible that the water uptake, at least initially, is more rapid. This means that, in the first step, the sulphuric acid droplet will attain a composition corresponding to some point on the lower part of the curve (r.h.)$'$. After this, ammonia is taken up until the point (r.h.)$'$, $(pNH_3)''$ is reached.

Case 2: The Combination (r.h.)$'$, $(pNH_3)'$

As will be seen from the diagram, there is no liquid-phase composition that can be in equilibrium with this combination in the gas phase. The sulphuric acid droplet introduced into this atmosphere is likely to take up water and ammonia until some point on the curve (r.h.)$'$ is reached. This curve is then followed to point $L21$. Now, however, the liquid phase has to disappear. In this process, the system will follow some course within the area $L21$–$S1$–$L12$, while $S1$, i.e. $(NH_4)SO_4$, is being precipitated.

Case 3: The Combination (r.h.)$''$, (pNH$_3$)$''$

This case is analogous to case 2. Phase $S2$ will be the final form, i.e. $(NH_4)_3H(SO_4)_2$.

Case 4: The Combination (r.h.)$''$, (pNH$_3$)$'$

In this case, the sulphuric acid droplet is likely to take up water and ammonia until some point on the curve (r.h.)$''$ is reached. The curve is then followed to point $L32$. The system is now still very far from equilibrium, and the liquid phase will therefore disappear. In this process, the system takes a course between lines $L32$–$S2$ and $L12$–$S1$, which may mean precipitation at the same time of phases $S2$ and $S1$. Eventually, some point on the line $S1$–$S2$ is reached. The location of the point is determined by kinetic conditions.

A USEFUL ALTERNATIVE FORM OF THE DIAGRAM

In certain cases, when determining the phase composition of the system in equilibrium, it may be helpful to plot log pNH$_3$ vs r.h. An example of such an r.h.–log pNH$_3$ diagram is given in Fig. 2a. It corresponds to the conditions presented in Fig. 1a.

The curve in Fig. 2a represents the combinations (r.h., pNH$_3$) located on the curve $L1$–$L2$–$L3$ in Fig. 1a. The points $L1$, $L2$, and $L3$ in Fig. 2a thus have the same physical meaning as the corresponding points in Fig. 1a.

The meaning of the areas introduced in Fig. 2a is easily seen:

A point located within the area below the curve $L1$–$L2$–$L3$ represents pure liquid phase. Thus, point 1 corresponds to case 1 above.

A point within the area $L2$–$L1$–$M1$ represents the pure phase $S1$. This is illustrated by point 2, corresponding to case 2 above.

A point within the area $L2$–$L3$–$M3$ represents the pure phase $S2$. Thus, point 3 corresponds to case 3 above.

A point within the area $L2$–$M1$–$M2$–$M3$ represents a mixture of the phases $S1$ and $S2$. This is illustrated by point 4, corresponding to case 4 above.

This diagram form is very useful for the study of how the phase composition of the system changes as a result of changes in r.h. and pNH$_3$. Such an application will be illustrated in the following section.

APPLICATION OF THE DIAGRAM TO ACTUAL ATMOSPHERIC CONDITIONS

In the foregoing the phase composition in the system $(NH_4)_2SO_4$–H_2SO_4–H_2O was studied as a function of the variables temperature (t), r.h., and pNH$_3$. In the real situation (referring here to the conditions in central Sweden), all these parameters are likely to show a more or less pronounced daily and seasonal variation. That this is the case for temperature and relative humidity is a well-known fact.

Using a method developed by Ferm (1979), 24-hr means of ammonia concentration are being measured as of spring 1977 in Sweden. A clear seasonal and a slight daily variation of the concentration following the air temperature have been established.

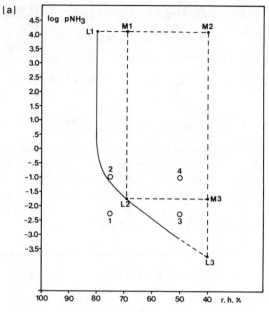

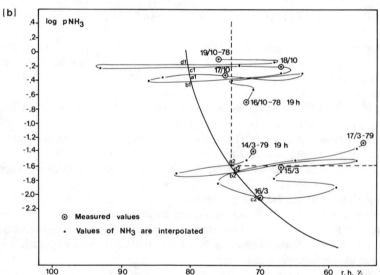

Fig. 2. R.h.–log pNH$_3$ diagrams. (*a*) Diagram at 25°C, corresponding to conditions in Fig. 1*a*; (*b*) diagram at 10°C, illustrating phase composition and transformation of fine particles as reflected by measured values of r.h. and pNH$_3$ at Rörvik during two three-day periods.

Let us now assume that equilibrium adjustment in the system in question is rapid relative to the existing 24-hr variation in *t*, r.h., and pNH$_3$. In such a case it would be possible to obtain a rather detailed picture of the variation with time of the phase composition. The prerequisite for this is access to phase diagrams within relevant temperature ranges and data for r.h. and pNH$_3$ in the form of sufficiently short-term mean values.

Today, some of these prerequisites are not at hand.

Due to the low concentration (approx. 0.001–10 ppb) of ammonia in the atmosphere, it is at present often necessary to use a sampling time (24-hr) which is somewhat too long. Furthermore, phase diagrams for other temperatures than 25°C, are not available. In central Sweden, the relevant temperature ranges are, however, primarily from $-5°$ to $+5°C$, from $+5°$ to $+15°C$, and from $+15°$ to $+25°C$.

In spite of these shortcomings, an attempt has been made to show how it is possible to follow the phase composition by means of an r.h.–log pNH$_3$ diagram. This attempt is illustrated in Fig. 2b. The data introduced in this figure were obtained through measurements at Rörvik, a clean-air area about 40 km south of Gothenburg, during three days in October 1978 and three days in March 1979.

On the first occasion the temperature was mostly between 8° and 12°C, whereas on the later occasion it ranged between $-6°$ and $+3°C$.

The values plotted in this figure are 6-hr means, the ammonia concentration being estimated from 24-hr means on the assumption that the 24-hr variation (seemingly small here) largely followed the temperature.

There are, as mentioned, no phase diagrams available for the temperature ranges concerned here. The solubility curve drawn into Fig. 2b was obtained by introducing solubility data valid for 10°C (Locuty and Laffitte, 1934) in Fig. 1a and assuming that the respective tie lines do not shift their position too much at a temperature drop from 25°C (30° for r.h.) to 10°C.

The solubility curve thus obtained can probably be used for the values from October but only as a rough approximation for the values from March. However, nothing else is at hand at present.

In spite of this, Fig. 2b at least gives a qualitative picture of what could have happened with the phase composition of the particles in question during the three-day periods involved.

Assume that sulphuric acid aerosol was introduced into the air around the sampling site on 16 October 1978 and remained there for the three following days. The diagram in Fig. 2b shows that, initially, this could result in the formation of a mixture of phases $S1$ and $S2$. As reaction in the solid state is likely to be slow here, nothing should happen until about 27 hr later, when point $a1$ has been passed and the solid particles have been transformed into liquid drops. About 12 hr later, point $b1$ is passed, which should lead to the crystallisation of phase $S1$. The particles remain in this state until point $c1$ has been passed, at which time they return to the liquid state. Later, when point $d1$ has been reached, phase $S1$ can recrystallise.

In the same manner, a sulphuric acid aerosol is assumed to have been introduced into the air on 14 March 1979. Here, too, phases $S1$ and $S2$ may crystallise initially, then transform to the liquid phase about 8 hr later when point $a2$ has been passed. After another 6 hr, point $b2$ will be passed, and phase $S2$ can crystallise. This state remains for about 32 hr, after which point $c2$ has been passed, and the particles transform to the liquid phase. After another few hours, point $d2$ will be passed, and phase $S2$ can recrystallise.

Thus, the diagram predicts that the particles from the October period should have a composition approximately corresponding to $(NH_4)_2SO_4$ and those from March corresponding to $(NH_4)_3H(SO_4)_2$. Analysis of daily particle samples taken on these days has given, for the ratio NH_4^+/H^+, very high values (22 and higher) for October samples and values from 2.8 to 1.7 for the March samples, which can be considered to be a fair agreement.

REFERENCES

Brosset, C. 1978. Water-soluble sulphur compounds in aerosols. Atmos. Environ. 12(1–3):25–38.

D'Ans, J. 1913. Zur Kenntnis der sauren Sulfate VII. Z. allg. anorg. Chem. 80: 235–245.

Ferm, M. 1979. Method for determination of atmospheric ammonia. Atmos. Environ., in press.

Lee, Y. H., and C. Brosset. 1979. Interaction of gases with sulphuric acid aerosol in the atmosphere. Presented at WMO Symposium on the long range transport of pollutants and its relation to general circulation including stratospheric/tropospheric exchange processes, Sofia, Bulgaria, 1–5 October.

Locuty, P., and P. Laffitte. 1934. Sur le système acide sulfurique–sulfate d'ammonium–eau. C. R. Acad. Sci. 199:950–952.

Tang, J. N., H. R. Munkelwitz, and J. G. Davis. 1978. Aerosol growth studies— IV. Phase transformation of mixed salt aerosols in moist atmosphere. J. Aerosol Sci. 9:505–511.

DISCUSSION

Jim Galloway: Sir, you mentioned the effect of kinetics on the dissolution and formation of solids. Could you be a bit more specific?

C. Brosset: Yes, sir, I could. Let's assume that the particle is absorbing water and ammonia. It's taking up water very rapidly and ammonia more slowly. I will draw your attention to the lines in Fig. 2. Ammonia will be taken up. If the water is being taken up much more rapidly, the reaction will more likely follow different lines than if ammonia is taken up more rapidly. It all depends on the relative absorption velocity of the various gases, and we have not yet measured those rates.

CHAPTER 19

THE DYNAMICS OF SECONDARY SULFUR AEROSOLS

Peter H. McMurry*

ABSTRACT

Significant amounts of sulfur-containing aerosol are formed in the atmosphere by gas-to-particle conversion of SO_2. The distribution of such secondary aerosols with respect to particle size is important in determining their environmental effects. In this paper, factors which determine the size distributions of evolving secondary aerosols are discussed. Evidence is presented which shows that growth mechanisms and hence size distributions of secondary sulfur aerosols vary in the atmosphere. Also, a theoretical analysis which quantitatively predicts the dynamic behavior of sulfur aerosols formed by gas-phase reactions is outlined.

INTRODUCTION

Sulfur-containing aerosols formed in the atmosphere by gas-to-particle conversion often comprise a substantial proportion of the submicron aerosol mass. These secondary aerosols are responsible for several of the undesirable effects associated with polluted air, including visibility reduction (White and Roberts, 1977; Trijonis, 1979; Cass, 1979), acid rain (Likens et al., 1979), and health effects (Amdur, 1971). Some of these effects have been discussed by other investigators at this symposium.

The ability of a given aerosol mass to contribute to such air pollution effects depends upon the distribution of that mass with respect to particle size. Whitby (1978) reported that secondary aerosols tend to accumulate in the submicron range and that atmospheric measurements for the volume mean diameters of these aerosols range from 0.15 to 0.5 μm. Such particles are particularly effective at scattering light and penetrating into the lungs (Lippmann, 1976).

In this paper, the physical and chemical formation mechanisms which influence secondary-aerosol size distributions will be discussed. Evidence is presented to show that chemical mechanisms for sulfur aerosol formation in the atmosphere vary with location and that this variation helps explain the range of mean particle sizes for observed secondary aerosols. A rigorous theoretical analysis which has been useful in analyzing the evolution of sulfur-containing aerosols formed by the homogeneous gas-phase reaction of SO_2 is also outlined.

*125 Mechanical Engineering, 111 Church St. S.E., University of Minnesota, Minneapolis, Minn. 55455.

GROWTH OF SECONDARY SULFUR AEROSOLS

Gas-phase sulfur species can be incorporated into particles by a variety of mechanisms. Homogeneous gas-phase oxidation of SO_2, for example, can lead to the formation of low-vapor-pressure species which condense on existing aerosol particles. Such condensable products can also form new particles (McMurry and Friedlander, 1979). Gas-to-particle conversion can also result from heterogeneous conversion of aerosol precursors such as SO_2 on or within existing aerosol particles or droplets. Theory predicts that secondary-aerosol mass formed by homogeneous chemical reactions tends to accumulate in smaller particles than that formed heterogeneously.

The size-distribution evolution of a growing secondary aerosol is determined primarily by the diameter dependence of particle growth rates (i.e., the growth law). Friedlander (1977) has discussed growth laws for secondary aerosols in some detail. Several theoretical growth laws which one might hypothesize for atmospheric aerosols are presented in Table 1. The diameter dependence of particle diameter

Table 1. Some simple theoretical growth laws

Mechanism	Volume growth rate	Diameter growth rate
Condensation of products from gas-phase reaction		
Free-molecule regime ($D_p \ll$ mean free path)	$\dfrac{dv}{dt} \sim D_p{}^2$	$\dfrac{dD_p}{dt} \sim D_p{}^0$
Continuum regime ($D_p \gg$ mean free path)	$\dfrac{dv}{dt} \sim D_p$	$\dfrac{dD_p}{dt} \sim D_p{}^{-1}$
Surface reaction	$\dfrac{dv}{dt} \sim D_p{}^2$	$\dfrac{dD_p}{dt} \sim D_p{}^0$
Droplet-phase reaction	$\dfrac{dv}{dt} \sim D_p{}^3$	$\dfrac{dD_p}{dt} \sim D_p + 1$

growth rates predicted by these theoretical growth laws is shown in Fig. 1. Note that for the case of condensational growth following gas-phase reactions, theory predicts that growth rates for particles large compared with the mean free path of air (0.07 μm) are negligible compared with growth rates for very small particles. As a result, secondary aerosols formed by this mechanism tend to accumulate in particles of about 0.15 μm, slightly larger than the mean free path of air. In contrast, diameter growth rates for particles growing by heterogeneous reactions within existing aerosol droplets increase with increasing diameter. Aerosols growing according to this growth law would tend to accumulate in larger aerosol particles.

Atmospheric data for aerosols growing by gas-to-particle conversion have been analyzed by McMurry and Wilson (1979) to determine the diameter dependence of particle growth rates. These investigators used data obtained with the University of

MODEL GROWTH LAWS

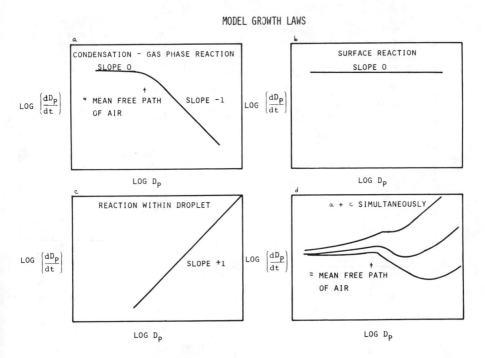

Fig. 1. Logarithmic plots showing the diameter dependence of particle diameter growth rates for several theoretical growth laws. (*a*) Applies to particle growth resulting from condensation of low-vapor-pressure products of gas-phase reactions. (*b*) Growth rates for particle growth from surface-limited heterogeneous chemical reactions. (*c*) Case of heterogeneous reactions which occur at a uniform rate throughout droplet volumes. If condensation and heterogeneous reactions were occurring simultaneously, the diameter dependence of particle growth rates would vary with particle diameter as shown in *d*.

Minnesota Mobile Laboratory in a variety of locations across the United States. Particle growth data were based on aerosol size distribution data measured with an electrical aerosol analyzer and an optical particle counter. By comparing observed growth rates with model growth laws such as those presented in Table 1 and Fig. 1, it was possible to infer possible mechanisms of aerosol formation for several data sets.

The diameter dependence of particle diameter growth rates for secondary-aerosol formation observed in several locations is shown in Fig. 2. Data obtained in the Great Smoky Mountains near Gatlinburg are shown in Fig. 2*a*. In this case, the diameter growth rate was approximately proportional to particle diameter, which is consistent with the hypothesis of particle growth by heterogeneous reactions of dissolved aerosol precursors within existing aerosol particles. The hypothesis that this new secondary aerosol contained sulfur is supported by data from a continuous aerosol sulfur monitor, which showed aerosol sulfur increasing simultaneously with submicron aerosol volume and the light scattering coefficient.

Particle growth rate data for smog-chamber aerosols measured by McMurry (1977) are shown in Fig. 2*b*. In these experiments, secondary sulfur aerosol was formed by photochemical reactions in the SO_2-NO_x-propylene system. Particle growth in this system was distinctly different from that observed in the Smokies

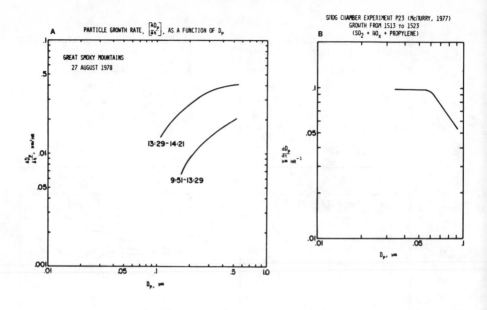

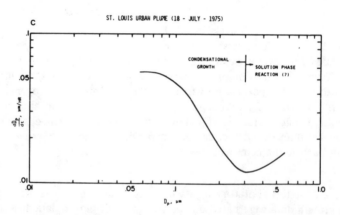

Fig.2 Logarithmic plots showing the diameter dependence of particle diameter growth rates from data for aerosols growing by gas-to-particle conversion. (*a*) Data obtained near Gatlinburg in the Great Smoky Mountains are consistent with the hypothesis of heterogeneous droplet-phase reactions (Fig. 1*c*). (*b*) In contrast, smog-chamber data for photochemical sulfur aerosol formation are consistent with the assumption of condensational growth (Fig. 1*a*), as would be expected if the secondary-aerosol products are initially formed in the gas phase. (*c*) Data obtained in the St. Louis urban plume. These data are qualitatively consistent with the assumption that secondary aerosols were formed simultaneously by condensation and droplet-phase reaction (see Fig. 1*d*).

(Fig. 2*a*). These data are consistent with the assumption of particle growth by condensation of low-vapor-pressure species from the gas phase.

Data obtained in the St. Louis urban plume (Fig. 2*c*) are qualitatively consistent with the assumption of particle growth by a combination of condensation (dominating small-particle growth) and heterogeneous chemical reactions. Whitby

(1979) concluded that most of the aerosol volume in this urban plume could be accounted for by the sulfur. This supports the argument that aerosol growth in Fig. 2c is dominated by sulfur aerosol formation.

The analysis of aerosol growth by McMurry and Wilson (1979) shows clearly that secondary aerosols grow in a variety of ways. A fundamental link between observed growth phenomena and emissions of aerosol precursors such as SO_2 requires a better understanding of atmospheric chemical reactions which lead to particle growth. In particular, this analysis suggests that heterogeneous chemical reactions of SO_2 are probably important and require more study.

DYNAMICS OF PHOTOCHEMICALLY FORMED SULFUR AEROSOLS

In the previous section, evidence was presented to show that secondary aerosols are formed in the atmosphere both by homogeneous gas-phase reactions and by heterogeneous reactions. In this section, a theoretical treatment which has been useful in predicting the dynamic behavior of aerosols formed by chemical reactions in the gas phase is outlined, and comparisons between theory and experiment are presented.

Condensable molecules formed by gas-phase reactions can deposit on existing aerosol particles or coagulate with other condensable molecules or molecular clusters. This process is shown schematically in Fig. 3. In general, once incorporated

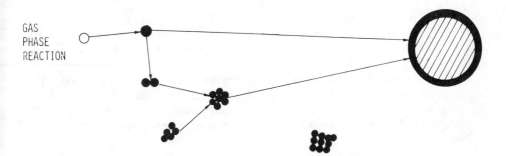

Fig. 3. Schematic diagram of aerosol formation when condensable molecules are formed by gas-phase reactions. In this case, products of gas-to-particle conversion can condense on existing aerosol particles or form new particles.

in a particle or molecular cluster, such molecules can reenter the vapor phase by evaporation. The problem of predicting the dynamics of such an aerosol is simplified dramatically if rates of molecular evaporation are sufficiently slow to be neglected. In this section it is shown that smog-chamber data for sulfur aerosols formed photochemically can be explained reasonably well if evaporation is neglected.

Using the numerical techniques developed by Gelbard (1979), the evolution of an aerosol size distribution growing at a constant volume rate by condensation and coagulation of low-vapor-pressure molecules formed in the gas phase has been solved theoretically. For the results presented in this paper, the system was assumed to be initially particle-free. Details of the dynamic equations for the evolving aerosol are discussed by McMurry and Friedlander (1979) and Gelbard (1979). In this paper, theoretical predictions are compared with data of Clark (1972).

Clark (1972) used a Whitby aerosol analyzer to measure size distributions of aerosols generated photochemically from a mixture of SO_2 in air. These experiments were conducted in a 17-m³ smog chamber, and in all cases the system was initially particle-free. Data were reported for rates of SO_2 photooxidation ranging from 6×10^5 to 2.6×10^7 molecules per cubic centimeter per second; these rates were varied by using a range of SO_2 concentrations. Clark's data are well suited for comparison with theoretical results because the rate of SO_2 removal was nearly constant.

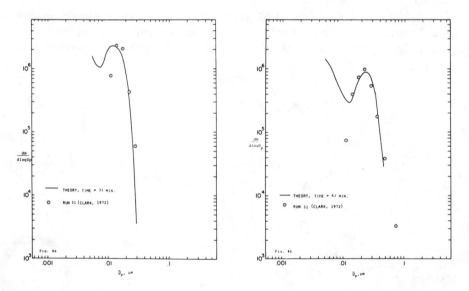

Fig. 4. Comparison of theory and experiments of Clark (1972) for size distributions of aerosols formed by photooxidation of SO_2 in a smog chamber. The two plots are for two different times after the start of gas-to-particle conversion. The rate of SO_2 photooxidation was 3.76×10^6 molecules per cubic centimeter per second.

Aerosol size distributions measured by Clark (1972) are compared with predictions of the theory outlined above in Fig. 4. Comparisons for times of 31 min and 62 min after the start of gas-to-particle conversion are presented. Agreement between experiment and theory is quite good.

In Fig. 5, the maximum aerosol number concentration for particles larger than 0.01 μm observed by Clark with a Whitby aerosol analyzer is shown as a function of the rate of SO_2 photooxidation. Data from all of the experiments reported by Clark are shown in this plot. The solid line is the theoretical result for this relationship based on experimental humidities and temperatures. Note that the theory tends to overpredict the maximum aerosol number concentration by about 50%, a relatively small error for this sort of analysis. Agreement between theory and experiment is worst at small reaction rates. This is expected, since the importance of the evaporation term in the theory and particle losses on the reactor walls both become increasingly important as reaction rates decrease.

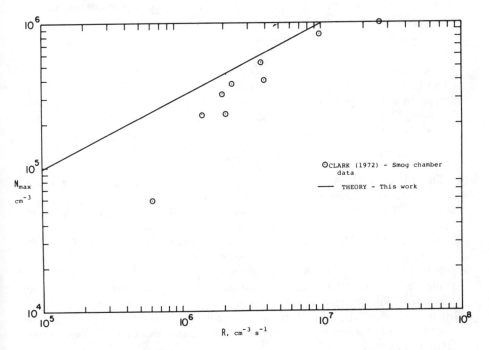

Fig. 5. Comparison of theory and data for the maximum number concentration of aerosol generated photochemically from SO_2 in an initially particle-free system, as a function of the rate of SO_2 photooxidation. Note that the theory tends to overpredict by about 50% except at low oxidation rates, where agreement is poorer.

The theory and data discussed above apply when aerosol formation is initiated in a particle-free system. In the atmosphere, an aerosol is always present. McMurry and Friedlander (1979) have shown that this preexisting aerosol has a profound effect on the dynamics of photochemically formed aerosols. Current work by the author on the dynamics of photochemically formed sulfate aerosols includes a rigorous analysis of the effects of this preexisting aerosol. The general theoretical approach outlined above, in which evaporation from particles or molecular clusters is ignored, is being applied in this analysis.

CONCLUSIONS

Field data for secondary atmospheric aerosol formation support the hypothesis that both homogeneous and heterogeneous chemical reaction mechanisms are important in converting SO_2 to aerosol sulfur. This conclusion is based on a limited data base obtained in a variety of locations across the United States.

A theoretical analysis which has successfully been applied to the analysis of photochemically formed smog-chamber aerosols has been discussed. The key simplifying assumption incorporated in the theory is that evaporation of condensed molecules from molecular clusters or particles is neligible. Agreement between experiment and theory supports the development of this approach for studying the behavior of photochemically formed sulfur aerosols in the atmosphere. This analysis will be useful in predicting ambient aerosol size distributions and diurnal number-concentration variations.

ACKNOWLEDGMENTS

This research was supported by EPA Grant R803851 and NSF Grant 78-05561. The contents of this paper do not necessarily reflect the views and policies of the Environmental Protection Agency or the National Science Foundation.

REFERENCES

Amdur, M. O. 1971. Aerosols formed by oxidation of sulfur dioxide—review of their toxicology. Arch. Environ. Health 23:459–468.

Cass, G. 1979. On the relationship between sulfate air quality and visibility with examples in Los Angeles. Atmos. Environ. 13:1069–1084.

Clark, W. E. 1972. Measurements of aerosols produced by the photochemical oxidation of SO_2 in air. Ph.D. thesis. Department of Mechanical Engineering, University of Minnesota, Minneapolis, Minn.

Friedlander, S. K. 1977. Smoke, Dust and Haze: Fundamentals of Aerosol Behavior. Wiley Interscience, New York.

Gelbard, F. 1979. The general dynamic equation for aerosols. Ph.D. thesis. Department of Chemical Engineering, California Institute of Technology, Pasadena, Calif.

Likens, G. E., R. F. Wright, J. N. Galloway, and T. J. Butler. 1979. Acid rain. Sci. Am. 241:43–51.

Lippmann, M. 1976. Size selective sampling for inhalation hazard evaluations. Pp. 287–310 in Liu, B. Y. H. (ed.), Fine Particles. Academic Press, New York.

McMurry, P. H. 1977. On the relationship between aerosol dynamics and the rate of gas-to-particle conversion. Ph.D. thesis. Department of Environmental Engineering Sciences. California Institute of Technology, Pasadena, Calif.

McMurry, P. H., and S. K. Friedlander. 1979. New particle formation in the presence of an aerosol. Atmos. Environ., in press.

McMurry, P. H., and J. C. Wilson. 1979. Growth laws for ambient aerosols. Paper presented at the ACS convention, September 10–14, Washington, D.C.

Trijonis, J. 1979. Visibility in the Southwest—an exploration of the historical data base. Atmos. Environ. 13:833–843.

Whitby, K. T. 1978. The physical characteristics of sulfur aerosols. Atmos. Environ. 12:131–160.

Whitby, K. T. 1979. Aerosol formation in urban plumes. Presented at New York Academy of Sciences conference on Aerosols: Anthropogenic and Natural—Sources and Transport, New York, N.Y.

White, W. H., and P. T. Roberts. 1977. On the nature and origins of visibility-reducing aerosols in the Los Angeles air basin. Atmos. Environ. 11:803–812.

DISCUSSION

John Nader, private consultant: Can the functions that you generated, relating rate of growth as a function of particle size to the mechanism, be applied to the gas stream of a source?

P. H. McMurry: Yes. In principle, the concept would apply to any case in which you have gas-to-particle conversion.

CHAPTER 20

MONTE CARLO SIMULATION OF DAILY SULFATE DISTRIBUTION IN THE EASTERN UNITED STATES: COMPARISON WITH SURE DATA AND VISIBILITY OBSERVATIONS

D. E. Patterson*

R. B. Husar*

C. Hakkarinen†

ABSTRACT

A long-range transport model is presented and applied to simulate daily sulfate concentrations over the eastern United States during August 1977. The model uses a Monte Carlo random-sampling technique to simulate horizontal dispersion as well as transformation and removal kinetics of individual sulfur emission quanta. On a daily basis, the spatial distribution of simulated SO_4^{2-} compares well with measured SO_4^{2-} on 12 days, with fair agreement on another 12 days. However, on 7 days there was a major disagreement. The measured sulfate data from the Sulfate Regional Experiment (SURE) and light extinction coefficients derived from routine visibility measurements show remarkable spatial and temporal agreement throughout the month. This suggests that the extensive visibility data base collected by the National Weather Service may be useful as a semiquantitative surrogate for sulfate measurements over the eastern United States.

INTRODUCTION

The lifetime of SO_2 in the atmosphere is normally about one day, corresponding to about 500 km transport distance from the source. Particulate SO_4^{2-}, however, typically resides in the atmosphere for three to five days and may be transported beyond 1000 km. Sulfate pollution thus transcends the scale of air quality control regions (AQCRs), states, and countries, making source identification difficult due to the superposition of multiple plumes of varying ages.

Diagnostic long-range transport modeling is an important tool for establishing the source identification of sulfur compounds. Such models utilize known S emission inventories, transport wind fields, and ambient SO_x concentration fields to estimate the rates of SO_2-to-SO_4^{2-} transformation and removal. The "best fit" rate constants between modeled and measured concentrations are inherently averages over all

*Center for Air Pollution Impact and Trend Analysis, Washington University, St. Louis, Mo. 63130.

†Electric Power Research Institute, Palo Alto, Calif. 94303.

sources and over the temporal-spatial scales of interest. This feature of models makes them attractive for regional-scale export-import budgeting (OECD, 1977) and for testing of control strategies. However, regional-scale models cannot predict high-resolution temporal or spatial scales such as diurnal or geographic variability in the rates. Nor can they normally obtain a good fit between *daily* modeled and measured concentration fields (e.g., OECD, 1977). Therefore regional models are forced into fitting monthly or seasonal average rates, which need not necessarily correspond to physically or chemically meaningful rate constants, due to spurious transport effects.

In the United States, the Sulfate Regional Experiment (SURE) conducted by the Electric Power Research Institute (EPRI) and Environmental Research and Technology, Inc., provides an excellent data base for testing long-range transport models with a temporal resolution of one day and spatial resolution of a few hundred kilometers. The purpose of this investigation is to compare the daily spatial patterns of SO_4^{2-} concentrations with those calculated by a simple Monte Carlo trajectory model and to examine the corresponding midday visibility contours as possible surrogates for sulfate fields.

DATA SOURCES AND DATA MANIPULATION

This study utilizes two data bases. The first is the daily average sulfate concentration at the 54 eastern U.S. SURE sites during August 1977 (Perhac, 1978), shown in Fig. 1a. The second data base consists of midday surface observations at about 200 National Weather Service sites in the eastern United States (Fig. 1b), which include visibility, relative humidity (RH), and surface wind direction (SWD) and speed (SWS). Visibility observations provide an estimate of haziness (the light extinction coefficient, b_{ext}) via the Koschmieder relationship, $b_{ext} = 24/visibility$, where visibility is in miles and b_{ext} has units of 10^{-4} m^{-1}. An RH correction is performed to normalize b_{ext} to 60% RH; the functional form will be described in a forthcoming paper.

The data for all reporting sites have been interpolated onto a 60 km × 60 km grid using an r^{-2} weighting: $V(ij) = \Sigma V(k) r^{-2}(kij)/\Sigma r^{-2}(kij)$, where $V(k)$ is the data of site k and $r(kij)$ is the distance from site k to grid point ij. About three to six sites influence each grid point. An additional smoothing is applied to the grid: (new value) = (4 × old value + sum of four surrounding grid values)/8. Finally, grid points lying outside the U.S. map boundaries are eliminated to avoid extrapolation problems.

The gridded data of all parameters except surface wind were used for contour plots and to define an eastern United States spatial average value, defined as the mean of all grids lying east of the north-south lines in Fig. 1a and b. Surface wind direction and speed were converted to eastward and northward wind speed components; each component was then gridded and smoothed independently. SURE sites were assigned to the location of the nearest National Weather Service site location for gridding calculations.

MODEL DESCRIPTION

Emissions

The distribution of quantized SO_x emissions used in this model application is shown in Fig. 1c. The density and position of dots on the map are in accordance with EPA's 1973 emission inventory by AQCR (EPA, 1976). With eastern United States annual SO_2 emission of 25 × 10^6 tons/year, each of the 204 emission quanta corresponds to 6.8 × 10^4 tons/year of SO_2, or 2.1 × 10^7 g of S per 3-hr time step. At

Fig. 1. Maps of data sources. (*a*) Location of SURE sulfate stations; solid vertical line is western boundary of computations for eastern U.S. average, and dashed lines delineate boundaries of region covered by SURE network. (*b*) Location of National Weather Service surface observation sites; solid vertical line is eastern U.S. average boundary. (*c*) SO$_2$ emission field used in model; each dot represents quantized annual emissions of 3.4×10^4 T of S per year, or 21 T S per 3-hr time step. (*d*) Snapshot of St. Louis, MO, plume on August 17, 1977. Model transport, including diffusion, is indicated by width of plume. Small dots are model SO$_2$, with SO$_4{}^{2-}$ denoted by larger dots.

pseudo steady state, continuous emission and transport winds cause about 5000 quanta to reside within the eastern United States.

Transport

Simulation of long-range transport requires accurate trajectory calculations. Unfortunately, the concept of "the trajectory" of an air parcel of any useful size during multiday transport has no physical realization, due to turbulent diffusion and wind shear and veer. We have chosen to use modified midday surface winds for horizontal transport. Pack et al. (1978) described trajectory comparisons in which modified surface winds predicted tetroon motion better than either geostrophic winds or measured upper-air winds. Comparison of the gridded surface-wind trajectories with upper-air trajectories by the method of Heffter et al. (1975) yielded the modification scheme WD = SWD + 20°; WS = SWS × 2.5. This surface-wind modification also provided best agreement to the observed motion of a large-scale hazy air mass during June 23–July 5, 1975. These modifications of surface winds to provide mean transport winds are not inconsistent with reports of Counihan (1975), Peterson (1966), Smith and Jeffrey (1975), and Maul (1978).

The spreading effect of vertical wind shear and veer in this Monte Carlo approach is simulated by superimposing a random perturbation on the position of the dots after each advecting time step. A pseudo diffusion coefficient, $K = 10^5$ m^2/sec, yielded sufficient dispersion to avoid excessive clustering of emissions downwind. This value of K is also in accordance with values used in long-range transport (Munn and Bolin, 1971). The diffusion is imposed as a displacement with random direction onto a circle of radius $\sqrt{2K \Delta t}$. With $\Delta t = 3$ hr, the displacement at each time step is 46 km, corresponding to a 4.3-m/sec wind with random orientation superimposed on the mean transport wind. The effect of this "diffusion" is illustrated in Fig. 1d.

Transport beyond the United States map boundaries ends the time history of a dot's motion, due to insufficient wind field information. Therefore no external inflow or recirculation from the Atlantic or Canada is permitted. However, transport within the United States west of the region and recirculation eastward is possible.

Transformation and Removal Kinetics

Three kinetic steps are imposed on the emissions after the transport calculations: SO_2 deposition, SO_2-to-SO_4^{2-} transformation, and SO_4^{2-} deposition. The kinetics are applied in 24-hr time steps. Rate constants k (in fraction per hour) correspond to stochastic decisions for each dot at each time: the probability of a given kinetic action $P = 1 - \exp(-k \Delta t)$. To avoid bias, SO_2 experiences the chance to be removed or converted to SO_4^{2-} in randomly chosen order. The Monte Carlo technique permits general variation of rates (i.e., probabilities), so that episodic wet removal and parameterized conversion can be readily included. In this application, transformation = 1%/hr, SO_2 removal = 2%/hr, and SO_4^{2-} removal = 1%/hr.

The separation of the transport and kinetic steps permits examination of multiple kinetic scenarios with minimal computational time. The resulting model calculations maintain the complete physical-chemical history for each dot at each step: origin, position, age, chemical form (SO_2 or SO_4^{2-}), and date (deposition or transport beyond borders). Therefore the model is capable of isolating both the regional impact of any given source area and the array of sources which impact on a given receptor area at any time.

MODEL APPLICATION TO AUGUST 1977

Daily Maps

Daily maps of measured and calculated sulfate concentration fields were prepared for August 1977, which was the first intensive sampling period of the SURE program. Figures 2–6 depict the daily SO_4^{2-} concentration (first column), the light extinction coefficient normalized to 60% RH (second column), the spatial distribution of quantized emission units (third column), and the interpolated surface wind field and sea-level pressure field (last column). Sulfate contours refer to daily averages, but all other parameters arise from local noon data. The simulation of the sulfate concentration field was initiated on July 26 to achieve a quasi steady state by August 1.

On the first day of the sampling program, sulfate levels above 10 μg/m^3 were observed only on the eastern seaboard, with New England experiencing 15–20 μg/m^3.

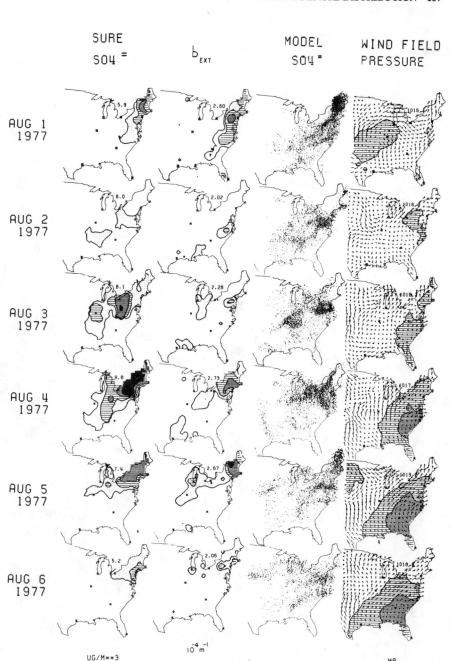

Fig. 2. Daily maps of 24-hr average SURE SO$_4^{2-}$ (first column), noon b_{ext} corrected to 60% RH (2nd column), modeled noon distribution of emitted sulfur quanta (3rd column), and unmodified noon surface wind field overlaid with the sea-level pressure (last column) for August 1-6, 1977. The stagnation over Illinois-Indiana on 2nd and 3rd caused high sulfate concentrations; the hazy air mass was transported to Canada and the Atlantic by August 6.

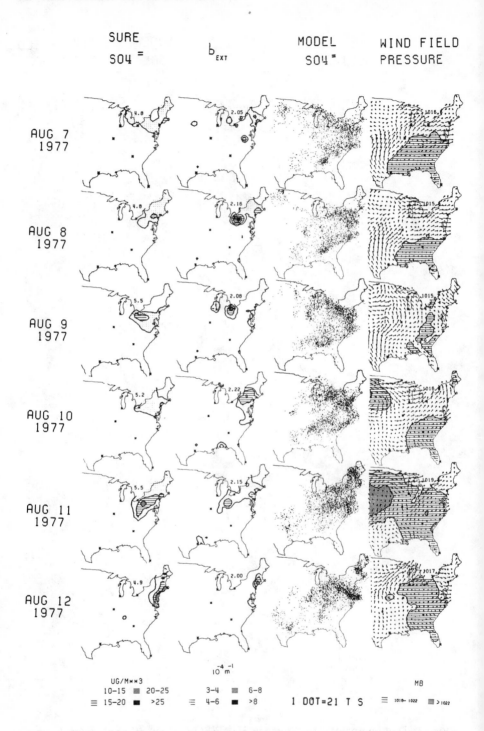

Fig. 3. As in Fig. 2, for the period August 7–12, 1977. Good regional ventilation on the back side of the high-pressure system centered in the southeast prevented buildup of sulfate until the 9th, when pockets of high concentration formed over the upper Ohio River valley. Model fields are in general agreement with measured sulfate and haze.

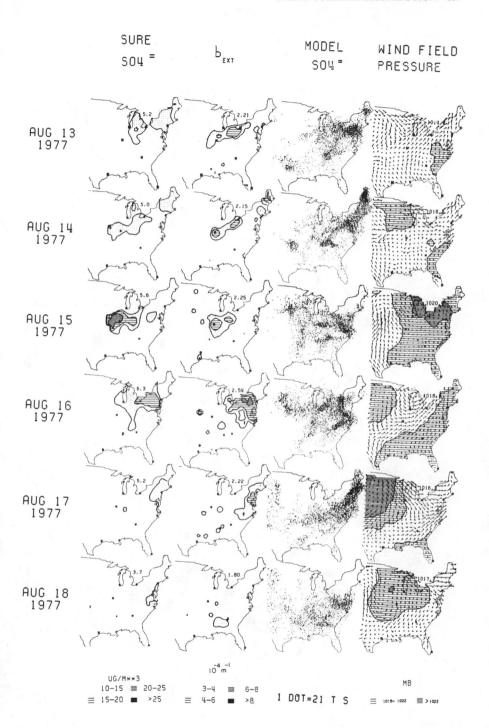

Fig. 4. On August 14–15, stagnation over Missouri and Illinois led to large areas of high sulfate and low visibility; the modeled field poorly matches the measured values. From August 16 to 18, the advancing Canadian front cleared the eastern United States.

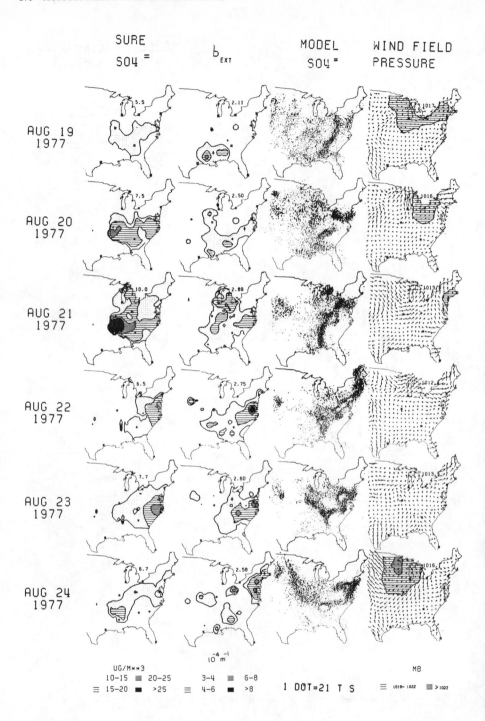

Fig. 5. From the 19th to the 21st, stagnation trapped most of the emissions within the U.S.; eastern U.S. average sulfate rose from 5.5 to 10 μg/m³. The model does not show the polluted area in South Carolina and Georgia which had recirculated from the Atlantic. During August 22–24, poor ventilation led to accumulation of emissions again.

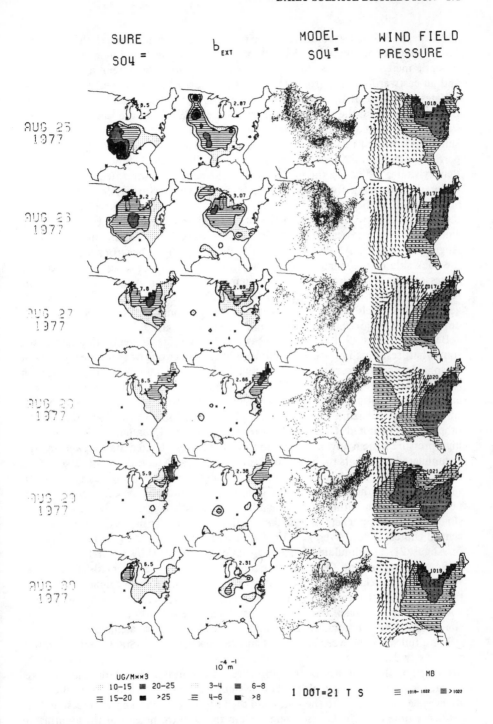

Fig. 6. The polluted air mass of the 25th is transported north and east on the back side of the stationary high centered over the eastern seaboard. On the 29th to 30th, poor ventilation in Illinois and Indiana was reflected in increasing sulfate, haziness, and simulated sulfate density.

The haze, with noon visibility less than 6 miles, was also confined to the seaboard. The simulated concentration field was most dense in the New England states. The synoptic meteorology was characterized by a high-pressure system in the Midwest and an eastward-moving front. On August 1, most of the observed and simulated sulfates were east of the cold front. By August 2, the front had passed over New England, sweeping the sulfates and haze to the Atlantic. In the same time period, stagnation centered around Illinois and Indiana resulted in the accumulation of emissions on August 2 and 3. By August 3, sulfate levels exceeded 20 $\mu g/m^3$ in large parts of Indiana and Ohio. The simulated high sulfate levels, however, appeared in Pennsylvania, indicating the inadequacy of the simulated wind field.

For the following five days, August 4–8, a high-pressure system resided over South Carolina, Georgia, and Alabama, causing stagnation within those states and the familiar clockwise circulation around the high. Consequently, the accumulated emissions in the Midwest were rapidly transported to New England and Canada on August 4 and to the Atlantic on August 5. By August 6–8, the strong ventilation on the back side of the stagnating high-pressure area reduced average eastern United States sulfate concentration below 5 $\mu g/m^3$ (from 9.8 on August 4), with a corresponding increase in mean visibility from 8 to 12 miles. The simulated SO_4^{2-} shows generally low levels between August 5 and 8, with some increase in concentration downwind of major sources, i.e., at the eastern seaboard.

Between August 9 and 13, the eastern United States ventilation decreased somewhat, resulting in pockets of sulfate and haze over one or two state areas, with reasonable spatial correspondence between simulated and measured levels. On August 14–15, stagnation over Missouri, Illinois, and Indiana resulted in an accumulation of sulfate and haze. For August 15, a spatial mismatch between SURE sulfate data and haziness on the one hand and the Monte Carlo simulation on the other should be noted. On August 16–18, a fast-moving Canadian front driven by a high-pressure system advanced over most of the eastern United States, sweeping the accumulated pollutants to the east and southeast to the Atlantic. The observed SO_4^{2-} levels, haziness, and Monte Carlo simulation all show the clearing of the region on the 18th. During the next three days, August 19–21, most of the emissions are trapped by stagnation within the boundaries of the United States. The Monte Carlo simulation shows pockets of emissions with increasing density drifting in the east-central United States. Average sulfate increased from 5.5 on August 19 to 10 $\mu g/m^3$ on August 21. The simulated concentrations on August 21 show a sulfate-free region in South Carolina and Georgia, in marked contrast to the observed haziness and sulfate data. In searching for the cause of this discrepancy, visible images of geostationary satellite pictures reveal a haze pall over the Atlantic, which, according to the wind field, drifted back toward the coastal states. Evidently the pall had left the simulation region some days before and was lost to further calculations.

For a full week, between August 19 and 26, eastern United States ventilation was poor, as evidenced by the low wind speeds from August 19 to 23, and was further decreased by the approaching high-pressure system on the 24th, 25th, and 26th. As usual, such approaching high-pressure systems move the emissions from the Ohio River valley region first to the south and then to the west, markedly increasing their residence time over the United States. From the 26th to the end of the month, the high-pressure system became stationary over the eastern seaboard, resulting in the swift northeasterly transport of the emissions accumulated in the Midwestern states. By August 30, the average sulfate level declined to about 6 $\mu g/m^3$ from the high of 9.2 $\mu g/m^3$ on August 26.

In the above analysis, extensive use has been made of daily weather maps and of visible geostationary satellite images. The comparison of measured sulfate, observed haziness, and simulated sulfate reveals an unexpectedly good correspondence between daily maps of sulfate and visibility reduction during August 1977. The spatial correspondence between measured and simulated SO_4^{2-} can only be considered as fair. Major synoptic events, such as large-scale stagnations or strong regional ventilation on the "back side" of high-pressure systems, were adequately simulated.

On 12 out of 31 days, "good" agreement is shown, in that calculated and measured high-concentration regions were within about 200 km of each other. Another 12 days show "fair" agreement, when the measured and simulated concentrations were within the same general area of the eastern United States but deviated by 300–500 km. The agreement is characterized as "poor" on 7 (23%) of the days, when high SO_4^{2-} levels appeared in either the observed or simulated fields without an appropriate match in the other set. Intuitively, the cause of spatial mismatch on poorly matching days is primarily attributed to improper wind fields used in the model. However, episodic wet removal or large-scale elevation of near-surface layers to above the mixing layer could also contribute to the mismatch. It is therefore evident that the transport module of this Monte Carlo model is marginally satisfactory. It would be desirable to know if other long-range transport models of the eastern United States (Henmi, 1979; Nieman et al., 1979; Shieh, 1977; Shannon, 1979; Meyers et al., 1979; Powell et al., 1979), utilizing different wind fields and computational procedures, also show similar discrepancies with daily SURE data.

Monthly Average Maps

The August 1977 monthly average concentration fields (Fig. 7) reveal that the highest measured SO_4^{2-} levels were in Pennsylvania, averaging more than 12 $\mu g/m^3$; the northeastern quadrant east of Missouri and north of Georgia shows more than 8 $\mu g/m^3$. With regard to sulfate maps, it should be noted that the contours are only meaningful within the SURE region. The lowest average visibility (less than 8.5 miles) is noted for Ohio, Pennsylvania, and the Boston-Washington megalopolis. Possibly, in the eastern megalopolis region, compounds other than sulfate contribute appreciably to visibility degradation. The average visibility is less than 10 miles in the sunbelt states of Louisiana, Mississippi, and Alabama, but the lack of SURE stations there prevents comparisons of SO_4^{2-} with b_{ext}.

The monthly average simulated SO_4^{2-} is shown in Fig. 7c and d. The superimposed daily sulfate quanta (21 tons of S per dot), Fig. 7c, show the highest density around Pennsylvania, with a smaller high-density region occurring in Tennessee and northern Alabama, as seen in the dot density contour map of Fig. 7d. The measured and simulated sulfate densities show two discrepancies. The simulated SO_4^{2-} pocket in Tennessee does not appear in the measured data, which may be due either to inadequacies of the model or to the incomplete spatial coverage of the SURE data base. Second, the SURE data base indicates only a factor of 2 variation of SO_4^{2-} (6–12 $\mu g/m^3$) within the region, while simulated contours predict a factor of 5–10 variation. The cause of this discrepancy could probably be attributed to model inadequacies: emission inventory and transport wind fields. It should be noted that other model predictions of the same time period as part of Project MAP3S (MacCracken, 1979) also show a good match to SURE data in Pennsylvania but, like the current Monte Carlo model, substantially underestimate the concentrations at the eastern and southern boundaries of the region. The density of sulfate quanta shown in Fig. 7d is in units of grams per square meter, i.e., vertical integral of S concentration. In order to convert those values to surface concentrations comparable with the SURE data, an assumption is required regarding the characteristic height of the sulfate layer. The characteristic height is the height at which the total vertical integral of SO_4^{2-} is accounted for, assuming uniform concentration equal to that at the surface.

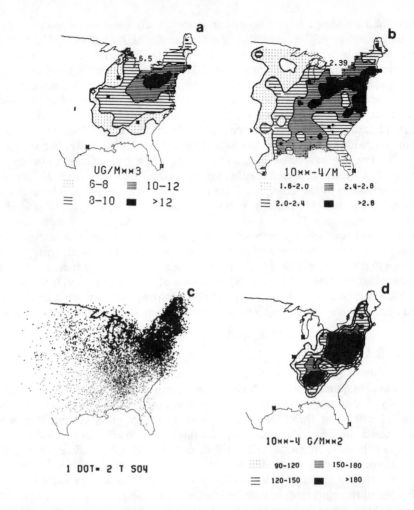

Fig. 7. August 1977 average distributions. (a) SO_4^{2-} as measured in the SURE project; (b) light extinction coefficient, b_{ext}, from visibility observations; (c) sulfate quanta from model; (d) gridded model SO_4^{2-}. If the characteristic height were 1500 m, model contours should be equivalent to SURE SO_4^{2-}. Apparently the model overestimates the high-sulfate regions and underestimates more remote sulfate concentrations, in part due to emission grid used.

Daily Variation of Eastern United States Averages

Another approach for the testing of "correlation" between SO_4^{2-}, b_{ext}, and simulated SO_4^{2-} is to examine their eastern United States averages from day to day, hence ignoring the spatial distribution. The daily variation of these quantities (Fig. 8) shows that in August 1977, sulfate varies between 3.7 and 10 $\mu g/m^3$, light extinction coefficient between 1.8 (13.3 miles) and 3.1 $\times 10^{-4}$ m^{-1} (7.7 miles). The covariation of SO_4^{2-} and b_{ext} averages is rather striking. The least-squares regression equation is $b_{ext} = 0.17 \times SO_4^{2-} + 1.3$, with correlation coefficient $r = 0.84$. The offset of $b_{ext} = 1.3$ is in part due to the fact that visibilities greater than 12–20 miles are not resolved over the eastern United States. The correlation coefficient of 0.84 implies

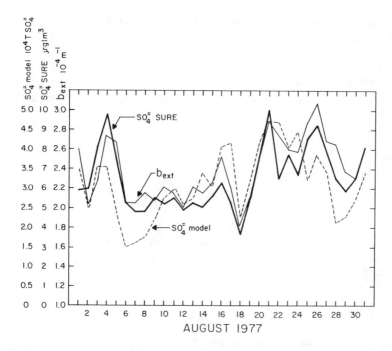

Fig. 8. Daily eastern U.S. averages for SURE SO$_4$$^{2-}$ and b_{ext}, with total eastern U.S. sulfate loading predicted in the simulation. If characteristic height were 1500 m, modeled and measured SO$_4$$^{2-}$ scales would be equivalent over the eastern U.S. area of 3.5 × 10^{12} m^2. The SURE SO$_4$$^{2-}$ and b_{ext} are highly correlated (r = 0.84). Simulated sulfate reflects the major synoptic events but exhibits major discrepancies on August 6–8, 16–17, and 28–29. The SURE model correlation is 0.55.

that 71% of the b_{ext} variance could be explained by SO$_4$$^{2-}$. It is important to recognize, however, that such a high correlation in this context does not establish a cause-effect relationship: when SO$_4$$^{2-}$ is high over the eastern United States, then it is likely that all other air pollutants are also accumulated, and their contribution is statistically incorporated into the apparent effect of sulfate.

The Monte Carlo simulated daily average sulfate concentration shows a pattern similar to those of measured SO$_4$$^{2-}$ and b_{ext}. The key features of the daily average sulfate variations are reflected in the simulation: the accumulation and clearing from August 2 to 6, the strong ventilation on August 18, followed by the accumulation between August 18 and 21. However, there are also strong deviations between the predicted and observed sulfate levels on August 6–8, on August 16–17, and on August 28–29. The cause of these discrepancies is not clear. The units of simulated sulfate levels in Fig. 8 are vertical integral of SO$_4$$^{2-}$ over the eastern United States. Some of the above variability could be due to day-to-day changes in mixing height.

SUMMARY

Data from the Sulfate Regional Experiment (SURE) provide an excellent data base for performance testing of regional-scale transport-transformation-removal models with a time resolution of one day. A long-range transport model is reported which utilizes a Monte Carlo or random-sampling technique to simulate horizontal dispersion and transformation and removal kinetics of individual SO_2 and SO_4^{2-} quanta. Over the month of August 1977, 12 days show good agreement (within 200 km) in the position of measured and simulated SO_4^{2-} palls; on 7 days there was severe mismatch, in that high-concentration palls existed in one set but not in the other. Another 12 days showed fair agreement, with measured and modeled palls within 300–500 km.

The spatial distribution and the daily variation of light extinction coefficient, obtained from routine airport visibility observations, and sulfate from SURE show remarkable correspondence. This suggests that the extensive visibility data base can serve as a semiquantitative surrogate for sulfate over the northeastern United States.

ACKNOWLEDGMENTS

This work was supported by EPA Grant No. R806606, William E. Wilson, Jr., Project Officer. We wish to thank Ms. Jan Holloway of CAPITA for preparing the maps and assisting throughout. The kind permission of EPRI to obtain and use the SURE data is gratefully acknowledged.

REFERENCES

Counihan, J. 1975. Adiabatic atmospheric boundary layers: a review and analysis of data from the period 1880–1972. Atmos. Environ. 9:871–905.

EPA. 1976. National Emissions Report (1973): National Emissions Data System (NEDS), EPA-450/2-76/007. U.S. Environmental Protection Agency, Research Triangle Park, N.C.

Heffter, J. L., A. D. Taylor, and G. J. Ferbert. 1975. A regional-continental scale transport, diffusion and deposition model. NOAA Tech. Memo. ERL ARL-50.

Henmi, T. 1979. Long-range transport model of SO_2 and sulphate and its application to the eastern United States. WMO Symposium on the Long Range Transport of Pollutants, WMO-No. 538. 321–328.

MacCracken, M. C. 1979. The multistate atmospheric power production pollution study—MAP3S. DOE/EV-0040 UC-11 UC-90. NTIS.

Maul, P. R. 1978. The effect of the turning of the wind vector with height on the ground level trajectory of a diffusing cloud. Atmos. Environ. 12:1045–1050.

Meyers, R. E., R. T. Cederwall, and J. A. Storch. 1979. Modeling sulfur oxide concentrations in the eastern United States: model sensitivity, verification and applications. In Preprints, Fourth Symposium on Turbulence, Diffusion and Air Pollution, American Meteorological Society, Boston. 673–674.

Munn, R. E., and B. Bolin. 1971. Global air pollution—meteorological aspects: a survey. Atmos. Environ. 5:363–402.

Nieman, B., A. Hirata, and L. Smith. 1979. Application of a regional transport model to the simulation of regional sulphate episodes over the eastern United States and Canada. WMO Symposium on the Long Range Transport of Pollutants, WMO-No. 538, 337–346.

OECD. 1977. The OECD program on long-range transport of air pollutants—measurements and findings. Organization for Economic Cooperation and Development, Paris.

Pack, D. H., G. J. Ferber, J. L. Heffter, K. Telegadas, J. K. Angell, W. H. Hoecker, and L. Mochta. 1978. Meteorology of long-range transport. Atmos. Environ. 12:425–444.

Perhac, R. M. 1978. Sulfate regional experiment in northeastern United States: the "SURE" program. Atmos. Environ. 12:641–647.

Peterson, K. R. 1966. Estimating low-level tetroon trajectories. J. Appl. Meteorol. 5:553–564.

Powell, D. C., D. J. McNaughton, L. L. Wendell, and R. Drake. 1979. A variable trajectory model for regional assessments of air pollution from sulfur compounds. Pacific Northwest Laboratory report PNL-2734.

Shannon, J. D. 1979. The advanced statistical trajectory regional model. In Proceedings, Fourth Symposium on Turbulence, Diffusion and Air Pollution, American Meteorological Society, Boston, 376–380.

Shieh, C. M. 1977. Application of statistical trajectory model to the simulation of sulfur pollution over northeastern United States. Atmos. Environ. 11:173–178.

Smith, F. B., and G. H. Jeffrey. 1975. Airborne transport of sulfur dioxide from the U.K. Atmos. Environ. 9:643–659.

DISCUSSION

R. D. S. Stevens, United Technology and Science, Inc., Downsview, Ontario: Have you looked at the relationship between haziness and nitrate?

D. E. Patterson: We have not, mostly because we do not have any good nitrate data. A number of people have done these correlations and found that sulfate shows a much stronger statistical correlation than nitrate. This may be due, in part, to problems associated with measuring ambient nitrate accurately.

SECTION V

EFFLUENT DELIVERY AND AIR MASS/LANDSCAPE INTERACTIONS

Co-Chairmen: S. E. Lindberg and R. P. Hosker

CHAPTER 21

EFFLUENT DELIVERY AND AIR MASS/LANDSCAPE INTERACTIONS

Chairmen: **Steven E. Lindberg**
Oak Ridge National Laboratory

R. P. Hosker
National Oceanic and Atmospheric Administration

OVERVIEW

Steven E. Lindberg*

R. P. Hosker†

The critical links between increased atmospheric emissions and their effects are the rates and mechanisms whereby atmospheric gases and particles are transported to biological, land, or water surfaces and made available to receptor organisms. Vegetation is a particularly important sink for atmospheric emissions. Because of the reactivity of their large surface area, vegetative canopies serve as very effective receptors for airborne materials delivered by both wet-fall and dry-deposition processes. It is important to recognize that before any effects occur, the airborne substances must somehow move from the surrounding envelope of air through a receptor boundary layer to the ultimate surface of interaction. Biological activity of the receptor may be very significant.

This session highlighted recent research on the interface between the transport medium and the ultimate receptor, with discussions of the physical and chemical aspects of wet and dry deposition to biological, land, and water surfaces. The authors emphasized recent results, interpretations, and experimental approaches relevant to wet and dry deposition of sulfur, its associated acidity, and trace elements.

Several authors addressed the general topic of dry deposition of gases and particles to soil and vegetation surfaces. Chamberlain summarized recent work on dry deposition of sulfur dioxide and described the advantages and drawbacks of various

---

*Environmental Sciences Division, Oak Ridge National Laboratory, P.O. Box X, Oak Ridge, Tennessee 37830.

†Air Resources Atmospheric Turbulence and Diffusion Laboratory, National Oceanic and Atmospheric Administration, 465 S. Illinois Avenue, Oak Ridge, Tennessee 37830

experimental techniques. Crop sulfur balance, total sulfur budget of the atmosphere, and the combination of atmospheric boundary-layer theory and observations of the gradients of meteorological variables are all viable approaches. In general, the resultant estimates of SO_2 dry deposition velocity fall in the range of 0.4 to 0.8 cm/sec, regardless of surface type, with a few exceptions. Values of 1 to 2 cm/sec can occur over luxuriant vegetation or high-pH damp soils, while still higher values, up to 8 cm/sec, may appear for a thoroughly wetted forest canopy. On the other hand, deposition velocities smaller by an order of magnitude or more may be found over dead vegetation or dry acid soil.

Hicks and Wesely discussed eddy flux measurements of deposition to a pine canopy with regard to the influence of diurnal cycles on turbulent transfer, surface properties, and pollutant fluxes. Deposition velocities varied widely during 24-hr cycles, so that surface fluxes of some pollutants ranged from negative to positive, with a value of zero not uncommon. Ozone flux was influenced by surface properties to a larger extent than sulfur or particle fluxes, although determination of nocturnal resistances for particles was complicated by the behavior of the canopy as both a sink and a source, as previously reported by these authors for sulfur.

The status of current experimental techniques was also addressed by Droppo, who emphasized studies in the atmosphere's constant-flux layer. Limitations of the flux-gradient and direct eddy flux measurement methods were described. Despite these, recent work has been able to show that particle mass transfer processes over natural surfaces cannot be considered in terms of a simple one-directional, single-rate flux. In particular, there is evidence that vegetation may sometimes serve as a source, rather than sink, for elements such as S, Ca, Fe, and occasionally Si. Recommendations for improvements in observation techniques are based on these results.

Sehmel presented a compilation of the results of more than 50 field experiments on both gas and particle deposition, carefully pointing out the differences due to particle size effects and gas speciation. Particle deposition velocities ranged over a factor of 10^5 from a minimum of 10^{-3} cm/sec, were primarily influenced by particle size, but were independent of atmospheric stability. Similarly, measured deposition velocities for major gases ranged over a factor of 10^5 from a minimum value of 10^{-4} cm/sec, with the range in deposition velocity for any given gas being at least an order of magnitude. These data were used to illustrate the need for development and application of predictive deposition velocity models.

Although there have been several field experiments aimed at direct measurement of deposition fluxes to surfaces in the terrestrial environment, there is a great need for similar determination of fluxes of particulate pollutants to natural water surfaces, according to the discussion by Hicks and Williams. The processes influencing pollutant exchange at the air-sea interface were reviewed in the context of developing a method suitable for measuring fluxes above water surfaces. Because of the complexities of the system, particularly those related to wave-induced continuous changes in the physical surface, this process is exceedingly difficult to both measure and model. The authors presented an argument for the use of eddy correlation, rather than surface accumulation or gradient methods based on their current experiments on particle and gas fluxes over Lake Michigan.

Two papers in this session dealt with some aspect of the wet-deposition phenomenon. Gatz documented the nature of the urban influence on pollutant deposition in order to identify sources and scavenging mechanisms. Urban emissions increased sulfur deposition magnitude and variability in the affected areas. The author concluded that sulfate wet deposition could be largely ascribed to nucleation scavenging of aerosol sulfate and not in-cloud scavenging of SO_2. Sulfate and the water-soluble fractions of Zn, Cd, and Pb exhibited similar deposition patterns, while dissimilar patterns occurred for particulate trace metals and both soluble and particulate soil elements—an observation with interesting implications for source and scavenging mechanism identification.

Dana presented an overview of wet deposition and scavenging. Methods for describing simple aerosol or gas interactions with precipitation are fairly well developed. Reasonable predictions of removal may be made if the pollutant atmospheric concentration, particle and/or hydrometeor size distributions, and gas properties such as solubility are known. Scavenging processes involving attachment, condensation, and cloud drop nucleation are less well understood, and modeling efforts are limited because current supporting measurements are inadequate. The utility of popular parameters such as scavenging coefficients and scavenging ratios depends on the pollutant in question and the complexity of the operative removal mechanisms.

A review of several Gaussian-plume-based transport and dry-deposition models was presented by Horst. The standard of comparison was the author's own surface-depletion model, which simulates effluent removal from the lower portions of a plume by a negative area source distributed over the ground surface. This results in a physically plausible gradual redistribution of material in the vertical as the pollutant moves downwind. The simpler Chamberlain source-depletion and the Csanady-Overcamp particle-reflection models were compared with this standard. A modified source-depletion model was introduced and shown to agree with the more complicated surface-depletion model. Unfortunately, data are unavailable to test these models adequately, particularly at large distances. Field experiments are essential to further model improvement; the present models suggest the variables which should be determined and reported.

The lack of a complete understanding of complex deposition processes was reflected in the consensus regarding the need for much more research in this area. For eddy flux measurement of sulfur transfer, sensors must be developed to differentiate between particulate and gaseous sulfur. Considerable work is also needed to define the processes which control particle deposition during diurnal cycles. The phenomenon of net upward fluxes of particles from vegetation requires considerable study to define its influence on measurements or calculations of longer-term pollutant fluxes to canopies. We need a better understanding of the relation between deposition fluxes measured using inert and biological surfaces in the field and those determined by eddy correlation methods, particularly in view of the possible existence of an "internal cycle" of material released from the canopy to the atmosphere and later redeposited on the canopy. In addition, there is a need for continued research on the role of the water surface as a particle source as well as on its role in controlling particle growth near the air-sea interface.

Application of either the gradient or eddy correlation methods to determination of surface fluxes will require an improvement in present-day analytical techniques to allow the resolution of concentration variations on the order of 1%. Future mechanistic studies perhaps should emphasize the eddy correlation technique because of its inherent noise-rejection features, rather than the more analytically demanding gradient method. Micrometeorological techniques may prove uneconomical for application in a monitoring mode. However, the present monitoring methods utilizing inert surrogate surfaces involve questionable assumptions. Research is needed to calibrate such surrogate surfaces against vegetation under both controlled and field conditions. The varying sensitivities of receptors to different forms of incident pollutants must also be examined.

The variability of results may be reduced if guidelines and limitations on experimental methods are more closely observed. Future studies should refer to past experience and the few available models for predictions of deposition velocity to determine micrometeorological and receptor surface variables and conditions which may influence the results. Results must be reported in detail.

Improvement of long-range transport and diffusion models largely awaits field data for comparison with existing models. Vertical as well as horizontal scans of plume depletion as functions of distance are needed. Such data will require new and undoubtedly expensive field studies.

Regarding wet deposition, the observation of behavioral similarities among several elements has interesting implications related to scavenging mechanisms and indicates the need for further study of the relationships between sulfur deposition and trace-metal deposition. There is presently a critical lack of data from large-scale field studies of dry- and wet-fall-only deposition rates of both organic and inorganic pollutants based on procedures specifically designed for contamination-free collection of trace constituents on an event basis. Information on the growth and removal of small particles (less than 0.1 μm) during precipitation is sparse. Nucleation of cloud droplets and their subsequent growth appears to be an efficient mechanism for removal of aerosols, but predictive capability and confirming measurements are largely unavailable. Modeling of wet scavenging of gases must proceed on a case-by-case basis, considering the solubility and reactivity of the pollutant. Mutual interference or reactions among multiple contaminants may affect these processes, but the mechanisms are poorly understood.

CHAPTER 22

DRY DEPOSITION OF SULFUR DIOXIDE

A. C. Chamberlain*

ABSTRACT

Measurements of sulfur dioxide (SO$_2$) deposition are reviewed to provide estimates of dry deposition velocity under various conditions. The methods of study are broken into several major categories; data obtained by crop sulfur balances (budgets), atmospheric sulfur budgets, and direct gradient measurements are emphasized. The special cases of SO$_2$ deposition to forests, snow, and bare soil are also described, and a few exceptions to the results are noted.

INTRODUCTION

The wet and dry deposition of oxides of sulfur was originally studied in relation to plant nutrition. Alway (1940) noted regions of sulfur deficiency in the Pacific Northwest, western Canada, and northern Minnesota, where winds from industrial sources seldom blow, and suggested that part of the sulfur requirements of plants might be supplied by direct absorption of sulfur dioxide by leaves. Bromfield and Williams (1974) state that sulfur deposited in dust-laden Harmattan winds in northern Nigeria is a useful source of sulfur to crops germinating at the start of the rainy season.

To assess the contribution of foliar absorption to the sulfur balance of crops, Thomas et al. (1943) grew plots of alfalfa on silica sand with added nutrients in a glasshouse. The plots were fumigated for an average of 900 hr with 0.1 ppm (280 μg m^{-3}) of SO$_2$. At harvest, the amount of sulfur in the crop was measured. The amount taken up by the roots was estimated from the known amounts supplied to the nutrient medium, and the balance was attributed to uptake from the atmosphere.

The amounts absorbed from the air were found not to be strongly related to pH or strength of the nutrient media, and when related to the area of the plots, fumigation time, and concentration of SO$_2$, the average rate of uptake corresponded to a velocity of deposition of 1 cm sec^{-1}. This work can be considered one of the starting points of the study of dry deposition of SO$_2$.

Another starting point is the paper of Meetham (1950) in which he attempted to calculate the amounts of smoke and SO$_2$ emitted and deposited in Great Britain. Meetham only had an indirect estimate of the effective height of the cloud of SO$_2$, which he needed to calculate the amount blown to sea. Taking this height as 280 m, Meetham calculated that 78% of the SO$_2$ emitted in Britain was deposited as wet and

---

*Atomic Energy Research Establishment, Harwell, Oxfordshire, England. OXII ORA

dry deposition. Recently Garland (1977) estimated an effective height of 1200 m from the results of captive balloon flights and calculated 30% deposition and 70% blown to sea. The mean effective height probably has increased in the last 30 years, as more SO_2 has been emitted from high stacks, but this increase is unlikely to account for the difference between the two estimates.

Meetham (1954) also considered the balance of SO_2 during the London fog of 5–9 December 1952, when stagnant air capped by an inversion at 150 m height covered London. From the estimated rate of emission of SO_2 and the concentrations reached (averaging 2.1 mg m^{-3} of SO_2), Meetham calculated that the mean life of an SO_2 molecule in the fog was 6 hr. Chamberlain (1960) showed that Meetham's estimates of dry deposition corresponded to velocity of deposition (v_d) values of 1.8 cm sec^{-1} over Britain as a whole and 0.7 cm sec^{-1} in London during the fog, and he proposed that experiments with ^{35}S should be done to measure the rate of uptake of SO_2 on various surfaces and in various conditions.

This was done at Harwell by Spedding (1969), who exposed barley leaves and also metallic and other surfaces to $^{35}SO_2$ and considered the mechanisms of deposition. For this purpose it is convenient to work in terms of resistances (Monteith, 1973). For a single leaf the total resistance is given by

$$r_t = \tfrac{1}{2} v_d^{-1} = r_a + r_s,$$

where r_a and r_s are the aerodynamic and surface (or stomatal) resistances. The factor $\tfrac{1}{2}$ enters because resistances are conventionally calculated in terms of the flux per unit plan area of leaf, whereas v_d refers to the area of both sides of the leaf. For canopies, both v_d and r_t are related to unit ground area, and r_t can be split into three components:

$$r_t = r_a + r_b + r_c$$

(Garland, 1977; Fowler and Unsworth, 1979). Here r_a is the aerodynamic resistance (equal to u/u_*^2 in neutral conditions), r_b is an additional boundary-layer resistance taking account of the difference between the molecular diffusivity of the diffusing gas and the kinematic viscosity of air, and r_c is the crop canopy resistance. To an approximation, r_c is equal to r_s divided by the leaf area index (LAI), which is the plane area of leaves per unit area of ground.

The advantage of this type of analysis is to separate the aerodynamic and boundary-layer resistances, which are determined by wind speed and by overall crop morphology, from the surface resistance, which is controlled by plant physiological variables.

Spedding (1969) found that the total resistance to uptake of SO_2 by barley leaves averaged 5.5 cm^{-1} sec when the stomata were open. When they were closed, r_t ranged from 33 cm^{-1} sec when the relative humidity was 80% to 250 cm^{-1} sec when it was 20%. Although the wind speed in Spedding's apparatus was only 20 cm sec^{-1}, the external resistance r_a was only about 1.2 cm^{-1} sec, so the stomatal or surface resistance was limiting. For a crop with LAI $= 4$, Spedding's results suggested a minimum crop canopy resistance $r_c = 5.5/4 = 1.4$ cm^{-1} sec, which agrees well with Fowler and Unsworth's (1979) measurements by the gradient method, reviewed below.

Interest in the dry deposition of SO_2 increased about 1970, when attention was given to modeling the long-range transport of SO_2. The methods of study can be considered in four categories: (a) sulfur balance in crops, (b) sulfur balance in atmosphere, (c) sulfur gradient in atmosphere, (d) analogy with transpiration of water.

SULFUR BALANCE IN CROPS

In this method the increase in sulfur content of a crop over a growth period is related to the SO_2 concentration, while the contribution of root uptake is assessed by monitoring the supply of S in nutrients, by parallel control experiments, or by adding ^{35}S tracer to the nutrient medium. This method has the advantage that deposition is measured over a long period, usually including a range of growth conditions. In chamber experiments, the wind speed over the crop is usually low, and this introduces an aerodynamic resistance which may be greater than in typical outdoor conditions.

The experiments of Thomas et al. (1943) have already been discussed. Olsen (1957) did similar chamber experiments with cotton plants but added ^{35}S tracer to the nutrient solution to enable the sulfur taken up by the roots to be estimated. He found that 30% of the sulfur content of healthy plants and 50% in sulfur-deficient plants was obtained from the atmosphere.

Lockyer et al. (1976) showed that SO_2 concentrations up to 200 μg m^{-3} over ten weeks increased the yield of ryegrass grown on sulfur-deficient soils and had no effect of yield on soils with adequate sulfur. At 400 μg m^{-3} of SO_2, there was some diminution of yield. In their chambers there were two to five air changes per minute. The mean velocity of deposition was 0.79 cm sec^{-1} (Cowling et al., 1979). With this value of v_d, the equivalent depth of air in the growth chamber depleted of SO_2 in 30 sec (corresponding to two air changes per minute) is 23 cm, so at this ventilation rate the supply of SO_2 may not have been quite sufficient to maintain the concentration in the canopy.

Balance studies have been done at the Grassland Research Institute, Hurley, and at Rothamsted Experimental Station in the United Kingdom using lysimeters [Department of the Environment (UK), 1976]. The total losses of sulfur in drainage and removal of crops were 43 and 58 kg $ha^{-1}year^{-1}$ respectively. Input in rain, estimated from funnel collections, was about 20 kg $ha^{-1}year^{-1}$ at both stations. However, these collections, made monthly, include SO_2 which has been deposited by gaseous diffusion to the funnel sides (Garland, 1977), and the true input in rainfall was probably not more than about 12 kg $ha^{-1}year^{-1}$. Hence about 30 to 45 kg of S per hectare per year must have been added by dry deposition to the vegetation in the lysimeters. As the SO_2 concentration was about 25μg m^{-3} at both stations, equivalent to 12.5 μg of S per cubic meter, v_d must have been about 1 cm sec^{-1}.

Experimental fumigation with $^{35}SO_2$ can be considered a variant of the sulfur-balance-in-crops method, with the advantage that foliar uptake can be distinguished readily from soil uptake. The disadvantage is that experiments tend to be episodic, usually in daytime, and in growth conditions not necessarily representative of seasonal averages. The first field experiment was carried out at the National Reactor Test Station (Hill, 1971). A source of $^{35}SO_2$ with stable SO_2 carrier was liberated over a ½-hr period over a field of alfalfa, averaging 1500 g fresh weight per square meter. After the fumigation, the amount of ^{35}S accumulated at 15 stations was determined by analysis of the alfalfa and related to the time-integral air concentration (dosage) measured by the adjacent air sampling equipment. The velocity of deposition was found to be independent of the concentration of SO_2 and averaged 2.3 cm sec^{-1}. The wind speed was not given.

Garland et al. (1973) and Owers and Powell (1974) did similar experiments over grassland. Garland found a mean v_d of 1.2 cm sec^{-1}. Mean wind speed at 1 m height was 4.0 m sec^{-1}. Analysis in terms of resistances gave a surface resistance of 0.6 cm^{-1} sec. Owers and Powell found a mean v_d of 1.3 cm sec^{-1} (wind speed at 2 m height equal to 3.2 m sec^{-1}).

Tracers are particularly useful in establishing the mechanisms of dry deposition. The work of Spedding, referred to earlier, established that stomatal aperture controlled deposition. Belot (1975) used SO_2 labeled with the stable isotope ^{34}S, and Garland and Branson (1977) used $^{35}SO_2$, to measure the uptake by pine needles while simultaneously measuring the transpiration of water. The amount of water transpired per unit area of leaf divided by the difference between the water vapour concentration in air within the leaf and in the ambient air is termed the conductance of the leaf and is analogous to the velocity of deposition. When allowance was made for the ratio of the molecular diffusivity of SO_2 (0.14 cm^2 sec^{-1}) to that of water vapour (0.25 cm^2 sec^{-1}), the conductance of SO_2 and water vapour were found to be equal over a wide range of stomatal opening (Fig. 1). Fowler and Unsworth (1979)

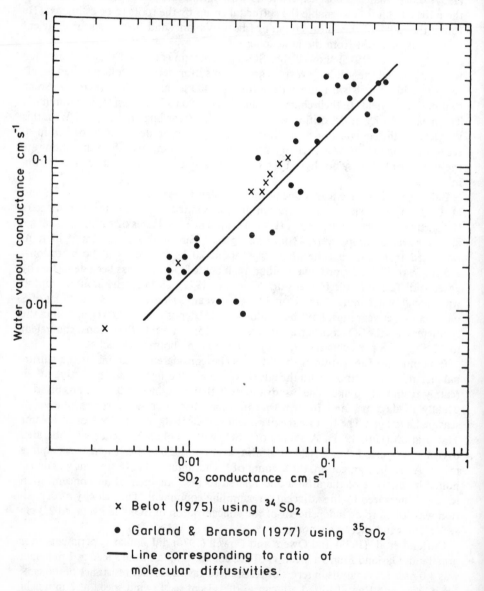

Fig. 1. Conductances for water vapour and SO_2 of Scots pine shoots.

have shown that conductances are not equivalent in wheat when stomata are closed, since there is apparently more cuticular absorption of SO_2 than there is cuticular transpiration of water vapour. Nevertheless, the analogy between transport of SO_2 and water vapour is useful for estimating the uptake of SO_2 by crops for which specific measurements have not been made.

SULFUR BUDGET IN THE ATMOSPHERE

The loss by deposition of SO_2 between two imaginary vertical boundaries in the atmosphere can be calculated from the difference in the advection flux of SO_2 and SO_4^{2-} across the two boundaries plus the input between them. The difficulties are that

a. the velocity of deposition is typically two orders of magnitude less than the wind speed, so long distances of travel are needed,

b. unless the upper boundary of the atmospheric layer is well defined (for example by an inversion), the upward escape of SO_2 through the top of the layer may be difficult to evaluate,

c. calculation of the fluxes across the boundaries requires integration of concentration multiplied by wind speed over large areas and is susceptible to small analytical errors,

d. it is necessary to assume that deposition of sulfate is negligible.

In the London fog of 1952, the upper boundary was well defined, and wind speeds were so low that advection could be neglected. Equating the deposition to the emission gave $v_d = 0.7$ cm sec^{-1} (Chamberlain, 1960).

Smith and Jeffrey (1975) sampled SO_2 and SO_4^{2-} by aircraft flights off the east coast of England in dry weather, computed the fluxes, and compared them with the inventory of emissions. In three flights in August and September 1973, fractions 0.57, 0.58, and 0.48 of the calculated source strengths were airborne as SO_2 at the sampling lines, and fractions 0.12, 0.09, and 0.20 were airborne as SO_4^{2-}. Hence the fractions deposited by dry deposition were 0.31, 0.33, and 0.32, with corresponding v_d values of 0.8, 0.8, and 0.6 cm sec^{-1}. Most of the loss was overland and in daytime. No allowance was made for SO_2 or SO_4^{2-} in the airstream incident on England. From aircraft measurements of the sulfur budget of the Labadie plume at 300 km on three occasions in July 1976, Gillani (1978) deduced average v_d values of 1.8 cm sec^{-1} (daytime) and 0.5 cm sec^{-1} (nighttime).

Under the present heading can also be considered experiments in which the confining box is real and not imaginary. The rate of loss of SO_2 from the air in the box to vegetation on its floor is compared with the rate of loss from leaks and from absorption on the walls when there is no vegetation in the box. Hill's (1971) box was a wind tunnel with recirculating airflow of about 2 m sec^{-1}. The alfalfa canopy had area 1.4 m^2, height 40 cm, leaf area index 10, and dry-weight density 350 g m^{-2}. Relative humidity and light intensity were controlled. The only possible criticism of the experimental arrangement is that there may have been an "oasis effect" inevitable in measuring the flux to a restricted area of crop. The rate of uptake of SO_2 (and several other gases) was measured by finding the rate of input of gas into the tunnel required to maintain a steady concentration. To maintain 5 pphm of SO_2, an input of 85 μl min^{-1} per square meter of alfalfa was required, which is equivalent to $v_d = 2.8$ cm sec^{-1}.

Hill's experiment showed a high rate of uptake obtained with a good flow of air over a luxuriant growth of crop. Milne et al. (1979) measured uptake of SO_2 by drought-resistant plants and arid soils, using a portable exposure box in which the air was stirred by a fan. They used the "die-away" method, in which a single release of gas is made and the concentration measured at various subsequent times. The die-away was found to be exponential, according to $\chi(t) = \chi_0 \exp(-\lambda t)$, where λ is the decay coefficient with units time^{-1}. The velocity of deposition was deduced from $v_d = \lambda V/A$, where V is the volume of the chamber and A the ground area of the vegetation within it. The aerodynamic resistance can be deduced from an experiment in which short (4 cm) grass from the Sydney area was artificially watered with dilute NaOH solution to make it a perfect sink. This gave $v_d = 0.9$ cm sec^{-1}, $r_a = 1.1$ cm^{-1} sec, which is a reasonable figure for open-air conditions also. Four experiments with the same grass, artificially watered but not wet at the time of the experiments, gave an average v_d of 0.46 cm sec^{-1}, and, assuming $r_a = 1.1$ cm^{-1} sec as before, this implies $r_c = 1.1$ cm^{-1} sec, which is comparable with the results found by Spedding using the ^{35}SO$_2$ tracer method with barley. When Milne et al. transferred their box to the arid region near Mount Isa in Queensland and erected it over an acid and sun-baked clay soil with thin canopies of drought-resistant species of grass, they found an average v_d of 0.1 cm sec^{-1}, implying r_c about 9 cm^{-1} sec. Thus both stomatal and cuticular resistances must have been very high in these conditions.

GRADIENT METHOD

The first demonstration of the gradient of SO$_2$ concentration over vegetation appears to be that of Gilbert (1968), in an investigation of the effect of atmospheric pollution on the growth of bryophytes near Newcastle on Tyne, United Kingdom. Gilbert noted that sensitive species survived in dense woodland and long grassland and deduced that this was because the surrounding vegetation absorbed the SO$_2$. To confirm this, he measured SO$_2$ at several levels of a lawn and found a 24% reduction between 1.8 and 0.15 m height. For smoke the reduction was only 3%.

Saito et al. (1971) measured SO$_2$ at heights of 2 m and 0.5 m (the canopy height) over long grass, and from the gradients of SO$_2$ and wind speed they deduced an average velocity of deposition of 1.6 cm sec^{-1}. However, the measurements were made when the wind blew from the direction of an iron foundry (distance not stated), and the average concentration at 2 m was 500 μg m^{-3}. It is possible, therefore, that the assumption of flux constant with height, essential to the gradient method, may not have been valid.

An extensive series of gradient experiments have been done by Garland (1977) and his colleagues. Wind speed, temperature, and SO$_2$ concentration were measured at several heights from 20 cm to 3 m above grassland, bare soil, and water surfaces. By fitting wind speeds on occasions with small temperature gradients to the logarithmic wind profile, the displacement height d and roughness length z_0 for each site were determined. The Richardson number for each run was estimated, and the dimensionless gradients for momentum (Φ_M) and heat (Φ_H) were estimated using the formulae of Dyer and Hicks (1970) and Webb (1970). It was assumed that the dimensionless gradient for SO$_2$ vapour transport

$$\Phi_V = \frac{k(z-d)u*}{F}\frac{d\chi}{dz}$$

was equal to Φ_H, and the flux F was deduced from

$$F = -k^2(z-d)^2\frac{du}{dz}\frac{d\chi}{dz}(\Phi_H\Phi_M)^{-1}.$$

The mean v_d values in daytime measurements were: short grass, 0.85 cm sec^{-1}; medium grass, 0.87 cm sec^{-1}; bare calcareous soil, 1.2 cm sec^{-1}. Using similar methods, Shepherd (1974) found v_d for medium grass (50–200 mm high) to average 0.8 cm sec^{-1} in daytime summer runs but only 0.3 cm sec^{-1} in autumn. Holland et al. (1974) also did a similar study in an area of wet moorland where cotton sedge grows on acid peat (pH 3.5). The mean v_d for SO$_2$ concentrations greater than 20 μg m^{-3} was 0.7 cm sec^{-1}. At lower SO$_2$ concentrations, v_d was apparently rather greater, though the measurements were rather inaccurate in these cases.

Garland (1977) and Shepherd (1974) showed that the canopy resistance r_c over grassland in summer ranged from 0.4 to 0.8 cm^{-1} sec. An experiment of Shepherd in autumn showed that r_c increased to 3 cm^{-1} sec.

Fowler and Unsworth (1979) used the gradient method to measure deposition of SO$_2$ to a wheat crop during two growing seasons and by night as well as by day. The canopy resistance was considered as stomatal and cuticular leaf resistances and a resistance to transport through the canopy to the soil, all in parallel. Writing these as r_{c_1}, r_{c_2}, and r_{c_3},

$$r_c^{-1} = r_{c_1}^{-1} + r_{c_2}^{-1} + r_{c_3}^{-1} .$$

With a green crop, there was typically a diurnal variation, with v_d reaching a maximum and r_c a minimum in the forenoon. The average daytime canopy resistance was 0.7 cm^{-1} sec. The variation and the magnitude of the surface resistance were compatible with stomatal aperture as the controlling factor. However, the resistance to uptake at night (r_c about 3 cm^{-1} sec) was less than the corresponding resistance for water vapour, probably because the cuticular resistance r_{c_2} was less for SO$_2$ than for water vapour. When the crop reached senescence shortly before harvest, r_c exceeded 9 cm^{-1} sec, and it is probable that most of the flux was being absorbed by the soil. The dying vegetation, when dry, took up little SO$_2$ itself and also reduced the uptake by the soil, compared with uptake to bare soil, by virtue of the resistance to transfer through the boundary layer of air within the canopy.

When the crop was wetted by dew at night, the canopy resistance fell, owing to increased cuticular uptake of SO$_2$, but this was observed to be a temporary effect in some instances. It is probable that the sorption of SO$_2$ in the water film on the leaf slowed down when the pH fell to about 3 (Brimblecombe, 1978).

Fowler and Unsworth (1979) estimated the total uptake of sulfur by the crop over the period 1 May–31 July as shown in Table 1. The sulfur content of the crop at harvest was 1.46 g m^{-2}, so the amount of S entering through the stomata (0.46 g m^{-2}) represented 32% of the total. This is the same as the percentage of S derived from the atmosphere by cotton plants in Olsen's (1957) experiments. It is assumed that the sulfur deposited on the cuticle was washed off by rain.

DEPOSITION TO FORESTS

It has been a matter of dispute for some years whether or not the water loss by evapotranspiration from forests is greater than that from grassland or arable crops in similar climatic conditions. The balance of opinion, in the United Kingdom at least, is that transpiration is not greatly different but that the loss by evaporation of intercepted water is much greater from forests. The reason why evapotranspiration is not greater is that canopy resistance appears to be greater for forests and this compensates for the lower aerodynamic resistance.

Garland (unpublished) attempted to measure deposition by the gradient method to a coniferous forest at Thetford, Norfolk, United Kingdom, but found the concentration gradients too small to obtain reliable measurement. Garland and Branson (1977), having shown by measurements with ^{35}SO$_2$ that the stomatal resistances for

Table 1. Components of resistance to uptake of SO_2 by a wheat crop and total sulfur deposition to crop

	Resistance (cm^{-1} sec)				$v_d = 1/r_t$ (cm sec^{-1})
	Aerodynamic, $r_a + r_b$	Stomatal, r_{c_1}	Cuticular, r_{c_2}	Total, r_t	
Day	0.5	1.4	2.5	1.4	0.7
Night	0.5	∞	2.5	3.0	0.3

Sulfur deposition during growth of crop,
(g of S per square meter; SO_2 concentration 50 μg m^{-3})

	Stomatal	Cuticular	Total
Day	0.46	0.26	0.72
Night		0.33	0.33
Total	0.46	0.59	1.05

Note that the stomatal canopy resistance $r_{c_1} = 1.4$ cm^{-1} sec was derived from measurements of resistance to water vapour of the crop. With $r_{c_2} = 2.5$ cm^{-1} sec in parallel, the total canopy resistance is 0.9 cm^{-1} sec, compared with 0.7 cm^{-1} sec deduced from the gradient measurements.
Source: Fowler and Unsworth, 1979.

uptake of SO_2 and transpiration of water by pine needles were inversely proportional to the molecular diffusivities of the two gases, used published data on the evapotranspiration of the Thetford forest and deduced that v_d (relative to 20 m reference height) probably varied during the day from 0.2 to 0.6 cm sec^{-1}, with a value possible as low as 0.05 cm sec^{-1} at night. When the canopy is wet, the aerodynamic resistance would be limiting, and v_d might be in the range 1 to 8 cm sec^{-1} depending on wind speed. However, unless there was continued rewetting of the leaves, the uptake would eventually be limited by the resulting lowering of the pH. Even so, Garland and Branson estimated that in typical U.K. climatic conditions the yearly deposition to the wet canopy might exceed that to the dry canopy by a factor of 5 or more.

DEPOSITION TO SNOW

Barrie and Warmsley (1978) found a deposition velocity of 0.25 cm sec^{-1} to snow by simultaneously measuring sulfur deposition and ambient concentration during a pollution episode in Alberta, Canada. Deposition velocities were also obtained at 40 other sites by relating the measured deposition rates to the calculated average ground-level concentration, and these ranged from 0.3 to 0.4 cm sec^{-1}. If snow-covered ground is an aerodynamically smooth surface and a perfect sink, then v_d can be calculated (see, for example, Chamberlain, 1968), and with average wind speeds at 2 m height of 2 to 3 m sec^{-1}, v_d values in the range 0.3 to 0.4 cm sec^{-1} can be expected. This is also near the average v_d found by Garland (1977) for deposition to fresh water by the gradient method. However, Whelpdale and Shaw (1974) found steeper gradients of SO_2 concentration over water than over snow. It is of interest to speculate the mode of transport into the surface layers. This may be by diffusion of SO_3^{2-} or SO_4^{2-} ions in the ice crystal lattices or by melting and refreezing of the surface layer.

DEPOSITION TO BARE SOIL

There have been a number of investigations of the uptake of SO_2 to bare soil in the field and in the laboratory, but the results are rather conflicting and difficult to interpret. Terraglio and Manganelli (1966) passed SO_2 through soil in glass columns and concluded that dry acid soils do not remove sulfur dioxide from the air but less acid soils can retain significant quantities. Moisture increases the amount retained. They also passed SO_2 over soils in small horizontal containers. With Nixon (New Jersey) soil, a mixture of silt and clay with pH 5.5, v_d increased from 0.23 to 0.55 cm sec^{-1} as the soil moisture increased from 0.52% to 10%, but with Lakewood soil, almost pure sand with pH 4.0, v_d was nil with 0.05% moisture and increased only to 0.17 cm sec^{-1} with 10% moisture. Lockyer et al. (1978), in similar experiments with a range of U.K. soils, found less effect of pH and water content. With soils at field capacity, v_d varied from 0.42 cm sec^{-1} (pH 4.5) to 0.65 cm sec^{-1} (pH 7.5), and the mean was 0.54 cm sec^{-1}. When the soil was oven-dried, the range was from 0.13 to 0.36 cm sec^{-1}, with mean 0.26 cm sec^{-1}. It is not possible to separate the aerodynamic and surface resistances, but, since the soils were sieved and leveled, it is likely that the moist alkaline soils were acting as perfect sinks. In both Terraglio and Manganelli's and Lockyer et al.'s (1978) experiments, the deposition to the soil was estimated from the loss of SO_2 from the gas stream in a flowing system.

Payrissat and Beilke (1975) used the "die-away" method in a stirred chamber. Soils with pH varying from 4.5 to 7.6 were used, and the experiments were done at two levels of relative humidity, about 45% and 80% (it is not clear what changes in soil moisture occurred because of the variation of relative humidity in the air). Velocity of deposition varied from 0.19 to 0.55 cm sec^{-1} at the lower relative humidity and from 0.38 to 0.60 cm sec^{-1} at the higher relative humidity. Payrissat and Beilke also estimated v_d with the soils replaced by activated charcoal as a perfect sink, and this enabled them to estimate the surface resistances of the soils, which ranged from 0.24 cm^{-1} sec for the most alkaline soil at high relative humidity to 3.8 cm^{-1} sec for the most acid soild at low pH. Also, using the die-away method, with a mobile chamber placed over bare soil in an arid region of Queensland, Australia, Milne et al. (1979) found $v_d = 0.04$ cm sec^{-1}, implying a surface resistance of more than 20 cm^{-1} sec.

Bromfield and Williams (1974) placed air-dried sieved soil (pH not stated) in Stevenson screens at three locations in the United Kingdom and estimated the deposition from the increase in soil sulfur over periods of 57 to 77 days. By reference to estimated ambient SO_2 levels, v_d was 0.4 cm sec^{-1}. The wind speed within the Stevenson screen would have been low, and by reference to the rate of uptake of SO_2 by lead peroxide candles, similarly placed, it can be estimated that v_d to a perfect sink would not have exceeded 0.6 cm sec^{-1}.

Using the gradient method over bare soil in Berkshire, United Kingdom, Garland (1977) found an average v_d of 1.2 cm sec^{-1}, which was higher than the average for either short or long grass or for water surfaces. The soil surface was sufficiently rough to give an aerodynamic resistance of order 1 cm^{-1} sec, and it appears that the calcareous soil acted as a perfect sink for SO_2 even in dry weather.

COMPARISON OF DRY DEPOSITION OF SO₂ TO DIFFERENT SURFACES

Resistances to evapotranspiration from various crops tend to be surprisingly similar, because crops with well-developed and aerodynamically rough surfaces (r_a a minimum) tend to have high surface resistances, and vice versa. Table 2 shows aerodynamic (r_a) and canopy (r_c) resistances for pine forest, potatoes, and lucerne (alfalfa) deduced by Sceicz et al. (1969) by considerations of energy balance. By the assumption that the canopy resistances to water vapour and SO₂ are inversely proportional to the molecular diffusivities (1.8:1), r_c, r_t, and v_d values for SO₂ are deduced.

It seems that a range of values from about 0.4 to 0.8 cm sec^{-1} covers many measurements and deduced values for the velocity of deposition of SO₂ to crops, bare soil, snow, and water surfaces, with the following exceptions:
1. Higher values, in the range 1 to 2 cm sec^{-1}, may be found with luxuriant, well-watered vegetation and with rough, moist soil of high pH.
2. Much higher values may apply to wet or snow-covered forests.
3. Lower values, by as much as an order of magnitude in some cases, apply where vegetation is dead or where bare arid soil is exposed.

Table 2. Resistances to evapotranspiration from crops and deduced resistances to deposition of SO₂

Crop	Resistance (cm^{-1} sec)				v_d (SO₂) (cm sec^{-1})
	r_a*	r_c(H₂O)*	r_c (SO₂)	r_t (SO₂)	
Pine forest	0.06	1.2	2.2	2.3	0.43
Potatoes	0.7	0.9	1.9	2.6	0.38
Lucerne	0.7	0.5	0.9	1.6	0.63

*From Sceicz et al., 1969.

REFERENCES

Alway, F. J. 1940. A nutrient element.
  J. Am. Soc. Agron. 32:913–921.

Barrie, L. A., and J. L. Warmsley. 1978. A study of sulphur dioxide deposition velocities to snow in northern Canada. Atmos. Environ. 12:2321–2332.

Belot, Y. 1975. Etude de la captation des polluants atmospheriques par les vegetaux. C.E.A., Fontenay-aux-Roses, France.

Brimblecombe, P. 1978. Dew as a sink for sulphur dioxide. Tellus 30:151–157.

Bromfield, A. R., and R. J. B. Williams. 1974. The direct measurement of sulphur deposition on bare soil. Nature 252:470–471.

Chamberlain, A. C. 1960. Aspects of the deposition of radioactive and other gases and particles, in *Aerodynamic Capture of Particles*, E. G. Richardson, editor. Pergamon Press, Oxford:pp. 63–88.

Chamberlain, A. C. 1968. Transport of gases to and from surfaces with bluff and wave-like elements. Q. J. R. Meteorol. Soc. 94:318–332.

Cowling, D. W., D. R. Lockyer, and A. W. Bristow. 1979. Uptake and utilisation of sulphur by forage plants, in Annu. Rep. Grassl. Res. Inst., 1978, pp. 22–23. Hurley, Berks, U.K.

Department of the Environment (U.K.). 1976. Effects of airborne sulphur roughness compounds on forests and freshwaters. London, HMSO.

Dyer, A. J., and B. B. Hicks. 1970. Flux-gradient relationships in the constant flux layer. Q. J. R. Meteorol. Soc. 96:715–721.

Fowler, D., and M. H. Unsworth. 1979. Turbulent transfer of sulphur dioxide to a wheat crop. Q. J. R. Meteorol. Soc., in press.

Garland, J. A. 1977. The dry deposition of sulfur dioxide to land and water surfaces. Proc. R. Soc. London, Ser. A 354:245–268.

Garland, J. A., and J. R. Branson. 1977. The deposition of sulphur dioxide to pine forest assessed by a radioactive tracer method. Tellus 29:445–454.

Garland, J. A., W. S. Clough, and D. Fowler. 1973. Deposition of sulphur dioxide on grass. Nature (London) 242:256–257.

Gilbert, O. L. 1968. Bryophytes as indicators of air pollution in the Tyne valley. New Phytol. 67:15–30.

Gillani, N. V. 1978. Project MISTT: mesoscale plume modelling of the dispersion, transformation and ground removal of SO_2. Atmos. Environ. 12:569–587.

Hill, A. C. 1971. Vegetation: a sink for atmospheric pollutants. J. Air Pollut. Control Assoc. 21:341–346.

Holland, P. K., J. Sugden, and K. Thornton. 1974. The direct deposition of SO_2 from atmosphere. Pt. 2. Measurements at Ringinglow Bog compared with other results. CEGB report NW/SSD/RN/PL/1/74.

Lockyer, D. R., D. W. Cowling, and L. H. P. Jones. 1976. A system for exposing plants to atmospheres containing low concentrations of sulphur dioxide. J. Exp. Bot. 27:397–409.

Lockyer, D. R., D. W. Cowling, and J. S. Fanlon. 1978. Laboratory measurements of dry deposition of sulphur dioxide on to several soils from England and Wales. J. Sci. Food Agric. 29:739–746.

Meetham, A. R. 1950. Natural removal of pollutation from the atmosphere. Q. J. R. Meteorol. Soc. 76:359–371.

Meetham, A. R. 1954. Natural removal of atmospheric pollution during fog. Q. J. R. Meteorol. Soc. 80:96–99.

Milne, J. W., D. B. Roberts, and D. J. Williams. 1979. The dry deposition of sulphur dioxide—field measurements with a stirred chamber. Atmos. Environ. 13: 373–379.

Monteith, J. L. 1973. Principles of Environmental Physics. Edward Arnold, London.

Olsen, R. A. 1957. Absorption of sulfur dioxide from the atmosphere by cotton plants. Soil Sci. 84:107–111.

Owers, M. J., and A. W. Powell. 1974. Deposition velocity of sulphur dioxide on land and water surfaces using a 35S tracer method. Atmos. Environ. 8:63–67.

Payrissat, M., and S. Beilke. 1975. Laboratory measurements of the uptake of sulphur dioxide by different European soils. Atmos. Environ. 9:211–217.

Saito, T., S. Isobe, Y. Nagai, and Y. Horibe. 1971. Transport of SO_2 gas to grass. J. Agric. Met. Tokyo 26: 177–180.

Sceicz, G., G. Enrodi, and S. Tajchman. 1969. Aerodynamic and surface factors in evaporation. Water Resour. Res. 5:380–394.

Shepherd, J. G. 1974. Measurements of the direct deposition of sulphur dioxide onto grass and water by the profile method. Atmos. Environ. 8:69–74.

Smith, F. B., and G. H. Jeffrey. 1975. Airborne transport of sulphur dioxide from the UK. Atmos. Environ. 9:643–659.

Spedding, D. J. 1969. Uptake of sulphur dioxide by barley leaves at low sulphur dioxide concentrations. Nature (London) 224:1229–1234.

Terraglio, F. P., and R. M. Manganelli. 1966. The influence of moisture on the adsorption of atmospheric sulfur dioxide by soil. Air Water Pollut. 10:783–791.

Thomas, M. D., R. H. Hendricks, T. R. Collier, and G. R. Hill. 1943. The utilization of sulphates and sulphur dioxide for the sulphur nutrition of alfalfa. Plant Physiol. 18:345–371.

Webb, E. K. 1970. Profile relationships: the log-linear range and extension to strong stability. Q. J. R. Meteorol. Soc. 96:67–90.

Whelpdale, D. M., and R. W. Shaw. 1974. Sulphur dioxide removal by turbulent transfer over grass, snow and water surfaces. Tellus 26:196–205.

DISCUSSION

V. P. Aneja, Northrup Services: What are the meteorological constraints, such as fetch, which must be met for successful field deposition experiments?

A. C. Chamberlain: The fetch needs to be one hundred times the height at which you make the measurements. Making measurements up to 2 meters above a crop of, say, wheat, one needs a fetch of several hundred meters. This is very much more difficult over forests, because one needs a much longer fetch, since one has to make measurements at greater heights and also because the turbulence is greater above the forest, so that the gradients that one sees are smaller. But over most crops, I think a uniform fetch of a few hundred meters is sufficient.

V. P. Aneja: This value of one hundred—what is the basis for it?

A. C. Chamberlain: There has been a great deal of work in this field in respect to evaporation of water, comparing alternate results to this gradient method. We've other methods of measuring transpiration—for example, the energy balance method, where essentially one measures all the solar energy entering the crop and apportions this among long-wave radiation, sensible heat flux, and transpiration carrying latent heat. From comparisons of such results, this hundred times rule is, I think, fairly well established. I dare say Mr. Hicks, from his experience in Australia, would say it should be more. I think the hundred times rule seems, as far as we can tell, to give a reasonably good estimate of the fetch required.

V. P. Aneja: One last question, if I may. Have you compared the results from field work with those from chamber experiments?

A. C. Chamberlain: In my written contribution this is done; I think the range, roughly 0.5 to 1 cm/sec, does cover the field, including chamber experiments, fairly well with two major exceptions. On dead vegetation or very dry low-pH soil, the deposition rate is very much less: 0.1 or even 0.01 cm/sec has been found in Australia with dry soil of low pH. The other exception involves Hill's measurements with a wind tunnel, as I have already mentioned, and one or two others where possibly there may have been a so-called oasis effect. In such cases the vegetated area is too small, so that the crop as a whole doesn't set up its own microclimate. The vegetation is fumigated from the side, as it were, as well as from above; in transpiration studies this is known as the oasis effect, because it corresponds to the fact that if you have a green oasis you get much more evaporation of water from it than you would from a widespread crop of the same type. In my opinion, subject to these exceptions, the gradient method does tie up with the chamber method reasonably well.

S. V. Krupa, U. Minnesota: I understand you and John Garland are working primarily on SO_2; are you also looking at sulfate deposition, for which the deposition rate is probably lower?

A. C. Chamberlain: No, we're not at the moment. John Garland concluded that his gradient method wouldn't detect sulfate profiles. I should mention that, in our location, we're not near a major source, and for various reasons we think that nearly all of our sulfate is in that fine-particle mode that was described earlier. Now it would take at least as long again to go into the reasons for thinking that a low velocity of deposition is in fact typical of these small particles. I can only refer you not only to our own work but to the work of Dr. Belot, in Paris, who is doing a very careful study of the means by which small particles do attach to leaves and who is also doing a computer study. I have a slide of this, but I don't think there's time to show it. At any rate, this does rather confirm that something like 0.5 micron is what we and many people find to be the sulfate mean particle size. It does seem to correspond to a very low deposition velocity, but I'm always ready to be proved wrong.

S. V. Krupa: The main concern I have relates to the fact that there has been some deposition work using Styrofoam particles; these are quite different from sulfate per se, because the sulfate aerosol carries a charge on it. This is quite different from using inert particles, and one also expects a different motion in the sense that the aerosol is more like a colloid.

A. C. Chamberlain: If the particle size were 0.01 micron, then I would expect that the charge would have a very large effect. However, in my experience covering about 25 years, we've never found a case where the electrical gradients which exist in the open air essentially affect the deposition of a charged particle of this 0.5-micron size. I don't know whether Dr. Davidson is still here, but I think we would agree that Dr. Friedlander's experience is the same. In the laboratory, especially if one uses material such as phosphates, one can generate electrical gradients or fields which will cause charged particles of almost any size to be deposited, but I don't believe that in a reasonably moist outside environment, where most surfaces are reasonable conductors, the necessary electrical fields can be generated.

CHAPTER 23

TURBULENT TRANSFER PROCESSES TO A SURFACE AND INTERACTION WITH VEGETATION

B. B. Hicks*

M. L. Wesely*

ABSTRACT

Strong diurnal variations are associated with both aerodynamic and surface resistances to the transfer of gaseous and particulate pollutants. Eddy flux measurements made over a loblolly pine plantation in July 1977 indicate that nocturnal aerodynamic resistances can frequently exceed 1 sec/cm and sometimes be considerably more than 10 sec/cm. Residual surface resistances to ozone transfer ranged from about 1.4 sec/cm during daytime to about 15 sec/cm at night. For small particles and sulfur compounds, similar daytime residual resistances were measured, but nocturnal resistances were typically about 50% of the ozone values. Exceedingly few nocturnal particle resistances were determined, however, because the trees then appeared to constitute more of a source than a sink.

INTRODUCTION

Turbulent transfer mechanisms above (and within) vegetal canopies and biological factors associated with the foliage both display strong diurnal cycles that combine to produce even stronger variations in the net flux of pollutants. In the case of sulfur dioxide, for example, field experiments agree that fluxes can be quite large in daytime, corresponding to deposition velocities about 1 cm/sec or more, but at night the closing of stomata and the damping of atmospheric turbulence often shut down the flux, almost completely. On these grounds alone, it is immediately apparent that the common assumption of an average SO_2 deposition velocity of 1 cm/sec over vegetation has serious deficiencies, even though it might well be appropriate for daytime conditions.

Less-soluble gases and particles will be subject to similar influences, although the relevance of stomata might not be so clearly evident. Here, we will consider diurnal variations in those turbulent and surface properties that combine to produce a net pollutant transfer, in order to demonstrate the sort of cycles that should be anticipated. The arguments will be supported by data obtained over a plantation of loblolly pine in North Carolina during July 1977.

*Atmospheric Physics Section, Radiological and Environmental Research Division, Argonne National Laboratory, Argonne, IL 60439.

THE RESISTANCE ANALOG

It is convenient to express the flux of some pollutant in terms of its concentration differences across layers of air that can be associated with clearly identifiable parts of the transfer process. The following individual "resistances" can be identified as having major effects.

Aerodynamic Resistance, r_a

The wind profile above a surface characterized by a roughness length z_0 and a displacement height d is usually written as

$$u_z = (u_*/k) \cdot \{\ln[(z - d)/z_0] - \psi_M\}, \tag{1}$$

where u_z is the wind speed at height z, u_* is the friction velocity, $k = 0.4$ is the von Karman constant, and ψ_M is a stability-dependent correction function. Following the obvious electrical analog, the aerodynamic resistance to momentum transfer is then written as

$$R_{am} = u_z/u_*^2 , \tag{2}$$

which is seen to be dependent upon stability to the extent inherent in (1).

In daytime convective conditions, it is known that sensible heat and water vapor transfer are controlled by similar eddy diffusivities, which can differ substantially from those applicable for the momentum case. Sinclair et al. (1975) have shown that the eddy diffusivity for carbon dioxide transfer also is the same as that for heat, rather than momentum, and it is often presumed that this applies to other atmospheric trace constituents as well. Accordingly, the following discussion will be based on consideration of the sensible heat flux, in unstable stratification.

A relation that is like (1) can be written for the transfer of heat:

$$T_0 - T_z = (H/\varrho c_p) \cdot \{\ln[(z - d_H)/z_H] - \psi_H\}/)ku_*), \tag{3}$$

where T_z is the temperature at height z, H is the sensible heat flux, ϱ is air density, c_p is the specific heat of air at constant pressure, d_H is a displacement height for heat, z_H is roughness height for heat, and ψ_H is another stability-dependent correction function. In this case, T_0 represents a "surface temperature" that is characteristic of the foliage. From (3), the total resistance to heat transfer between the surface and the air can be derived as

$$r_H = \{\ln[(z - d_H)/z_0] + \ln(z_0/z_H) - \psi_H\}/(ku_*). \tag{4}$$

This can be expressed in terms of the aerodynamic resistance to momentum transfer as

$$r_H = r_{am} + [\psi_M - \psi_H + \ln(z_0/z_H)]/(ku_*), \tag{5}$$

where, as in (3), the displacement heights for heat and momentum are assumed equal. It is clear that the property $\ln(z_0/z_H)$ is directly related to the surface itself, whereas the stability correction functions modify the aerodynamic behavior. Thus, it appears that trace pollutants respond to a modified aerodynamic resistance (if it is assumed to be equal to that for sensible heat) which can be written as

$$r_{aH} = r_{am} + (\psi_M - \psi_H)/(ku_*). \tag{6}$$

In daytime, convection enhances thermal transfer, and hence $\psi_H > \psi_M$, with the obvious consequence that $r_{aH} < r_{am}$. In stable stratification, studies have shown that

$\psi_M \cong \psi_H$, and hence the aerodynamic resistance to pollutant transfer is not detectably different from u/u_*^2.

The properties ψ_M and ψ_H have been the subject of extensive experimental investigation. Dyer (1974) has summarized the available evidence; his recommendations have been followed here.

Figure 1 shows how r_{aH} varied through the course of a week during which direct measurements of the fluxes of sensible heat and momentum were made above a pine plantation in North Carolina. The site in question was a large area of loblolly pine, operated as a forest meteorology research site by Duke University. During July 1977, an intensive series of eddy flux measurements was made by application of the covariance technique, which evaluates the vertical flux of a quantity with density C as the time average of instantaneous products $\overline{w'C'}$, where w is the vertical wind speed and primes indicate deviations from mean values. Sensible heat fluxes were evaluated as $\overline{w'T'}$ and friction velocities as $(-\overline{u'w'})^{1/2}$, where overbars represent time averages. The trees' tops were about 17 m high, and eddy correlation sensors were mounted at about 23 m height. The site and the instrumentation routinely operated there are described in detail by Arnts et al. (1978).

Large diurnal variations are evident in Fig. 1, with nocturnal aerodynamic resistances exceeding 10 sec/cm on one night and approaching 1 sec/cm on most others.

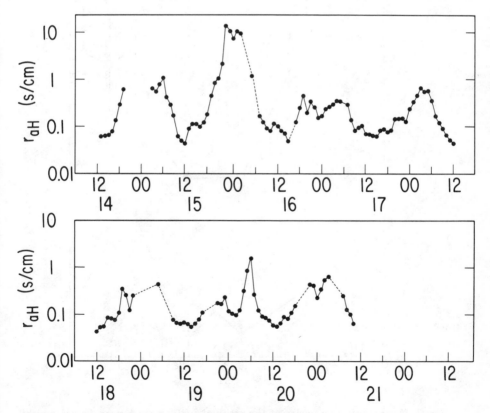

Fig. 1. Aerodynamic resistance to sensible heat transfer, as indicated by eddy correlation measurements above a pine plantation during the period 14–21 July 1977. Dates and times (EST) are shown.

Surface Resistance, r_s

In the case of sensible heat transfer, the term $[\ln(z_0/z_H)]/(ku_*)$ in (5) represents an additional resistance which can be associated with the effects of molecular diffusion in the immediate vicinity of the plant surfaces (see Garratt and Hicks, 1973, for example); field experiments over natural surfaces show that $\ln(z_0/z_H)$ should have been about 2.0 in the case of the pine forest of present interest. For the transfer of water vapor, it is well known that yet another resistance plays an important role, associated with the extent of stomatal closure, which in turn is strongly influenced by other factors such as the radiation intensity, plant water stress, etc. (see Norman, 1979, for example). Wesely et al. (1978) present experimental data that indicate a stomatal influence on the rate of uptake of ozone by maize, and they expand these considerations to include mesophyll and cuticular resistances. While the relevance of these properties to the case of the fluxes of soluble gases is obvious, it is not clear that partially soluble gases and particles can be addressed in a similar way.

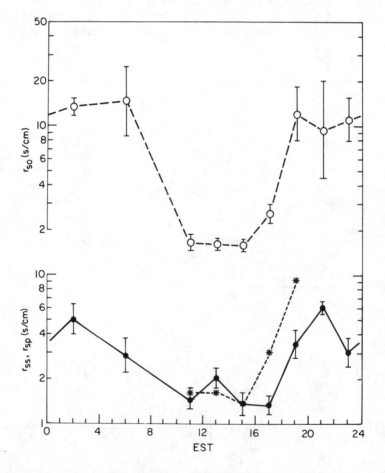

Fig. 2. Average diurnal cycles of the residual surface resistances determined by direct measurement of the eddy fluxes. Open circles, ozone; dots, airborne sulfur; asterisks, small particles. Error bars indicate ± one standard error.

Figure 2 illustrates the average diurnal cycles of the residual surface resistances calculated from eddy fluxes and air concentrations of the three quantities measured during the July 1977 experiment. In each case, fluxes F and concentrations C are used to determine total resistances $R = C/F$, from which aerodynamic resistances r_{aH} [computed as in Eq. (6)] are subtracted in order to determine residual "surface resistances" for ozone (r_{so}), small particles (r_{sp}), and atmospheric sulfur (r_{ss})*. Ozone fluxes were measured by means of eddy correlation employing an ethylene-reaction ozone concentration sensor especially modified to have a time constant shorter than 1 sec (Cook and Wesely, 1977). A modified aerosol "charger"† was used to measure fluxes of particles probably in the size range 0.1–0.4 μm diameter (see Husar et al., 1976; Wesely and Hicks, 1977; Wesely and Hicks, 1979). Sulfur fluxes were evaluated by use of a modified flame photometric sulfur detector, which in this particular case was incapable of differentiating between gaseous and particulate sulfur components.‡ It should be noted, in passing, that the questions of surface uptake being addressed have required that attention be focused on data for which the surface was a sink for the quantity in question; thus occasions on which upward fluxes were detected and some other cases apparently strongly influenced by them have been omitted. For this reason, relatively few values of r_{sp} can be computed for nighttime conditions, since then the pine plantation appeared to be a strong source of small particles. It also appeared that a small upward flux of sulfur sometimes occurred at night, but the explanation of these observations has remained elusive, and their practical significance continues to be uncertain.

Daytime values of r_{so}, r_{ss}, and r_{sp} evident in Fig. 2 appear to average about 1.4 sec/cm, which is what might be expected for stomatal resistance. For the case of ozone fluxes to maize, Wesely and Hicks (1979) have investigated the dependence of surface resistance on a range of external properties and find support for the hypothesis that stomata exert considerable control in this particular case. Garland (1977) and Wesely and Hicks (1977) have suggested similar behavior for sulfur dioxide, but there seems no obvious reason for small particles to respond to such a biological mechanism. The possibility that the similarity in surface residual resistances is fortuitous can be addressed by looking for a wind-speed dependence, which, if found, would indicate the importance of some mechanism other than stomatal.

Table 1 lists the results of a correlation analysis between each of the surface resistances and u_*, performed logarithmically so as to minimize the effects of extremes resulting from errors of measurement. In the cases of ozone and sulfur, no significant correlation with u_* is found. However, a significant correlation is evident in the case of particles, corresponding to a power-law relationship

$$r_{sp} \propto u_*^{\,p}, \tag{7}$$

with the value of the exponent p being 1.0 ± 0.3. Thus, there appears to be a mechanistic difference between the cases of particle and ozone deposition to the pine plantation reported on here. Daytime deposition of sulfur appears to behave as for

---

*Note that the surface residual resistances reported here would differ slightly if the term $\ln(z_0/z_C)$, analogous to $\ln(z_0/z_H)$ in Eq. (5), were taken into account. For the present comparative purposes, this particular complexity has been avoided.

†Supplied by Washington University of St. Louis.

‡Provided and operated by the Aerosol Research Branch, ESRL, U.S. EPA.

Table 1. Results of a correlation analysis, performed logarithmically, between daytime (1100–1600) surface residual resistances and friction velocity

	r_{so}	r_{ss}	r_{sp}
Number of observations, n	35	25	28
Average value of $\ln(r_s)$, x_1	0.50	0.45	0.41
Standard deviation, ς_1	0.33	0.50	0.50
Average value of $\ln(u_*)$, x_2	3.83	4.02	3.82
Standard deviation, ς_2	0.25	0.15	0.29
Correlation coefficient, r	0.25	0.13	0.56
	±0.16	±0.20	±0.13
Exponent of power law, p	0.33	0.43	0.98
	±0.22	±0.67	±0.27

Notes: (a) The standard error on r can be approximated as $(1 - r^2)/\sqrt{n}$.
(b) The value of p is calculated as $r\varsigma_1/\varsigma_2$.
(c) The standard deviation of p is then $(\varsigma_1/\varsigma_2) \cdot \sqrt{(1 - r^2)/n}$.
(d) Calculations have retained more significant digits than are tabulated.

ozone, even though much of the sulfur was present in particulate form. Further consideration of this matter is not warranted, since the data are highly scattered and because the possibility of an artificial correlation resulting from errors in measurement of properties like u_* cannot be disproved.

Nighttime ozone and sulfur residual surface resistances differ greatly, however. Previously, Wesely and Hicks (1979) have shown that ozone uptake can be very slow at night, especially when dewfall occurs and the surface is wet. However, the deposition of atmospheric sulfur is likely to be enhanced in these circumstances. While it is not clear that this is the cause of the differences seen in Fig. 2, it is worth while to note that the average ozone and sulfur surface resistances at night do indeed conform with this picture. After dusk, values of r_{so} quickly rise to more than 10 sec/cm, and they appear to continue to rise slowly through to about dawn (averages are 11.0 sec/cm at 2300 hr, 13.5 sec/cm at 0200 hr, and 14.7 sec/cm at 0600 hr). On the other hand, sulfur values increase more slowly and appear to peak at about 6.1 sec/cm at 2100 hr, dropping to 2.9 sec/cm by 2300 hr.

DISCUSSION AND CONCLUSIONS

It is clear that the diurnal cycles evident in Figs. 1 and 2 can be combined to produce deposition velocities that vary strongly throughout a day. It is equally clear that the selection of a particular "cardinal" value for the deposition velocity of any specific pollutant is a matter which is likely to be more subjective than objective, since the processes involved will probably be influenced by meteorological, physical, chemical, and biological factors, even in the case of particles. The data presented here show that aerodynamic factors are likely to control the surface fluxes of pollutants on many occasions, especially on summer nights, when air above the height of measurement used here is likely to be stability-decoupled from the constraints normally imposed by surface drag. In these highly stratified flows, surface fluxes of all pollutants carried aloft might as well be taken to be zero.

The specific processes that contribute to the deposition of particles remain unclear; the present finding that r_{sp} varies approximately linearly with u_* might be indicative of a mechanical effect within the canopy, perhaps related to resuspension or emission of small particles from the trees. It is also possible that the result is a consequence of experimental error, perhaps due to some undetected instrumental problem that might cause dynamic pressure fluctuations, for example, to contaminate particle concentration signals. More data are certainly required to generate confidence in the present $r_{sp} \sim u_*$ relationship.

Ozone fluxes appear to be controlled more often by surface properties than either sulfur or particle fluxes. This is particularly evident at night as a consequence of the low solubility of ozone and the related small uptake rate whenever foliage is wet. During the daytime, ozone fluxes appear to be stomatally controlled, as has been demonstrated elsewhere for other types of vegetation.

The inability of the sulfur sensor to respond independently to either particulate or gaseous sulfur was a severe drawback of this experiment. This problem, which was associated with the introduction of critical pressure drops whenever filters or denuding tubes were inserted in the sampling system, was solved later, and a series of field experiments has since been performed. The data obtained in these later experiments will be reported elsewhere in due course.

ACKNOWLEDGMENTS

This work was supported jointly by the U.S. Department of Energy and the U.S. Environmental Protection Agency.

REFERENCES

Arnts, R. R., R. S. Seila, R. L. Kuntz, F. L. Mowry, K. R. Knoerr, and A. C. Dudgeon. 1978. Measurements of α-pinene fluxes from a loblolly pine forest. Proceedings, 4th Conference on Sensing of Environmental Pollutants, Am. Chem. Soc., pp. 829–833.

Cook, D. R., and M. L. Wesely. 1977. Modification of an ozone sensor to permit eddy-correlation measurement of vertical flux. Argonne National Laboratory Radiological and Environmental Research Division Annual Report, ANL-77-65, Part IV, pp. 107–112.

Dyer, A. J. 1974. A review of flux-profile relationships. Boundary-Layer Meteorol. 7:363–372.

Garland, J. R. 1977. The dry deposition of sulphur dioxide to land and water surfaces. Proc. R. Soc. London, Ser. A 354:245–268.

Garratt, J. R., and B. B. Hicks. 1973. Momentum, heat and water vapour transfer to and from natural and artificial surfaces. Q. J. R. Meteorol. Soc. 99:680–687.

Husar, R. B., E. S. Macius, and W. P. Dannevik. 1976. Measurement of dispersion with a fast response detector. Proceedings Third Symposium on Atmospheric Turbulence, Diffusion and Air Quality, Am. Meteorol. Soc., pp. 293–298.

Norman, J. 1979. Modeling the complete crop canopy. Chapter 3.6 of "modification of the Aerial Environment of Crops," Am. Soc. Agric. Eng., pp. 249–277.

Sinclair, T. R., L. H. Allen, Jr., and E. R. Lemon. 1975. An analysis of errors in the calculation of energy flux densities above vegetation by a Bowen-ratio profile method. Boundary-Layer Meteorol. 8:129–139.

Wesely, M. L., J. A. Eastman, D. R. Cook, and B. B. Hicks. 1978. Daytime variations of ozone eddy fluxes to maize. Boundary-Layer Meteorol. 15:361–373.

Wesely, M. L., and B. B. Hicks. 1977. Some factors that affect the deposition rates of sulfur dioxide and similar gases on vegetation. J. Air Pollut. Control Assoc. 27:1110–1116.

Wesely, M. L., and B. B. Hicks. 1979. Dry deposition and emission of small particles at the surface of the earth. Proceedings Fourth Symposium on Turbulence, Diffusion, and Air Pollution, Am. Meteorol. Soc., pp. 510–513.

DISCUSSION

Orie Loucks, Institute of Ecology: Could you clarify the distinction between aerodynamic resistance and residual surface resistance; specifically, what part of it is stomatal resistance?

B. B. Hicks: The aerodynamic resistance is that part of the transfer, or is a characteristic of that part of the transfer that is determined by atmospheric conditions. It includes the effect of convection—convective outcrops in the daytime—and it includes the effect of stratification at night. The stomatal resistance is part of the residual that I have been talking about.

Orie Loucks: Can you further explain what part of the resistance is a strict surface resistance segregated from the stomatal resistance?

B. B. Hicks: For SO_2, for example, recall the slide that Dr. Chamberlain showed which illustrated several different resistances from a level above the canopy down to the ground. One of those would represent resistance describing transfer into the soil beneath the canopy; another one would represent cuticular resistance into the leaves but not with the stomata; another one would represent stomatal resistance. The smallest of those resistances is going to determine to a large extent the deposition process because they are parallel. In practice, we find that in daytime, when the stomates are open, that this pathway is very efficient. For ozone our data indicate almost precisely the same resistance as for water vapor. What happens when the stomata close down and we have to estimate the deposition velocity to our canopy, which is now not absorbing the gases as fast as it can? I'd like to simply point out that I would go with cuticular resistance most of the time, in addition to something perameterizing the soil transfer. But be very careful about dewfall, because once the canopy gets wet the situation is vastly different. Finally I would emphasize that at night it is quite likely that you'll begin to get a controlling influence by the aerodynamic processes.

Leonard Newman, Brookhaven Lab: Bruce, did you say that stomatal resistance is not important to deposition of particulates?

B. B. Hicks: Yes, but what I meant to say is that the relevance of the physical picture isn't clear. For example, there's a theory that hygroscopic particles will rapidly grow when they encounter a humidity gradient. You can picture the situation in that way, but the precision of the analogy isn't clear. I hope I didn't mislead anyone else.

A. C. Chamberlain, Atomic Energy Research Establishment: Can you comment further on the phenomenon of particle emission from canopy surfaces?

B. B. Hicks: We have a picture slowly developing in our minds that, in fact, many surfaces reprocess particles. We certainly see evidence of particles being emitted from a number of the surfaces over which we work. For example, with acid sulfate particles, ammonium bisulfate might adsorb onto a surface. Then at night a gust of wind through the canopy can resuspend the particle. The entire process can be exceedingly complicated, as far as we can tell.

CHAPTER 24

EXPERIMENTAL TECHNIQUES FOR DRY-DEPOSITION MEASUREMENTS*

J. G. Droppo†

ABSTRACT

Recent experimental approaches for measurement of dry surface fluxes of sulfur compounds are reviewed and discussed with emphasis on surface-layer studies. Evidence indicates that natural surfaces are sources as well as sinks for sulfur. Field results are related to requirements and implications in future studies of surface sulfur fluxes.

INTRODUCTION

The motivation for recent studies of surface pollutant fluxes derives from a need for better quantification of atmospheric pollutant budgets. For example, models of long-range sulfur transport are severely limited by the state of the art for definition of surface source and sink relationships.

The understanding of any subject is no better than experimental observations that provide verification of theories and models. The basis for dry deposition remains relatively uncertain despite a multitude of recent field and laboratory experiments. Each has provided glimpses of the complexity of the processes, but in total they do not provide a complete picture. Recent field results do provide a basis for evaluation of experimental techniques for future research.

A major point from recent studies is that dry removal of air pollutants should not be expected under all conditions. Natural surfaces are sources as well as sinks for air pollutants. To allow for this, "surface pollutant flux" or "surface flux" is proposed in lieu of terms implying either a surface sink or source for dry surface pollutant fluxes.

There are a variety of viable research approaches for studying pollutant flux rates. Each of these has contributed to the overall understanding of surface flux rates. No one approach has proven significantly better than any other. Each approach has indigenous shortcomings that limit interpretation and/or application of results. A detailed discussion of the advantages and limitations of the experimental approaches is given in Droppo and Hales (1974).

This paper reviews recent experimental efforts and relates these to implications for application of experimental approaches.

---

*This paper is based on work performed under the U.S. Department of Energy Contract EY-76-C-06-1830.

†Pacific Northwest Laboratory, Richland, Washington 99352.

EXPERIMENTAL APPROACHES

A conceptual framework for the experimental approaches sorted by the scale of application is given in Fig. 1. These are discussed below with selected examples of recent applications.

Starting at the left, the studies with the greatest horizontal extent and relatively short time scale are listed as "transport budget." These approaches involve computing the surface flux rates for individual plumes based on pollutant budgets. Dry surface removal rates are computed from pollutant budgets. One variation of this approach incorporates the concentration changes of a tracer material for a more direct determination of surface flux rates; this allows the possibility of expansion of the scale of applicability of the approach.

As an example of the total budget approach, Gillani (1978) used data for the plume from a coal-fired plant to compute dry-deposition velocities for sulfur dioxide; he found peak values between 1.0 and 1.5 cm/sec. Young et al. (1976) estimated dry removal rates for several pollutants, including sulfur dioxide, using a budget approach on the plume from a large metropolitan area.

The "surface layer flux" approaches are based on measurements of the vertical fluxes of pollutants in the constant flux layer (as defined by momentum fluxes). The techniques involve direct application of methods developed for studying momentum, heat, moisture, and carbon dioxide fluxes. Recent applications will be given in a separate section below.

The "surface flux" approaches are studies of fluxes directly over the surfaces; they differ from "surface layer flux" approaches mainly by their exclusion of atmospheric eddy transport. In reality, many "surface flux" studies include a component of the eddy transport, but the emphasis is clearly on the flux near the receptor or source.

Examples are chamber, box, wind tunnel, and stomatal response studies. These include both laboratory and field studies. For example, Beckerson and Hofstra (1978) found evidence in laboratory studies that the combined effects of sulfur dioxide and ozone were different from the effects of the gases alone in terms of stomatal resistances. Although these studies were at damage-level concentrations, the implication is that sulfur dioxide surface fluxes cannot be considered independently of other atmospheric pollutants.

A surface flux field study has provided valuable information on natural sulfur source flux from salt marches on the coast of North Carolina (Aneja et al., 1979). They were able to determine surface flux rates of H_2S plus COS and $(CH_3)S$ using a flux reactor chamber technique.

Using a well-stirred-chamber method, Milne et al. (1979) have obtained sulfur dioxide removal rates that are generally less than those reported over similar surfaces in less arid regions. The technique attempts to minimize atmospheric eddy transport by providing sufficient mixing. Small removal rates were found with an approach that should provide an upper range of transport rates. These results underline the importance of considering both type and condition of surface.

"Receptor uptake budgets" are studies of incremental changes in pollutant concentrations within the receptors as a result of atmospheric exposure. The very small change in concentrations attributed to pollutant fluxes seriously limits practical application of this method for most pollutants. This can be overcome by either using a unique tracer with a low background concentration or conducting budget experiments over a sufficiently long time period to produce significant changes.

For example, the component input by dry and wet deposition of atmospheric sulfur to two European forests was measured over a period of six years (Mayer and

SCALES OF EXPERIMENTAL TECHNIQUES FOR DETERMINING SURFACE POLLUTANT FLUXES

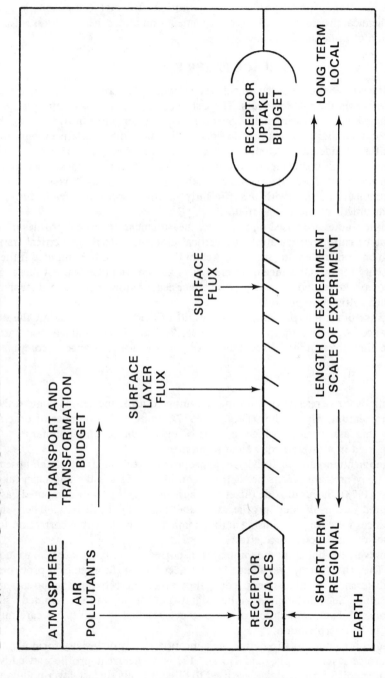

Fig. 1. Scales of experimental techniques for determination of dry surface pollutant flux rates.

Ulrich, 1978). Their results in terms of total fluxes indicate that the fluxes over forests are considerably greater than over unforested areas.

The approaches illustrated in Fig. 1 cover the range of application from computation of regional air quality to determination of dose rates on receptors. The experiments range from dependence on definition of complex atmospheric physical and chemical processes to no direct dependence on atmospheric processes for computation of pollutant flux rates.

SURFACE-LAYER FLUX STUDIES

Surface-layer flux research on sulfur compounds has comprised a major fraction of efforts over the past few years. These studies have been in two major groups: profile and eddy flux measurements. Surface flux rates for sulfur dioxide have been the topic of a number of recent profile field studies, providing a relative wealth of data on sulfur dioxide fluxes compared with just a few years ago. However, in comparison with sulfur dioxide, the reported studies of sulfur aerosol surface fluxes are fewer and have fewer data points. Both profile and eddy flux studies require measurements at sites with a sufficiently uniform horizontal fetch during time periods practically devoid of trends.

Profile studies are based on the direct measurement of vertical profiles of sulfur dioxide or sulfur particulates. The vertical gradient reflects the vertical transport rate to or from underlying surfaces. Actual flux rates could be computed if the eddy diffusivity for the substance of interest were known. In practice this value is generally not available, and an analogy with either heat or momentum eddy diffusivities is assumed in flux computations.

It is useful for interpretation of the results of field studies to separate the surface resistance, R_s, from atmospheric resistance, R_a. The ratio of ambient concentration to the flux gives the total resistance, R_t. The surface resistance is computed by:

$$R_s = R_t - R_a. \tag{1}$$

The atmospheric resistance is normally computed using the aerodynamic resistance for momentum. This author (Droppo, 1977), along with a number of others, has made this assumption, which is strictly only valid if the pollutant is being transported in a fashion equivalent to momentum.

If momentum eddy diffusivity is appropriate, then the effect of the absence of bluff body pressure effects needs to be considered. This is the approach taken by Fowler (1978). Sulfur dioxide fluxes over agricultural crops were computed based on measured sulfur dioxide profiles and thermal eddy diffusivity. Then surface resistances were computed using aerodynamic resistances, with a correction for the absence of bluff-body force effects.

However, the pollutant may not be transported in a fashion identical to momentum. Then the "surface resistance" will reflect both surface removal processes and any difference in atmospheric eddy flux processes between the pollutant and momentum. Such differences can be a significant fraction of the computed "surface resistance." The computation of a resistance residual is useful, but care must be taken in its interpretation.

Studies are under way to define the appropriate eddy diffusivity for pollutants in the surface layer (Droppo and Doran, 1979). Concurrent profile and eddy flux measurements of ozone fluxes are used to directly study surface-layer pollutant turbulent transport. Preliminary results indicate that ozone is transported at least as effectively as the heat eddy diffusivity implies.

The advent of remote sensing by correlation spectroscopy provides a new tool for remote atmospheric sulfur dioxide studies (Millán and Hoff, 1978). In a profile study by Platt (1978), a system using tower-mounted mirrors and differential optical spectroscopy provided a means of monitoring sulfur dioxide profiles over an extended time period. He found agreement within 20% with short-term comparisons using chemically determined profiles. His analysis assumed that the sulfur dioxide eddy diffusivity was equal to momentum eddy diffusivity. An interesting discussion of this paper by Brimblecombe (1979) and Platt (1979) mentions a fairly large frequency of cases where the SO_2 concentration decreased with height and suggests reasons for this behavior.

Studies of sulfur aerosol fluxes are inherently more difficult than those of sulfur dioxide because of the much smaller relative flux rates. Recent studies tend to confirm the previously postulated order-of-magnitude differences between these fluxes. Obtaining equivalent experimental accuracy for sulfur aerosols is made more difficult for both the profile and eddy flux approaches by the smaller ratio between fluxes and the natural concentrations.

One recent study of sulfur aerosols used integrated gradients measured over extended time periods, compared with those normally used in profile studies (Everett et al., 1979). They found larger values for particulate sulfur dry removal than previously reported. Deposition velocities averaging about 1.4 cm/sec were nearly equal to the momentum transfer velocities. These are greater than most other published data, which are on the order of 0.1 to 0.2 cm/sec (Droppo, 1977; McMahon and Denison, 1979; Garland, 1978). Sehmel (1976) points out that extrapolation of his wind-tunnel rates for particle deposition to very rough surfaces does produce relatively large predicted deposition velocities. Sulfur aerosol fluxes clearly require more research to better define the processes.

Eddy flux studies are based on the computation of fluxes from measured time series of vertical velocity and sulfur concentrations. This method provides a direct measurement of the sulfur flux. The only serious limitation is the requirement for a sensor that accurately indicates sulfur concentration fluctuations. Sufficiently fast-response sulfur monitors have become available in the past few years (Hicks and Wesely, 1978; Hadjitofi and Wilson, 1979). The first results of application in eddy flux studies are encouraging.

Hicks and Wesely (1978) measured both total sulfur (sulfur gas and particulate) and particles (0.05 to 0.10 μm) over a pine forest using the eddy correlation approach. They found that the total sulfur fluxes were downward during the daytime. However, the total sulfur flux was upward at a much smaller rate during nightime. The canopy was a source of small particles during the afternoon. These results raise the problem of explaining such a sulfur flux reversal.

The application of the eddy flux technique to the flux of sulfur aerosols has an additional limitation beyond the normal spectral response and related requirements. Sulfur does not occur in a single chemical compound in the atmosphere. Instead, there are multiple forms for which measurement techniques often have significantly different sensitivity (Robinson, 1978). The eddy correlation method requires that the signal from the sulfur monitor accurately reflect actual sulfur concentration fluctuations. Unequal sensitivity to different sulfur compounds (that may have fluxes in different directions) could easily result in spurious sulfur flux values.

PNL PARTICULATE SULFUR FLUX STUDIES

Our profile surface-layer studies at PNL have produced results that are pertinent to the topic of this paper (Droppo et al., 1976; Droppo, 1977; Droppo and Doran,

1979). These studies entail the acquisition of approximately 1-hr nuclepore filter samples taken at three log-spaced heights. The volume and temperature of air drawn through each filter were monitored to allow accurate flow corrections. The material on the filters was analyzed using a specially developed x-ray fluorescence procedure to maximize the accuracy of the gradient values. Three levels of detailed micrometeorological data were also recorded to define eddy diffusivities. The results from three sites are presented here: Centralia, grassland; Hanford, arid vegetation; and Sangamon, cornfield.

The Hanford tests over arid surfaces during low winds produced results that most closely agreed with classical models of aerosol removal. In addition, the gradient accuracy appeared to be the best in the Hanford tests. Figure 2 shows one set of sulfur

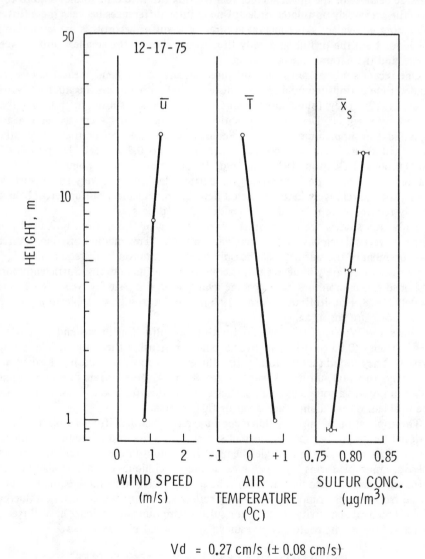

Fig. 2. Particulate sulfur, wind, and temperature profiles at Hanford. Error bars illustrate ±1% resolution.

concentration, wind, and temperature profiles at Hanford. Three levels were chosen as the minimum needed to both define the gradient and provide an intermediate point to assess the accuracy of resolution of the gradient. Deviations from a linear fit with logarithmic height will reflect both experimental variability and natural curvature. Hanford tests indicated resolution on the order of ±1%. Tests at other sites showed more variability. Our current estimate is that the gradient accuracy ranges between ±2% and ±5%. Absolute accuracy for sulfur concentrations is not as good as the relative accuracy.

Table 1 contains the two best Hanford tests with near-background sulfur aerosol concentrations. The deposition velocities of a few tenths of a centimeter per second agree with values extrapolated from wind-tunnel tests. This may reflect the fact that the arid surfaces most closely match the wind-tunnel surfaces. These are computed with momentum eddy diffusivities. The atmospheric resistances are momentum transfer resistances, and the surface resistances are residual resistances as discussed above.

Table 1. Summary of results for dry deposition of sulfur aerosols

Test	V_d (cm/sec)	R_t (sec/cm)	R_a (sec/cm)	R_s (sec/cm)	$\bar{u}$ (17 m) (m/sec)
1	0.10	10.0	2.3	7.7	1.13
3	0.27	3.7	0.44	3.3	1.28

The data taken over nonarid vegetation show quite a different trend. Daytime particulate sulfur gradients consistently indicate a surface source term. Figures 3 and 4 show the sulfur profiles for the Centralia and Sangamon sites, respectively. These are presented to illustrate trends in profiles.

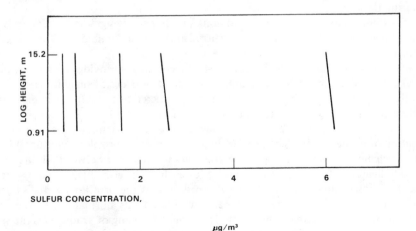

MAY 1975 GRADIENTS OVER GRASSLAND

Fig. 3. Centralia particulate sulfur profiles.

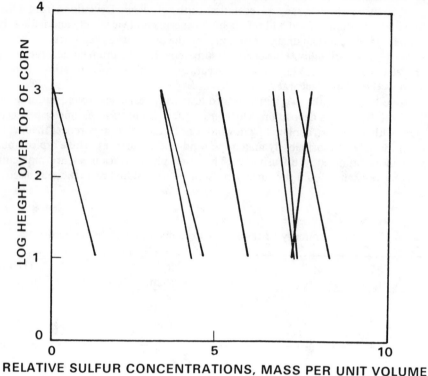

Fig. 4. Sangamon particulate sulfur profiles.

Centralia daytime tests consistently had negative gradients—the profiles at lower concentrations have about the same percentage change as profiles at higher concentrations. Much of the consistency is likely the result of the similar midday conditions during these tests.

Sangamon daytime tests also had consistently negative gradients. The only test that suggested a surface sink effect occurred at the highest ambient sulfur concentrations.

The Sangamon profile with the lowest concentrations provides an interesting insight to particulate surface and sink processes. The order-of-magnitude change in sulfur concentration over the profile may be expected as a result of limited vertical dispersion for the early morning time of this test.

The x-ray fluorescence analysis provides comparable accuracy of relative concentration of elements other than sulfur. Figure 5 shows the profiles of sulfur and three other elements. The ordering of the gradients is consistent between all three levels and indicates different vertical flux rates for each of the elements, including a reversal to a downward flux for silicon. The ordering of the profiles was the same for other tests. Figure 6 shows a comparison of gradients normalized to the upper heights with this and two other tests. The order of sorting of gradients is the same; the direction of the silicon gradient has reversed in one run, however.

Uncertainties in the gradients have precluded computation of specific flux rates. However, the order of magnitude of these gradients indicates that the source term is

JULY 1976 GRADIENTS OVER CORN

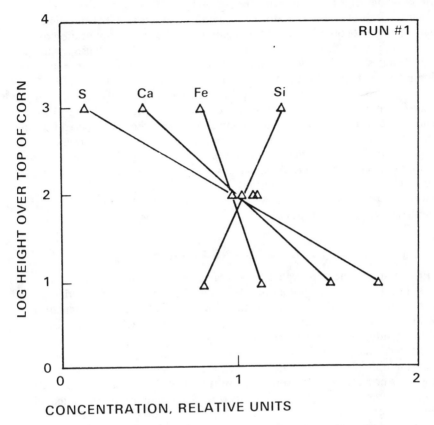

CONCENTRATION, RELATIVE UNITS

Fig. 5. Normalized Sangamon particulate profiles for various elements.

NORMALIZED GRADIENTS OVER CORN - JULY 1976

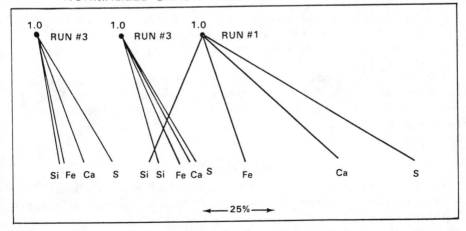

Fig. 6. Normalized particulate gradients for three Sangamon tests.

equivalent in magnitude (but opposite in sign) to removal rates predicted by deposition velocities of a few centimeters per second.

Our analysis of sulfur and ozone flux data indicates that it is critically important that the ambient conditions meet the requirements for proper application of profile or eddy flux methods, both for the meteorological variables and also for the pollutant of interest. Our experience is that failure to operate only under appropriate conditions injects very large variability into the results. In detailed studies of ozone fluxes by both eddy flux and profile methods, an acceptable level of variability that allowed comparison of methods was obtained only when all tests with bimodal winds, significant trends, bad fetches, etc., were eliminated. Computed flux values were not used as criteria. The strict criteria eliminated cases with extreme as well as what appeared to be reasonable flux values. The result was that the total resistance values for ozone were in a relatively narrow range (270 ± 160 sec/cm for grassland, 290 ± 150 sec/cm for arid vegetation). These narrow ranges are highly unusual for reported deposition resistances; we attribute this primarily to the use of stringent criteria for acceptance of tests.

CONCLUDING REMARKS

Profile studies have provided considerable information on sulfur dioxide flux rates. Resolution of the appropriate characterization of eddy diffusivities is needed for both sulfur dioxide and particulate sulfur profile studies to eliminate uncertainties in flux computations and to allow computation of surface resistances that do not include residual atmospheric terms.

The particulate source and sink processes over natural surfaces cannot be considered as a simple unidirectional single-rate flux. There is evidence that natural surfaces may be either sinks or sources of sulfur particulates. The nonequivalent normalized gradients for different elements (Fig. 5 and 6) illustrate the potential complexity of the fluxes.

Studies of particulate sulfur fluxes need to be specific to sulfur compound. In addition, account needs to be taken of varying sensitivity of monitors to different forms of particulate sulfur.

Based on the recent research results, future sulfur surface flux studies need to incorporate more detailed monitoring of the ambient meteorological, surface, and air pollutant concentrations.

In surface-layer studies, the variability of results using either profile or eddy flux methods may be considerably reduced by strict adherence to meeting required ambient conditions. Experimental methods should not be applied out of the range of their valid time and space scales.

The state of the art in surface flux measurements has not yet been able to meet the information need; sulfur sink and source surface processes are only incompletely understood. The way is open for improvement of current or development of new experimental approaches.

REFERENCES

Aneja, V. P., J. H. Overton, L. T. Cupitt, J. L. Durham, and W. E. Wilson. 1979. Direct measurements of emission rates of some atmospheric biogenic sulfur compounds. Tellus 31:174–178.

Beckerson, D. W., and G. Hofstra. 1979. Stomatal responses of white bean to O_3 and SO_2 singly or in combination. Atmos. Environ. 13:533–535.

Brimblecombe, P. 1979. Discussion: dry deposition of SO_2. Atmos. Environ. 13:1064–1065.

Droppo, J. G. 1977. Proof test of a system to measure the dry deposition of sulfur aerosol in the plume from a Northwest fossil fuel power plant. In Pacific Northwest Laboratory Annual Report of 1976 to the ERDA Assistant Administrator for Environmental Safety, Part 3, Atmospheric Sciences, BNWL-2100 PT 3. Pacific Northwest Laboratory, Richland, Washington 99352.

Droppo, J. G., and J. C. Doran. 1979. Measurements of surface layer turbulent ozone flux processes. In Preprint, Fourth Symposium on Turbulence, Diffusion and Air Pollution, Reno, Nevada. American Meteorological Society, Boston, Mass.

♦ Droppo, J. G., and J. M. Hales, 1974. Profile methods of dry deposition measurement. In Proceedings of Symposium on Atmosphere-Surface Exchange of Particulate and Gaseous Pollutants, Richland, Washington, Sept. 4–6, 1974, R. J. Engelmann and G. A. Sehmel, coordinators (U.S. ERDA CONF-740921) pp. 192–211. Available from NTIS, Springfield, Va. 22161.
Available from NTIS, Springfield, Va. 22161.

Droppo, J. G., D. W. Glover, O. B. Abbey, C. W. Spicer, and J. Cooper. 1976. Measurement of Dry Deposition of Fossil Fuel Pollutants. Final Report under EPA Contract 68-02-1747, Battelle, Pacific Northwest Laboratories, Richland, Wash.

Everett, R. G., B. B. Hicks, W. W. Berg, and J. W. Winchester. 1979. An analysis of particulate sulfur and gradient data collected and Argonne National Laboratory. Atmos. Environ. 13:931–934.

Fowler, D. 1978. Dry deposition of SO_2 on agricultural crops. Atmos. Environ. 12:369–373.

Garland, J. A. 1978. Dry and wet removal of sulphur from the atmosphere. Atmos. Environ. 12:349–362.

♣Gillani, N. V. 1978. Project MISTT: mesoscale plume modeling of the dispersion, transformation, and ground removal of SO_2. Atmos. Environ. 12:569–588.

Hadjitofi, A., and M. J. G. Wilson. 1979. Fast response measurements of air pollution. Atmos. Environ. 13:751–760.

Hicks, B. B., and M. L. Wesely. 1978. Recent Results for Particle Deposition Obtained by the Eddy-Correlation Method, Environmental Research contribution No. 78-12. Argonne National Laboratory, Argonne, Ill.

Mayer, B., and B. Ulrich. 1978. Input of atmospheric sulfur by dry and wet deposition to two Central European forest ecosystems. Atmos. Environ. 12:375–377.

McMahon, T. A., and P. J. Denison. 1979. Review paper—empirical atmospheric deposition parameters—a survey. Atmos. Environ. 13:571–585.

Millán, M. M., and R. M. Hoff. 1978. Remote sensing of air pollutants by correlation spectroscopy—instrumental response characteristics. Atmos. Environ. 12:853–864.

Milne, J. W., D. B. Roberts, and D. J. Williams. 1979. The dry deposition of sulphur dioxide—field measurements with a stirred chamber. Atmos. Environ. 13:373–379.

Platt, U. 1978. Dry deposition of SO_2. Atmos. Environ. 12:363–367.

Platt, U. 1979. Author's reply to discussion: dry deposition of SO_2. Atmos. Environ. 13:1065.

Robinson, J. W. 1978. Discussion: total sulfur aerosol concentration with an electrostatically pulsed plume photometric detector system. Atmos. Environ. 13:1064.

Sehmel, G. A. 1976. Predicted dry deposition velocities, in Proceedings of Atmospheric-Surface Exchange of Particulate and Gaseous Pollutants, CONF-740921. Department of Commerce, Springfield, Va.

Young, J. A., T. M. Tanner, C. W. Thomas, N. A. Wogman, and M. R. Peterson. 1976. Concentrations and Rates of Removal of Contaminants from the Atmosphere In and Downwind of St. Louis, PNL Annual Report for 1975 to USERDA Division of Biomedical and Environmental Research, Part 3, Atmospheric Sciences, BNWL-2000 PT3, UC-11. Pacific Northwest Laboratory, Richland, Wash. 99352.

DISCUSSION

V. P. Aneja, Northrop Services: I just finished doing a velocity of deposition experiment myself, so I fully appreciate the complexity and problems you face. My question is, do you believe the numbers in the sulfur profiles you found?

J. G. Droppo: I can't do anything other than believe them because of their consistency.

V. P. Aneja: Well, I didn't expect to see reversed gradients; I expected to see removal. Does this fit? Are there estimates of the experimental variability? I believe this is ±50%, okay?

J. G. Droppo: I believe that they may be better than that, but I wouldn't support them any more than that. I do believe that there are reverse gradients over vegetation, and they may be within ±50%, but the gradient values may be double or half of what I showed. I do believe they go that way; I can't explain it any other way—it was very consistent in both experiments.

M. Lippmann, New York University Medical Center: Have you done or considered doing any size-selective aerosol sampling, the implication being that the aerosol generated locally from the surface should be larger in size than the sulfate aerosol depositing from regional transport?

J. G. Droppo: Yes, I believe the logical next step is to have the particle flux as a function of size distribution, because bacteria could be coming up from the surface and have sulfur compounds in them, and they would, of course, be in a larger size range than the sulfur aerosol that you normally think of as deposited.

B. D. Zak, Sandia Lab: If I remember correctly, it's Barringer Research in Canada that offers, as a commercial service, a survey technique which depends upon the fact that vegetation incorporates the minerals in the soil in which it is growing to some extent and then aerosolizes the minerals. I believe that one can go to Barringer and hire a helicopter and search for uranium or whatever, so . . . I believe so. I heard a talk, must have been three or four years ago, and I believe it's Barringer, but I can't swear to it.

J. G. Droppo: That's very interesting. I know of people that search for uranium with aircraft; I didn't realize that they were depending upon the plants to aerosolize. I assumed it was a mechanical or radiological sensor or something.

D. F. Gatz, Illinois State Water Survey: I wonder if you could comment on your filter sampling technique. You had data there for silicon, calcium, and iron, which are large particles; if you're not careful to be isokinetic in your sampling, you could introduce errors that would affect your results.

J. G. Droppo: That's a good point. I did the analysis for the sulfur particles to make sure I was isokinetic, but that was for less than 0.5 micron—I'm okay in that size range. I really have not considered the other sizes. I agree that if you aren't careful in the large ranges, you could get inertial effects, yes.

C. I. Davidson, Carnegie-Mellon University: I wonder if you could get an idea of the importance of soil resuspension by going back to the dates that you took the data and trying to get a handle on what the surface moisture levels and wind speeds close to ground were. There have been some experiments that have demonstrated the existence of a resuspension layer close to the ground which can spread out for many miles or even tens of miles in very dry soil conditions and appreciable wind speeds. It should be possible to do a correlation by looking at your data in comparison to meteorological parameters.

J. G. Droppo: It sounds very interesting. I think one thing I haven't looked at is the composition of the soil at the different sites. One set at Hanford was extremely dry. I agree that that is one way to study the problem, yes.

W. Graustein, Yale University: There was a paper about five or six years ago in *Science* by Fish, who studied the electrical generation of aerosols from plant tissues. He found that, in these electric fields, that particles in the size range of about three-tenths to one-half micron came flying off the vegetation. He didn't do any chemical studies on these, but certainly all the literature indicates the particles on these surfaces are loaded with nutrient elements. Some work that I've done has indicated that potassium, in particular, seems to be given off from foliage. I wonder if you have any analyses for potassium amongst your data?

J. G. Droppo: Yes, we have potassium, but the accuracy wasn't high enough to really include it, so I really didn't look at it. Your remark about charged particles, I think, involved the point made earlier that charged particles are a problem or could possibly be a problem. Dr. Chamberlain responded that the mobility is usually very low, but then if you have a very large metal tower altering local fields, who knows?

CHAPTER 25

MODEL PREDICTIONS AND A SUMMARY OF DRY DEPOSITION VELOCITY DATA*

George A. Sehmel†

ABSTRACT

Literature values of independent measurements of dry deposition velocities are summarized as a function of particle diameter and gas speciation. In most of the experiments reported in the literature, there are uncertainties that have hindered the development of general predictive deposition velocity models. However, one model discussed here (Sehmel and Hodgson, 1978) offers a more useful approach for predicting particle dry deposition velocities as a function of particle diameter, friction velocity, aerodynamic surface roughness, and particle density.

INTRODUCTION

Independent measurements of dry deposition velocities, v_d, for particles and gases have been reported in the literature for many field experiments. However, there has been limited cross comparison between results. In the following pages, some comparisons between experimental results are made for both particle and gas deposition. For particles, comparisons are made for experimental results in which the particle diameter range is indicated. For gases, ranges of deposition velocities are listed for 16 gases. More detailed comparisons are made for sulfur dioxide and iodine deposition. These comparisons show that there are large ranges in deposition velocity measurements. These ranges have hindered the generalization of results. Consequently, the literature is devoid of predictive deposition velocity models based solely upon experimental field data. Deposition velocity, K_{1m}, predictions based upon surface mass transfer measured in a wind tunnel are suggested to currently offer the most useful model for predicting dry deposition velocities, and some of these model predictions are presented.

DEPOSITION VELOCITY

The deposition velocity is used to describe transfer across the air-surface interface in atmospheric diffusion and transport models. Chamberlain and Chadwick (1953) defined the deposition velocity as the ratio of the deposition flux divided by the air-

---

*This paper is based on work performed under the U.S. Department of Energy Contract EY-76-C-06-1830.

†Pacific Northwest Laboratory, Richland, Washington.

borne pollutant concentration per unit volume at some height above a deposition surface. Usually, the deposition velocity is reported in units of centimeters per second. The particle deposition velocity is greater than or equal to the gravitational settling velocity, v_t.

Historically, the deposition velocity, v_d, has been defined as a deposition flux, F, divided by an airborne concentration, X. That is, for a depositing polydispersed aerosol and sometimes for a monodispersed aerosol,

$$v_d = - F/X. \tag{1}$$

(The minus sign is required since the flux downward is a negative quantity, whereas the deposition velocity is defined as positive.) For a monodispersed particle concentration, C, the deposition velocity, K_{1m}, is defined as

$$K_{1m} = - F/C. \tag{2}$$

The deposition velocity nomenclature K, rather than v_d, reflects that deposition velocities are a function of particle diameter. The subscript 1m indicates a 1-m reference height for the pollutant concentration.

EXPERIMENTAL PLUME DEPLETION

The dry deposition velocity can be used in atmospheric transport models to predict plume depletion. To date, experimental particle plume depletion measurements include up to 97% removal across a distance of 3200 m (Simpson, 1961), 65% between 45 and 91 m (Sehmel and Hodgson, 1976), and 80% at a distance of 200 m (Start, 1970a, 1970b). Data for iodine, a gas, indicate that plume depletion is up to 3% at a distance of 380 m (Zimbrick and Voilleque, 1969). This experimental data base is too limited to generalize plume depletion characteristics for particles and gases. To predict plume depletion for general situations, transport and diffusion meteorological models must be used that include a surface mass transfer coefficient such as the dry deposition velocity.

DRY DEPOSITION VELOCITIES FOR PARTICLES

Experimental dry deposition velocities for many materials and various deposition surfaces have been summarized by Sehmel (1979), who listed deposition velocities for different materials as a function of particle diameter. The range of experimental deposition velocities for each experiment was presented rather than the "average" deposition velocities. This range is critical in emphasizing the uncertainties in many deposition experiments and in our ability to predict deposition mass transfer rates. The deposition velocities reported range over five orders of magnitude from a minimum of 10^{-3} cm/sec to 180 cm/sec.

Since few measurements for deposition to water surfaces have been made (Slinn et al., 1978), the literature results were listed by Sehmel as a function of particle diameter for land surfaces. These particle deposition velocity data are organized graphically in Fig. 1. The ranges of deposition velocities for each set of experimental conditions are shown as a function of particle diameter range. The dashed lines are for field experiments determined with polydispersed aerosols. In contrast, the solid lines are data obtained with much narrower size distributions. In general, the data show the following:

1. The deposition velocities in any individual experiment range over several orders of magnitude.

PARTICLE DEPOSITION VELOCITIES MEASURED IN THE FIELD

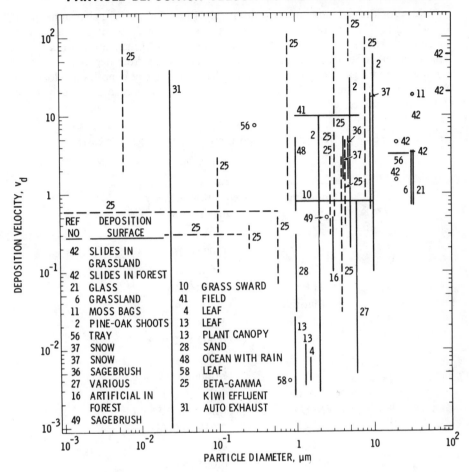

Fig. 1. Particle deposition velocities measured in the field. Numbers refer to references for cited data.

2. A minimum deposition velocity exists and is approximately 10^{-2} cm/sec for particle diameters in the range of 0.1 to 1 μm diameter.

Most data gained through field experiments, however, cannot be extrapolated to other situations because the variables have not been adequately controlled. Deposition results are often reported when a particle diameter distribution was investigated rather than a single diameter or a narrow particle diameter range. Although the particle diameter greatly influences deposition velocities, other variables do too, for instance, wind speed. Convair (1960) has shown that field deposition velocities increase with wind speed, even though in these field tests the deposition velocities were independent of atmospheric stability. Sehmel and Hodgson's (1978) model supports these test results by predicting that the deposition velocity is nearly independent of atmospheric stability.

DRY DEPOSITION VELOCITIES FOR GASES

Values for the experimental deposition velocities of gases were also summarized by Sehmel (1979). The ranges of deposition velocities for different gases are shown in Table 1. The range of velocities is at least one order of magnitude for each gas. For nitrogen oxides (NO and NO_x), the range is actually from negative to positive numbers. For all gases other than krypton (Kr), the range of deposition velocities is from 10^{-4} to 7.5 cm/sec. For Kr the maximum deposition rate is much lower, 2.3×10^{-11} cm/sec.

Table 1. Summary of deposition velocity ranges for gases

From Sehmel, 1979

Depositing gas	Deposition velocity range (cm/sec)
SO_2	0.04 to 7.5
I_2	0.02 to 26
HF	1.6 to 3.7
ThB	0.08 to 2.6
Fluorides	0.3 to 2.4
Cl_2	1.8 to 2.1
O_3	0.002 to 2.0
NO_2	1.9
NO	Minus to 0.9
PAN	0.8
NO_x	Minus to 0.5
H_2S	0.015 to 0.38
CO_2	0.3
$(CH_3)_2S$	0.064 to 0.28
CH_3I	10^{-4} to 10^{-2}
Kr	2.3×10^{-11} max

The deposition velocities for iodine (I_2) and sulfur dioxide (SO_2) in Table 1 have been studied extensively since those gases could be effluents from nuclear and non-nuclear industries respectively. Even though I_2 and SO_2 have been studied extensively, their deposition velocities are so widely scattered that they cannot be confidently predicted. The deposition velocity of iodine ranges three orders of magnitude, from 0.02 to 26 cm/sec; the deposition velocity of sulfur dioxide ranges two orders of magnitude, from 0.04 to 7.5 cm/sec.

The deposition velocity data for I_2 and SO_2 are further organized in Figs. 2 and 3. In each figure, the data have been organized or ranked according to the maximum deposition velocity reported in each experiment. Commonly, deposition velocities are a function of experimental conditions and show a wide range even for the same types of deposition surfaces. This wide range is most evident for grass and water surfaces. Although a 1-cm/sec deposition velocity is often assumed for gases, Figs. 2 and 3 show that 1 cm/sec for I_2 and SO_2 may have an uncertainty range from 10^{-1} to

IODINE DEPOSITION VELOCITY SUMMARY

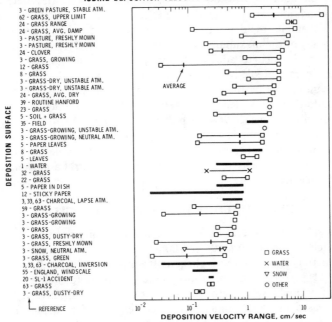

Fig. 2. Iodine deposition velocity summary.

SO₂ DEPOSITION VELOCITY SUMMARY

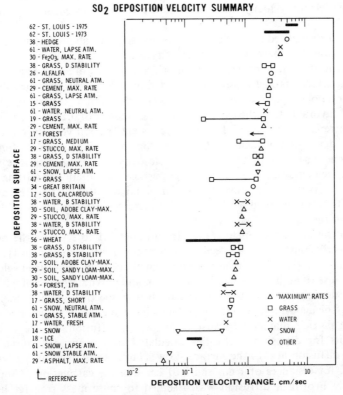

Fig. 3. SO₂ deposition velocity summary.

10 cm/sec. In addition, maximum SO_2 deposition velocities determined for laboratory conditions are less than some experimental conditions measured in the field. As an example, the deposition velocities reported over St. Louis (Fig. 3) are greater than any of the maximum rates shown within the figure. Obviously, there is much yet to be learned to adequately interpret these data and to develop more precise and accurate predictive deposition velocity models.

Even though predicting the deposition velocity of particles is uncertain, there is some indication that the deposition velocities for gases may depend upon atmospheric stability (Bunch, 1968). Whelpdale and Shaw (1974) have also shown a dependency upon atmospheric stability. They report that the deposition velocities to grass and snow for neutral conditions are greater than for stable conditions. In contrast, the deposition velocity to a water surface for a neutral atmosphere is greater than for a stable atmosphere.

PREDICTED DRY DEPOSITION VELOCITIES

Sehmel and Hodgson (1978) have predicted deposition velocities, K_{1m}, as a function of particle diameter from 10^{-2} to 100 μm, of friction velocities from 10 to 200 cm/sec, of aerodynamic roughness heights from 10^{-3} to 10 cm, of particle densities from 1 to 11.5 g/cm^3, and of atmospheric stabilities (unstable and stable atmospheres) for Obukhov's lengths from -10 to $+10$ m, respectively. These predictions were based on wind-tunnel-derived correlations for surface mass-transfer resistances for depositing particles. Predictions indicate that deposition velocities can range over several orders of magnitude from about 10^{-3} up to 20 cm/sec. Moreover, the deposition velocities increase as roughness height increases, usually increase as friction velocity increases, and are nearly independent of atmospheric stability.

When it is recognized that field deposition experiments abound in uncertainties, it is encouraging that these deposition velocity predictions are in the same range as those determined in field experiments. In fact, the validity of Sehmel and Hodgson's deposition velocity model has been supported in a field experiment (Sehmel et al., 1973). For a polydispersed aerosol, the predicted deposition velocity of 0.17 cm/sec compared favorably with the experimental measurement of 0.21 cm/sec. For a concentration reference height of 1 m and a constant friction velocity of 30 cm/sec, predicted deposition velocities, K_{1m}, are shown in Fig. 4 as a function of aerodynamic surface roughness and particle density. In all cases, the predicted deposition velocities are greater than the particle's gravitational settling velocity. Thus, the particle density must be kept in mind when predicting the deposition velocity. The gravitational settling velocity increases proportionally with particle density and the square of particle diameter.

Only in the particle diameter range from about 0.1 to 1 μm is the deposition velocity nearly constant for a selected surface roughness, particle density, and friction velocity. For particle diameters larger than about 1 μm, deposition velocities increase because of an increase in eddy diffusion and gravitational settling. For large particles, deposition velocities approach their respective gravitational settling velocities. The deposition velocity of particles less than about 0.1 μm increases with decreasing particle diameter because of Brownian diffusion. Figure 4 shows the lower limits for deposition velocities calculated from transport caused only by Brownian diffusion next to the surface. Lower limits are shown, on the left side of the figure, for distances of 1 cm and 0.01 cm. For the calculation of these lower limits, only Brownian diffusion was assumed to cause mass transfer below the respective distance, whereas both Brownian diffusion and atmospheric diffusion

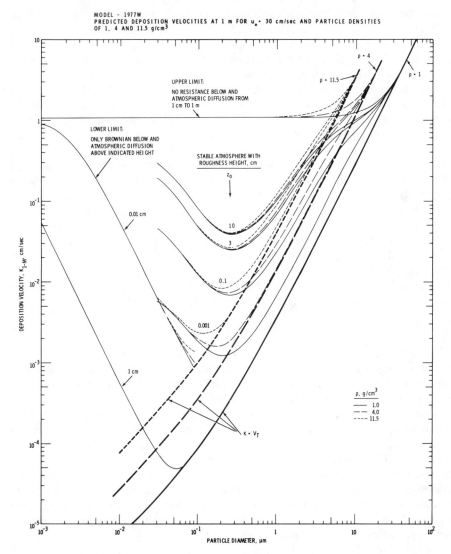

Fig. 4. Predicted deposition velocities at 1m for u_* = 30cm/sec and particle densities of 1, 4, and 11.5 g/cm³.

were assumed to cause mass transfer for greater distances. These lower limits are a function of each distance. Thus, Brownian diffusion is a controlling factor in deposition velocities. When the controlling diffusion distance decreases from 1 cm to 0.01 cm adjacent to the deposition surface, the lower limit for deposition velocities increases nearly two orders of magnitude. Obviously, predicted deposition velocities for small particles are dependent upon Brownian diffusion adjacent to the deposition surface.

The dry deposition velocity upper limit for each particle density is also shown in Fig. 4. For these calculations, the surface resistance to mass transfer within the 1 cm adjacent to the surface was assumed to be zero. Between 1 cm and 1 m, deposition velocities were calculated by including only the effects of atmospheric diffusion (Sehmel, 1970, 1973) and gravitational settling. For particle diameters less than

1 μm, this upper limit is nearly constant and decreases from 1.1 cm/sec at 1 μm to 1.08 cm/sec at 10^{-3} μm. To simplify calculations, the ground surface area rather than the total canopy surface area was used as the deposition surface. For this upper limit, deposition velocities increase for particle diameters greater than 2 μm and approach their respective terminal settling velocities.

Further predictions by Sehmel and Hodgson (1978) show that instability increases the value of deposition velocities, although the increase is small compared with the effects of particle diameter, friction velocity, and aerodynamic surface roughness. The deposition velocity of particles with diameters outside the range from 0.1 to1 μm always increases with an increase in friction velocity. There is a minimum deposition velocity of 0.025 cm/sec; however, that velocity occurs for a friction velocity in the range of 20 to 30 cm/sec. It is not known whether this minimum velocity is physically real or only reflects uncertainties in the prediction techniques.

To predict deposition velocities of particles, estimates of both the aerodynamic surface roughness height, z_0, and air friction velocity, u_*, are needed. Aerodynamic surface roughness is about 15% of the vegetation and physical roughness height (Plate, 1971), although this simple relationship does not attempt to describe any changes in surface roughness that occur as wind speed changes, such as a field of long grass becoming smooth during high wind speeds. The magnitude of the friction velocity ranges from 3 to 15% of the air velocity at a height of 2 m.

CONCLUSIONS

Mass transfer across the air-surface interface is especially important for toxic materials, regardless of their size. For mass transfer across the air-surface interface, airborne concentrations can be predicted from atmospheric diffusion and transport models that include a deposition velocity boundary condition.

Generalizing of early field experiments has been limited in that experimental variables have not been adequately controlled; for example, often the particle size distribution has been either not known or not reported. However, deposition velocities are a function of particle diameter. As a result, most field-determined mass transfer data should be interpreted with qualification when generalization is attempted. The deposition velocities determined in these early field experiments ranged from 10^{-3} up to 20 cm/sec, a range also predicted by Sehmel and Hodgson's empirical model. For predictive generalizations, Sehmel and Hodgson's deposition velocity predictions are recommended for use.

REFERENCES

(Numbers refer to references in Figs. 1–3)

1. Allen, M. D., and R. D. Neff. 1975. Measurements of deposition velocity of gaseous elemental iodine on water. Health Phys. 28:707–715.
2. Belot, Y., and D. Gauthier. 1975. Transport of Micronic Particles from Atmosphere to Foliar Surfaces, Heat and Mass Transfer in the Biosphere. Part 1: Transfer Processes in the Plant Environment, pp. 583–591.
3. Bunch, D. F. (ed.) 1968. Controlled Environmental Radioiodine Tests. Progress Report Number Three, IDO-12063. Idaho Operations Office, USAEC. Available from NTIS.
4. Cataldo, D. A., E. L. Klepper, and D. K. Craig. 1976. Fate of plutonium intercepted by leaf surface: leachability and translocation to seed and root tissues. Proceedings of Transuranium Nuclides in the Environment, International Atomic Energy Agency, Vienna, pp. 291–301.

5. Chamberlain, A. C. 1960. Aspects of the deposition of radioactive and other gases and particles. Int. J. Air Pollut. 3:63–88.
6. Chamberlain, A. C. 1966. Transport of lycopodium spores and other small particles to rough surfaces. Proc. R. Soc. 296:45–70.
7. Chamberlain, A. C., and R. C. Chadwick. 1953. Deposition of airborne radioiodine vapor. Nucleonics 8:22–25.
8. Chamberlain, A. C., and R. C. Chadwick. 1966. Transport of iodine from atmosphere to ground. Tellus 18:226–237.
9. Cline, J. F., D. O. Wilson, and F. P. Hungate. 1965. Effect of physical and biological conditions on deposition and retention of ^{131}I on plants. Health Phys. 11:713–717.
10. Clough, W. S. 1973. Transport of particles to surfaces. Aerosol Sci. 4:227–234.
11. Clough, W. S. 1975. The deposition of particles on moss and grass surfaces. Atmos. Environ. 9:1113–1119.
12. Convair. 1960. Fission Products Field Release Test II, Report NARF 60-l0T (FZK-9-149; AFSWC-TR-60-26), U.S. Air Force Nuclear Aircraft Research Facility, Division of General Dynamics Corp., Ft. Worth, Texas. Available from National Technical Information Service, U.S. Dept. of Commerce, Springfield, Va.
13. Craig, D. K., B. L. Klepper, and R. L. Buschbom. 1976. Deposition of various plutonium-compound aerosols onto plant foliage at very low wind velocities. Proceedings of the Atmosphere-Surface Exchange of Particulate and Gaseous Pollutants—1974 Symposium, Richland, Wash., September 4–6, 1974. Energy Research and Development Administration Symposium Series CONF-740921: pp. 244–263. Available from National Technical Information Service, U.S. Dept. of Commerce, Springfield, Va.
14. Dovland, H., and A. Eliassen. 1976. Dry deposition on a snow surface. Atmos. Environ. 10:783–785.
15. Droppo, J. G., D. W. Glover, O. B. Abbey, C. W. Spicer, and J. Cooper. 1976. Measurement of Dry Deposition of Fossil Fuel Plant Pollutants, EPA-600/4-76-056. EPA, Research Triangle Park, N.C.
16. Fritschen, L. J., and R. Edmonds. 1976. Dispersion of fluorescent particles into and within a Douglas fir forest. Proceedings of the Atmosphere-Surface Exchange of Particulate and Gaseous Pollutants—1974 Symposium, Richland, Wash., September 4–6, 1974. Energy Research and Development Administration Symposium Series CONF-740921: pp. 280–301. Available from National Technical Information Service, U.S. Dept. of Commerce, Springfield, Va.
17. Garland, J. A. 1976a. Dry deposition of SO_2 and other gases. Proceedings of the Atmosphere-Surface Exchange of Particulate and Gaseous Pollutants—1974 Symposium, Richland, Wash., September 4–6, 1974. Energy Research and Development Administration Symposium Series CONF-740921: pp. 212–227. Available from National Technical Information Service, U.S. Dept. of Commerce, Springfield, Va.
18. Garland, J. A. 1976b. Discussions—dry deposition on a snow surface. Atmos. Environ. 10:1033.
19. Garland, J. A., D. H. F. Atkins, C. J. Readings, and S. J. Caughey. 1974. Deposition of gaseous sulphur dioxide to the ground. Atmos. Environ. 8:74–79.
20. Gifford, F. A., and D. H. Pack. 1962. Surface deposition of airborne material. Nucl. Safety 3(4):76–89.
21. Gregory, P. H. 1950. Deposition of airborne particles on trap surfaces. Nature 66:489.

22. Hawley, C. A., Jr., and E. H. Markee, Jr. 1964. Controlled Environmental Radioiodine Tests, CONF-765. Proceedings of the 2nd Conference on Radioactive Fallout, Germantown, MD, November 1964: pp. 821–835.

23. Heinemann, K., K. J. Vogt, and L. Angeletti. 1976a. Deposition and biological half-life of elemental iodine on grass and clover. Proceedings of the Atmosphere-Surface Exchange of Particulate and Gaseous Pollutants—1974 Symposium, Richland, Wash., September 4–6, 1974. Energy Research and Development Administration Symposium Series CONF-740921: pp. 136–152. Available from National Technical Information Service, U.S. Dept. of Commerce, Springfield, Va.

24. Heinemann, K., M. Stoeppler, K. J. Vogt, and L. Angelletti. 1976b. Studies on the Deposition and Release of Iodine on Vegetation, ORNL-tr-4313. Available from National Technical Information Service, U.S. Dept. of Commerce, Springfield, Va.

25. Henderson, R. W., and R. V. Fultyn. 1965. Radiation Measurements of the Effluent from the Kiwi B4D-202 and B4E-301 Reactors, LA-3397. Los Alamos Scientific Lab., Los Alamos, N.M.

26. Hill, A. C. 1971. Vegetation: a sink for atmospheric pollutants. J. Air Pollut. Control Assoc. 21(6):341–346.

27. Hobert, M., K. J. Vogt, and L. Angeletti. 1976. Studies on the Deposition of Aerosols on Vegetation and Other Surfaces, ORNL-TR-4314. Available from Oak Ridge Nationald Laboratory, Oak Ridge, Tenn.

28. Islitzer, N. F., and R. K. Dumbauld. 1963. Atmospheric diffusion-deposition studies over flat terrain. Int. J. Air Water Pollut. 7:999–1022.

29. Judeikis, H. S., and T. B. Stewart. 1976. Laboratory measurement of SO_2 deposition velocities on selected building materials and soils. Atmos. Environ. 10:769–776.

30. Judeikis, H. S., and A. G. Wren. 1977. Deposition of H_2S and dimethyl sulfide on selected soil materials. Atmos. Environ. 11:1221–1224.

31. Little, P., and R. D. Wiffen. 1977. Emission and deposition of petrol engine exhaust Pb—I. Deposition of exhaust Pb to plant and soil surfaces. Atmos. Environ. 11:437–447.

32. Markee, E. H., Jr. 1967. A Parametric Study of Gaseous Plume Depletion by Ground Surface Absorption. AECL-2787. USAEC Meteorgological Information Meeting, C. A. Mauson, ed., Chalk River Nuclear Laboratories, Atomic Energy of Canada Limited, Chalk River, Ontario, Canada, September 11–14, 1967: pp. 602–614.

33. Markee, E. H., Jr. 1971. Studies of the Transfer of Radioiodine Gas to and from Natural Surfaces, ERL-ARL-29. Environmental Research Laboratories, Air Resources Laboratories, Silver Spring, Md.

34. Meetham, A. R. 1950. Natural removal of pollution from the atmosphere. J. R. Meteorol. Soc. 76:359–371.

35. Megaw, W. J., and R. C. Chadwick. 1956. Some Field Experiments on the Release and Deposition of Fission Products and Thoria, AERE-HP-M-114. Available from National Technical Information Service, U.S. Dept. of Commerce, Springfield, Va.

36. Nickola, P. W., and G. H. Clark. 1976. Field measurement of particulate plume depletion by comparison with an inert gas plume. Proceedings of the Atmosphere-Surface Exchange of Particulate and Gaseous Pollutants—1974 Symposium, Richland, Wash., September 4–6, 1974. Energy Research and Development Administration Symposium Series CONF-740921: pp. 74–86. Available from National Technical Information Service, U.S. Dept. of Commerce, Springfield, Va.

37. Nickola, P. W., and R. N. Lee. 1975. Direct measurements of particulate deposition on a snow surface. Pacific Northwest Laboratory Annual Report for 1974, Atmospheric Sciences. BNWL-1950-3. Battelle, Pacific Northwest Laboratory, Richland Wash.: pp. 185–188.
38. Owers, M. J., and A. W. Powell. 1974. Deposition velocity of sulphur dioxide on land and water surfaces using a ^{35}S tracer method. Atmos. Environ. 8:63–67.
39. Parker, H. M. 1956. Radiation exposure from environmental hazards. Proceedings of the International Conference on the Peaceful Uses of Atomic Energy, August 8–20, 1955, Geneva, United Nations: 13:305–310.
40. Plate, E. J. 1971. Aerodynamic Characteristics of Atmospheric Boundary Layers, TID-25465. U.S.A.E.C. Available from National Technical Information Service, U.S. Department of Commerce, Springfield, Va.
41. Porch, W. M., and J. E. Lovill. 1976. Aerosol Deposition and Suspension during a Texas Dust Storm, UCRL-78011. Lawrence Livermore Laboratory, Livermore, Cal.
42. Raynor, G. S. 1976. Experimental studies of pollen desposition to vegetated surfaces. Proceedings of the Atmosphere-Surface Exchange of Particulate and Gaseous Pollutants—1974 Symposium, Richland, WA, September 4–6, 1974. Energy Research and Development Administration Symposium Series CONF-740921: pp. 264–279. Available from National Technical Information Service, U.S. Dept. of Commerce, Springfield, Va.
43. Sehmel, G. A. 1970. Particle deposition from turbulent air flow. J. Geophys. Res. 75:1766–1781.
44. Sehmel, G. A. 1973. Particle eddy diffusivities and deposition velocities for isothermal flow and smooth surfaces. Aerosol Sci. 4:125–138.
45. Sehmel, G. A. 1979. Deposition and Resuspension Processes, PNL-SA-6746. Battelle, Pacific Northwest Laboratory, Richland, Wash.
46. Sehmel, G. A., S. L. Sutter, and M. T. Dana. 1973. Dry deposition processes. Pacific Northwest Laboratory Annual Report for 1972, Atmospheric Sciences, BNWL-1751, Pt 1. Battelle, Pacific Northwest Laboratory, Richland, Wash., pp. 43–49.
47. Sehmel, G. A., and W. H. Hodgson. 1976. Field deposition velocity measurements to a sagebrush canopy. Pacific Northwest Laboratory Annual Report for 1975, Atmospheric Sciences, BNWL-2000-3. Battelle, Pacific Northwest Laboratory, Richland, Wash., pp. 89–91.
48. Sehmel, G. A., and W. H. Hodgson. 1978. A Model for Predicting Dry Deposition of Particles and Gases to Environmental Surfaces, PNL-SA-6721. Battelle, Pacific Northwest Laboratory, Richland, Wash.
49. Shepherd, J. G. 1974. Measurements of the direct deposition of sulphur dioxide onto grass and water by the profile method. Atmos. Environ. 8:69–74.
50. Silker, W. B. 1976. Air to sea transfer of marine aerosol. Proceedings of the Atmosphere-Surface Exchange of Particulate and Gaseous Pollutants—1974 Symposium, Richland, Wash., September 4–6, 1974. Energy Research and Development Administration Symposium Series CONF-740921: pp. 391–398. Available from National Technical Information Service, U.S. Dept. of Commerce, Springfield, Va.
51. Simpson, C. L. 1961. Estimates of Deposition of Matter from a Continuous Point Source in a Stable Atmosphere, HW-69292. Hanford Atomic Products Operation, Richland, Wash.
52. Simpson, C. L. 1964. Deposition Measurements at Hanford. BNL-914. Conference on AEC Meteorological Activities, May 19–22, 1964, Brookhaven National Laboratory, C-42.

53. Slinn, W. G. N., L. Hasse, B. B. Hicks, A. W. Hogan, D. Lal, P. S. Liss, K. O. Munnich, G. A. Sehmel, and O. Vittori. 1978. Some aspects of the transfer of atmospheric trace constituents past the air-sea interface. Atmos. Environ. 12:2055–2087.

54. Start, G. E. 1970a. Comparative diffusion and deposition of uranine dye, molecular iodine gas, and methyl iodide gas. Proceedings of the Fifth Annual Health Physics Society Midyear Topical Symposium, Idaho Falls, Idaho, November 3–6, 1970: Vol. II, pp. 463–473.

55. Start, G. E. 1970b. Diffusion and deposition studies, in ERLTM-ARL 20, Air Resources Laboratories, Silver Springs, Md.

56. Stewart, K. 1963. The particulate material formed by the oxidation of plutonium. C. M. Nichols, ed., Progress in Nuclear Energy, Series IV, Technology and Safety, Vol. 5. Pergamon Press, Macmillan Company, New York: pp. 535–579.

57. Stewart, N. G., and R. N. Crooks. 1958. Long-range travel of the radioactive cloud from the accident at Windscale. Nature 182:627–628.

58. Unsworth, M. H., and D. Fowler. 1976. Field measurements of sulphur dioxide fluxes to wheat. Proceedings of the Atmosphere-Surface Exchange of Particulate and Gaseous Pollutants—1974 Symposium, Richland, Wash., September 4–6, 1974. Energy Research and Development Administration Symposium Series CONF-740921: pp. 342–353. Available from National Technical Information Service, U.S. Dept. of Commerce, Springfield, VA.

59. Van der Hoven, I., ed. 1970. Atmospheric Transport and Diffusion in the Planetary Boundary Layer, ERLTM-ARL-20. Air Resources Laboratories, Silver Springs, Md.

60. Vaughan, B. E. 1976. Suspended particle interactions and uptake in terrestrial plants. Proceedings of the Atmosphere-Surface Exchange of Particulate and Gaseous Pollutants—1974 Symposium, Richland, Wash., September 4–6, 1974. Energy Research and Development Administration Symposium Series CONF-740921: pp. 228–243. Available from National Technical Information Service, U.S. Dept. of Commerce, Springfield, Va.

61. Vogt, K. J., K. Heinemann, W. Matthes, G. Polster, M. Stoeppler, and L. Angeletti. 1973. Propagation of Pollutants in the Atmosphere and Stress on the Environment, BNWL-TR-204. Available from National Technical Information Service, U.S. Dept. of Commerce, Springfield, Va.

62. Voilleque, P. G., D. R. Adams, and J. B. Echo. 1970. Transfer of krypton-85 from air to grass. Health Phys. 19:835.

63. Whelpdale, D. M., and R. W. Shaw. 1974. Sulphur dioxide removal by turbulent transfer over grass, snow, and water surfaces. Tellus 26:196–204.

64. Young, J. A. 1978. The rates of change of pollutant concentrations downwind of St. Louis. Pacific Northwest Laboratory Annual Report for 1977 to the DOE Assistant Secretary for Environment, Atmospheric Sciences, PNL-2500PT3. Battelle, Pacific Northwest Laboratory, Richland, Wash., pp. 1.35–1.39.

65. Zimbrick, J. D., and P. G. Voilleque. 1969. Controlled Environmental Radioiodine Tests at the National Reactor Testing Station, Progress Report No. 4, IDO-12065. AEC Operations Office, Idaho Falls, Id.

DISCUSSION

A. C. Chamberlain: I agree with what George Sehmel has said. His curves for deposition velocity are very well founded. I wonder about this dreadnought of 10 meters. A forester in Great Britain has roughly estimated 1 meter. Is your area a very large forest, and isn't it perhaps more typical to consider smaller areas?

G. A. Sehmel: The prediction is from a correlation of data we have for the surface mass transfer resistance. As far as I know, we really don't have a good estimate of the available surface leaf index and whether the vegetation is wet or dry. This is an area that is unknown, as far as I am concerned. Surface leaf index and available surface area are not known as well as the question of whether particles or gases deposit on the vegetation or on the underlying ground surface.

Pal S. Arya, North Carolina State University: I wonder why the Brownian motion should increase the deposition velocity of the small particles.

G. A. Sehmel: The assumption we make is that the surface is a perfect sink. If there is some concentration, high Brownian diffusion could control the deposition. Brownian diffusion coefficients are very high compared to diffusion coefficients of particles for particle diameters ≤ 0.1 μm. Just looking at the Brownian diffusion coefficient as the controlling mechanism adjacent to the surface results in the type of curve discussed here.

CHAPTER 26

TRANSFER AND DEPOSITION OF PARTICLES TO WATER SURFACES

B. B. Hicks*

R. M. Williams†

ABSTRACT

Consideration of the nature of open water in moderate and high winds leads to the expectation that breaking waves and foaming white water might play a critical role in determining the rate at which particles are deposited to open water surfaces. Corresponding models of the deposition process can be derived. These indicate that particle deposition velocities are not likely to exceed 1 cm/sec except in high winds. Micrometeorological formulations of flux-gradient relationships in the air immediately above the surface indicate that experimental verification of such predictions of relatively small deposition velocities will be exceedingly difficult to obtain. For application of the familiar gradient method it appears necessary to resolve concentration differences of the order of 1%, and concentration fluctuations that must be measured in the eddy correlation method will be of similar magnitude.

INTRODUCTION

Research on the processes involved in the deposition of particles to water surfaces has been marked by a number of apparently contradictory but otherwise convincing contributions. We have, on the one hand, results of radioactive fallout budget studies which indicate that small atmospheric particles enter the oceans at rates similar to those over land (see Volchok et al., 1970). On the other hand, we have strong evidence from wind tunnel studies that particle deposition velocities to water surfaces are exceedingly low (Moller and Schumann, 1970; Sehmel and Sutter, 1974). Both viewpoints find field data in sympathy with them, and so the matter remains largely unresolved.

The paucity of relevant field observations indicates the difficulties involved in obtaining direct experimental evidence in the real world. Much related information has been obtained by exposing surrogate collection surfaces of various geometric configurations. In recent years, the use of such pots, plates, and pans has fallen into some disfavor, as it has become apparent that they cannot be expected to provide the

*Atmospheric Physics Section, Radiological and Environmental Research Division, Argonne National Laboratory, Argonne, Ill. 60439.

†Ecological Sciences Section, Radiological and Environmental Research Division, Argonne National Laboratory, Argonne, Ill. 60439.

same uptake characteristics as a natural water surface, especially in the presence of breaking waves. As the use of artificial collection surfaces diminishes, interest in the alternative micrometeorological methods of flux measurement appears to be increasing. The present purpose is to review some of the processes that have recently been suggested as important factors in the exchange of pollutants between air and water. Methods for investigating the surface flux directly will be considered, and an experimental program to measure particle and trace gas fluxes over Lake Michigan will be described.

LIMITING PROCESSES

Most studies of particle air-water interchange concede that the rate of transfer is controlled by surface properties, such as the presence of a thin layer of molecular diffusion which for very small particles is sometimes characterized by a surface resistance $r_s = (Sc)^{2/3}/ku_*$ (see Slinn et al., 1978, for example). Here, Sc is the Schmidt number, $k = 0.4$ is the von Karman constant, and u_* is the friction velocity. The resistance r_s is very large, typically attaining values in excess of 10^3 sec/cm. In comparison, aerodynamic resistances can be approximated as $(C_f u_*)^{-1}$, where $C_f = 0.035$ is the appropriate friction coefficient. Clearly, the surface molecular boundary-layer resistance would usually dominate.

A number of other processes need to be considered, such as thermophoresis, diffusiophoresis, inertial impaction, and the Stefan velocity. Most contemporary models focus on matters such as these, which combine to reinforce the belief that the net transfer of small particles is largely controlled by near-surface behavior. Recently, however, Slinn (1979) has considered the effect of the humidity profile above a natural water body and its influence on the size of small (and presumably hygroscopic) particles. As a consequence of the rapid growth of such particles following an increase in humidity, Slinn proposes a highly efficient surface capture mechanism and a corresponding low value of the surface resistance r_s. It remains to be seen whether field data support this prediction or the earlier and more widely accepted assumption that r_s is quite large.

Almost without exception, existing models of particle behavior above water surfaces focus on details of particle physics and chemistry. It is of some interest, therefore, to look at the problem from the oceanographic viewpoint, as an exercise in air-sea exchange. The concept of a spatially and temporally continuous layer of molecular diffusion above open water surfaces is especially difficult to accept, since in the case of open ocean the global average wind speed is about 8 m/sec, with breaking waves covering about 3% of the surface. It has been suggested that these highly agitated and foaming areas provide a preferred route by which airborne particles (and presumably other atmospheric trace contaminants as well) can become entrained in subsurface water (Hicks and Wesely, 1978). While it is clear that the transfer of all properties will be affected by the presence of breaking waves, it is not immediately obvious how much the net fluxes will be altered, nor can we say in what circumstances the downward flux will be overwhelmed by emissions from the broken water surface. The residual surface resistance associated with a broken water surface (r_{bs}) might indeed be near zero, as intuition suggests, but experimental support is lacking. Likewise, it seems probable that the surface resistance associated with unbroken water (r_{ss}) will be considerably greater, although influenced to an unknown extent by the chemistry and morphology of the particles involved. In this latter context, surface tension seems likely to present a strong impediment to particle transfer, unless some mechanism is available to overcome or bypass it. Possible mechanisms for reducing the average effects of surface tension could be physical, as discussed above for the case of breaking waves, or chemical, such as when the particles of interest are wettable or hygroscopic (Slinn, 1979).

The downward flux of particles into a water body can therefore be considered in terms of three contributing resistances: r_a, the aerodynamic resistance, which is common to all properties; r_{ss}, the smooth-surface resistance, which is probably large for small particles and largest if they are insoluble; and r_{bs}, the broken-surface resistance, which is likely to be small for most particles and trace gases. The critical factor, then, is the apportionment between r_{ss} and r_{bs}, which can be addressed by use of well-known relations between the wind speed and the proportion of the surface covered by white water (see Kraus, 1972, for example). The model of particle transfer that results is considerably less complicated than many that have been proposed in recent years. For example, there is no particle size effect, even though in practice such a dependence might result from surface tension and molecular effects.

Figure 1 illustrates the wind speed dependence of the deposition velocities that result from the assumption of a range of values of r_{ss} and r_{bs}. Calculations have assumed neutral stratification; generalization to more realistic circumstances presents no great difficulties.

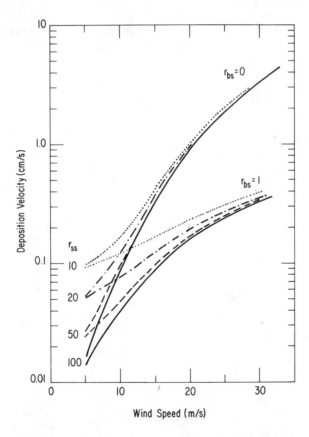

Fig. 1. Predicted desposition velocities for the case of particles over a natural water surface with breaking waves. Two alternative conditions of surface resistance above the broken water surface are considered ($r_{bs} = 0$ and 1 sec/cm) and four considerably higher values of that above broken water ($r_{ss} = 10$, 20, 50, and 100 sec/cm.

EXPERIMENTAL CONSIDERATIONS

There are certainly far fewer reports of experimental measurements of particle deposition velocities to natual water surfaces than there are models and predictions. In itself, this is a convincing demonstration of the practical difficulties associated with the direct measurement of particle fluxes. It is informative to consider the magnitudes of vertical gradients and turbulent fluctuations that correspond to the deposition velocities presented in Fig. 1. In this way, experimental guidelines can be developed in a manner that permits ready comparison of the alternative techniques and selection of the method most likely to be successful. Manipulation of standard micrometeorological relations yields the result

$$\Delta C/C = (v_d/ku_*) \cdot \ln(z_2/z_1), \tag{1}$$

which relates the concentration difference ΔC between heights z_2 and z_1 to the deposition velocity v_d. Once again, neutral stratification is assumed, but generalization presents no substantial difficulties. For purposes of illustration, Fig. 2 has been constructed by use of the deposition velocities presented in Fig. 1, for the case $r_{ss} = 20$ sec/cm. The calculations indicate that the concentration difference over a twofold height interval (e.g., 3 to 6 m, 5 to 10 m, etc.) should be expected to be less than 1% in winds below 15 m/sec, even if $r_{bs} = 0$. If it is intended to test the model presented here, it is evident therefore that experiments would be required to resolve concentration differences that would not normally be expected to exceed 1% over tractable height differences.

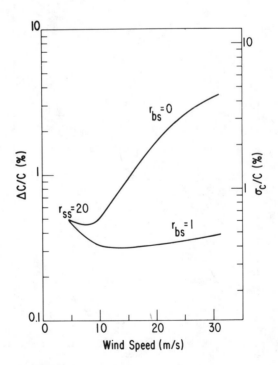

Fig.2.   The magnitude of gradients over twofold height intervals and of turbulent variations in concentration of pollutants deposited according to the desposition velocities of Fig. 1. The case $r_{ss}=20$ sec/cm has been selected for illustration.

Further algebraic manipulation of standard micrometeorological equations leads to an expression that permits estimation of the magnitude of turbulent concentration fluctuations:

$$\varsigma_c/C = v_d/(r\alpha u_*),\tag{2}$$

where ς_c is the standard deviation of the concentration C, r is the correlation coefficient between variations in C and in the vertical wind component, and α is a constant. Experiments have indicated that r is likely to be about 0.35 and α about 1.3 in near-neutral conditions; both will increase as instability increases. The magnitude of the ratio ς_c/C is quite similar to that of the gradient $\Delta C/C$ for the case illustrated in Fig. 2.

Equations (1) and (2) can be applied to test the feasibility of any micrometeorological gradient or turbulence variance measurement technique, provided appropriate estimates of u_* and required accuracies of measurement of v_d are available. In practice, the requirement for extreme accuracy in the measurement of concentration tends to rule against the successful application of these particular methods. As a direct consequence, recent experimental work over Lake Michigan has concentrated on tests of the alternative eddy-correlation method, which evaluates the vertical flux as the covariance $\overline{w' C'}$, in which primes denote deviations from mean values and the overbar indicates a time average.

The eddy-correlation method of direct flux measurement (e.g., Scrase, 1930) has the attraction of a basic immunity to random noise generated by the pollutant concentration sensor, since it is only those concentration fluctuations associated with changes in the vertical wind component that contribute to the net flux. However, it is required that the sensor respond accurately and rapidly to these fluctuations; time constants less than 1 sec are normally required. Initial field trials have been conducted with a rapid-response electrical mobility sensor and a nephelometer. The former sensor detects small particles typically in the 0.05–0.25 μm size range; it was used in earlier studies conducted over land and reported by Wesely et al. (1977). The latter detector responds to particles in the optically active size range, from about 0.2 to 0.8 μm. Other sensors routinely deployed include an anemometer to measure the horizontal wind speed, so as to measure the friction velocity from the appropriate covariance ($u_*^2 = -\overline{u' w'}$), a microbead thermistor temperature sensor to permit evaluation of the sensible heat flux from the covariance $\overline{w' T'}$, and an ozone sensor for use in related investigations of the uptake of ozone at natural water surfaces.

A fixed tower erected about 10 km off shore near Chicago has served as a focal point for Argonne's studies of dry deposition to open water surfaces. Figure 3 shows this tower. In normal operation, sensors for eddy correlation application are mounted at a height of about 8 m. The experiments are both technically and logistically demanding; in practice an experimental team is required to monitor signals carefully throughout each period. Nevertheless, data obtained so far have failed to provide definitive evidence that the deposition velocities of small particles follow the predictions of models like that illustrated in Fig. 1. In a feasibility study conducted in November 1978, small-particle deposition velocities were not significantly different from zero in average winds of about 2 m/sec (see Williams et al., 1978), but later experiments have indicated more scatter than might otherwise be anticipated.

An independent program to investigate particle fluxes over Lake Michigan by use of the gradient method has been reported by Sievering et al. (1979). As expected, these data are also highly scattered, and their interpretation is not clear. Although Sievering et al. conclude that their data indicate rather large deposition velocities, typically in the range 0.2–1 cm/sec, an independent analysis of their entire data set yields averages that are not significantly different from zero. As in the case of the preliminary eddy correlation studies, wind speeds were low throughout the gradient measurement experiments except for one occasion of 7.9 m/sec.

Fig. 3.   The fixed tower located on a sandbar about 10 km off shore from Chicago. This tower has been the site of preliminary tests of eddy-correlation approaches to the general problem of particle flux measurement.

CONCLUSIONS

It is clear that there is a great need for direct measurements of the fluxes of particulate pollutants into natural water bodies, but it is equally clear that existing technical capabilities will not permit close investigation of this matter. For application of the familiar gradient method, it appears necessary to resolve concentration differences that might be expected to be about 1%. For eddy correlation purposes, concentration fluctuations of similar magnitude must be measured. Since the size of the signal to be detected is so closely tied to the corresponding deposition velocity, it seems likely that a considerable effort will need to be expended in order to decide which of the various models that have been proposed provides the best description of the phenomenon of particle fluxes to large exposed water surfaces. Consideration of the technical difficulties involved and of the inherent noise-rejection features of the eddy correlation method suggests that greater emphasis should be placed on the use of this particular technique, thereby avoiding the need to resolve exceedingly small differences in concentration (as would be required in gradient studies) or to assume that some selected collection material is representative of the natural surface.

The question of small-particle emission associated with breaking waves has been avoided here. This and the related matter of particle growth in near-saturated air need to be addressed in future theoretical and experimental studies.

ACKNOWLEDGMENTS

This work has been supported by the U.S. Department of Energy and the U.S. Environmental Protection Agency.

REFERENCES

Hicks, B. B., and M. L. Wesely. 1978. On the deposition of submicron particles to open water surfaces. Argonne National Laboratory Radiological and Environmental Research Division Annual Report, January–December 1978, in press.

Husar, R. R., E. S. Macius, and W. P. Dannevik. 1976. Measurements of dispersion with a fast response detector. Proceedings, Third Symposium on Atmospheric Turbulence, Diffusion and Air Quality, Am. Meteorol. Soc., Boston, pp. 293–298.

Kraus, E. B. 1972. Atmosphere-Ocean Interaction. Oxford University Press, Oxford.

Moller, U., and G. Schumann. 1970. Mechanisms of transport from the stmosphere to the earth's surface. J. Geophys. Res. 75:3013–3019.

Scrase, F. J. 1930. Some characteristics of eddy motion in the atmosphere. Meteorological Office Geophysical Memoirs, Air Ministry, London, No. 52.

Sehmel, G., and S. L. Sutter. 1974. Particle deposition rates on a water surface as a function of particle diameter and air velocity. Battelle Pacific Northwest Laboratory, BNWL-1850, Part 3, pp. 171–174.

Sievering, H., M. Dave, D. A., Dolske, R. L. Hughes, and P. McCoy. 1979. An experimental study of lake loading by aerosol transport and dry deposition in the southern Lake Michigan basin. U.S. Environmental Protection Agency Report EPA-905/4-79-016.

Slinn, W. G. N., L. Hasse, B. B. Hicks, A. W. Hogan, D. Lal, P. S. Liss, K. O. Munnich, G. A. Sehmel, and O. Vittori. 1978. Some aspects of the transfer of atmospheric trace constituents past the air-sea interface. Atmos. Environ. 12:2055–2087.

Slinn, W. G. N. 1979. Modeling of atmospheric particulate deposition to lakes. Presented at the ACS Symposium on "Atmospheric Input of Pollutants to Natural Waters," Sept. 9–14, Washington, D.C.

Volchok, H. L., M. Feiner, H. J. Simpson, W. S. Broecker, V. E. Noshkin, V. T. Bowen, and E. Willis. 1970. Ocean fallout—the Crater Lake experiment. J. Geophys. Res. 75:1084–1091.

Wesely, M. L., B. B. Hicks, W. P. Dannevik, S. Frisella, and R. B. Husar. 1977. An eddy-correlation measurement of particulate deposition from the atmosphere. Atmos. Environ. 11:561–563.

Williams, R. M., M. L. Wesely, and B. B. Hicks. 1978. Preliminary eddy-correlation measurements of momentum, heat and particle fluxes to Lake Michigan. Argonne National Laboratory Radiological and Environmental Research Division Annual Report, January–December 1978, Part III, in press.

DISCUSSION

Cliff Davidson, Carnegie-Mellon University: Bruce, I agree that it's a very difficult problem we're trying to tackle. I wonder if under certain circumstances the gradient method might be the solution? You can always use a large-scale approach with several replicate samplers for a long period of time to achieve higher accuracy, while you can't quite do that with the eddy correlation method.

B. B. Hicks: You're quite right. The replication capability with gradients is real. But I would like to point out and emphasize that your suggestion allows you to improve the statistics in a concrete way. In other words, given an accuracy of measurement that might be ∼10%, and taking 100 different profiles, you're down to the 1% mark. This requires the assumption of random behavior, but I'm not sure that these things always are random.

A. C. Chamberlain: My impression was that, with regard to the effects of breaking waves, both momentum measurements on the sea and the evaporation measurements you suggested, have indicated that it was the capillary waves which provide most of the surface roughness. This means that you don't increase evaporation very much until you get not merely breaking waves but what's called a white sea.

B. B. Hicks: I confirm that. I'm reminded that the experiments addressing that question are very difficult because you never know where the interface is. I don't want anyone to get the idea that I'm advocating the breaking-wave theory very strongly; it's simply a way of thinking of the problem.

CHAPTER 27

AN URBAN INFLUENCE ON DEPOSITION OF SULFATE AND SOLUBLE METALS IN SUMMER RAINS

Donald F. Gatz*

ABSTRACT

Recent observations of abnormally acidic precipitation have raised questions regarding the distribution of acidic precipitation, its time trend, its sources of acidity, and the relevant physical and chemical processes involved in its formation. This paper attempts to answer some of these questions based on the content of sulfate and other materials in the atmosphere and in samples of summer convective rainfall from mesoscale sampling networks near St. Louis.

These and other observations in the literature agree that rainfall deposits locally emitted sulfur at short distances downwind of cities. This causes enhanced deposition and concentration of sulfur in local rainfall and increases the local variability of these parameters relative to that of rainfall volume and crustally derived materials, on both daily and seasonal scales.

Airborne sulfate concentrations vary by a factor of at least 2–4 from urban to rural areas on individual days. This is similar to the observed variation of sulfate deposition or concentrations in rain. Thus, there may be no need to invoke extensive SO_2 scavenging in rain systems to explain the observed enhancements; nucleation scavenging of atmospheric sulfate appears adequate.

INTRODUCTION

Observations of abnormally acidic precipitation in northern Europe and the northeastern United States have raised questions regarding (1) the geographical distribution of acid precipitation, (2) its trend with time, (3) the sources of its acidic materials, and (4) the physical and chemical processes involved in its formation.

The purpose of this paper is to attempt at least partial answers to some of these questions for the St. Louis, Missouri, area. Specifically, this paper describes the urban influence on sulfate deposition in rain at St. Louis and interprets the deposition data in terms of possible sources and scavenging mechanisms of sulfur in precipitation.

Hales and Dana (1979) measured sulfate, nitrate, and ammonium ion concentrations in rain near St. Louis during the summers of 1972 and 1973 as part of Project METROMEX. They concluded that (1) there was sufficient airborne particulate sulfate to account for the sulfate in rain, but (2) for several reasons, the elevated sulfate concentrations found in the rain downwind of the city could only be explained by a mechanism in which locally emitted sulfur dioxide was scavenged in the cloud.

*Atmospheric Sciences Section, Illinois State Water Survey, Urbana, Ill. 61801.

Also as part of METROMEX the Illinois State Water Survey measured the areal deposition of both soluble and insoluble fractions of a number of elements or ions in rain during the summers of 1972 to 1975 (Gatz et al., 1978). The analyses included Li, Na, Mg, K, Ca, Fe, Zn, Pb, and Cd. These elements represent both natural and man-made sources. Soluble-sulfate deposition was measured in four of the events sampled in 1974 and 1975. Further, the data set includes measurements of sulfate and metal concentrations in ground-level air, which allows the testing of a key feature in the reasoning of Hales and Dana (1979) that led to their conclusion that sulfur dioxide scavenging was significant.

These data are analyzed to show the urban effect on sulfate deposition and its variability. In addition, the data are analyzed to test whether they are consistent with the conclusions of Hales and Dana (1979) regarding the importance of local sources and sulfur dioxide scavenging.

EXPERIMENTAL METHODS

The methods used to collect and analyze both precipitation samples (Gatz et al., 1978) and aerosol samples (Gatz, 1978) have been given in detail previously. They are outlined here only briefly.

Precipitation Samples

In METROMEX, rain samples were collected in the two different sampling networks shown in Fig. 1. The 1900-km² 80-collector rectangular network extending eastward from St. Louis was used between 1972 and 1974, and the 2200-km² 85-collector network covering Alton and Wood River, Illinois, was used in 1975. Rain was collected in precleaned open wide-mouth polyethylene bottles mounted atop metal fence posts. Samples were ordinarily changed daily.

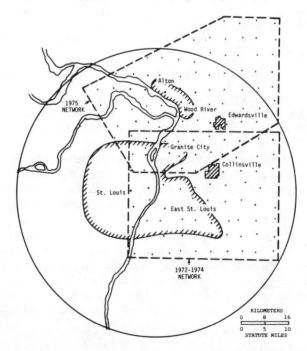

Fig. 1. Rain chemistry sampling networks in METROMEX, 1972–1975.

In the laboratory, sample bottles were weighed with and without the water sample to determine sample mass and volume. The samples were analyzed for pH and conductivity and then filtered through preleached 0.45-μm-pore-diameter Millipore type HA filters to remove insoluble particulate matter. The filtrate was acidified to pH 2 and stored in polyethylene bottles. Filters were dried, folded, and stored in plastic vials.

Filtrates were analyzed for cations directly using standard flame or flameless atomic absorption spectrophotometry. Sulfate was determined by an automated colorimetric barium–methylthymol blue method on a Technicon Autoanalyzer II system (Peden et al., 1979) after neutralizing the samples with ammonium hydroxide. Filters were dissolved in strong acids and analyzed using the same methods.

The raw analytical data were computer-corrected for blanks, dilutions, and dry deposition, and the data were edited to remove obvious outliers caused by contamination. Dry-deposition corrections were based on sample duration and the median dry-deposition rate at each sampling site. Analyses of dry-deposited material in the sample collectors were made on 17 to 32 days, depending on the element, in the 1972–1974 network and on 5 days in 1975. Dry-deposition corrections were not applied to sulfate, Pb, or Cd but were expected to be less than 10% for most rains, as they were for the other elements.

Aerosol Samples

Airborne particulate matter was collected on preweighed 37-mm Nuclepore 0.8-μm-pore-diameter filters, exposed face down under 25-cm-diam polyethylene funnel rain shields, about 1 m above grass or a flat roof. The locations of the sampling sites in 1974 and 1975 are shown in Fig. 7 (which is discussed and appears later in the text). Sample durations varied from 6 to 12 hr, but most collections were 12 hr.

In the laboratory, loaded filters were reweighed to determine sample mass, after conditioning at 47% relative humidity for 24 hr to allow the moisture absorbed on the particulate matter to come to equilibrium.

Elemental analysis of the filters was performed at Crocker Nuclear Laboratory, University of California, Davis, using ion-excited x-ray fluorescence (Flocchini et al., 1972, 1976).

RESULTS

Deposition Patterns

Consider first some previous results which show which of the various soluble and insoluble constituents of rain have similar deposition patterns. Tables 1–4 are loadings tables from factor analyses performed separately on four rain events in which sulfate was measured (Gatz, 1979). The loadings are correlation coefficients between the variables and the factors. They show which variables have similar deposition patterns (those with high loadings on the same factors). In addition, they show how similar a variable's deposition pattern is to that of the factor. High loadings (0.86 or higher) indicate very similar patterns; moderate loadings (0.71 to 0.85) show that the patterns are moderately similar. Loadings less than 0.50 are usually not shown; they explain little of the variance, and the tables are easier to read without them.

In each rain event, factor analysis grouped the sulfate deposition pattern with that of soluble pollutant metals and with rainfall, while soluble soil elements, insoluble soil elements, and insoluble pollutants grouped on different factors. Only in the 1

Table 1. Loadings table from factor analysis on deposition data from storm of
2 August 1974

	Factor 1	Factor 2
Soluble K	0.90	
Soluble Ca	0.81	
Soluble Li	0.78	
Soluble Mg	0.71	
Soluble SO_4^{2-}		0.85
Soluble Zn		0.83
Rainfall		0.79
Soluble Na	(0.49)	(0.48)
Variance explained %	56	15

August 1975 case did one of the soluble pollutant elements occur on a different factor from sulfate. Note, however, that loadings were often in the moderate range for
at least some of the variables in the "soluble pollutants" factor.

Table 2. Loadings table from factor analysis on deposition data from storm of
13 July 1975

	Factor 1	Factor 2	Factor 3	Factor 4
Insoluble Fe	0.91			
Insoluble K	0.86			
Insoluble Li	0.64			
Soluble SO_4^{2-}		0.77		
Soluble Zn		0.72		
Soluble Cd		0.62		
Rainfall		0.55		
Soluble Ca			0.92	
Soluble K			0.88	
Soluble Mg			0.61	
Soluble Li				0.76
Insoluble Zn				−0.65
Variance explained, %	32	15	13	9

Source: Gatz, D. F. 1979. Associations and spatial relationships among rainwater constituents deposited
in mesoscale sampling networks by individual storms near St. Louis. Paper presented at Commission on
Atmospheric Chemistry and Global Pollution Symposium on the Budgets and Cycles of Trace Gases and
Elements in the Atmosphere, Boulder, Col., August, 1979.

Table 3. Loadings table from factor analysis on deposition data from storm
of 19 July 1975

	Factor 1	Factor 2	Factor 3	Factor 4	Factor 5
Soluble SO_4^{2-}	0.91				
Rainfall	0.86				
Soluble Zn	0.72				
Soluble Pb	0.66				
Soluble Mg		0.86			
Soluble Ca		0.84			
Soluble K		0.70			
Soluble Li		0.64			
Insoluble Fe			0.91		
Insoluble K			0.86		
Insoluble Li			0.72		
Insoluble Mg				0.95	
Insoluble Ca				0.94	
Insoluble Zn					0.75
Insoluble Pb					0.74
Variance explained, %	21	17	13	12	9

Table 4. Loadings table from factor analysis on deposition data from storm of
1 August 1975

	Factor 1	Factor 2	Factor 3	Factor 4
Insoluble Fe	0.96			
Insoluble Li	0.96			
Insoluble K	0.96			
Insoluble Zn	0.50	(0.49)		
Soluble K		0.92		
Soluble Mg		0.84		
Soluble Ca		0.59		
Soluble SO_4^{2-}			0.78	
Soluble Cd			0.73	
Rainfall			0.73	
Soluble Zn				0.80
Soluble Li				0.77
Variance explained, %	33	17	15	10

Source: Gatz, D. F. 1979. Associations and spatial relationships among rainwater constituents deposited
in mesoscale sampling networks by individual storms near St. Louis. Paper presented at Commission on
Atmospheric Chemistry and Global Pollution Symposium on the Budgets and Cycles of Trace Gases and
Elements in the Atmosphere, Boulder, Col., August, 1979.

With this indication from factor analysis that the soluble pollutants and rainfall have similar deposition patterns, it is of interest to compare the actual deposition patterns. This is done for the four events in Figs. 2–5. Figure 2 shows results for the

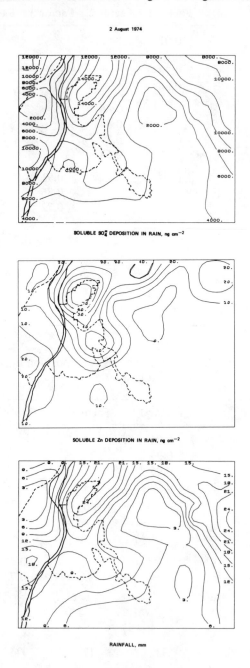

2 August 1974

SOLUBLE SO₄ DEPOSITION IN RAIN, ng cm⁻²

SOLUBLE Zn DEPOSITION IN RAIN, ng cm⁻²

RAINFALL, mm

Fig. 2. Distribution of rainfall and constituent deposition for rain of 2 August 1974.

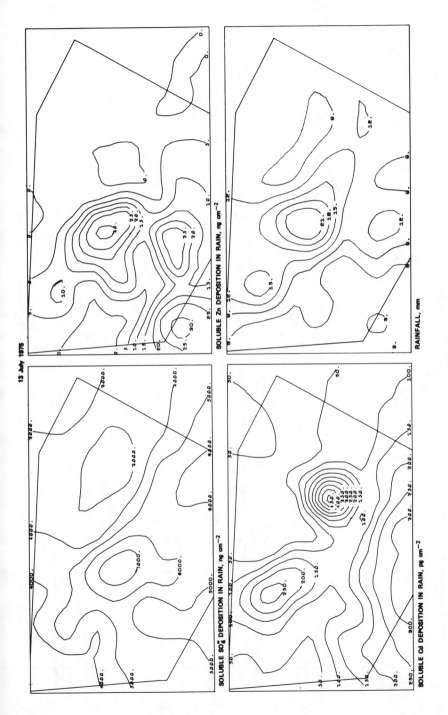

Fig. 3. Distribution of rainfall and constituent deposition for rain of 13 July 1975.

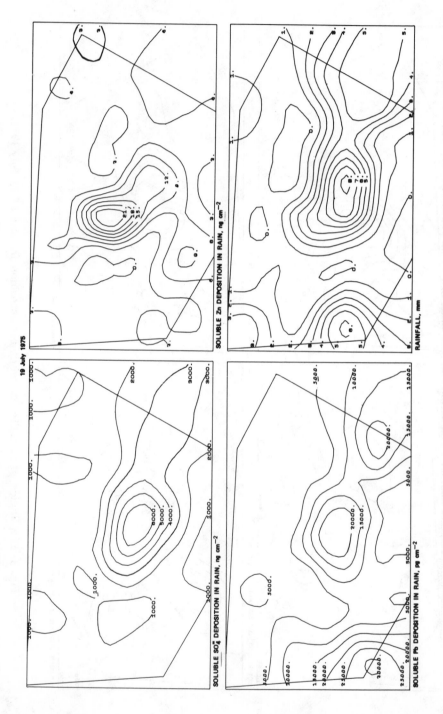

Fig. 4. Distribution of rainfall and constituent deposition for rain of 19 July 1975.

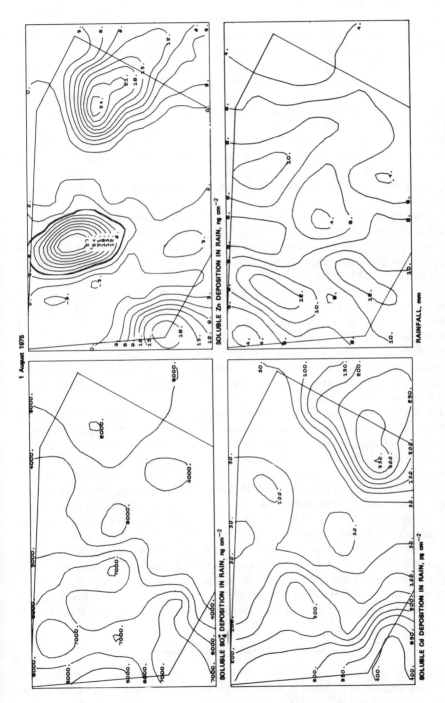

Fig. 5. Distribution of rainfall and constituent deposition for rain of 1 August 1975.

rain of 2 August 1974, which deposited an average of 11.3 mm of rain in the network. The figure shows good agreement between the deposition patterns of soluble sulfate, soluble Zn, and rainfall. (After this, mention of sulfate or any of the metals means the soluble portion, unless specified otherwise.) This agrees with the loadings shown in Table 1. Ordinary product-moment correlation coefficients for sulfate with the other variables were: sulfate-rainfall, 0.94; sulfate-Zn, 0.53.

Figure 3 shows results for the storm of 13 July 1975, which deposited an average of 10.2 mm of rain. As a group the loadings were smaller in this case than the previous one (0.55 to 0.77), and this is reflected in the comparison of deposition patterns. Generally high element and ion deposition occurred over or near the industrial areas of Alton–Wood River and Granite City, Illinois, and this was accompanied by a similar rainfall pattern. However, some individual differences between patterns may be seen. For example, a maximum in Zn deposition occurred near Granite City, while a corresponding Cd maximum was displaced somewhat to the east. Product-moment correlations for this event were: sulfate-rainfall, 0.41; sulfate-Zn, 0.59; and sulfate-Cd, 0.41.

Similar results were found for the event of 19 July 1975, which gave an average of 2.0 mm of rain over the network. Figure 4 shows that there was general agreement between the main features of the various deposition patterns but differences in some details. The product-moment correlations were: sulfate-rainfall, 0.76; sulfate-Zn, 0.56; and sulfate-Pb, 0.52.

The event of 1 August 1975 gave an average of 7.3 mm of rain over the network and was the only one in which the soluble-Zn deposition did not occur on the same factor as sulfate and rainfall. Nevertheless, its deposition pattern is included along with the others in Fig. 5, where it clearly shows itself to be different. As in the other cases, sulfate, Cd, and rainfall display general similarity in the area of maximum deposition in the southwestern portion of the network, but they differ elsewhere in minor ways. The product-moment correlations were: sulfate-rainfall, 0.43; sulfate-Cd, 0.41; and sulfate-Zn, 0.19.

The general agreement between sulfate deposition and soluble-Zn deposition for the four events discussed above suggests that the long-term average deposition pattern (i.e., over many events) for soluble Zn is similar to that of sulfate, which was measured in only four events. The average soluble-Zn deposition (nanograms per square centimeter per event) patterns for the 1972–1974 and 1975 networks are shown in Fig. 6 along with average rainfall patterns for the same respective events. Both Zn and rainfall depositions have been normalized (i.e., divided by their respective network mean values). This facilitates comparisons, because both Zn and rainfall depositions appear in the same nondimensional units. The figure shows maximum soluble-Zn deposition very close to the suspected sources, which include an electrolytic Zn refiner, two municipal incinerators, a brass plant, and a secondary copper smelter, and not related to the rainfall pattern. Since sulfate deposition was similar to soluble-Zn deposition in three of the four events presented earlier, it is reasonable to expect a similar but not identical urban influence for sulfate deposition.

Spatial Variability of Deposition

The urban influence on the spatial variability of Zn deposition and, by analogy, on sulfate deposition is shown in Table 5 in terms of the relative standard deviation. Data for rainfall and the soluble portion of elements for which the earth's crust is a major source are shown for comparison. Spatial variabilities are presented for daily deposition, as well as for integrated deposition over all days in which the various

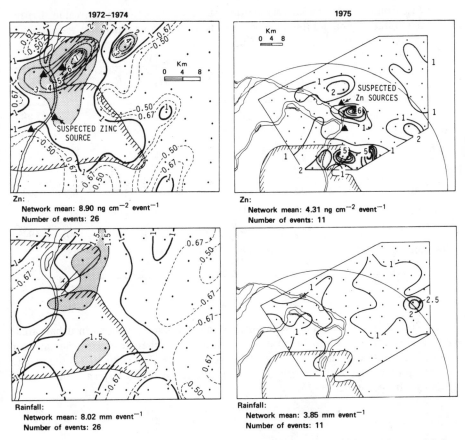

Fig. 6. Distribution of median deposition per event for Zn and rainfall in multiple storms. Values are normalized to their respective network means.

elements were measured. The table shows median relative standard deviations near 100% for rainfall and the earth's crust elements on a daily time scale, while that for Zn is over 175%. On the seasonal time scale, rain and the crustal elements had spatial variabilities of 30–45%, while that for Zn was 125%. Thus, as expected, the influence of the city on Zn deposition, and probably sulfate deposition, is to increase the spatial variability on both daily and seasonal time scales.

Patterns of Concentrations in Air

Since deposition of materials in rain is influenced by their concentrations in air, it is of interest to examine the distribution of sulfate and pollutant elements in the St. Louis area. Figure 7 shows median values at 12 locations where measurements were taken during the summers of 1973 to 1975. Not all sites were sampled each summer, however. Please note that the concentrations shown in Fig. 7 represent *total* concentrations, not just the soluble portion. Sulfate concentrations were calculated from measurements of elemental sulfur, assuming that particulate sulfur was 100% sulfate. Sulfur in rain is virtually 100% soluble, but that is not true for the pollutant metals.

Table 5. Summary of spatial variabilities (standard deviation/mean value, %) for precipitation and soluble element wet deposition over daily and seasonal time scales

	Rainfall	Li	Na	Mg	K	Ca	Zn
Number of days	41	27	20	20	26	20	26
Variability, individual days							
Minimum	17	37	56	58	54	50	105
Mean	112	112	98	100	147	92	195
Median	100	101	84	92	126	86	176
Maximum	316	224	193	222	281	160	494
Variability, all days	35	38	32	44	45	35	125

Sulfate, Zn, and Pb all show the highest median concentrations in urban areas and lowest concentrations in rural areas. Ratios of highest to lowest medians for sulfate, Zn, and Pb were 2.3, 5.8, and 8.0, respectively. The value of 2.3 for sulfate agrees quite closely with the urban/rural sulfate ratios in the eastern United States reported by Altshuller (1976).

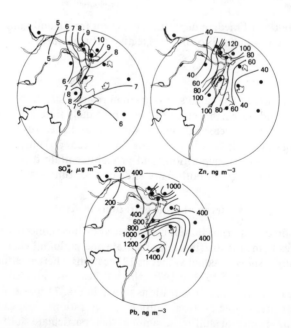

Fig. 7. Distribution of median airborne pollutant concentrations for measurements made during summers, 1973–1975.

Examinations of spatial patterns of airborne concentrations on many individual days from the summers of 1973–1975 (not shown) confirm that the same patterns as seen in Fig. 7 also occurred on individual days. The maximum in the Pb distribution east of St. Louis is an anomaly in the data caused by two sites near heavily traveled roads. The true maximum was probably over the St. Louis urban area, where the greatest traffic density occurs.

DISCUSSION

The two matters of most concern in this paper are (1) the urban influence on deposition and its variability and (2) the implications of the data regarding sources and scavenging mechanisms for sulfur in rain near St. Louis. We shall discuss them in this order.

Urban Influence

The four rain events in which sulfate was measured show sulfate deposition peaks associated with both (1) heavy rain areas and (2) urban-industrial source areas, independent of rainfall. The results over many events, using soluble Zn as an analog of sulfate, suggest that on seasonal time scales the deposition maxima associated with rainfall maxima in single events tend to "average out" away from the city, just as rainfall itself does, while the local maxima near the sources persist.

This is confirmed and expressed quantitatively by the calculated variability results (Table 5). Over single events, rain and crustal element deposition had relative standard deviations near 100%, while that of Zn was greater by a factor of 1.75. On the seasonal scale, the spottiness in the rainfall and its contribution to the crustal element variability was considerably smoothed, so that their relative standard deviations decreased to 30–45%. The local Zn anomaly persisted, however, so that overall the Zn relative standard deviation was still 125%, a factor of 3 to 4 higher than rain and the crustal elements.

Sources and Scavenging Mechanisms

Now we examine the implications of the data regarding sources and scavenging mechanisms of the locally deposited sulfur. Hales and Dana (1979) found that there was more than enough sulfate aerosol present in the St. Louis atmosphere to account for all the sulfate in the rain solely by scavenging of sulfate particles. Yet, because of the locally enhanced sulfate concentrations in rain downwind of the city and the perceived "highly uniform" atmospheric sulfate concentrations, local sulfur dioxide was presumed to have been converted to sulfate by a gas-phase oxidation in the precipitation system. We will examine this argument in the light of data presented earlier.

Two aspects of the data appear to be the most pertinent. First, we will compare the observed variabilities of sulfate concentrations in air with those in rain, and second we will consider the association between the deposition patterns of sulfate and the other soluble pollutants, such as Zn, Pb, and Cd.

Airborne sulfate distributions

Hales and Dana (1979) did not give a quantitative definition of "highly uniform" airborne sulfate concentrations, but there is evidence in the literature that sulfate concentrations differ by a factor of at least 2 between urban and rural areas. For example, Altshuller (1976) reported ratios of mean urban and rural sulfate concentrations near 2 for much of the eastern United States. Further, it should be noted that these urban concentrations were probably conservative, since they were measured near the center of an urban or industrial area. On individual days, larger concentrations would be expected some distance downwind after conversion of some of the sulfur dioxide to sulfate.

The distribution of median airborne sulfate concentrations measured in METROMEX (Fig. 7) illustrates the variability of sulfate concentrations in the St. Louis area. This distribution exhibits a maximum/minimum ratio of 2.3.

For individual days, the ratios were greater. Plotted distributions of airborne sulfate for the four rain events (Figs. 2–5) are not shown because of space limitations, but their respective maximum/minimum sulfate ratios, based on our data, were 7.9, 4.2, 3.0, and 3.5, respectively. Similar ratios were obtained by the EPA Regional Air Pollution Study (RAPS) for total sulfate concentrations on the last three of these days, based on particulate sulfur measurements by ten dichotomous virtual impactors in the St. Louis area (Dzubay, 1979).

These urban/rural ratios *are* more uniform than those of some of the other pollutants, such as Zn or Pb, for which distant sources are relatively less important than for sulfate. This may be seen in Fig. 7. The maximum/minimum ratios for the median Zn and Pb concentrations in Fig. 7 are 5.8 and 8.0, respectively, and ratios on individual days would ordinarily be greater. Nevertheless, the urban/rural sulfate ratios are clearly greater than about 2 or 3 on individual days.

The next question is, "How will such urban/rural differences in sulfate aerosol be reflected in the sulfate in rain?" The answer to this is, unfortunately, not at all clear.

One may examine either concentrations in rain (micrograms of sulfate per milliliter) or deposition to the surface (micrograms of sulfate per square centimeter), but both are strongly affected by rainfall amount. Hales and Dana (1979) have made a step in the right direction in their attempt to normalize sulfate concentrations to 1 cm of rain. The distribution of such values should show clearly any urban influence on sulfate concentrations in rain, apart from effects caused by rainfall amount. The variability of these normalized concentrations, then, is the proper one to compare against the variability of airborne sulfate. Hales and Dana, however, listed only nonnormalized data at individual collection sites. The nonnormalized maximum/minimum concentration ratio for their storm of 23 July 1973 was 4.5. Since the highest concentrations usually occur in small samples and the lowest in large samples, a normalization to unit rainfall should reduce the maximum/minimum ratio. Thus the maximum/minimum ratio in the normalized data should be no more than 4.5. This ratio is the same order as those found for sulfate in air in the St. Louis area. Since there is a well-known process, namely nucleation, by which sulfate aerosol is efficiently scavenged, there appears to be no compelling reason in the 23 July 1973 storm for concluding that sulfur dioxide scavenging must have occurred.

The sulfate depositions in the four events presented earlier have not been normalized to unit rainfall, but again, ratios of such data should be "worst cases," since normalization of *depositions* to unit rainfall should decrease the high values, which occur with the heaviest rainfall (Figs. 2–5), and increase the low values, which occur with the lightest rainfall. Maximum/minimum ratios of sulfate deposition for the events shown in Figs. 2–5 were, respectively, 22.7, 3.4, 32.4, and 5.0. Thus, even for such "worst cases," two of the four events had ratios of the same order as those of airborne sulfate.

The ratios of rainfall for the samples with the maximum and minimum sulfate deposition were, respectively 39.8, 17.4, 31.8, and 1.2 in the four rains. This suggests that the two cases of high sulfate maximum/minimum deposition ratios were caused by large variations in rainfall. Again, there appears to be no strong reason to require sulfur dioxide scavenging in the cloud if, as has been shown previously by Hales and Dana (1979), there is sufficient airborne sulfate to account for the sulfate in rain.

These data are similar to those of Hales and Dana (1979) in that they show local enhancement of sulfur in rain near the city. Nothing in these data disagrees with the conclusion of Hales and Dana that this observed enhancement was caused by deposition of sulfur emitted from local sources.

Association between deposition patterns

Factor analyses showed that sulfate deposition patterns group consistently with those of other soluble pollutants and rainfall. The loading values from the factor analyses suggested, and a comparison of actual deposition patterns showed, that soluble pollutants and rainfall depositions were generally similar in their major features but differed in some details.

It appears very significant that sulfate deposition patterns are similar to soluble pollutant metals and rainfall but not to insoluble metals or soluble soil elements. These relationships should be determined by the distribution of sources and by the processes of scavenging and/or precipitation formation. Source distribution can probably explain why pollutant distributions are different from those of soil elements, but the reason why soluble and insoluble materials have different deposition patterns would appear related to scavenging and/or precipitation formation processes.

One could speculate that the association between the deposition patterns of sulfate and soluble metals implies that the sulfur is scavenged in the same way as the metals, namely, by particulate (sulfate) scavenging. Other explanations, however, involving sulfur dioxide scavenging by cloud droplets with high catalyst concentrations, can also be envisioned. Whatever mechanisms are eventually proposed, however, they must now be consistent with the observations reported here.

CONCLUSIONS

Observations reported here and others in the literature agree that rainfall deposits at least some locally emitted sulfur at short distances downwind of the sources. This causes enhanced deposition and variability in the affected areas.

Airborne sulfate concentrations near St. Louis vary by a factor of 2–4 from urban to rural locations on individual days. This variation is roughly the same as that observed in either concentration or deposition of sulfate in rain. Thus, there is no need to invoke in-cloud scavenging of sulfur dioxide to explain enhanced sulfur in rain near the city, since there is ample aerosol sulfate and an available scavenging mechanism (nucleation) to account for it.

The observation of consistently similar deposition patterns for sulfate and soluble pollutant metals places some interesting constraints on possible scavenging mechanisms and suggests that more detailed study of this group of materials would be fruitful for understanding the processes involved.

ACKNOWLEDGMENTS

I thank Mr. S. A. Changnon, Jr., and Mr. R. G. Semonin for general supervision of this work, Dr. G. J. Stensland for helpful discussions of the material, Mr. R. K. Stahlhut for programming the computer plots, and Mr. M. E. Peden, Mrs. F. F.

McGurk, and Ms. L. M. Skowron for sample analysis. This work was supported by the U.S. Department of Energy, Division of Biomedical and Environmental Research, under Contract Number EY-76-S-02-1199.

REFERENCES

Altshuller, A. P. 1976. Regional transport and transformations of sulfur dioxide to sulfates in the U.S. J. Air Pollut. Control Assoc. 26(4):318–324.

Dzubay, T. 1979. Personal communication.

Flocchini, R. G., T. A. Cahill, D. J. Shadoan, S. J. Lange, R. A. Eldred, P. J. Feeney, and G. W. Wolfe. 1976. Monitoring California's aerosols by size and elemental composition. Environ. Sci. Technol. 10:76–82.

Flocchini, R. G., P. J. Feeney, R. J. Sommerville, and T. A. Cahill. 1972. Sensitivity versus target backings for elemental analysis by alpha-excited x-ray emission. Nucl. Instrum. Methods 100:379–402.

Gatz, D. F. 1978. Identification of aerosol sources in the St. Louis area using factor analysis. J. Appl. Meteorol. 17(5):600–608.

Gatz, D. F. 1979. Associations and spatial relationships among rainwater constituents deposited in mesoscale sampling networks by individual storms near St. Louis. Paper presented at the Commission on Atmospheric Chemistry and Global Pollution Symposium on the Budgets and Cycles of Trace Gases and Aerosols in the Atmosphere, Boulder, Col., August 1979.

Gatz, D. F., R. G. Semonin, and M. E. Peden. 1978. Deposition of Aerosols. In Ackerman, B., et al., Summary of METROMEX, Vol. 2. Causes of Precipitation Anomalies. Bulletin 63, Illinois State Water Survey, Urbana, pp. 345–381.

Hales, J. M., and M. T. Dana. 1979. Precipitation scavenging of urban pollutants by convective storm systems. J. Appl. Meteorol. 36:294–316.

Peden, M. E., L. M. Skowron, and F. M. McGurk. 1979. Precipitation sample handling, analysis, and storage procedures. Research Report 4, Contract No. EY-76-S-02-1199, U.S. Department of Energy, Illinois State Water Survey, Urbana, C00-1199-57.

DISCUSSION

Ron McLean, Domtar Research, Montreal, Canada: You seem to be suggesting that the sulfate deposition was being influenced by heavy metals. Can you give us a rough idea of the proportion of the ion balance that the heavy metals were contributing?

D. F. Gatz: I don't have any specific numbers, but it's quite small, maybe a few percent at the most. I don't want to suggest at this point that the metals were acting as a catalyst for sulfur dioxide scavenging. We could generate quite a discussion on that point. The fact that these deposition patterns were similar to each other but, at the same time, quite different from the patterns of several other constituents suggests that there are some constraints on what could possibly be causing the deposition and what the scavenging mechanisms might have been. I'm not suggesting any details about that at this point.

John Miller, NOAA: If we were setting up an urban monitoring network, wouldn't your work be rather useful to show that there is quite a bit of differentiation just downwind of the city, so that one site probably wouldn't be enough for monitoring in urban areas? Do you have a comment?

D. F. Gatz: It depends what you are monitoring, of course. No, I wouldn't ever just pick one site, because there's so much variation. I wish we had more than four cases to look at from our Metromex data, but I think the fact that the zinc is highly variable suggests that sulfate may be as well.

Chuck Hakkarinen, Electric Power Research Institute: I notice you used a bulk collector for this measurement program. Could you comment on what fraction of zinc and other metals in the collector were deposited with the rain and what fraction may have been deposited as dry fall during the time the collectors were in the field before the rain occurred?

D. F. Gatz: The zinc data were corrected for dry deposition based on analyses of several days of collections without rain. The fraction that the dry deposition contributes to the total varies considerably, depending on how much it rains. On the heavy rain days it contributes just a few percent. Looking at cases where there was very light rain, 1–2 mm, you might get a 50% contribution, or you might dominate the entire deposition by dry. For the cases I have discussed here, I think probably 10% or less of the total deposition is dry fall.

Bill Graustein, Yale University: I was interested to note that in the elements you said were soil-derived, there was a clustering of magnesium, potassium, and calcium, while sodium was in a separate group. Do you have any idea as to what the mechanism would be that fractionates these elements in the soil?

D. F. Gatz: I don't want to place too much confidence in the sodium measurements. They do behave strangely, as indicated by factor analyses of other cases. We may have had some problems with sodium contamination, so I don't think the data warrant any statements about fractionation.

G. S. Henderson, University of Missouri: You mentioned rainfall affecting your deposition pattern. I've recently read some articles which suggested that downwind of St. Louis the particles themselves were causing changes in the rainfall distribution patterns. Would you care to comment on this and suggest what kind of element might be responsible?

D. F. Gatz: Rainfall is definitely affected by the city but probably not by the pollution from the city. I have a paper that will be appearing in the *Journal of Applied Meteorology* in which we tried to determine statistically whether the pollutants could in any way explain the rainfall anomaly. The result is no, it can't. The explanation may involve urban heat or roughness.

CHAPTER 28

OVERVIEW OF WET DEPOSITION AND SCAVENGING*

M. Terry Dana†

ABSTRACT

Methods for describing simple pollutant aerosol or gas interactions with hydrometeors are well developed, and reasonable predictions of removal may be made if the pollutant gas-phase concentration and particle size spectrum (or gas properties) are known. Scavenging processes involving attachment, condensation, or cloud-drop nucleation are less well known, and current models are limited in that supporting measurements are inadequate. The scavenging coefficient and scavenging ratio, the main parameters for describing precipitation scavenging, have usefulness which varies depending on the pollutant in question and the complexity of the removal processes involved.

INTRODUCTION

The study of the removal of pollutants from the atmosphere by precipitation has a comparatively short history as a branch of atmospheric science, but its importance—if not its success at definitively solving the requisite problems—has grown rapidly as we are coming to recognize the limited capacity of the atmosphere for absorbing and processing wastes. Securing an understanding of pollutant (aerosol or gas)–hydrometeor interactions is a difficult enough problem if the pollutants are inert chemically, but in fact contaminants are modified by contact with natural aerosols and water vapor, undergo chemical reactions in the gas and liquid phases, and are capable of influencing the formation of precipitation or storms, frequency of precipitation, and, ultimately, climate. Therefore, a description of what happens between release of a pollutant and its eventual "disappearance" or deposition is often complicated and may involve results drawn from several scientific disciplines. There have been considerable advances in describing basic microphysical processes, especially the interactions between particles and gases and hydrometeors, but, on the scale of storms and pollutant clouds or plumes, progress has been limited by impaired understanding of plume diffusion and lack of suitable in-cloud measurements of pertinent parameters. The chemical behavior of important pollutants such as sulfur dioxide and its reaction product sulfate is also poorly understood as they exist in the real atmosphere.

---

*Based on work performed under U.S. Department of Energy Contract EY-76-C-06-1830.
†Pacific Northwest Laboratories, Richland, Wash. 99352.

Because of the uncertainties involved in transport, diffusion, and chemical behavior of a particular pollutant release situation, modelers of atmospheric impact have generally sought and utilized grossly simplified parameterizations for precipitation removal. Perhaps because there is no formal discipline associated with precipitation scavenging, and thus a lack of recognized definitive sources for practically useful results, these "quick fix" parameterizations are often outdated and more naive than necessary for inclusion in the relevant models.

A thorough review of wet deposition and scavenging in the context of these proceedings is impossible; the attempt here is to mention some of the areas of basic knowledge and to indicate areas of needed research. There are several recent surveys of atmospheric deposition which are useful in updating the common wisdom, and, at the least, this limited review aims to introduce these to the reader. For detailed reviews of deposition, including wet deposition and scavenging, the reader should consult Slinn, 1979; Slinn et al., 1978; Hales, 1978; Slinn, 1977; Hidy, 1973; Hales, 1972a; and Engelmann, 1968.

In this paper, the most basic terminology will be used. Although "precipitation scavenging" as a field of study is not particularly susceptible to rampant jargonizing, it has its share of specialized terms; these are unfortunately often misunderstood and are used contradictorily by various authors. Thus, the special terms are left to their inventors (some among the authors above), but areas of greatest potential hazard regarding usage and interpretation will be pointed out. In this regard, the title of this paper, while not originally intended to serve this purpose, illustrates the nonsynonymity of "wet deposition" and "scavenging." "Scavenging" here will mean simply the collection or absorption of something by (into) something else. "Precipitation scavenging" means collection of something by precipitation, and since meteorological precipitation by definition involves deposition on the earth, removal from the atmosphere to the ground is implied. For practical purposes "wet deposition" is the same as precipitation scavenging, although one could find, if pressed, instances of wet deposition without precipitation.

AEROSOL SCAVENGING

Description of the spatial and temporal behavior of an aerosol in the atmosphere begins with a gas-phase continuity equation (Hales, 1972a; Hidy, 1973):

$$\frac{\partial c_y}{\partial t} = -\nabla \cdot c_y \bar{v} - w + r_y , \tag{1}$$

where c_y = pollutant concentration, mass/volume,

t = time,

$\bar{v}$ = (three-dimensional) velocity vector,

w = scavenging removal term,

r_y = chemical reation term.

The scavenging term represents the amount of pollutant per unit volume of space removed from the gas phase to the liquid phase per unit time; it is related to, and may define, the scavenging coefficient Λ by

$$w = \Lambda c_y . \tag{2}$$

In a well-mixed volume (divergence term zero) and with no chemical reactions, (1) reduces to

$$\frac{dc_y}{dt} = -\Lambda c_y \, ,$$

(3)

and subsequently the familiar exponential decay of the quantity of pollutant with time. It should be pointed out that this result is for specific conditions, and in the present terminology the removal is only to the liquid phase, not necessarily to deposition on the ground.

Theoretical evaluation of the microphysical scavenging coefficient may be obtained for the case of collection of aerosol particles by raindrops. For raindrops of radius R and aerosol particles of radius a, the monodisperse particle scavenging coefficient* is given by

$$\Lambda(a) = \int_0^\infty dR \, N(R)v_z z(R)\pi R^2 E(a, R),$$

(4)

where $N(R)$ is the number of raindrops of radius R per unit volume, v_z is their terminal velocity, and $E(a, R)$ is the collection efficiency as defined in Fig. 1. Slinn (1976a) has suggested a semiempirical expression for E (retention efficiency assumed unity), an adaptation of which (Dana and Hales, 1976) is illustrated in Fig. 2 along with some measured values. Since the expression for rainfall rate,

$$J = \int_0^\infty dR \, N(R)v_z(R)\tfrac{4}{3} \cdot \pi R^3 \, ,$$

(5)

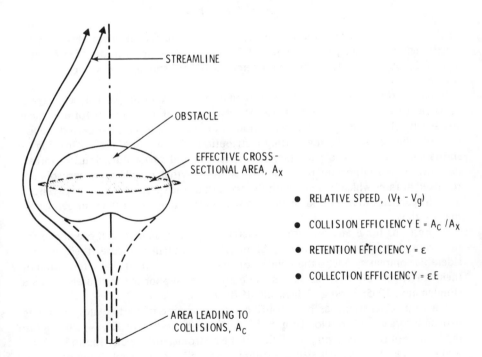

Fig. 1. Schematic of hydrometeor-pollutant collection geometry.

---

*Barring evaporation of the raindrops, this might be called a precipitation scavenging coefficient.

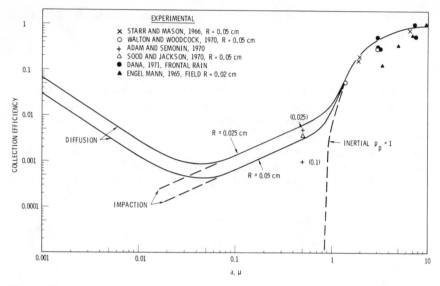

Fig. 2 Theoretical collection efficiencies for aerosols (retention efficiency unity) and selected measured values.

is similar, an approximate expression for scavenging coefficient may be derived:

$$\Lambda(a) \cong \frac{\alpha J}{R_m} E(a,\, R_m) \; .$$

(6)

In Eq. (6), R_m is a mean radius parameter and α is a numerical factor chosen to fit experimental washout results. Slinn (1977) uses the mass mean radius for R_m and $\alpha = 0.5$; other rain spectrum parameters and corresponding α values may be substituted.

Equation (6) provides a useful approximation for raindrop-particle scavenging coefficient for monodisperse particles. Determination of a coefficient for a particle size spectrum requires a second integration of Eq. (4) over a. This has been done for aerosols characterized by log-normal distributions (Dana and Hales, 1976); Fig. 3 shows some results for a particular rain spectrum and various distribution spread parameters. An important result of this analysis is that polydisperse aerosols are removed at rates which increase with the breadth of the spectrum and are scavenged as if the particles were monodisperse of a size considerably larger than the geometric mean radius.

Expressions for E and $\Lambda(a)$ for ice-crystal scavenging, similar to those above for raindrops, have been suggested by Slinn (1976a), but in this case, there is considerable complexity due to the number of ice-crystal types (and characterization of their scavenging-effective size), and there are fewer supporting measurement data (Englemann, 1965; Sood and Jackson, 1970, 1972).

Thus far, discussion has been confined to scavenging of particulates by falling hydrometeors outside of clouds; use of Eq. (6) or integration of Eq. (4) can lead to predictions of the scavenging coefficient if the hydrometeor size distribution is fairly well known and the particle size spectrum is very well known. Application of Eq. (3) on the micro scale, or a macro counterpart under suitable conditions (notably, Λ space- and time-independent) to a suitable model for the air concentration (plume) can lead to estimates of plume depletion by scavenging. The wet deposition can be estimated from an expression for the wet flux F to the ground,

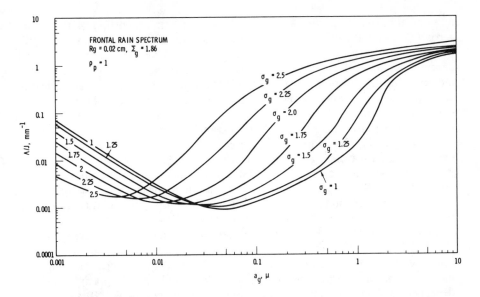

Fig. 3. Theoretical mass scavenging coefficients for unit-density particles, normalized to unit rainfall rate. a_g and R_g are the geometric number mean radii of the log-normal particle and raindrop size spectra; ς_g, Σ_g are the geometric standard deviations, respectively.

$$F = \int_0^h \Lambda c_y \, dz, \tag{7}$$

where h is an appropriate height in the atmosphere above which scavenging is insignificant (e.g., for c_y represented by a Gaussian plume, h can be taken as infinity).

Precipitation scavenging of particles depends importantly on the scavenged particle size spectrum; thus it is crucial to be able to characterize the particle size at times critical to the scavenging mechanism. The size spectrum of a particular pollutant release may be reasonably defined at the release point, but subsequently the aerosol may become attached to other particles or plume water droplets, or it may grow by condensation of water vapor. Moreover, the particle size spectrum may be poorly known at all times; for example, a reaction product such as sulfate, resulting from oxidation of sulfur dioxide, is expected to range widely in size and propensity to attach or grow. Radke et al. (1976) measured collection efficiencies for particles over a broad size range in an industrial plume; these values were at least a factor of 10 higher than those shown in Fig. 2. Slinn (1976b) has applied a scavenging model including attachment and condensation to suggest scavenging coefficients compatible with these measurements. This model relies on parameters which have not been measured and on assumed growth times which limit its general applicability at this time.

It follows that pollutant aerosols become involved with clouds and precipitation-formation processes. Thus, in addition to the uncertainties listed above for "below-cloud" scavenging, one must consider the mechanisms by which aerosols are incorporated into the cloud liquid phase and the process of removal of the cloud water as precipitation.

Among the many proposed mechanisms for attachment of aerosol particles to cloud liquid water (Junge, 1963; Mason, 1971; Dingle and Lee, 1973), it appears that nucleation is the most important; it has been estimated that over 50% of the total mass of atmospheric particulates, and likely, a greater proportion of the sulfate aerosol (Weiss et al., 1977), exists as condensation nuclei under normal storm conditions (Junge, 1963). Calculations of the time required for growth of droplets, which depends on aerosol surface properties and solubilities plus temperature 'and molecular diffusivity of water (Junge, 1963; Mason, 1971), show that nucleation and growth proceed slowly enough to provide a rate-limiting step in the precipitation scavenging process. Also, it appears that the process of pollutant nucleation scavenging parallels the process of formation of precipitation (i.e., process of removal of water); this correspondence can be used to simplify modeling of the overall in-cloud scavenging mechanism (Klett, 1977; Scott, 1978).

In the model proposed by Scott (1978), it is assumed that a sizable portion of the atmospheric sulfate particles act as condensation nuclei. The resulting cloud droplets are included in precipitation formed in three precipitating storm categories: storms requiring ice-growth (Bergeron) process, warm-rain synoptic scale storms, and convective storms. The results, in the form of "washout" or scavenging ratio (see definition of scavenging ratio in a later section), are shown in Fig. 4 along with some data (Perkins et al., 1970; Davis, 1972). The results are compatible with ratios measured recently (Scott and Laulainen, 1979) and exhibit precipitation rate dependences similar to previous observations (Georgii and Beilke, 1966).

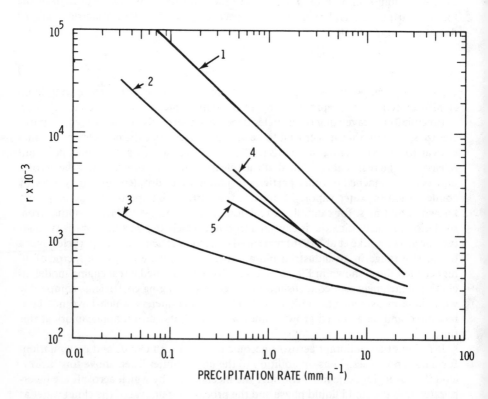

Fig. 4. Sulfate aerosol scavenging ratio predicted for three types of precipitation: (1) convective storms, (2) warm rain, and (3) ice-growth-initiated. Curves 4 and 5 are observed ratios at coastal Washington state for two experiment series. Source: Scott, B. C., 1978.

GAS SCAVENGING

Derivation of useful descriptors of gas scavenging mechanisms follows the same general procedure as for aerosols, above, with some notable differences. A material balance for the liquid phase is required in addition to Eq. (1) for the gas phase (Hales, 1972a). Linkage between these two is provided by w, and via assumptions similar to the aerosol case, a counterpart to Eq. (3) can be obtained:

$$\frac{dc_y}{dt} = \frac{3K_y}{ac}(c_y - H'c_x) , \qquad (8)$$

where K_y is a mass transfer coefficient, c is the total gas-phase concentration $[(R'T)^{-1}$, where R' is the gas constant], c_x is the liquid-phase gas concentration, and H' is a solubility parameter. Comparison of Eq. (8) with Eq. (3) reveals that description of gas scavenging by a scavenging coefficient is generally unsatisfactory and that the rate of change of liquid-phase concentration may be positive or negative, depending on the relative liquid- and gas-phase concentrations and the solubility. For vertical rainfall and linear solubility (H' not a function of concentration), Eq. (8) may be solved analytically for simple plumes (Hales et al., 1973). This procedure is useful for gases which form simple solutions, such as water vapor, tritium compounds, and N_2O; under proper conditions, more strongly dissolved gases may be treated with a suitable linearization of the solubility. Generally, however, nonlinearly soluble gases like SO_2, I_2, and NH_3 require numerical solution of Eq. (8); computer codes have been developed for SO_2 scavenging calculations (Dana et al., 1973; Hill and Adamowicz, 1977). Such calculations using a Gaussian plume model have compared favorably with measurements in power-plant and small-scale controlled-source experiments (Dana et al., 1975, 1976).

When the gas- and liquid-phase concentrations of the pollutant gas are in equilibrium,

$$c_y = H'c_x , \qquad (9)$$

and calculations of gas scavenging become considerably simpler. Criteria for predicting equilibrium conditions have been developed (Dana et al, 1975); these show that in the case of SO_2, equilibrium conditions exist generally beyond about 20 stack heights from a tall-stack source. It follows that equilibrium scavenging applies for SO_2 scavenging on a regional scale, and precipitation concentrations can be estimated as a function only of gas-phase concentration, precipitation pH, and temperature, the variables on which H' depends. Regional estimations of SO_2 scavenging and comparisons with recent measurements may be found in Hales and Dana (1979) and Dana (1979).

Gases which are reactive in either the gas or liquid phase offer complications similar to those expressed above for in-cloud scavenging of particles. In the case of gas-phase reactions, uncertainty arises as to a description of the air concentration c_y relevant to scavenging applications, and the model for c_y must account for this loss of gaseous pollutant prior to the onset of scavenging. Liquid-phase oxidation mechanisms, which are of primary current interest in connection with sulfate formation from SO_2, can be enormously complicated, and the predicted rates of oxidation, can be widely variable. Several reviews consider the liquid-phase oxidation of SO_2 (Harrison et al., 1976; Levy et al., 1976); despite considerable study of various mechanisms, e.g., by O_2 in pure aqueous solution (Fuller and Crist, 1941) or by inclusion of trace materials such as NH_3, O_3, H_2O_2, and metals (Levy et al., 1976; Penkett et al., 1977), little consensus has been reached regarding primary mechanisms or useful rate expressions for scavenging applications (Hales, 1978). In a cloud environment, the solubility of SO_2 is such that, at equilibrium, a relatively small fraction of the SO_2 occurs in the aqueous phase (see "Scavenging Ratios," below). Thus oxidation must be rapid in order for much of the sulfate to be formed in the liquid phase. In the absence of fast oxidation in cloud droplets, the equilibrium solubility expression, Eq. (9), can be used as a steady-state approximation to the liquid-phase SO_2 concentration. For the case of raindrops falling through plumes, where Eq. (9) does not apply, the formalism for calculation of scavenging including oxidation does exist (Dana et al., 1976; Hill and Adamowicz, 1977); what are required for reliable results from such calculations are realistic parameterizations of liquid-phase conversion rates.

SCAVENGING RATIOS

A useful approach to precipitation scavenging prediction and characterization of the scavenging properties of various pollutants involves the use of scavenging ratios. The scavenging ratio is merely the ratio of precipitation-phase to gas-phase concentration, with the concentrations normally evaluated at ground level. Various units are used by various authors; perhaps the best formulation is dimensionless:

$$r = \frac{\text{precipitation concentration (mass/volume precipitation*),}}{\text{gas-phase concentration (mass/volume air)}} \qquad (10)$$

or

$$r = \frac{c_x}{c_y} . \qquad (11)$$

Compared with the scavenging coefficient (Λ) as a descriptor of scavenging, r is relatively easy to measure and represents an integral result of the scavenging process over a whole storm system, including any relevant mechanisms noted above†. The relationship to the hydrometeor-particle interaction scavenging coefficient Λ follows from equating the wet flux to the ground, from Eq. (7), to $c_x J$ and dividing by the ground-level air concentration c_{y0}:

$$r = \frac{c_{x0}}{c_{y0}} = \frac{1}{J c_{y0}} \int_0^h \Lambda c_y \, dz . \qquad (12)$$

If Λ is invariant with height, the integral over z may be calculated using some model for the air concentration c_y. The simplest case would be a uniformly mixed layer of height h with $c_y = c_{y0}$. Hence, Eq. (12) becomes

$$r = \frac{\Lambda h}{J}, \qquad (13)$$

or

$$r = \frac{\alpha h}{R_m} E(a, R_m) , \qquad (14)$$

applying Eq. (6).

The scavenging ratio is eminently suitable for equilibrium gas scavenging applications, since Eq. (9) the solubility factor is directly relatable to the scavenging ratio. For unitary consistency, the scavenging ratio for gases can be defined as $r = H = (H')^{-1}$, where H is a dimensionless solubility parameter (Henry's law constant for the case of gases forming simple solutions). It should be emphasized that at equilibrium, r for a gas is a function of local conditions; the ground-level measured value of r may be unrelated to the value of r in the cloud above, especially if aqueous-phase

---

*Rain equivalent in the case of frozen forms.

†Use of scavenging ratio terminology avoids some of the semantic pitfalls of the scavenging coefficient, but it also may obscure the mechanisms involved.

reactions occur subsequent to deposition. One may be tempted to apply the approach leading to Eq. (13) to "back out" a value of Λ for equilibrium gas scavenging, but this is not particularly useful (except for unitary convenience for parameterization in transport and deposition models), since the height h involved is fictitious owing to the irrelevance of the previous history of the phase transitions of the gas.

Engelmann (1970), Slinn (1976a), and Slinn et al. (1978) have compiled values of r measured and calculated for various species. Table 1 lists some of these, mainly drawn from the latter reference. Table 2 lists H for SO_2, calculated from the solubility relationship of Hales and Sutter (1973). These data show the broad range of

Table 1. Measured scavenging ratios for selected species

Species	r	Remarks
Aerosols		
Pollen	$0.65-3.8 \times 10^6$	15–50 μm diameter
Pb	0.06×10^6	
Ca	0.06×10^6	
Ca	0.29×10^6	
Mn	0.31×10^6	Generally micrometer dimension or less
Cl	1.1×10^6	
Na	0.98×10^6	
Gases[a]		
CO	37.3	At 15°C, fresh water
H$_2$S	32.2	At 15°C, fresh water
N$_2$O	1.3	At 15°C, fresh water
CO$_2$	0.98	At 15°C, fresh water
HT[b]	0.018	At 25°C, calculated
HTO[b]	4.7×10^4	At 25°C, calculated

[a]Solubility parameters H equal to r under equilibrium conditions.
[b]Source: Hales, J. M., 1972b.

Table 2. Calculated solubility H for SO_2

Temperature (°C)	pH	H at indicated air concentration			
		1 ppb	3 ppb	10 ppb	30 ppb
0	4	1700	1700	1600	1500
	5	1.6×10^4	8.4×10^4	5.4×10^4	3.4×10^4
25	4	2400	2400	2400	2300
	5	2.2×10^4	1.9×10^4	1.5×10^4	1.0×10^4

solubilities and thus scavenging potential of the gases; generally, however, the scavenging ratios for typical gases are at least an order of magnitude less than for micrometer-sized particles under normal atmospheric conditions. Aerosol scavenging ratios tend to range from 10^5 to 10^6; these values are compatible with estimations of total accumulation of aerosols by cloud droplets and thus to precipitation using typical values of aerosol and cloud drop concentrations (Slinn et al., 1978).

CONCLUSIONS

The calculation procedures for predicting scavenging and precipitation scavenging are well established for at least the hydrometeor-pollutant interactions, given the distribution of the pollutant in the gas phase and the cloud drop liquid phase and the particle size distributions (or gas solubility properties). These are of course very big "givens." For below-cloud particle scavenging, parameterization of raindrop size is quite reliable owing to the extensive work in characterizing rain spectra; for the case of ice crystals, it is more complicated but manageable. Although the collection efficiencies for particles are strongly dependent on particle size, we are fortunate that fairly broadly polydisperse aerosols (most are) tend to act like large particles whose efficiencies approach unity. Although confirming experimental measurements are very few, it is likely that many particles of $a < 0.1$ μm are hygroscopic and grow to efficiently removed sizes under precipitation conditions. Nucleation appears to be an efficient process for effectively enlarging a large portion of the mass of atmospheric particles. The rates of growth and nucleation are not well known, however, and are dependent on particle surface and chemical properties and on cloud conditions which are difficult to measure and predict.

Prediction of gas scavenging from plumes (and to some extent in ostensibly well-mixed pollutant layers) depends heavily on an accurate model for the air concentration c_y which must include gas-phase reactions when necessary. Otherwise, equilibrium scavenging predictions based on solubilities are easily done for gases whose solubility has been studied. These are relatively few, however, and applications to specific gases should include a careful consideration of the adequacy of solubility knowledge.

Individual steps in the precipitation scavenging process are generally manageable, given key data. Full analyses of complicated many-step scavenging situations—e.g., sulfate formation in gas or liquid phase, inclusion in cloud droplets, formation of precipitation, collection of cloud droplets, and precipitation deposition—can only be roughly approximated, however, since many mechanisms and pollutant properties are still poorly understood and the supporting meteorological data are often inadequate.

REFERENCES

Adam, J. R., and R. G. Semonin. 1970. Collection efficiencies of raindrops for submicron particles. In Engelmann, R. J., and W. G. N. Slinn (ed.), Precipitation Scavenging (1970). U.S. Atomic Energy Commission, Oak Ridge, Tenn.(NTIS: CONF-700601).

Dana, M. Terry. 1971. Washout of soluble dye particles. In Pacific Northwest Laboratory Annual Report for 1970 to the USAEC Division of Biology and Medicine, Vol. II: Physical Sciences, Part 1. Atmospheric Sciences, BNWL-1551 (II.1). Battelle, Pacific Northwest Laboratories, Richland, Wash. PP. 63–67.

Dana, M. Terry. 1979. SO₂ vs. sulfate wet deposition in the eastern United States. Submitted to J. Geophys. Res.

Dana, M. Terry, and J. M. Hales. 1976. Statistical aspects of the washout of polydisperse aerosols. Atmos. Environ. 10:45–50.

Dana, M. Terry, J. M. Hales, W. G. N. Slinn, and M. A. Wolf. 1973. Natural Precipitation Washout of Sulfur Compounds from Plumes, EPA-R3-73-047. Battelle, Pacific Northwest Laboratories, Richland, Wash.

Dana, M. Terry, J. M. Hales, and M. A. Wolf. 1975. Rain scavenging of SO_2 and sulfate from power plant plumes. J. Geophys. Res. 80(30):4119–4129.

Dana, M. Terry, D. R. Drewes, D. W. Glover, and J. M. Hales. 1976. Precipitation scavenging of fossil-fuel effluents, EPA-600/4-76-031. Battelle, Pacific Northwest Laboratories, Richland, Wash.

Davis, W. E. 1972. A model for in-cloud scavenging of cosmogenic radionuclides. J. Geophys. Res. 77:2159–2165.

Dingle, A. N., and Y. Lee. 1973. An analysis of in-cloud scavenging. J. Appl. Meteorol. 12:1295–1302.

Engelmann, R. J. 1965. Rain scavenging of zinc sulphide particles. J. Atmos. Sci. 22:719.

Engelmann, R. J. 1968. The calculation of precipitation scavenging. In Slade, D. H. (ed.), Meteorology and Atomic Energy. U.S. Atomic Energy Commission, Oak Ridge, Tenn. Available from NTIS as TID-24190.

Engelmann, R. J. 1970. Scavenging prediction using ratios of concentrations in air and precipitation. In Engelman, R. J., and W. G. N. Slinn (ed.), Precipitation Scavenging (1970). U.S. Atomic Energy Commission, Oak Ridge, Tenn. (NTIS: CONF-700601).

Fuller, E. C., and R. H. Crist. 1941. The rate of oxidation of sulfite ions by oxygen. J. Am. Chem. Soc. 63:1644.

Georgii, H-W., and S. Beilke. 1966. Atmospheric Aerosol and Trace Gas Washout. Air Force Cambridge Research Laboratories (NTIS No. AD 634 907).

Hales, J. M. 1972a. Fundamentals of the theory of gas scavenging by rain. Atmos. Environ. 6:635–659.

Hales, J. M. 1972b. Scavenging of Gaseous Tritium Compounds by Rain, BNWL-1659. Battelle, Pacific Northwest Laboratories, Richland Wash.

Hales, J. M. 1978. Wet removal of sulfur compounds from the atmosphere. Atmos. Environ. 12:389–399.

Hales, J. M., and M. Terry Dana. 1979. Regional-scale deposition of sulfur dioxide by precipitation scavenging. Atmos. Environ. 13:1121–1132.

Hales, J. M., and S. L. Sutter. 1973. Solubility of sulfur dioxide in water at low concentrations. Atmos. Environ. 7:997–1001.

Hales, J. M., M. A. Wolf, and M. Terry Dana. 1973. A linear model for predicting the washout of pollutant gases from industrial plumes. AIChE J. 19(2):292–297.

Harrison, H. T., T. V. Larson, and P. V. Hobbs. 1976. Oxidation of sulfur dioxide in the atmosphere: a review. IEEE Annals No. 75:1–7.

Hill, F. B., and R. F. Adamowicz. 1977. A model for rain composition and the washout of sulfur dioxide. Atmos. Environ. 11:917–927.

Hidy, G. 1973. Removal of gaseous and particulate pollutants. In Rasool, S. I. (ed.), Chemistry of the Lower Atmosphere. Plenum Press, New York.

Junge, C. E. 1963. Air Chemistry and Radioactivity. Academic Press, New York.

Klett, J. 1977. Precipitation scavenging. In Rainout Assessment: The ACRA System and Summaries of Simulation Results, LA-6763. Los Alamos Scientific Laboratory, Los Alamos, N.M.

Levy, A., D. R. Drewes, and J. M. Hales. 1976. SO_2 Oxidation in Plumes: A Review and Assessment of Relevant Mechanistic and Rate Studies, EPA-450/13-76-022. Battelle, Pacific Northwest Laboratories, Richland, Wash. 99352.

Mason, B. J. 1971. The Physics of Clouds. Clarendon, Oxford.

Penkett, S. A., B. M. R. Jones, and K. A. Brice. 1977. Rate of Oxidation of Sodium Sulfite Solutions by Oxygen, Ozone, and Hydrogen Peroxide and Its Relevance to the Formation of Sulfate in Cloud and Rainwater, AERE-R 8584.

Perkins, R. W., C. W. Thomas, J. A. Young, and B. C. Scott. 1970. In-cloud scavenging analysis from cosmogenic radionuclide measurements. In Engelmann, R. J., and W. G. N. Slinn (ed.), Precipitation Scavenging (1970), U.S. Atomic Energy Commission, Oak Ridge, Tenn. (NTIS: CONF-700601).

Radke, L. F., E. E. Hindman, and P. V. Hobbs. 1976. A case study of rain scavenging from a kraft process paper mill plume. In Semonin, R. G., and R. W. Beadle (eds.), Precipitation Scavenging (1974). ERDA, Oak Ridge, Tenn. (NTIS No. CONF-741003).

Scott, B. C. 1978. Parameterization of sulfate removal by precipitation. J. Appl. Meteorol. 17(9):1375–1389.

Scott, B. C., and N. S. Laulainen. 1979. On the concentration of sulfate in precipitation. J. Appl. Meteorol. 18(2):138–147.

Slinn, W. G. N. 1976a. Precipitation scavenging: some problems, approximate solutions and suggestions for future research. In Semonin, R. G., and R. W. Beadle (eds.), Precipitation Scavenging (1974). ERDA, Oak Ridge, Tenn. (NTIS No. CONF-741003).

Slinn, W. G. N. 1976b. Rain scavenging of active aerosol particles from a plume. In Pacific Northwest Laboratory Annual Report for 1975 to the USERDA Division of Biomedical and Environmental Research, Part 3. Atmospheric Sciences, BNWL-2000 PT3. Battelle, Pacific Northwest Laboratories, Richland, Wash.

Slinn, W. G. N. 1977. Some approximations for the wet and dry removal of particles and gases from the atmosphere. Water, Air, Soil Pollut. 7:513–543.

Slinn, W. G. N. 1979. Precipitation scavenging. In Atmospheric Sciences and Power Production—1979. U.S. Department of Energy, Oak Ridge, Tenn. In press.

Slinn, W. G. N., L. Hassee, B. B. Hicks, A. W. Hogan, D. Lal, P. S. Liss, K. O. Munnich, G. A. Sehmel, and O. Vittori. 1978. Some aspects of the transfer of atmospheric trace constituents past the air-sea interface. Atmos. Environ. 12:2055–2087.

Sood, S. K., and M. R. Jackson. 1970. Scavenging by snow and ice crystals. In Engelmann, R. J., and W. G. N. Slinn (ed.), Precipitation Scavenging (1970). U.S. Atomic Energy Commission, Oak Ridge, Tenn. (NTIS: CONF-700601).

Sood, S. K., and M. R. Jackson. 1972. Scavenging Studies of Snow and Ice Crystals, IITRI C 6105-18. IIT Research Institute, Chicago, Ill.

Starr, J. R., and B. J. Mason. 1966. The capture of airborne particles by water drops and simulated snow crystals. Q. J. R. Meteorol. Soc. 92:490.

Walton, W., and A. Woodcock. 1970. The suppression of airborne dust by water sprays. In Aerodynamic Capture of Particles. Pergamon, Oxford.

Weiss, R. E., A. P. Waggoner, R. J. Charlson, and N. C. Ahlquist. 1977. Sulfate aerosol: its geographical extent in the mid-western and southern United States. Science 195:979–981.

CHAPTER 29

A REVIEW OF GAUSSIAN DIFFUSION-DEPOSITION MODELS

Thomas W. Horst*

ABSTRACT

The assumptions and predictions of several Gaussian diffusion-deposition models are compared. A simple correction to the Chamberlain source-depletion model is shown to predict ground-level airborne concentrations and dry-deposition fluxes in close agreement with the exact solution of Horst.

INTRODUCTION

It is often necessary to know the dry deposition from one or a few specific sources of pollution. The spatial distribution of airborne contamination is then not simply a function of height determined by the deposition flux at the surface. It also depends on the atmospheric transport and diffusion between source and receptor. This is predicted by diffusion models, which typically depend on source height, receptor location, wind speed, and atmospheric stability. A diffusion-deposition model accounts, in addition, for the loss of airborne contamination caused by deposition between source and receptor.

The spatial distribution of airborne contamination from a point source is determined by a conservation-of-material equation, subject to the deposition boundary condition. Direct solution of this equation is complicated by the need to parameterize the turbulent diffusion of the contaminant, the simplest scheme being the assumption of gradient transfer (K theory). A direct solution is commonly avoided by using an empirical formula, the Gaussian plume model, to describe observations of point-source diffusion.

Gaussian plume diffusion models are attractive since they are computationally simple, they can directly incorporate measurements of the plume dimensions, and they apply to both vertical and lateral diffusion and to both ground-level and elevated releases. However, it is difficult to modify them for gravitational settling and deposition because they do not explicitly model the vertical transport process.

Gravitational settling and deposition are much easier to include in a direct solution of the conservation equation, as typified by a gradient-transfer model. K-theory models can also directly incorporate physical properties of the atmosphere, such as vertical profiles of wind speed and eddy diffusivity. However, K theory is strictly applicable only to vertical diffusion from a ground-level source and should not be applied to lateral diffusion or to vertical diffusion in the vicinity of an elevated source.

---

*Pacific Northwest Laboratory, Richland, WA 99352.

Further, a realistic vertical distribution of wind speed and eddy diffusivity usually requires a numerical solution of the equations. Some of these restrictions can be lifted by using closure schemes more sophisticated than K theory but only at the expense of additional equations requiring numerical solution.

At this time, Gaussian plume models are usually chosen for environmental assessment. Given the proper data base, they are no less accurate than K-theory models and are computationally much simpler. Except for the largest particles, vertical atmospheric transport is dominated by diffusion rather than settling and deposition. Thus, some approximation is permissible when including the latter processes in the Gaussian plume model.

This paper compares several Gaussian diffusion-deposition models. The author's surface-depletion model is described first because, assuming a Gaussian description of the diffusion, it is an exact solution to the diffusion-deposition problem for nonsettling particles and is used as a standard of comparison for the other, computationally simpler, models. The original Chamberlain model is described next, followed by the simple Csanady/Overcamp model. Finally, the author proposes a simple correction to Chamberlain's source-depletion model which yields results which closely approximate the exact solution.

GAUSSIAN DIFFUSION MODEL

The basic form of the Gaussian diffusion model for a point source of contamination at height h is

$$C_0(x, y, z) = \frac{Q_0 e^{-y^2/2\varsigma_y^2}}{2\pi u \varsigma_y \varsigma_z} \{\exp[-(h - z)^2/2\varsigma_z^2] + \exp[-(h + z)^2/2\varsigma_z^2]\} . \quad (1)$$

C_0 is the airborne contamination (mass/volume) at a height z and at a distance x downwind and y crosswind of the source, Q_0 is the rate of contaminant emission (mass/time), and u is the mean wind speed. The lateral and vertical spread of the plume, ς_y and ς_z, are functions of downwind distance and atmospheric stability and are commonly determined from empirical graphs (Gifford, 1976). Since dry deposition does not appreciably alter the crosswind distribution, the discussion below deals with the crosswind-integrated form of Eq. (1),

$$\overline{C_0}^y(x, z) = \int_{-\infty}^{+\infty} C_0(x, y, z)dy = Q_0 D(x, z, h) , \quad (2)$$

where

$$D(x, z, h) = \frac{1}{\sqrt{2\pi} u \varsigma_z} \{\exp[-(h - z)^2/2\varsigma_z^2] + \exp[-(h + z)^2/2\varsigma_z^2]\} . \quad (3)$$

SETTLING vs NONSETTLING CONTAMINANTS

Aerosol particle sizes may be divided into two categories that depend on the relative contributions of gravitational settling and turbulent diffusion to vertical atmospheric transport. For gases and nonsettling particles, those smaller than approximately 10 μm radius, settling may be neglected compared with diffusion. These will be the chief concern of this review since they include the respirable particles and since they are transported farther from the source. The author's surface-depletion model is an accurate correction to the Gaussian plume model for the deposition of gases and nonsettling particles.

A rigorous modification of the Gaussian plume diffusion model to account for gravitational settling has not been found. A common procedure is to replace the source height in (1) or (3) by

$$h' = h - v_s x/u ,\qquad (4)$$

where v_s is the particle settling velocity. This model conserves mass and is probably a reasonable approximation for $v_s x/uh < 1$. A model based completely on gradient-transfer theory or other direct solution of the conservation equation would probably account more accurately for gravitational settling.

HORST SURFACE-DEPLETION MODEL

The author's surface-depletion model (Horst, 1977) is based on the linearity of the conservation-of-material equation which describes atmospheric transport and diffusion. This allows the superposition of solutions such as Eq. (1) to account for a number of sources at different locations and the use of image sources to satisfy boundary conditions. The deposition flux to the surface,

$$F_d(x, y) = v_d(z_*)C(x, y, z_*) ,\qquad (5)$$

is represented as a material sink, i.e., a ground-level source for diffusion of the material deficit resulting from deposition at that point. The airborne contamination at any location is calculated as the sum of the nondepositing diffusion from the original source at $(0, 0, h)$ plus the diffusion from the (negative) surface sources which account for the loss of airborne material by deposition upwind of that location.

The crosswind-integrated solution is

$$\overline{C}^y(x, z) = Q_0 D(x, z, h) - \int_0^x v_d(z_*)\, \overline{C}^y(x', z_*)\, D(x - x', z, 0)\, dx' .\qquad (6)$$

The integrand of (6) is the crosswind-integrated deposition flux $v_d\overline{C}^y$ at location x' multiplied by the diffusion function for a source at ground level and for a receptor at height z and downwind a distance $x - x'$. Since the depleted airborne contamination near ground level $\overline{C}^y(z_*)$ is in the integrand, (6) must initially be solved at $z = z_*$, the reference height for deposition. This is a problem if $z_* = 0$ because $D(x - x', 0, 0)$ is not well behaved as $x' \to x$. However, z_* need never equal zero, since when the deposition velocity v_d is determined experimentally from Eq. (5), C is commonly measured at some finite height above the surface.

Unfortunately, the integral in (6) usually requires a numerical solution. This is particularly tedious since the integrand is a function of x, the upper limit of integration. Hence the integral from 0 to $x + \Delta x$ cannot be computed by simply extending a previously determined integral from 0 to x. Nevertheless, (6) is available when an exact solution is required; and it permits testing, with identical diffusion assumptions, of simpler solutions for the deposition boundary conditions.

CHAMBERLAIN SOURCE-DEPLETION MODEL

Chamberlain (1953) proposed one of the simpler and most commonly used diffusion-deposition models. His model accounts for the deposition of airborne material by appropriately reducing the source strength as a function of downwind distance:

$$\overline{C}^y(x, z) = Q(x)D(x, z, h) .\qquad (7)$$

Conservation of mass requires that

$$\frac{dQ(x)}{dx} = -v_d(z_*) \, \overline{C}^y(x, z_*) , \qquad (8)$$

and hence,

$$Q(x) = Q_0 \exp\{ -\int_0^x v_d(z_*)D(x', z_*, h) \, dx' \} . \qquad (9)$$

Note that although the source-depletion model correctly evaluates the deposition flux in terms of the airborne contamination near the surface, $\overline{C}^y(z_*)$, it artificially assumes that deposition is a loss at the source $(0, 0, h)$ rather than at the surface. However, since the integrand in (9) does not depend either on the depleted airborne contamination $\overline{C}^y$ or on the upper limit of the integral, the source-depletion model is much easier to apply than the surface-depletion model.

Horst (1977) has compared the Chamberlain source-depletion model with the surface-depletion model. The results are a function of atmospheric stability, source height, downwind distance from the source, and the ratio v_d/u. Figure 1 compares model predictions of the ground-level airborne contamination $\overline{C}^y(z_*)$ for three Pasquill stability categories and a source height of 10 m. The predictions are also for a

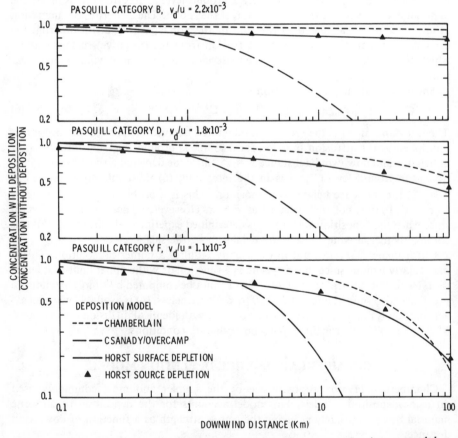

Fig. 1. The reduction of the ground-level airborne contamination caused by deposition, as a function of distance downwind of the source. $h = 10$ m, $v_d/u_* = .03$, $z_* = 3$ cm.

single particle size, and consequently the ratio of the deposition velocity to the friction velocity, v_d/u_*, is 0.03 for all calculations. However, v_d/u varies due to the dependence of the ratio u_*/u on atmospheric stability. Figure 2 similarly compares predictions of the suspension ratio S, the fraction of pollutant remaining airborne.

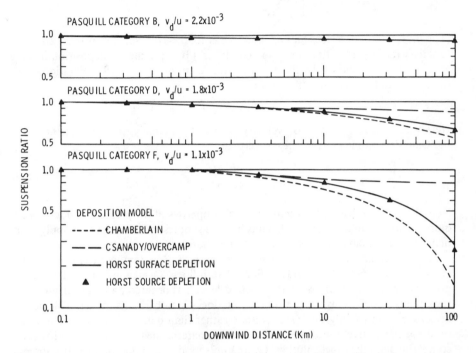

Fig. 2. The fraction of contamination remaining airborne as a function of distance downwind of the source. $h = 10$ m, $v_d/u_* = 0.03$, $z_* = 3$ cm.

The errors of the Chamberlain model are caused by its retention of the Gaussian plume shape. This, in effect, introduces artificial vertical mixing which instantaneously distributes the losses due to deposition throughout the entire vertical extent of the plume, rather than selectively depleting the portion of the plume near ground level. Thus, at downwind locations close to the source, the Chamberlain model overpredicts $\overline{C}^y(z_*)$. Because the deposition is therefore overpredicted, the model is biased in the opposite sense for locations far from the source. At all distances, the total deposition between source and receptor is overestimated, and the suspension ratio is consequently underpredicted.

In Figs. 1 and 2, Pasquill category B represents unstable atmospheric conditions, D represents neutral atmospheric conditions, and F represents stable conditions. The differences between the two models increase with increasing stability. This is expected because the vertical mixing due to real atmospheric processes decreases with increasing stability and hence the effects of the artificial mixing become more noticeable. The dependence on source height is not shown, but it may be approximated from Figs. 1 and 2 by multiplying the downwind distance by the ratio of a new source height to the 10-m source height. Finally, since without deposition the models are identical, the model differences will also increase with v_d/u.

CSANADY/OVERCAMP SOURCE-DEPLETION MODEL

Csanady's (1955) model for the deposition of settling particles has been extended by Overcamp (1976) to the case of nonsettling contaminants. This model accounts for deposition by reducing the strength of the conventional image source,

$$\bar{C}^y(x, z) = \frac{Q_0}{\sqrt{2\pi}\, u\varsigma_z} \left\{ \exp\left[\frac{-(h' - z)^2}{2\varsigma_z^2} \right] + \alpha(x, z)\exp\left[\frac{-(h' + z)^2}{2\varsigma_z^2} \right] \right\}. \tag{10}$$

[The $h' + z$ term describes an image source at $(0, 0, -h')$ which is introduced in (1) to satisfy the no-deposition boundary condition.] By equating the deposition flux (5) at $z_* = 0$ to the sum of the fluxes due to turbulent diffusion and gravitational settling, Overcamp finds $\alpha(x, z = 0)$ to be

$$\alpha_0(x) = 1 - 2v_d/(v_s + v_d + v_t) . \tag{11}$$

Here v_t is a turbulent diffusion velocity defined by Csanady to be the rate at which the Gaussian plume expands in the vertical,

$$v_t = \frac{uh'}{\varsigma_z} \frac{d\varsigma_z}{dx} . \tag{12}$$

Note that this model depends only on local properties of the plume and does not require computation of an integral. Thus it is the simplest of the models to apply, particularly if an analytical expression is available for $\varsigma_z(x)$.

In Figs 1 and 2, Overcamp's model for a nonsettling contaminant, $v_s = 0$, is compared with the preceding models. For all stabilities, Overcamp's model correctly predicts greater reductions in the ground-level airborne contamination and correspondingly higher suspension ratios than does Chamberlain's model. This occurs because it reduces only the image source, rather than both real and image sources, resulting in relatively more depletion of the material near the surface. However, comparing with the exact solution, Overcamp's model still underpredicts the air concentration reduction at short distances and greatly overpredicts the reduction at large distances. For $h = 10$ m the crossover occurs at $x \cong 1$ km; for $h = 100$ m this distance is increased by a factor of 5–10. Due to the excessive depletion of $\bar{C}^y(x, z_*)$ at large downwind distances, the deposition ceases and the suspension ratio is unrealistically predicted to approach a constant value.

Horst (1978) relates the model differences seen in Figs. 1 and 2 to the assumptions of Overcamp's model. Briefly, Overcamp's model underpredicts $\bar{C}^y(x, z_*)$ at short distances because, similar to Chamberlain's model, the material loss caused by deposition at $(x_d, z = 0)$ appears to occur at $(x = 0, -h)$. Downwind of x_d, therefore, the effect of that deposition is underestimated at ground level because the loss has been diluted by diffusion over the distance $(x^2 + h^2)^{1/2}$ rather than the smaller distance $x - x_d$. At large distances, this effect is overridden by the model's neglect of the additional downward diffusion flux produced by the reduced ground-level airborne contamination. Equation (12) only accounts for diffusion in the absence of deposition, and thus, at large distances where $v_t < v_d$, the depleted ground-level contamination is not effectively replenished by airborne contamination from higher levels.

HORST SOURCE-DEPLETION MODEL

The errors of source-depletion models are caused by their inadequate prescription of the deposition-modified vertical distribution of airborne contamination. The author has developed a modified source-depletion model which compares quite well with the surface-depletion model but is computationally comparable with the Chamberlain model. It accounts for the loss of airborne contamination in the same manner as the Chamberlain model, by reducing the source strength as a function of downwind distance [Eq. (8)]. However, a second correction $P(x, z)$ is applied to account for the change in the vertical contamination distribution caused by deposition,

$$\overline{C}^y(x, z) = Q(x) D(x, z, h) P(x, z) . \tag{13}$$

Substitution of (13) into (8) gives

$$Q(x) = Q_0 \exp\{ -\int_0^x v_d(z_*) D(x', z_*, h) P(x', z_*) dx'\} . \tag{14}$$

$P(x, z)$ is derived from a gradient-transfer model. Without deposition, the airborne contamination near the ground level is independent of height. In order to approximate $P(x, z)$ it will be assumed that this is true at all heights, i.e., $D \cong (uH)^{-1}$, where H is a mixing depth. For $\varsigma_z/h \gg 1$, this approximation is exact at ground level if $H = \sqrt{\pi/2} \, \varsigma_z$. Thus

$$\overline{C}^y(x, z) \cong \frac{Q(x)}{u\,H} P(x, z) , \tag{15}$$

and conservation of mass requires that

$$\frac{1}{H} \int_0^H P(x, z) \, dz = 1. \tag{16}$$

With D independent of height, gradient-transfer theory predicts that deposition of nonsettling particles produces a vertical distribution equal to

$$\overline{C}^y(x, z) = \overline{C}^y(x, z_*) [1 + v_d R(z, z_*)] , \tag{17}$$

where the resistance R is

$$R(z, z_*) = \int_{z_*}^z \frac{dz'}{K(z')} \tag{18}$$

and $K(z)$ is the vertical eddy diffusivity. For the Gaussian plume model the resistance becomes

$$R(z, z_*) = \frac{1}{u} \sqrt{\frac{2}{\pi}} \int_{x(z_*)}^{x(z)} \frac{dx'}{\varsigma_z(x')} . \tag{19}$$

Finally, combining (15), (16), and (17) gives

$$P(x, z_*) = \left[1 + \frac{v_d}{H} \int_0^H R(z, z_*) \, dz \right]^{-1} . \tag{20}$$

For particles with a finite settling speed v_s, Eqs. (17) and (20) are modified by replacing the quantity $v_d R$ with

$$\frac{v_d - v_s}{v_s} [1 - \exp(- v_s R)] . \tag{21}$$

In Figs. 1 and 2 the author's source-depletion model is compared with the previous models. Its predictions compare very well with the exact solution, particularly for $\varsigma_z/h > 1$, where most of the deposition occurs. Since the model assumes $D = (u\,H)^{-1}$, its predictions are poorest for $\varsigma_z/h < 1$, where the ground-level contamination from the elevation plume is not appreciable.

CONCLUSIONS

Four Gaussiuan diffusion-deposition models have been compared. The author's surface-depletion model is an exact solution but is applicable only to nonsettling contaminants and is computationally complex. The source-depletion models are applicable to both settling and nonsettling contaminants. The Overcamp/Csanady model is the easiest to use, and, at short downwind distances, its accuracy for nonsettling contaminants is intermediate between the Horst and Chamberlain models. At large downwind distances, however, Overcamp's model is unrealistic, and Chamberlain's model is better. Chamberlain's model is intermediate in computational requirements, but with the addition of the author's profile correction, its predictions closely approximate the exact solution.

Two cautions are in order at this point. The Gaussian diffusion model is basically empirical and thus is in general agreement with the available data. Adequate data are not presently available to validate plume-depletion models, however. Therefore the accuracy of the diffusion-deposition models has been tested only against a mathematically correct modification of the Gaussian plume model, not against real data. Second, the exact solution is applicable only to nonsettling contaminants, and thus the models for settling particles have not been tested.

REFERENCES

Chamberlain, A. C. 1953. Aspects of travel and deposition of aerosol and vapor clouds. Atomic Energy Research Establishment HP/R 1261, Harwell, Berkshire, England.

Csanady, G. T. 1955. Dispersal of dust particles from elevated sources. Aust. J. Phys. 8:545-550.

Gifford, F. A. 1976. Turbulent diffusion-typing schemes: a review. Nucl. Saf. 17: 68-86.

Horst, T. W. 1977. A surface depletion model for deposition from a Gaussian plume. Atmos. Environ. 11:41-46.

Horst, T. W. 1978. Comments on "A General Gaussian Diffusion-Deposition Model for Elevated Point Sources." J. Appl. Meteorol. 17:415-416.

Overcamp, T. J. 1976. A general Gaussian diffusion-deposition model for elevated point sources. J. Appl. Meteorol. 15:1167-1171.

DISCUSSION

B. D. Zak, Sandia Lab: In all of the models that you've discussed, you presume no limitation on vertical mixing. Could you comment on how a limitation on vertical mixing would change the results that you've obtained?

T. W. Horst: It would have very little effect on the comparison of the models. For one thing, the diffusion curves that I used here for stable atmospheric conditions do go to a limited value, so that the plume is limited in size at large distances. Also, it would be very easy to incorporate an inversion effect, for example, by putting a reflection at the inversion. I don't think it would have much effect on the model comparisons, although it certainly would affect the absolute values that you calculated from each model.

C. H. Goodman, Southern Company Services: In terms of applying such a model to simulate a single source or a group of sources close together, what are the uncertainties in your mind regarding how far out one could use the model, considering model limitations, knowledge of the vertical meteorology, wind shear, and so on?

T. W. Horst: The limitations are not in the deposition models, the limitation is in the Gaussian diffusion model.

C. H. Goodman: But you show plots out to a hundred kilometers—do you think there could be ways to correct the limitations in any of the models or the assumptions in the Gaussian model, so that you could apply it to make an environmental assessment of a given arrangement of sources?

T. W. Horst: Yes, but the place to make the correction is in the appropriate dispersion parameters, or perhaps by applying something mentioned earlier—as you go further downwind, a lid on the vertical mixing will have more and more effect. The correction must come in the diffusion model. I think the question you asked is independent of the particular deposition model. In fact, as you go further and further downwind and the material is more and more uniformly mixed in the boundary layer, then the modified source depletion model I have shown here should become more and more correct. But the place to correct is in the diffusion representation, and that's really not what I'm addressing here.

I forgot to make one point—that the Gaussian diffusion model is basically empirical through its dependence on the sigma z's and sigma y's. It therefore agrees with what data are available. However, these data have been extrapolated up to a hundred kilometers, and the reliability of that is certainly highly in question. Unfortunately, data are *not* available to test plume depletion models. Thus I tested these deposition models *only* by testing them against the mathematically correct modifications of the Gaussian diffusion model, and I have not been able to test them against real data.

SECTION VI

PROCESS-LEVEL EFFECTS OF SULFUR DEPOSITION

Co-Chairmen: D. W. Johnson and J. M. Skelly

CHAPTER 30

PROCESS-LEVEL EFFECTS OF SULFUR DEPOSITION

Chairmen: **Dale W. Johnson**
Oak Ridge National Laboratory

John M. Skelly
Virginia Polytechnic Institute and State University

OVERVIEW

Dale W. Johnson*

One useful way to gain understanding of ecosystems is to conceptually dissect them into a series of components and interconnecting processes. Within this framework, researchers identify those components and processes considered most important to the question at hand and design experiments to evaluate their behavior. Though sometimes criticized as narrow or "reductionist," this approach has been quite productive in acid rain effects studies.

Within Session V, there was considerable discussion of the effects of sulfur deposition on sulfur nutrition of terrestrial ecosystems. Krupa and Chevone noted increased foliar sulfur levels in both crop and tree species growing near a coal-fired power plant in Minnesota, probably as a result of increasing dry-fall inputs. Wet-fall sulfate inputs, which did not increase along with SO_2 emissions from the nearby power plant, were therefore thought to be due to regional-scale sulfur burdens. Noggle discussed the positive influence of gaseous sulfur on crop nutrition where soils are not capable of supplying adequate sulfur for crop needs. Using radioactive tracer techniques, Noggle demonstrated that atmospheric sources can provide substantial proportions of crop sulfur requirements. Atmospheric sources of sulfur are of particular importance to crops at present, since the use of sulfur-containing fertilizers has declined markedly. Turner reviewed the close relationships between nitrogen and sulfur in forest ecosystems and the potentially positive influences of atmospheric sulfur deposition on sulfur-deficient forests in Australia and the northwestern United States. Benefits can include both increased forest productivity and prevention of disease outbreaks associated with chronic sulfur deficiencies. Increasing use of nitrogenous fertilizers in commercial forests lands of the northwestern and southeastern United States are likely to create increasing demands upon the sulfur reserves of these ecosystems.

*Environmental Sciences Division, Oak Ridge National Laboratory, P.O. Box X, Oak Ridge, Tennessee 37830.

In contrast to these findings, several other investigators demonstrated the potentially detrimental effects of acid deposition often associated with sulfur deposition. Evans and Lewin found reduced yield in pinto beans and soybeans as a result of irrigation at pH 2.5. At a pH of 3.5, yields of pinto beans were reduced also, but yields of soybeans increased above control (pH 5.7) levels. The authors concluded that further research must be conducted before judgments can be made as to which specific pH levels are damaging to vegetation.

Cronan found that natural soil leaching processes had been overwhelmed by atmospheric sulfuric acid deposition in some forest soils in New Hampshire. In addition to accelerating the losses of nutrient cations from forest ecosystems, acid deposition has apparently also triggered the release of aluminum from soil horizons that would normally immobilize this element. Such releases of aluminum could have severe detrimental effects on fish populations in acidified lakes, as demonstrated by Schofield. Schofield noted that some lakes in the Adirondack region of New York have lost their bicarbonate buffering capacity and are devoid of fish life. Other lakes are less acidic but subject to episodic inputs of acid and soluble aluminum during snowmelt periods. Aluminum created toxic conditions for fish at pH levels not otherwise harmful. Hendry described some initial results of an Adirondack lake study and reviewed the complex effects of acidification on primary productivity in lakes. Of particular interest is the reduced availability of phosphorus due to aluminum complexation as the pH approaches 5.5, since phosphorus is often the limiting nutrient in lakes subject to acidification.

Atmospheric sulfur deposition can have either beneficial or adverse effects on terrestrial ecosystems. The beneficial effects of sulfur deposition on sulfur-deficient sites have been ignored too long by acid rain researchers, and they must be included in the final integrated assessment of sulfur pollution effects. Further research is needed in plant sulfur nutrition, especially in several commercially important tree species for which very limited data are currently available. Research is also needed to determine the extent of potentially sulfur-deficient crop and forest lands, the current role of atmospheric sulfur deposition in alleviating these deficiencies, and the potential costs associated with removing this inadvertent source of sulfur fertilizer. Similarly, the potential contribution of increasing nitrogen deposition toward ameliorating the far more widespread occurrence of nitrogen deficiency deserves investigation.

On the other hand, there is little chance of the acid commonly associated with sulfur deposition producing beneficial effects in either terrestrial or aquatic ecosystems. Acid inputs appear to be increasing in many parts of the world, yet the levels required to produce adverse effects on terrestrial ecosystem productivity remain a matter of speculation. The effects of acid deposition on both foliar and soil leaching processes deserve further study, particularly in ecosystems where cation nutrients are in short supply (e.g., potassium-deficient forest ecosystems in New York). Aquatic ecosystems in regions with poorly buffered soils are particularly susceptible to acidification, and the appearance of toxic levels of aluminum in some of these systems is certainly cause for alarm. There is a clear need for continuing research into mechanisms of aluminum and hydrogen ion transport to aquatic ecosystems and the complex effects of these inputs on the aquatic biota.

CHAPTER 31

SULFUR ACCUMULATION BY PLANTS; THE ROLE OF GASEOUS SULFUR IN CROP NUTRITION

J C Noggle*

ABSTRACT

Sulfur (S) requirements to maintain high crop production range from 10 to 40 kg/ha · year. The relationship between soil S supply and crop yield is presented, and the need for supplemental S to maintain high crop yields is discussed. A decline in the use of fertilizers that contain S has placed a greater dependency on the atmosphere as a source of supplemental S to meet the needs of plants.

A technique was developed and used to measure the amount of atmospheric S accumulated by crops grown in the field. Amounts of atmosphere-derived S in cotton grown 3 and 16 km from a coal-fired power plant were 44 and 26 percent, respectively, of the total S content in the plant tissue. Results from this study and from collection of wet and dry deposition of atmospheric S provide evidence that a significant quantity of S is transferred from the atmosphere to the agroecosystem.

INTRODUCTION

It has long been established that sulfur (S) is an essential plant nutrient needed for the synthesis of proteins and other organic compounds within vegetative tissue. In the production of food and fiber, a knowledge of the amount of S required for optimum plant growth and how to supply it in the most efficient manner is of prime importance.

Before we consider the sources of S to crops, let us review briefly the relationship between S supply and crop yield as illustrated in Fig. 1. As the supply of S is increased, the crop yield increases to some constant level, at which another growth factor limits a further increase in yield. To obtain maximum yield under a given set of environmental conditions, total S supply must at least equal the total S requirement of the crop, identified as point A on the S supply axis.

The actual value for point A is a function of both the amount of S that must accumulate in the plant tissue for maximum yield and the efficiency of S uptake from the source. The amount of S absorbed by crops ranges from 10 to 40 kg/ha · year (Ensminger and Jordan, 1958). Crops such as corn (*Zea mays* L.) and wheat (*Triticum aestivum* L.) use a lower amount of S than crops such as cabbage (*Brassica oleracea* L.), alfalfa (*Medicago sativa* L.), and cotton (*Gossypium hirsutum* L.). Crop requirements for S and phosphorus (P) are similar, but much higher amounts of nitrogen (N) and potassium (K) are needed. The "critical" S concentration in the

---

*Tennessee Valley Authority, Muscle Shoals, Alabama 35660.

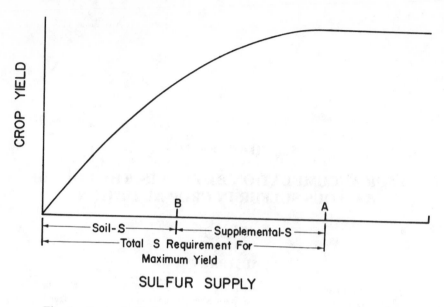

Fig. 1. General relationship between sulfur supply and crop yield.

plant tissue is an important parameter in determining the S requirement of crops. It is defined as the minimum concentration of S that can occur in the plant tissue without a reduction in growth and will vary between plant species. Total plant S requirement is the product of the critical concentration and dry-matter production.

The uptake efficiency of S is the percentage of the supply that is accumulated by the plant in one growing season. Although S uptake efficiency has not been studied extensively under field conditions, it is probably similar to nitrogen efficiency, which usually ranges from 50 to 60%. This means that the supply of available S should be about two times the plant uptake for maximum yields. Sulfur supplied in excess of the amount needed to meet the crop requirement cannot be utilized effectively by that particular crop, but it can be returned to the soil in crop residue and utilized by a subsequent crop.

The soil is the primary source of S for plants, and the amount supplied is identified as point B on the supply axis (Fig. 1). The instantaneous S supply in soil includes a "pool" of soluble sulfate in the soil solution and sulfate adsorbed on clay surfaces, both of which can be measured by chemical analysis. Although the size of the sulfate pool can be measured at a point in time, it is in a constant state of change. Soluble sulfate can be readily leached from sandy soils, but adsorbed sulfate must be replaced by another anion before it will be moved out of the root zone by the downward movement of water. Increasing the soil pH by liming reduces the number of positive charges on the clay and thereby transfers adsorbed sulfate to the soil solution. Sulfur in the harvested portion of crops is usually removed from the farm. If the size of the sulfate pool were infinite, there would be no concern, but Bardsley and Kilmer (1963) investigated the sulfate status in several soils of the southeastern United States and reported that many of them were low in sulfate sulfur. These soils contain only one or two times the annual crop requirement of S. Losses of sulfate by leaching and crop removal require that the sulfate pool be replenished from some source to maintain high crop yields.

The total supply of S in soil also includes the S in the soil organic matter and crop residues. This organic S is in a chemical form that cannot be used directly by plants, but the decomposition of the organic material by microorganisms oxidizes organic S compounds to the sulfate form. Although the rate of mineralization of S from soil organic matter has not been studied extensively under field conditions, Bettany et al. (1974) found that 30 to 70% of the S in alfalfa plants grown in the greenhouse was derived from mineralized S. Since the soils used in their study contained 2 to 6% organic matter, the data suggest that only soils high in organic matter content can supply an appreciable percentage of the crop's S requirements from the decomposition of organic matter.

Since the amount of S supplied annually by a particular soil is finite, the position of point *B* is fixed, and its position relative to point *A* depends on the total S requirement of the crop. For example, a given amount of soil S could supply the total requirement for a crop such as wheat but only a part of the requirement for a crop such as cotton. Also, total S requirement depends upon the yield level of the crop; during the past three decades, crop yields have increased dramatically due to improved varieties and production practices. A given S supply that met the crop needs 30 years ago would fall far short of meeting the S requirements for today's higher yields.

When the supply of available soil S is less than the total S requirement, supplemental S is needed to maintain maximum crop yield. The need for supplemental plant nutrients usually brings to mind fertilizers, and for many years farmers have applied various fertilizer materials to add supplemental N, P, K, and micronutrients to their crops. Prior to about 1950, ordinary superphosphate was the basic ingredient of the commercial fertilizers used to supply N, P, and K to crops. Ordinary superphosphate was prepared by reacting phosphate rock with sulfuric acid, with a resulting composition of 8 to 9% P and about 14% S. The use of this material usually added adequate S along with the other plant nutrients to meet the needs of the crops. The production of superphosphate was subsequently changed to a process of reacting rock phosphate with phosphoric acid to yield concentrated superphosphate with a composition of 20 to 22% P and less than 1% S. The application of other fertilizer materials such as ammonium sulfate, potassium sulfate, and potassium magnesium sulfate also added S to the soil, but the use of these materials has declined in recent years.

The recent trend to apply fertilizers with a relatively high N, P, and K content and essentially no S should have resulted in increased occurrences of S deificiency, but widespread shortages have not been observed. According to Beaton et al. (1971), the only crops that show S deficiencies are usually found on highly leached soils remote from industrial areas.

During the period that the use of S-containing fertilizers declined, the combustion of fossil fuels has increased, resulting in the emission of large quantities of sulfur compounds into the atmosphere. Sulfur in the atmosphere exists as sulfur dioxide (SO_2) and hydrogen sulfide (H_2S) in a gaseous form and as sulfate in association with aerosols and particulate matter. Sulfur is transferred from the atmosphere to agroecosystems by direct sorption of SO_2 by plants and soil, by dry particulate deposition, and in rainfall. Numerous investigators have reported that atmospheric SO_2 is absorbed directly by plant foliage. Fried (1949) used ^{35}S-labeled SO_2 to demonstrate that alfalfa plants can absorb SO_2 through the leaves and convert it into organic S compounds. Cowling et al. (1973) found beneficial effects of SO_2, such as increased yield and S content, in perennial ryegrass that was grown with an inadequate supply of S to the roots. Olsen (1957) showed that cotton plants grown in an atmosphere containing 0.01 to 0.05 ppm SO_2, although supplied with adequate sulfate in a nutrient solution, still absorbed about 30% of the plant S from the atmosphere. Several investigators (Alway et al., 1938; Payrissat and Beilke, 1975;

Roberts and Koehler, 1965; and Yee et al., 1975) have reported direct adsorption of SO_2 by soils. Jordan et al. (1959), who extensively surveyed S in precipitation and dry particulate deposition in Southern states, reported values of 10 to 30 kg/ha · year; these values are similar in magnitude to crop requirements.

These observations introduce an interesting question, Has the atmosphere replaced commercial fertilizers as a significant source of S for crops? For atmospheric S to affect crop production significantly, supplemental S must be needed, and the S that is transferred from the atmosphere to the agroecosystem must contribute a relatively large proportion of these supplemental needs.

The usual procedure to determine the need for a supplemental plant nutrient is to apply the nutrient and note the presence or absence of a yield response. With S, this approach has been frustrating. Soils that tested low in available S and showed a yield response to S application in greenhouse experiments have failed to show a yield response in the field. Many researchers have explained the lack of the expected response in the field by stating that atmospheric S may have provided sufficient S to prevent a deficiency in those plants growing on plots receiving no fertilizer S. This conclusion was adequate for their purposes because they were only concerned with establishing the need to add fertilizer S, but the evidence is not sufficient to convince those outside of agriculture that atmospheric S has an economic significance to crop production.

In 1976, the Tennessee Valley Authority initiated an investigation to evaluate the economic significance of the atmosphere as a source of S for crop production. The study includes both establishing the need for supplemental S and also quantifying the transfer of atmospheric S to the agroecosystem, but only the accumulation of SO_2 by plants grown in the field will be discussed in this paper.

EXPERIMENTAL PROCEDURE

Hartsell fine sandy loam (fine-loamy, siliceous, thermic, Typic Hapludults) soil with a very low S content was collected in De Kalb County, Alabama. The total soil needed for the study was thoroughly mixed and weighed into 45-kg portions for subsequent inclusion into plastic-lined, 10-gal galvanized cans. The cans were used as growth containers to ensure that soil S supply was the same for plants at all sites. Supplementary plant nutrients and 348 μCi of $^{35}SO_4^{2-}$ were mixed with each 45-kg portion of soil.

Growth containers were installed in the field at all locations shown in Table 1 and in filtered-air greenhouses at the Muscle Shoals and Crossville sites.

Table 1. Locations of experimental growth containers

Site designation	Location
Widows Creek	4 km southeast of the TVA Widows Creek coal-fired power plant near Stevenson, Alabama
Crossville	70 km south-southwest of Widows Creek power plant at the Auburn University Experiment Station near Crossville, Alabama
Colbert	3 km north of the TVA Colbert coal-fired power plant near Barton, Alabama
Muscle Shoals	16 km east of the TVA Colbert coal-fired power plant

Three replicates of cotton [*Gossypium hirsutum* (L.) var. Deltapine 16] and soybeans [*Glycine max* (L.) Merr. c.v. Essex] were planted at each site. Deionized water was added to growth containers in the field and greenhouses to maintain adequate soil moisture.

The entire above-ground portions of the plants were harvested on September 14, dried at 70°C for 72 hr in a forced-air oven, weighed, and ground to pass through a 20-mesh screen. The total S concentration in the plant material was measured by the Leco furnace method (Todemann and Anderson, 1971), and the ^{35}S content was determined on the same sample using liquid scintillation counting techniques.

RESULTS AND DISCUSSION

The technique used to measure accumulation of atmospheric S by plants provides a direct method for calculating the amount of plant S derived from the soil. Basically it consists of determining the specific acitivty (S.A.) of available soil S and using this value to calculate soil-derived S in plants grown at any location. Atmosphere-derived S was the difference between total plant S and soil-derived S.

The method chosen to measure the S.A. of soil S was based on the assumption that plants accumulated radioactive $^{35}SO_4^{2-}$ and available stable sulfate from the soil in the same ratio as they occurred in the sulfate pool. Therefore, when the soil was the only source of S, the S.A. of plant S and soil S were the same.

The total S and radioactive ^{35}S content of cotton grown in greenhouses with charcoal-filtered air at Muscle Shoals and Crossville were used to determine the S.A. of available S in the soil (Eq. 1):

S.A. soil S (cpm/mg S) = S.A. plant S (cpm/mg S) =

$$\frac{\text{total radioactivity in plants (cpm)}}{\text{Total plant S (mg)}} . \tag{1}$$

The calculations for mean S.A. of S in cotton plants at the two sites are shown in Table 2. The total dry-matter production at Muscle Shoals was higher than at

Table 2. Determination of specific activity of sulfur in cotton grown in greenhouses with charcoal-filtered air

Site	Oven-dry weight (g)	Sulfur concentration (%)	Total plant S (mg)	Total radioactivity (cpm $\times$ 10^6)	Specific activity (cpm $\times$ 10^3/mg S)
Muscle Shoals	238	0.33	785	20.97	26.71
Crossville	166	0.47	780	21.51	27.58
Mean					(27.15)

Crossville, but the calculated values for S.A. of soil S were similar, indicating that size of plants did not affect the S.A. of soil S.

To quantify the amount of S accumulated from the atmosphere, we must first determine the amount of soil-derived S in the plants grown in the field (Eq. 2):

$$\text{soil-derived S (mg)} = \frac{\text{total radioactivity in field plants (cpm)}}{\text{S.A. of soil S (27.15} \times 10^3 \text{ cpm/mg S)}} \tag{2}$$

and then calculate atmosphere-derived S (Eq. 3):

$$\text{atmosphere-derived S (mg)} = \text{total plant S (mg)} - \text{soil-derived S (mg)} . \quad (3)$$

The radioactivity and total S in cotton were used with Eqs. 2 and 3 to calculate the source of S in plants grown at each of the four field sites (Table 3). Cotton accumulated S from the atmosphere at all sites except Crossville, which is remote from an SO_2 source. The calculated value for soil-derived S at Crossville was greater than the measured total S in the plants and resulted in a negative value for atmosphere-derived S; however, it was not significantly different from zero. Forty-five percent of the S in cotton at Colbert and smaller proportions at Muscle Shoals and Widows Creek were accumulated from the atmosphere. The taller stacks and S emission control devices on the Widows Creek power plant may have influenced atmospheric-S accumulation at that site.

Table 3. Sources of sulfur in cotton grown in ambient air at four sites

Site	Oven-dry weight (g)	Sulfur concentration (%)	Total plant S (mg)	Total radioactivity (cpm $\times$ 10⁶)	Soil-derived S (mg)	Atmosphere-derived S (mg)	(%)
Crossville	98	0.26	255	7.08	261	−6	
Widows Creek	144	0.31	446	10.87	400	46	10
Muscle Shoals	90	0.42	378	7.57	279	99	26
Colbert	199	0.36	716	10. 80	398	318	44

The sources of S in soybeans grown in the field in 1978 and in fescue grown in 1977 were calculated in the same manner, and the results are shown in Table 4. Although soybeans accumulated less S from the atmosphere than cotton, the proportion of total plant S derived from the atmosphere was slightly higher in the soybeans. About one-third of the total plant S in fescue grown near the power plants was derived from the atmosphere.

Cotton did not show a growth response as a result of the accumulation of atmospheric S. Apparently, there was sufficient S in the 45 kg of soil to prevent S deficiency. The highest dry-matter production of cotton was in the greenhouse at Muscle Shoals, and the plants contained 0.33% S (Table 2), which is higher than the critical concentration of 0.20% reported by Jordan and Bardsley (1958). An evaluation of the economic significance of atmospheric S does not depend on a positive yield

Table 4. Sources of sulfur in soybeans and fescue grown in ambient air

Site	Soybeans				Fescue			
	Total plant S (mg)	Soil-derived S (mg)	Atmosphere-derived S (mg)	(%)	Total plant S (mg)	Soil-derived S (mg)	Atmosphere-derived S (mg)	(%)
Crossville	93	84	9	10				
Widows Creek	117	83	34	29	115	76	39	34
Muscle Shoals	145	76	69	48				
Colbert	270	133	137	51	86	57	29	34

response during one cropping of the soil. As previously discussed, most soils contain sufficient S in the sulfate pool to meet the S requirements for one or more years. The prime consideration concerns the contribution of the atmosphere in replenishing the sulfate pool so that high crop production can be maintained. Results from these experiments indicate that SO_2 contributed a significant amount of S to plants grown in the vicinity of coal-fired power plants. Much of this S would be returned to the soil in crop residues. Since these crops were grown in the field for only three months during a period of relatively low rainfall, atmospheric S that would ordinarily accumulate in soil by rainfall and dry deposition throughout the year was not available to these plants. Experiments now in progress will be maintained in the field for a sufficient period of time to measure all forms of S transferred from the atmosphere to the agroecosystem.

As previously discussed, the importance of atmospheric S to crop production depends on the need for supplemental S and the relative amount accumulated from the atmosphere. Preliminary observations from studies of soils with less than 1% organic matter indicate that the rate of S mineralization is sufficient to supply only about one-third of the crop requirement annually and show that supplemental S is needed to maintain maximum crop yields. The results of the study reported here and consideration of wet- and dry-deposition data provide evidence that a significant quantity of S is transferred from the atmosphere to the agroecosystem.

Results from these experiments and from current studies of the sulfur-supplying capacity of selected soils will be used to evaluate the economic significance of atmospheric S on crop productivity in the Tennessee River valley.

REFERENCES

Alway, F. J., A. W. Marsh, and W. J. Methley. 1938. Sufficiency of atmospheric sulfur for maximum crop yields. Soil Sci. Soc. Am., Proc. 2:229–238.

Bardsley, C. E., and V. J. Kilmer. 1963. Sulfur supply of soils and crop yields in the southeastern United States. Soil Sci. Soc. Am., Proc. 27:197–199.

Beaton, J. D., S. L. Tisdale, and J. Platou. 1971. Crop Response to Sulfur in North America. The Sulfur Inst. Tech. Bull. 18.

Bettany, J. R., J. W. B. Stewart, and E. H. Halstead. 1974. Assessment of available soil sulfur in a ^{35}S growth chamber experiment. Can. J. Soil Sci. 54:309–315.

Cowling, D. W., L. H. P. Jones, and D. R. Lockyer. 1973. Increased yield through correcting of sulfur deficiency in ryegrass exposed to sulfur dioxide. Nature 243:479–480.

Ensminger, L. E., and H. V. Jordan. 1958. The role of sulfur in soil fertility. Adv. Agron. 10:408–434.

Fried, M. 1949. The absorption of sulfur dioxide by plants as shown by the use of radioactive sulfur. Soil Sci. Soc. Am., Proc. 13:135–138.

Jordan, H. V., and C. E. Bardsley. 1958. Response of crops to sulfur on Southeastern soils. Soil Sci. Soc. Am., Proc. 22:254–256.

Jordan, H. V., C. E. Bardsley, Jr., L. E. Ensminger, and J. A. Lutz, Jr. 1959. Sulfur Content of Rainwater and Atmosphere in Southern States. U.S.D.A. Tech. Bull. 1196.

Olsen, R. A. 1957. Absorption of sulfur dioxide from the atmosphere by cotton plants. Soil Sci. 84:107–111.

Payrissat, M., and S. Beilke. 1975. Laboratory measurements of sulfur dioxide by different European soils. Atmos. Environ. 9:211–217.

Roberts, S., and F. E. Koehler. 1965. Sulfur dioxide as a source of sulfur for wheat. Soil Sci. Soc. Am., Proc. 29:696–698.

Todemann, A. R., and T. D. Anderson. 1971. Rapid analysis of total sulfur in soils and plant material. Plant Soil 35:197–200.

Yee, M. S., H. L. Bohn, and S. Miyamats. 1975. Sorption of sulfur dioxide by calcareous soils. Soil Sci. Soc. Am., Proc. 39:268–270.

DISCUSSION

Lance Evans, Brookhaven Lab: J C, I'd like to comment on your first slide, if I may. It might be misleading to a lot of people in the audience who are not familiar with agriculture. In most places within the United States right now there are no sulfur deficiencies that we can pick up and such that the distance between A and B would probably be very short if at all in most of the areas in the United States. I think we should also recognize that plants will take up nutrients in an amount over that which is necessary, what is called an agricultural luxury consumption, if it's available. It doesn't mean it's actually used; it just means that it's taken up; it may be used if necessary.

J C Noggle: I'm glad you made that point, because that's one that I didn't make. Any sulfur that would be taken up in excess of the requirements will not be used by that particular crop, but it would be useful to go back into the sulfate pool and prevent its depletion, but it is a very good point. I think I even failed to point out the two criteria for sulfur to be beneficial, if you like. One is, there must be a need for supplemental sulfur. The other is, the amount that is transferred from the atmosphere to the plants must be a significant amount of this need. The thing I tried to point out, and it's really difficult to point out, is the fact that we don't have a lot of soils that respond to sulfur. My whole contention is, what we're doing is replenishing the sulfate pool for next year. We are always keeping it up by replenishing it this year for next year, and if we quit doing that, then I guarantee you you'll see tremendous sulfate shortages in those deficiencies. But there are soils already (for instance, Alabama) receiving recommended rates of 10 pounds of sulfur per acre.

Lance Evans: J C, I'd like to state, however, that very few crops have a requirement for sulfur that's above what is ordinarily in the soil right now. In fact, the highest requirement for crops is something in the order of 20 kg per hectare actual sulfur removed by crops for soybeans and a few other plants. In crops like barley and corn the removal of the crop material itself is only about 5 kg per hectare. There are some estimates that the amount of sulfur deposited per hectare at the average over the United States right now is on the order of 40 kg per hectare.

Ellis Cowling, North Carolina State University: I would just like to comment further on what Dr. Noggle has said and what Lance Evans has said. There are 15 essential elements that are required by plants. All 15 are absorbed through the soil and through the atmosphere, so the same general principles that Dr. Noggle showed in that first slide pertain to all the essential elements that are required by plants for growth, and I would like to remind those of you who are not biologists or foresters that it is a very common practice to add supplemental fertilizers in agriculture and a very uncommon practice to do this in either rangelands or forestlands, which in total comprise 50% of the total land area of the United States, and that in those, atmospheric sources plus mineralization have provided the nutrients that have permitted plants to grow ever since God created them. So atmospheric uptake of nutrients is a very important thing for us to understand; it is an important aspect of the general phenomenon of atmospheric deposition and plant nutrition. The most important point that I would like to emphasize is simply that what Dr. Noggle has said pertains to all the essential elements that are required by plants for growth. The amount by which the nutrients are supplied from atmospheric sources will vary with the element and the particular plant involved, but atmospheric sources make an important contribution, and if we looked at the 15 elements on the average in the southeastern United States, about half of all the elements required for sustained-yield forestry are now provided by atmospheric deposition.

J C Noggle: And probably a significant amount relative to the requirement with the exception of nitrogen. The plant requirements for nitrogen are so high that the amount from deposition of atmospheric sources is insignificant for their needs. However, considering sulfur and all of the other nutrients, I think that the amount transferred from the atmosphere is a significant amount of the requirement.

Danny Jackson, Battelle Columbus Laboratories: I was just wondering if you had considered the effect of atmospheric absorption of SO_2 by the soil and what effect that would have on the specific acitivity of the sulfur in soil following exposure to the atmosphere in the field.

J C Noggle: Well, I personally have not studied the absorption of SO_2 by soil, but other people have of course shown that soils do absorb SO_2. In this particular technique it really makes no difference whether the sulfur is going through the leaves or through the soil or into the soil or into the roots. There still is a change in the specific activity of the soil, and therefore uptake would be measured.

Danny Jackson: What I'm leading to is that your calibration of the specific activity that you made on that soil in the greenhouse will change if the soil absorbs a significant quantity of the SO_2 out of the atmosphere. Therefore your calibration that you had made would no longer, possibly no longer, be valid.

J C Noggle: But the ones in the greenhouse don't have SO_2. It's a filtered-air greenhouse.

Danny Jackson: I think you still misunderstand. You made the calibration in the greenhouse, and that calibration may change if the atmospheric SO_2 is put in the soil out in the field.

J C Noggle: You've got my technique down; the whole principle of this technique is the dilution of specific activity in the field.

CHAPTER 32

EFFECTS OF SIMULATED ACID RAIN ON GROWTH AND YIELD OF SOYBEANS AND PINTO BEANS*

Lance S. Evans†‡

Keith F. Lewin‡

ABSTRACT

In order to assess the degree of damage that acid rain has or might have on plants, experiments were performed to determine the change in seed yield of soybeans and pinto beans after exposure to simulated rain of pH 5.7, 3.1, 2.9, 2.7, and 2.5. Moreover, the effects of simulated acid rain were determined on a variety of other experimental parameters to understand further how plants respond to this environmental stress. Simulated acid rain of pH 3.1 and below decreased the dry mass of seeds, leaves, and stems of pinto beans. On a percentage mass basis the decrease in seed yield was comparable with reductions in biomass of leaves and stems. The decrease in yield of pinto beans by simulated acid rain was attributed to both (1) a decrease in the number of pods per plant and (2) a decrease in the number of seeds per pod. In soybeans, simulated acid rain decreased the dry mass of both stems and leaves. Seed yield also decreased after treatment with rain of pH 2.5. However, an increase in seed yield occurred when plants were exposed to rain of pH 3.1. A larger dry mass per seed was responsible for the larger dry mass of seed per plant.

INTRODUCTION

Simulated acid rain has been shown to injure plants if the pH is below 3.4 (Jonsonn and Sundberg, 1972; Wood and Bormann, 1974, 1975, 1977; Ferenbaugh, 1976; Shriner, 1977; Evans, Gmur, and Da Costa, 1977, 1978; Evans and Curry, 1979). Since plant foliage is affected by pH levels below 3.4 and acidic solutions reduce the reproductive ability of ferns even above pH 3.4 (Evans and Bozzone, 1977; Evans and Conway, unpublished results) we questioned if seed production of higher plants could be reduced by exposure to simulated acid rain.

Some recent evidence (Evans and Curry, 1979) suggests that broad-leaved (dicotyledonous) herbaceous plants may be the most sensitive of all angiosperms and gymnosperms. For our experiments we used two agronomic crops that are also

*Research supported in part by United States Department of Energy Contract No. EY-76-C-02-0016 and in part by Associated Universities Contract No. 469167. By acceptance of this article, the publisher and/or recipient acknowledges the United States Government's right to retain a non-exclusive, royalty-free license in and to any copyright covering this paper.

†Laboratory of Plant Morphogenesis, Manhattan College, The Bronx, N.Y. 10471.

‡Land and Freshwater Environmental Sciences Group, Department of Energy and Environment, Brookhaven National Laboratory, Upton, N.Y. 11973.

broad-leaved herbs: soybeans and pinto beans. Since these plants and their legume relatives are also economically and nutritionally important, they appeared to be good test organisms.

Experiments were performed to determine alterations in (1) the rate of leaf growth, (2) masses of dry (*a*) seeds, (*b*) stems, (*c*) leaves, (*d*) pods, and (*e*) prematurely abscised leaves in mature plants after exposure to simulated rain of pH levels from 5.7 to 2.5.

MATERIALS AND METHODS
General Plant Growth Techniques

All experiments were performed with either soybeans (*Glycine max*) cv. Amsoy 71 or pinto beans (*Phaseolus vulgaris*) cv. "Univ. of Idaho 111" (all seeds purchased from Agway, Inc., P.O. Box 539, 205 Marcy Avenue, Riverhead, New York 11901). Seeds were planted in 15-cm-diam clay pots and were allowed to germinate and grow in a greenhouse. Seeds were planted in a soil mixture composed of two parts of manure-enriched soil, one part of peat moss, one part of perlite, and two parts of coarse sand (volume basis). One hundred ten milliliters of ground limestone was added to each 99 liters of the above mixture. Soybeans were inoculated with *Rhizobium japonicum* (R.P. 152-6, North American Plant Breeders Association, Princeton, Illinois 61356), and pinto beans were inoculated with *Rhizobium phaseoli* (R.P. 132-4, North American Plant Breeders Association, Princeton, Illinois 63156). Plants had numerous effective nodules as noted visually by the presence of leghemoglobin.

The greenhouse (7.35 m $\times$ 7.6 m = 55.9 m^2 floor space) was closed to outside air contamination at all times and was equipped with activated charcoal air filters to remove oxidants. An air conditioning unit (45,000 cal kg [International Steam Table] per hour, Carrier Air Conditioning Corporation, Carrier Parkway, Syracuse, New York 13221) was provided so that the greenhouse temperature would not exceed 27°C.

In all experiments a plastic hood was fitted around the stem at the base of each plant to cover the soil. During experiments, plants were irrigated under the covers, so test solutions were not washed from leaves. Irrigation water was never applied to foliage.

Rainfall Application Procedures

Experimental plants were sprayed in a fully enclosed chamber (Evans, Gmur, and Da Costa, 1977) with light provided. Plants were placed on a turntable (1.7 m in diameter) which revolved at 2.5 rpm. Simulated acid rain was distributed from a set of four nozzles, 1.5 m above the level of the turntable. Each nozzle delivered raindrops of 0.353 mean volume radius (Evans, Gmur, and Da Costa, 1977). A vibrator on the nozzle apparatus assured uniform delivery of drops.

Only one rainfall with an effective rainfall rate of 7.2 mm hr^{-1} was given daily. The duration of the rainfall treatment varied among experiments from 6 to 20 min but was uniform for any particular experiment. A simulated rain solution (Evans, Gmur, and Da Costa, 1977) was created to simulate anion and cation contents of rain present in northeastern United States. The pH of the rain was adjusted with H_2SO_4 or NaOH to obtain appropriate levels.

Pinto Beans: Yield

Thirty-five plants were divided into five groups. Each group was exposed to simulated rain at a pH level of 5.7, 3.1, 2.9, 2.7, or 2.5 for 20 min daily. Fifteen-day-old plants received rainfall 5 days weekly until the pods were mature 65 days later. Forty-five rainfalls were applied.

Soybeans: Yield

One hundred forty-four plants were divided into three groups. Each group was exposed to simulated rain at a pH level of 5.7, 3.1, or 2.5 for 10 min daily. Fourteen-day-old plants received rainfall 5 days weekly until all pods were mature 112 days later. Seventy-eight rainfalls were applied.

The increase in length of middle trifoliate leaves of soybeans and of pinto beans was measured. The leaf length measurements were made after leaves had reached their maximum size.

Statistical analyses were performed on all data to view if means were significantly different at the 0.05 or 0.01 confidence level. Either the Student-Newman-Keuls test or the Dunnett's T test was used (Steel and Torrie, 1960).

RESULTS

Pinto Beans: Yield

Experiments were performed to determine if simulated acid rain has an effect on growth, development, and yield of pinto beans. Plants were exposed to simulated rain at pH levels of 5.7, 3.1, 2.9, 2.7, and 2.5 for 20 min daily, 5 days per week. Forty-five rainfalls were applied over the growing period of 83 days.

When pinto beans were exposed to simulated rain of pH 3.1 or below, the dry mass per plant was lower than for plants exposed to rain of pH 5.7 (Table 1).

Table 1. Effect of simulated acid rain on dry mass quantity of various plant parts of pinto beans[a]

Experimental parameter	pH 5.7	pH 3.1	pH 2.9	pH 2.7	pH 2.5
Dry mass of stems and leaves at harvest, g per plant	15.0[1] [b,c] ±3.0	11.4[1] ±2.5 (0.76)[d]	10.8[1] ±2.3 (0.72)	11.6[1] ±2.8 (0.77)	10.4[1] ±3.6 (0.69)
Dry mass of seeds at harvest, g per plant	17.8[1] ±2.2	14.9[1] ±1.8 (0.84)	12.8[1] ±1.2 (0.72)	12.6[2] ±1.8 (0.71)	10.9[2] ±0.9 (0.61)
Number of seeds at harvest, number per plant	67[1] ±7	52[1] ±6 (0.78)	47[1] ±5 (0.70)	50[1] ±8 (0.75)	38[2] ±2 (0.57)
Number of pods at harvest, number per plant	14[1] ±1	12[1] ±1 (0.86)	12[1] ±1 (0.86)	11[1] ±1 (0.79)	11[1] ±0 (0.79)
Dry mass of leaves abscised before harvest, g per plant	4.6[2] ±0.4	5.0[2] ±0.4 (1.09)	6.3[1] ±0.8 (1.37)	5.5[1] ±0.5 (1.20)	6.6[1] ±0.7 (1.43)

[a]Thirty-five plants in each of two experiments were divided into five groups. Each group was exposed to simulated rain at pH levels of 5.7, 3.1, 2.9, 2.7, or 2.5 for 20 min daily. Fifteen-day-old plants received rainfalls 5 days weekly until all pods were mature (plants had senesced) 65 days later. A total of 45 rainfalls were applied. The results of both experiments were grouped together for statistical analyses.

[b]Mean and standard error.

[c]Dunnet's T-test analysis was performed on the data to demonstrate significant (0.05) differences between mean of control (pH 5.7) and all other mean values. Means followed by the same superscript value are not significantly different. Values followed by a different superscript are significantly (0.05) different.

[d]Values in parentheses are proportions of the mean values obtained at rain of pH 5.7.

Although the mean dry mass of leaves and stems of plants exposed to pH 5.7 was not significantly (0.05) greater than from plants exposed to lower pH levels, the mean values of plants exposed to these lower pH levels were between 69 and 77% of those of plants exposed to rain of pH 5.7.

Concomitant with this decrease in dry mass of leaves and stems at low pH, there was a decrease in dry mass of seeds. Plants exposed to simulated rain of pH 2.5 and 3.1 had dry seed masses between 61 and 84%, respectively, of the dry mass of plants exposed to pH 5.7. Statistical analyses show that the dry mass of seeds per plant was significantly lower at pH 2.5 and 2.7 than at pH 5.7 at the 0.05 level. This decrease in seed mass with a decrease in rain pH was also reflected by a lower number of seeds per plant. No differences in the dry mass of individual seeds occurred among the experimental treatments. The average dry mass per seed was 0.27 g.

The number of seeds per plant decreased with a decrease in rain pH. This decrease in seed number could be due to a decrease in the number of seeds per pod and/or the number of pods per plant. Both the number of seeds per pod and the number of pods per plant decreased with a decrease in rain pH.

The mass of leaves abscised before harvest was significantly (0.05) lower for plants exposed to either pH 5.7 or pH 3.1 compared with plants exposed to pH 2.5. This increase in mass of abscised leaves at low pH was also reflected by an increase in the number of abscised leaves at low pH (Table 2). In pinto beans the mass of leaves abscised before harvest was a substantial portion of the entire plant biomass produced. At pH levels of 5.7, 3.1, 2.9, 2.7, and 2.5 about 11, 14, 18, 16, and 21%, respectively, of the entire plant biomass abscised before harvest.

Table 2. Number of abscised leaves per plant of pinto beans exposed to simulated rain of various pH levels

Days after planting	Number of rainfalls	Number of abscised leaves				
		pH 5.7	pH 3.1	pH 2.9	pH 2.7	pH 2.5
57	28	0 ± 0[a]	1 ± 0	1 ± 0	2 ± 0	3 ± 0
62	31	3 ± 1	3 ± 1	4 ± 0	6 ± 1	6 ± 1
66	35	4 ± 1	6 ± 1	6 ± 1	7 ± 1	8 ± 1
73	40	5 ± 1	7 ± 1	6 ± 1	8 ± 1	9 ± 1

[a]Mean and standard error of 7 plants per pH level in one typical experiment.

Pinto Beans: Leaf Growth Rates

Experiments were performed to determine if simulated acid rain affects the growth rate of pinto bean leaves. The lengths of mature middle trifoliate leaflets are shown in Table 3. The data demonstrate that leaf numbers 2 through 8 have shorter leaflets if exposed to rainfall of either pH 2.5 or 2.7 than to pH 5.7. Leaflet lengths of leaf numbers 9 through 15 were not affected by rain pH. Regression analysis was performed between leaflet length and leaflet area. The correlation coefficient was 0.88 and the slope of the line was 1.75 for leaflet lengths greater than 45 mm. From this analysis, each millimeter of leaflet length represents 1.75 cm^2 of leaflet area. The reduction in the length of the middle leaflet of leaf number 4 from 111 mm at pH 5.7 to 100 mm at pH 2.7 represents a 15% (6.3 cm^2) reduction in leaf area. Similarly, the 103-mm length of leaflet number 7 at pH 2.7 is 17% smaller than for leaflets exposed to pH 5.7. This 15–17% reduction in leaflet area at pH 2.7 compared with that at 5.7 is consistent with the 35% decrease in mass of leaves and stems between plants grown at these two pH levels (Table 1).

Table 3. Effect of simulated acid rain on the length of middle leaflets of mature leaves of pinto beans[a]

Leaf number[b]	Length (mm)				
	pH 5.7	pH 3.1	pH 2.9	pH 2.7	pH 2.5
1	115[1c]	126[1]	121[1]	[d]	[d]
2	118[1]	117[1]	110[2]	103[2]	[d]
3	116[1]	113[1]	110[1]	108[1]	98[2]
4	111[1]	112[1]	111[1]	100[2]	94[2]
5	112[1]	112[1]	108[1,2]	100[2]	87[3]
6	109[1]	113[1]	110[1]	99[2]	97[2]
7	117[1]	120[1]	117[1]	103[2]	96[2]
8	124[1]	129[1]	125[1]	110[2]	99[3]
9	121[1]	128[1]	122[1]	117[1]	101[2]
10	113[1]	122[1]	121[1]	115[1]	108[1]
11	109[1]	109[1]	120[1]	110[1]	110[1]
12	112[1,2]	108[1,2]	119[1]	103[2]	107[1,2]
13	102[1]	97[1]	109[1]	97[1]	103[1]
14	91[1]	89[1]	101[1]	84[1]	91[1]
15	81[1]	75[1]	85[1]	78[1]	82[1]

[a]The 57-day-old plants were exposed to twenty-eight 20-min rainfalls before they were scored.
[b]Trifoliate leaves were numbered from the bottom to the top of the plant.
[c]Numerical values in superscripts represent the significant values. Values not followed by the same number differ significantly (P = <0.05); means followed by the same number are not significantly different. The Student-Newman-Keuls test was applied to the data.
[d]Insufficient samples were available.

Soybeans: Yield

Experiments were performed to determine if simulated acid rain affects growth, development, and yield of soybeans. One hundred forty-four plants were divided into three groups. Plants were exposed to simulated rain at pH levels of 5.7, 3.1, and 2.5 for 10 min daily, 5 days per week. Seventy-eight rainfalls were applied over the growing period of 126 days. When plants were exposed to simulated rain at pH 3.1 and 2.5 the dry mass of stems decreased 15 and 29%, respectively, compared with plants exposed to rain of pH 5.7 (Table 4). In a similar manner the dry mass of leaves decreased 5% and 14% for pH levels of 3.1 and 2.5, respectively, below plants exposed to pH 5.7.

In contrast to the decrease in leaf and stem biomass, the mass of seeds per plant increased 11% in plants exposed to pH 3.1 compared with plants exposed to pH 5.7. Plants exposed to pH 2.5 had an 11% decrease in seed mass compared with plants exposed to pH 5.7. These differences in total seed mass were also reflected in commercial seed yield, as well. Commercial seed yield is defined as the mass of seeds that will not pass through a mesh screen with 5-mm-diam holes. This was a somewhat artificial designation. However, seeds more than 5 mm in diameter would surely represent a marketable product.

The mean numbers of seeds more than 5 mm in diameter at pH levels of 5.7, 3.1, and 2.5 were 169, 165, and 150, respectively. Mean mass of seeds of plants exposed to pH 3.1 was slightly greater (0.13 g per seed) than that of plants exposed to rain of pH 5.7 (0.11 g per seed). Plants exposed to rain of pH 2.5 had a mean seed mass average of 0.11 g per seed, the same as plants exposed to pH 5.7. The greater seed yield of plants exposed to pH 3.1 compared with pH 5.7 or pH 2.5 may be attributed to a greater mass per seed.

Table 4. Effect of simulated acid rain on dry mass or quantity of various plant parts of soybeans[a]

Experimental parameter	pH 5.7	pH 3.1	pH 2.5
Dry mass of stems at harvest, g per plant	24.0[1] [b] ±0.5	20.4[2] ±0.4 (0.85)[c]	17.0[3] ±0.3 (0.71)
Dry mass of leaves at harvest, g per plant	40.7[1] ±0.6	38.6[2] ±0.6 (0.95)	35.0[3] ±0.6 (0.86)
Dry mass of seeds at harvest, g per plant	19.7[2] ±0.5	21.8[1] ±0.4 (1.11)	17.6[3] ±0.4 (0.89)
Dry mass of seeds at harvest (commercial yield),[d] g per plant	18.9[2] ±0.5	21.0[1] ±0.4 (1.11)	17.1[3] ±0.4 (0.90)
Number of seeds at harvest, number per plant	192[1] ±3	179[2] ±3 (0.93)	169[3] ±3 (0.88)
Number of seeds at harvest (commercial yield), number per plant	169[1] ±3	165[1] ±3 (0.98)	150[2] ±3 (0.89)
Number of pods at harvest, number per plant	92[1] ±1	84[2] ±1 (0.91)	81[2] ±1 (0.88)
Dry mass of seeds and pods at harvest, number per plant	31.2[1] ±0.6	32.8[1] ±0.6 (1.05)	26.6[2] ±0.5 (0.85)
Dry mass of leaves abscised before harvest, g per plant	3.3[1] ±0.3	2.6[1] ±0.2 (0.79)	1.6[2] ±0.2 (0.48)
Dry biomass produced above ground, g per plant	99.2[1] ±1.5	94.5[2] ±1.1 (0.95)	80.2[3] ±1.3 (0.81)

[a] Plants were exposed to 78 rainfalls of 10 min each during the 126-day period from planting to harvest.
[b] Numerical values in superscripts above the mean and standard error represent the significance values. Values not followed by the same number differ very significantly ($P = \leqslant 0.01$); means followed by the same number are not significantly different. The Student-Newman-Keuls test was applied to the data.
[c] Numbers in parentheses are proportions of the mean values obtained at rain of pH 5.7.
[d] Seeds of the "commercial yield" did not pass through a mesh screen with holes 5 mm in diameter.

The number of pods per soybean plant decreased as the pH of the simulated rain decreased. Plants exposed to simulated rain of pH 3.1 and 2.5 had 91 and 88%, respectively, of the number of pods of plants exposed to pH 5.7 rain. Plants exposed to simulated rain of pH levels 5.7, 3.1, and 2.5 had averages of 1.8, 2.0, and 1.9 seeds (commercial yield) per pod, respectively. Even though the total number of pods per plant was lowest in plants exposed to both pH 3.1 and pH 2.5, the average number of seeds per pod was greatest at these low pH levels.

The mass of leaves abscised before harvest was highest in plants exposed to pH 5.7 rain. The lower the rainfall pH the lower the mass of abscised leaves. In soybeans, the plant biomass that abscised before harvest was small compared with the total biomass produced. The dry mass of leaves abscised before harvest comprised only 3.3, 2.8, and 2.0% of the entire biomass produced at simulated rain pH levels of 5.7, 3.1, and 2.5, respectively. A significantly (0.05) lower mass of leaves abscised from plants exposed to pH 2.5 than from plants exposed to either 3.1 or 5.7.

Soybeans: Leaf Growth Rates

The effect of simulated rain pH on the length of middle leaflets of trifoliate leaves is shown in Table 5. A 0.92 correlation coefficient between leaflet length and leaflet area was observed. Each millimeter of leaflet length above 40 mm is correlated with an area of 1.25 cm². From the equation of the line, the difference between 64 mm and 82 mm leaflet length of leaf number 2 at pH levels of 2.5 and 5.7, respectively, reflects a 47% decrease in leaflet area. When similar correlations were made of middle leaflets from leaves numbers 5, 9, and 13, reductions of 50, 53, and 46% were obtained, respectively. Analyses were performed to determine the amount of leaf area reduction at pH 3.1 compared with the reduction at pH 5.7. For leaf numbers 2, 5, 9, and 13, the percent area reductions were 0, 10, 13, and 14, respectively.

Table 5. Effect of simulated rain pH on the length of middle leaflets of mature leaves[a] of soybeans

	Length (mm)		
Leaf number	pH 5.7	pH 3.1	pH 2.5
1[b]	65[1c]	63[1]	58[2]
2[b]	82[1]	82[1]	64[2]
3[b]	95[1]	90[2]	76[3]
4[b]	110[1]	108[1]	77[2]
5[d]	119[1]	112[2]	81[3]
6[d]	130[1]	121[2]	91[3]
7[d]	140[1]	130[2]	91[3]
8[d]	154[1]	141[2]	104[3]
9[e]	165[1]	149[2]	108[3]
10[e]	176[1]	154[2]	105[3]
11[e]	169[1]	152[2]	112[3]
12[e]	167[1]	149[2]	105[3]
13[e]	163[1]	146[2]	99[3]
14[e]	155[1]	138[2]	102[3]

[a] Trifoliate leaves were numbered from the bottom to the top of the plants. The leaves had reached maximum enlargement before they were measured.
[b] The 35-day-old plants were exposed to fourteen 10-min rainfalls before they were measured.
[c] Numerical values in superscripts above the mean values represent significance levels. Values not followed by the same number differ significantly (P = <0.05); means followed by the same number are not significantly different. The Student-Newman-Keuls test was applied to the data.
[d] The 51-day-old plants were exposed to twenty-six 10-min rainfalls before they were measured.
[e] The 73-day-old plants were exposed to forty 10-min rainfalls before they were measured.

DISCUSSION

Pinto bean and soybean plants respond differently to simulated acid rain. However, this does not mean that they do not have some similarities in their responses. Previous results (Evans and Curry, 1979) suggest that the broad-leaved herbaceous plants are the most sensitive of all angiosperms and gymnosperms. So, even though the foliage of broad-leaved herbaceous plants may be most sensitive, the responses among these broad-leaved plants may differ. In this way, each plant species may cope with the stress applied by acid rain in a particular manner.

The different responses between pinto beans and soybeans are most dramatically expressed by the fact that the dry mass of soybeans seeds increased while the yield of pinto beans decreased when plants were exposed to simulated acid rain of pH 3.1 compared with exposures to pH 5.7. The yield of pinto bean decreased at each pH of 3.1 and below. This decreased yield resulted from both (a) a decrease in the number of pods per plant and (b) a decrease in the number of seeds per pod. Thus the difference between seed yields of pinto bean and soybean is accounted for by the fact that soybeans exposed to simulated rain of pH 3.1 produced a larger dry mass per seed than plants exposed to pH 5.7.

In both plant species the number of pods per plant decreased at a rain of pH 3.1 and below compared with plants exposed to rain of pH 5.7. This decrease in number of pods per plant may result if an alteration in the sequential steps required during flowering, pollination, pod set, and pod development occurs. Simulated acid rain seems to have a negative influence on at least one of these processes. This negative influence is similar to the influence that acidic conditions have on fertilization in ferns (Evans and Bozzone, 1977). Ferns may be much more sensitive to acidity than legumes because unicellular, ciliated spermatozoids of ferns may have a direct contact to acidic solutions for several hours. The experimental results with soybeans, pinto beans, and bracken fern suggest that flowering, fertilization, and associated events that contribute to the development of viable progeny may be sensitive to acid rain.

Exposure to simulated acid rain caused different responses in pinto bean and soybean with respect to leaf abscission before harvest. Less leaf abscission occurred with exposure to rainfalls of low pH in soybeans. However, since the biomass of foliage abscised before harvest averaged less than 3% of the entire biomass of soybean plants, the overall pH effect on foliar abscission in soybeans may be negligible. Pinto bean plants exposed to low-pH rainfalls had a greater amount of foliar abscission than plants exposed to pH 5.7. In pinto beans the mass of leaves abscised was between 14 and 31% of the total plant biomass at harvest. The greater premature leaf abscission in pinto bean at low-pH rainfalls may be responsible, in part, for the decreased yield.

Simulated acid rain decreased the rate of leaf expansion in leaves of both plant species. However, the two species differed with regard to which leaves were most injured. In pinto beans, leaf expansion was most affected in leaf numbers 2 through 8, while leaves 9 through 15 were essentially unaffected. In contrast, leaf numbers 5 through 14 were most affected by simulated acid rain in soybeans, although leaf numbers 1 through 4 of soybeans were also affected. The relative importance of these differences in overall plant responses to reductions in leaf enlargement by simulated acid rain is unknown at present. Since lower leaf numbers of pinto beans are most affected by acidic solutions, this decreased leaf expansion may contribute to the early abscission of leaves. In turn, both processes may be responsible for the decrease in seed yield. In soybeans, the relatively small effect of acidic solutions on leaf expansion of the lower leaves may not have been sufficient to accelerate leaf abscission.

Long-term rainfall measurements taken at Brookhaven National Laboratory (Nagle, 1975) for the summer months (June through September) show an average of about 8.5 cm per month. In our experiments the rainfall rates were 3.1 and 3.9 cm per month for pinto beans and soybeans, respectively. These rates are less than half that normally found. Nagle has documented that 42% of all summer days had some precipitation. In our experiments, 54% and 62% of all days had precipitation for pinto beans and soybeans, respectively. In this way our rainfalls were only slightly more frequent than rainfalls at Brookhaven National Laboratory.

These investigations have attempted to determine the change in seed yield of two agronomic crops of economic importance after exposure to simulated acid rain. Moreover, the effects of simulated acid rain were determined on a variety of other experimental parameters in addition to seed yield in an attempt to understand further how plants respond to this environmental stress. Many other environmental influences may modify the responses of plants to one particular environmental agent (Brun, 1978). Additional experiments must be performed before a sufficient level of knowledge will be reached where it will be possible to assess the extent of damage to vegetation caused by specific rainfall pH levels.

REFERENCES

Brun, W. A. 1978. Assimilation. Pp. 45–76 in Norman, A. G. (ed.), Soybean Physiology, Agronomy, and Utilization. Academic Press, New York.

Evans, L. S., and D. M. Bozzone. 1977. Effect of buffered solutions and sulfate on vegetative and sexual development in gametophytes of *Pteridium aquilinum*. Am. J. Bot. 64:897–902.

Evans, L. S., and T. M. Curry. 1979. Differential responses of plant foliage to simulated acid rain. Am. J. Bot. 66:953–062.

Evans, L. S., N. F. Gmur, and F. Da Costa. 1977. Leaf surface and histological perturbations of leaves of *Phaseolus vulgaris* and *Helianthus annuus* after exposure to simulated acid rain. Am. J. Bot. 64:903–913.

Evans, L. S., N. F. Gmur, and F. Da Costa. 1978. Foliar responses of six clones of hybrid poplar to simulated acid rain. Phytopathology 68:847–856.

Ferenbaugh, R. W. 1976. Effects of simulated acid rain on *Phaseolus vulgaris* L. (Fabaceae). Am. J. Bot. 63:283–288.

Jonsonn, B., and R. Sundberg. 1972. Has the acidification by atmospheric pollution caused a growth reduction in Swedish forests? A comparison of growth between regions with different soil properties. In B. Bolin (ed.), Supporting Studies to Air Pollution across National Boundaries. The Impact on the Environment of Sulfur in Air and Precipitation. Sweden's case study for the United Nations Conference on The Human Environment. Royal Ministry of Foreign Affairs, Royal Ministry of Agriculture, Stockholm.

Nagle, C. M. 1975. Climatology of Brookhaven National Laboratory 1949 through 1973. Brookhaven report 50466.

Shriner, D. S. 1977. Effects of simulated rain acidified with sulfuric acid on host-parasite interactions. Water, Air, Soil Pollut. 8:9–14.

Steel, R. G., and J. H. Torrie. 1960. Principles and Procedures of Statistics. New York, McGraw-Hill.

Wood, T., and F. H. Bormann. 1974. The effects of an artificial mist upon the growth of *Betula alleghaniensis* (Britt.). Environ. Pollut. 7:259–268.

Wood, T., and F. H. Bormann. 1975. Increases in foliar leaching caused by acidification of an artificial mist. Ambio 4:169–171.

Wood, T., and F. H. Bormann. 1977. Short-term effects of a simulated acid rain upon the growth and nutrient relations of *Pinus strobus,* L. Water, Air, Soil Pollut. 7:479–488.

DISCUSSION

S. V. Krupa, University of Minnesota: Lance, could you make a comment about the frequency of occurrence of rainfall with a pH of 3.1 or under? Would you also comment as to why in your first slide you did not really illustrate a linear decrease of biomass with pH?

L. S. Evans: The percentage of rainfalls during the summertime with a pH of 3.1 or lower would probably be on the order of, I'd say, between 10 and 20% of the time at Brookhaven, based upon what we know. I think the second part of your question had to do with the nonlinearity of the response between 5.7, 3.1, 2.9, and so on; is that correct?

S. V. Krupa: Yes, actually when the pH dropped you had no difference in the percent biomass lost. You had **lower** biomass loss in a pH value lower than the one preceding it.

L. S. Evans: This is the variability between treatments, because the treatment pH's are very close together. There were only seven plants per treatment in the pinto bean experiments. The nonlinearity down in that area is probably not too unusual, because the treatment pH's were so close together, 0.2 pH unit apart. I don't think we have very much information about what the dose-response curve would be in those areas.

Al Heagle, North Carolina State: You've made some comparison between species, with the results that you got, and I wonder if you can do this when your numbers of rains were different and the durations of rains were different. Would you like to comment on this?

L. S. Evans: We think in general that the response of soybeans was different than that of pinto beans. In general I think that pinto beans are more sensitive to the rain solutions than are soybeans. As you suggest, though, it's difficult to say that with the data that I've shown here. If you look at the number of hairs on a leaf surface, there is essentially no difference between the two species. In the field, I don't know if you would find a difference. What I was trying to point out, and if I didn't say it this way then I said it wrong before, there's a difference in the response of these plants, the way they respond and the whole plant growth response, between pinto beans and soybeans, to the simulated rain solutions. I don't think with the data that we have that there's that much difference in the total sensitivity of pinto beans and soybeans. I don't think that we can say that there's that much of a difference in their sensitivity levels.

Patricia Irving, Argonne National Laboratory: Do you think that your work is a realistic assumption of what happens in the real world as far as acid rain is concerned? Your applications were five days a week; the only significant differences occurred at very low pH's that are not occurring naturally.

L. S. Evans: The amount of rainfall that we gave, the amount of rainfall in a three-month period, would be comparable to the growth of pinto beans. At BNL over the last 25 years the amount of rainfall would be 25.6 cm. We actually gave a rainfall of 10.8 cm. The results are comparable for a four-month period—I have them in front of me if you'd like to have those—so the rainfall rate that we gave was less than half of that you would find normally at BNL. You would also find that during the summer-month period that 42% of the days would have precipitation. In our experiments we gave pinto beans 54% days with exposure and soybeans 62% days with exposure. This is a little bit higher, but the amount of rainfall per hour was comparable to what one would find in nature, so we think that our rainfall applications, the actual water applied, is close to reality.

Patricia Irving: At such low pH's?

L. S. Evans: The pH levels were low, but, as I said in the beginning, we were interested in establishing growth-rate responses. We've got to have pH levels that will give us an effect to compare to other pH levels that may not give us an effect, so that we can look at a dose-rate response.

I'd like to clarify something that I said before and make sure that I keep straight. About 10–20% of the rain events at Brookhaven would have a pH of 3.7 or lower, and 5% or less of the rain events would have a pH of about 3.1 to 3.2.

CHAPTER 33

VEGETATION: EFFECTS OF SULFUR DEPOSITION BY DRY-FALL PROCESSES

S. V. Krupa*

B. I. Chevone*

J. L. Bechthold†

J. L. Wolf†

ABSTRACT

Atmospheric inputs of sulfur into a terrestrial ecosystem were studied relative to a coal-fired power plant. Dry- and wet-fall inputs into cornfields in the vicinity of the sulfur source were calculated to be 0.82 and 4.41 kg/ha during the crop season. High wet-fall inputs of sulfur seemed to be due to regional transport and import of SO_4 into the area rather than due to the point source. Wet-fall inputs of sulfur did not increase linearly during the four-year study, while dry-fall input did. Foliar sulfur concentration in vegetation also increased temporally. It is suggested that this increase in foliar sulfur is due to the incremental dry fall of atmospheric sulfur.

INTRODUCTION

Deposition of atmospheric sulfur through dry-fall processes involves direct transfer to, or collection of, gaseous and particulate sulfur species on land and water surfaces. Estimates regarding the importance of dry vs wet processes of atmospheric sulfur removal range from a ratio of 60:40% to 40:60% (National Academy of Sciences, 1978).

Sulfur dioxide (SO_2) and particulate sulfate (SO_4), in general, constitute the most abundant sulfur species in the atmosphere. Reduced forms of sulfur such as hydrogen sulfide (H_2S) are most likely deposited to the ground after oxidation to SO_2 and SO_4. Therefore, information presented in the following sections is restricted to SO_2 and SO_4.

Results of investigations on the dry deposition of SO_2 and SO_4 on vegetation have been previously summarized by several authors (Chamberlain, 1975; Droppo, 1974; Garland, 1977, 1978; Lindberg et al., 1979; Sehmel and Hodgson, 1974). The objective of this paper is to review the results of a four-year study of atmospheric sulfur inputs and effects on terrestrial vegetation in the vicinity of a coal-fired power plant, with emphasis on dry fall within the natural variability of measured background parameters.

---

*Department of Plant Pathology, University of Minnesota, St. Paul, Minn. 55108.

†Northern States Power Company, Minneapolis, Minn. 55401.

LOCATION OF THE POINT SOURCE AND ITS CHARACTERISTICS

Northern States Power Company's (NSP) SHERCO coal-fired power plant is located in Sherburne County, near Becker, Minnesota. The facility is surrounded by a highly agricultural and irrigated area with flat terrain, approximately 65 km northwest and downwind (summer months) from an urban complex (Minneapolis–St. Paul). The power plant became operational in 1976.

During summer months, meteorological conditions have produced periodic ozone episodes in the study area due to transport of the urban plume (Laurence et al., 1977). The power plant is equipped with a 200-m stack and employs advanced pollutant scrubber technology (Table 1), resulting in the abatement of 99.9% of particulates and about 60% of the SO_2 in the flue gas. Net SO_2 emissions from the point source during 1978 were approximately 59 metric tons per day.

Table 1. Northern States Power Company coal-fired power plant: source description and characteristics

Location: Becker, Sherburne County, Minnesota (93° 55' W; 45° 20' N)

Characteristics of surrounding area	Power generating capacity (MW)	Sulfur content of coal (%)	Ash content of coal (%)	Flue gas concentration of SO_2 (ppm)	Flue gas concentration of NO_x (ppm)	Scrubber characteristics
Summer months: downwind from urban plume; high relative humidity; flat terrain, agricultural (irrigated)	1400 (stack height 200 m), plume exit velocity 90 m/sec	0.8	8.4	250 (6.55 × 10⁵ $\mu g/m^3$)	400	Wet scrubber followed by limestone $(CaCO_3)$ scrubber; heated plume (82°C)

Adapted from Hegg, D. A., and P. V. Hobbs. 1980.

SULFUR CHEMISTRY OF THE POINT SOURCE; PLUME AND GROUND-LEVEL SO₂ CONCENTRATIONS

Hegg and Hobbs (1979) investigated the sulfur chemistry of the plume using aircraft monitoring (Table 2). Sulfate concentrations in the center of the plume ranged from 0.7 to 4.96% (0.88 ± 0.30 to 7.86 ± 0.35 $\mu g/m^3$) relative to SO_2 levels (55 to 1100 $\mu g/m^3$). The conversion rate of SO_2 to SO_4 in the plume was calculated to be 0 to 2.2% ± 1.3% hr⁻¹ (Hegg and Hobbs, 1980).

Ground-level SO_2 concentrations were continually measured over a three-year period at eight locations in the vicinity of the point source based upon predicted isopleths. These data were modeled according to Larsen (1977), and a summary is presented in Table 3. Highest measured average ground-level SO_2 concentrations were 150 $\mu g/m^3$/0.5 hr, 136 $\mu g/m^3$/hr, 121 $\mu g/m^3$, and 107 $\mu g/m^3$/8 hr, occurring 0.01% of the time annually.

Air quality (SO_2) was also modeled using U.S. EPA's climatological dispersion model (CDM) applied to a single point source (Fig. 1). Predicted isolines of average ground-level SO_2 range from 0.25 to 2.0 $\mu g/m^3$ during June–September.

AN ESTIMATE OF ATMOSPHERIC SULFUR INPUTS INTO CORNFIELDS IN THE VICINITY OF THE POINT SOURCE DURING THE CROP SEASON

Dry-fall and wet-fall inputs of sulfur into cornfields in the vicinity of the point source were estimated for the crop season, May 15–September 15 (Table 4). Sulfur

Table 2. Sulfur dioxide and sulfate concentrations in the plume[a]

Date	Range (km)	Concentration of SO$_2$ (μg/m^3)[b]	Concentration of SO$_4$ (μg/m^3)[c]
June 17, 1978	3.2	1090	1.90 ± 0.40
	17.6	346	6.01 ± 0.77
	20.0	275	4.70 ± 0.67
June 21, 1978	3.2	686	5.16 ± 0.34
	9.6	320	2.38 ± 0.30
	32.0	369	2.72 ± 0.31
	64.0	55	2.73 ± 0.31
	Ambient	26	0.88 ± 0.30
June 22, 1978	3.2	1079	7.87 ± 0.35
	19.2	1108	7.77 ± 0.30
	32.0	482	3.70 ± 0.36
	48.0	79	1.56 ± 0.37
	Ambient	34	1.37 ± 0.36

[a]Adapted from Hegg, D. A., and P. V. Hobbs. 1980.
[b]These are "point" measurements (obtained from bag sampler) near the center of the plume. The error in these measurements is ±1.3 μg/m^3. Actual measured values in parts per billion were converted to micrograms per cubic meter using reference conditions: 25°C; 760 mm Hg.
[c]These values were derived from ion-exchange chromatography of Teflon filters.

Table 3. Average ground-level sulfur dioxide concentrations (μg/m^3) relative to the point source[a]

Averaging time (hr)	Percent frequency of occurrence (hr/year)			
	0.01	0.1	1.0	10
0.5	121.8 ± 16.5 (107.2–145.7)[b]	73.9 ± 11.3 (63.1–93.5)	25.2 ± 6.0 (18.9–36.2)	0
1.0	112.9 ± 15.5 (112.9–136.2)	67.9 ± 10.7 (57.6–86.5)	21.2 ± 5.8 (15.7–31.4)	0
3.0	99.0 ± 13.9 (86.7–120.8)	57.9 ± 9.4 (48.7–74.7)	15.2 ± 5.6 (9.4–23.6)	0
8.0	86.5 ± 12.6 (74.9–106.6)	50.0 ± 8.4 (40.6–63.9)	9.7 ± 5.8 (3.4–19.4)	0
24.0	72.0 ± 10.7 (61.8–90.1)	38.5 ± 7.3 (31.4–51.4)	3.9 ± 5.5 (0–14.9)	0

[a]Values were derived from three years of measured SO$_2$ concentrations at 8 locations based on isopleths. Model used was according to Larsen, R. I. 1977. An air quality data analysis system for interrelating effects, standards, and needed source reductions: Part 4. A three parameter averaging-time model. J. Air Pollut. Control Assoc. 27(5):454–459.
[b]Values in parentheses represent concentration range.

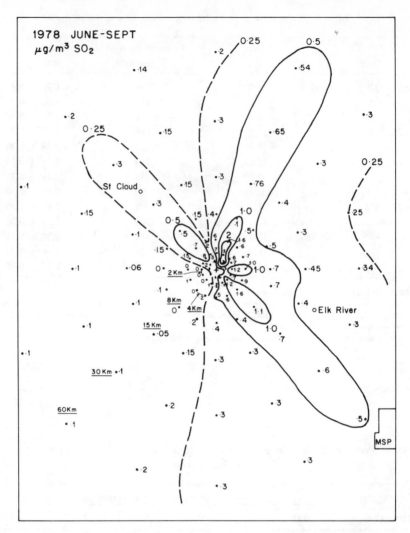

Fig. 1. Climatological dispersion model (CDM) showing isolines of average ground-level SO₂ (μg/m³) in relation to the point source ($+$) during June–September, 1978.

Table 4. An estimate of point-source sulfur to a cornfield in the study area during The crop season (May 15–September 15)[a]

Dry fall		
	Foliage	0.52 kg/ha
	Soil	0.30 kg/ha
Wet fall		4.41 kg/ha
	Total	5.23 kg/ha
Ratio of	Dry fall	
	to wet fall	16:84%

[a]Calculations are based on three-year summertime average measurements of SO₂ and SO₄²⁻ in rain.

input to the field through wet-fall processes was about five times the input due to dry fall. Because of the increments in the power generating capacity of the point source, daily SO_2 emissions in metric tons have been increasing since 1976, reaching a maximum in 1978 (33, 54, and 59 metric tons during 1976, 1977, and 1978, respectively). Ambient SO_2 concentrations measured during our study years, however, are not sufficient to account for much of the SO_4^{2-} concentrations found in rain in this area (3–10 mg/liter). High sulfur inputs through wet fall in the study area, therefore, seem to be attributable to long-distance transport of SO_4.

Several recent measurements of ozone and SO_4 concentrations in polluted air masses, outside the source regions, show a strong correlation between elevated ozone and SO_4 particle concentrations (Isaksen et al., 1978). Trajectory analyses indicate that elevated ozone concentrations in our study area were found to be associated with air masses originating from a south and east direction (PES, 1978; Lyons et al., 1978). Similarly, the source of moisture and cloud condensation nuclei for the rainfall in central Minnesota originates from a south and east quadrant a great portion of the time during summer months (W. A. Lyons, personal communication). If ozone is used as a marker for understanding sulfate transport, it can be concluded that a large fraction of SO_4^{2-} in wet fall in our study could have originated because of regional transport. This is supported by the predominance of neutralized SO_4^{2-} or incorporation of aged SO_4 aerosol into our rain samples (Gardner, 1979; Krupa et al., 1979).

Minnesota oxidant analysis and modeling indicate that 50–60% of ozone under episodic conditions is imported into Minnesota from areas upwind (PES, 1978). These features, coupled with low concentrations of atmospheric SO_2 in our study, may explain the overriding influence of wet fall over dry fall.

VEGETATION RESPONSE TO ATMOSPHERIC SULFUR INPUTS

No visible SO_2-induced vegetational injury was observed over an 80-km radius around the point source during the four years of power plant operation. This is consistent with the measured ground-level SO_2 concentrations during that time (Table 3). Foliar sulfur concentrations were determined in several plant species in 48 permanent field plots located in eight different directions from the stack at distances of 4 km to 80 km. All vegetation species studied were sampled twice during the growth season (June and August) for foliar sulfur analyses. Nonparametric statistical analysis was used to evaluate changes in foliar total sulfur content in the populations of a given plant species. Typical graphic representations of temporal changes in the foliar total sulfur content for two selected plant species (corn and birch) are presented in Figs. 2 and 3. Significant increases in foliar sulfur were found in corn in 1977 and 1978 compared with 1975 (preoperational phase of the power plant) and 1976 (one-half generating capacity of the power plant); this was not observed with birch.

Corn is an irrigated crop in our study area, whereas birch occurs as a natural stand. Thus there are distinct differences in stomatal resistance and hence SO_2 diffusivity between the two plant species. Further, corn is fertilized with nitrogen (NH_3). Corn has a greater surface area and higher photosynthetic and transpiration rates as compared with birch. These differences between the two plant species may account for the differences in temporal changes in foliar sulfur content.

A plot-by-plot analysis over the four years showed increased foliar S ($R^2 = 0.87$) in 17 of the 48 permanent plots independent of the plant species in the plot. Increases in foliar total sulfur ranged from 11 to 80% (Table 5), with the exception of red pine (*Pinus resinosa*). The unusually high increase (200%) in red pine foliar

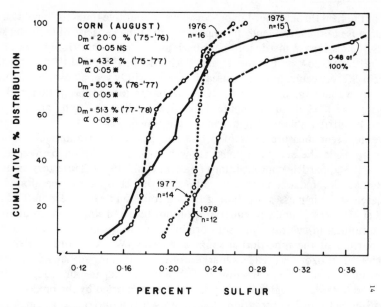

Fig. 2. Smirnov-Kolmogorov nonparametric statistical analysis of foliar total sulfur concentrations in corn in certain study plots located radially at 4 to 80 km from the point source. NS, statistically not significant; asterisk statistically significant at 95% level. The graph shows temporal variations in foliar total sulfur from 1976 to 1978.

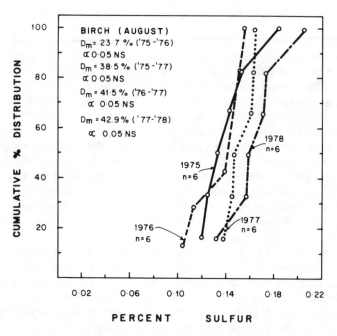

Fig. 3. Smirnov-Kolmogorov nonparametric statistical analysis of foliar total sulfur concentrations in birch in study plots located radially at 4 to 80 km from the point source. NS, statistically not significant. The graph shows temporal variations in foliar total sulfur from 1976 to 1978.

Table 5. Temporal variations in foliar sulfur concentrations in selected plant species in permanent study plots located at various distances and directions from the source

Plot location (km and direction from source)	Plant species	Percent foliar [S], 1975–1978	Percent increase 1975–1978
48N	Bog willow	0.18–0.20[a]	11.1
32N	Corn	0.17–0.24	41.2
8N	Paper birch	0.14–0.16[a]	14.0
16NE	Paper birch	0.13–0.17[a]	30.8
80E	Alfalfa	0.30–0.48	60.0
48E	Trembling aspen	0.20–0.23	15.0
32E	Paper birch	0.12–0.16[a]	33.3
4E	Soybean	0.21–0.30	42.9
32SE	Alfalfa	0.24–0.41	70.8
8SE	Green ash	0.20–0.24	20.0
4SE	Red pine	0.09–0.27[b]	200.0[b]
80S	Soybean	0.21–0.33[b]	60.0[b]
4SW	Corn	0.16–0.22[b]	37.5[b]
16W	Elm	0.14–0.16	14.3
4W	Corn	0.20–0.25	25.0
32NW	Elm	0.13–0.16	23.1
8NW	Corn	0.20–0.36	80.0

[a]Boron (B) was used as a marker for the point-source plume. Foliar B concentration increased significantly along with S.

[b]1975–1977 data only. Crop species were rotated in the study plots during summer, 1978. With red pine, the sampling trees died due to drought, followed by bark beetle infestation.

sulfur is not readily explainable. Several factors, such as the proximity of the plot location (6.5 km) to the point source or physical (drought) and biological stress (beetles), might have had significant roles. Further, red pine is a conifer capable of year-round sulfur accumulation. The geographic distribution of study plots showing increased foliar sulfur concentrations is consistent with modeled isopleths of ground-level SO_2 concentrations (Fig.4). Total foliar sulfur concentrations increased in these plots independent of soil type (sand—sandy loam—silt loam—peat). Additionally, the soils in this area can be considered as deficient to marginal in terms of available sulfur content (7.0 ppm).

Increased foliar sulfur content (Table 5), observed in all the plant species in our study, however, is still within the norm of foliar sulfur concentrations reported for these plant species (Table 6). It seems that vegetation in the study area is responding to soil sulfur deficiencies through increased atmospheric sulfur inputs, as indicated by an increase through changes from the lower extremity of the normal range of foliar sulfur to a higher concentration.

As previously mentioned, sulfur inputs into the ecosystem through wet-fall processes dominate the dry fall. However, wet-fall inputs have not shown a significant linear increase from 1976 to 1978 and have remained within a set range (4.41 ± 1.5 kg/ha) from year to year, even though the range of SO_4^{2-} concentrations in rain

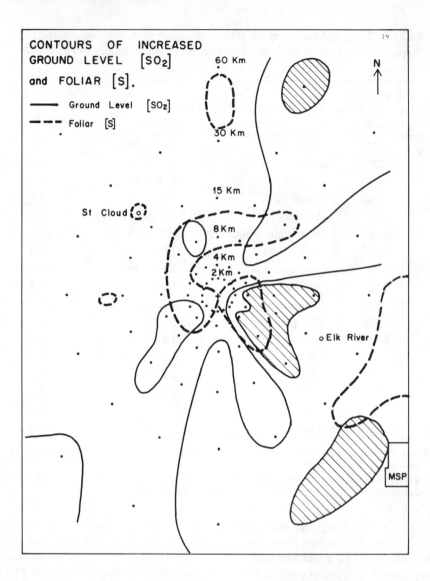

Fig. 4. Modeled contours of increased ground-level SO₂ concentrations relative to the point source (+). Contours depict increments of SO₂ concentrations during 1977 or 1978 compared with 1976. Contours enclosing the hatched areas depict increments in SO₂ levels during both 1977 and 1978, with the highest increment in 1978. Dotted lines represent geographic areas with significantly increasing foliar total sulfur concentrations in vegetation in 1975 compared with 1978.

varied. Since the available sulfur content in the soil did not appear to be a controlling factor, the only significant variable has been increments in dry fall (SO₂). Therefore, it seems highly probable that dry-fall input of sulfur has resulted in increased foliar sulfur content in vegetation. While no yield losses have been noted at this time, continued pollutant quantification and biomonitoring will indicate long-term subtle vegetational effects due to tissue sulfur accumulation through atmospheric dry-fall inputs of sulfur.

Table 6. Normal concentration or range in percent foliar sulfur reported for various plant species evaluated in the vicinity of the point source

Plant species	Percent foliar S (normal concentration or range)	Reference
Paper birch (*Betula papyrifera*)	0.11–0.24	Linzon et al., 1979; Linzon, 1978; Rennie and Halstead, 1977; Kubota et al., 1970
Trembling aspen (*Populus tremuloides*)	0.15–0.22	Linzon et al., 1979; Kubota et al., 1970
Red pine (*Pinus resinosa*)	0.33	Rennie and Halstead, 1977
Willow sp. (*Salix* sp.)	0.23	Kubota et al., 1970
Ash sp. (*Fraxinus* sp.)	0.26	Linzon et al., 1979
Elm sp. (*Ulmus* sp.)	0.17	Linzon et al., 1979
Corn (*Zea mays*)	0.10–0.30	Jones and Eck, 1973
Soybean (*Glycine max*)	0.12–0.52	Ohlrogge, 1960
Alfalfa (*Medicago sativa*)	0.20–0.57	Bickoff et al., 1972

ACKNOWLEDGMENT

The authors wish to thank the following individuals for their valuable assistance and advice: Dr. R. J. Kohut, Mr. Robert Munter, Mr. D. W. Gardner, Mr. Sid Nystrom, Mr. Dave Heberling, and Mr. Dan DeShon. Special thanks are due Dr. George McVeil for his work in pollutant dispersion modeling.

REFERENCES

Bickoff, E. M., G. O. Kohler, and D. Smith. 1972. Chemical composition of herbage. In C. H. Hanson (ed.), Alfalfa Science and Technology. Am. Soc. Agr. Inc., Madison, Wis.

Chamberlain, A. C. 1975. The movement of particles in plant communities. In J. L. Monteith (ed.), Vegetation and the Atmosphere. Academic Press, vol. 1, pp. 155–203.

Droppo, J. G. 1974. Dry deposition processes on vegetation canopies. In Coord, R. J. Engelmann, and G. A. Sehmel (eds.), Atmosphere-Surface Exchange of Particulate Pollutants. Proceedings of a Symposium Held at Richland, Washington, 1974. Battelle, Pacific Northwest Laboratories. Pp. 104–111.

Gardner, D. W. 1979. The chemistry of rain in the vicinity of sulfur dioxide point source in central Minnesota. M.S. thesis, University of Minnesota, in preparation.

Garland, J. A. 1977. The dry deposition of sulphur dioxide on land and water surfaces. Proc. R. Soc. London, Serv. A 354:245–268.

Garland, J. A. 1978. Dry and wet removal of sulphur from the atmosphere. Atmos. Environ. 12:349–362.

Hegg, D. A. and P. V. Hobbs. 1980. Measurements of gas to particle conversion in the plumes from five coal-fired electric power plants. Atmos. Environ. 14: 99–116.

Isaksen, I. S. A., E. Hesstvedt, and Ö. Hov. 1978. A chemical model for urban plumes: Test for ozone and particulate sulfur formation in St. Louis urban plume. Atmos. Environ. 12:599–604.

Jones, J. B., and H. V. Eck. 1973. Plant analysis as an aid in fertilizing corn and grain sorghum. In L. M. Walsh and J. D. Begton (eds.), Soil Testing and Plant Analysis. Soil Sci. Soc. Am. Inc., Madison, Wis. Pp. 349–364.

Krupa, S. V., D. W. Gardner, M. R. Coscio, Jr., and J. L. Bechthold. 1979. Coulometric evaluation of rainfall chemistry in central Minnesota relative to plant pathogenic air pollutants (abstract). Ann. Am. Phytopathol. Soc. Washington, D.C.

Kubota, J., S. Riegers, and V. A. Lazar. 1970. Mineral composition of herbage browsed by moose in Alaska. J. Wild. Manage. 34(3):565–569.

Larsen, R. I. 1977. An air quality data analysis system for interrelating effects, standards, and needed source reductions: Part 4. A three-parameter averaging-time model. J. Air Pollut. Control Assoc. 27(5):454–459.

Laurence, J. A., F. A. Wood, and S. V. Krupa. 1977. Possible transport of ozone and ozone precursors in Minnesota. Ann. Am. Phytopathol. Soc. 31.

Lindberg, S. E., R. C. Harriss, R. R. Turner, D. S. Shriner, and D. D. Huff. 1979. Mechanisms and rates of atmospheric deposition of selected trace elements and sulfate to a deciduous forest watershed. Oak Ridge National Laboratory, Environmental Sciences Division. Publ. No. 1299.

Linzon, S. N. 1978. Effects of air-borne sulfur pollutants on plants. In J. O. Nriagu (ed.), Sulfur in the Environment, Part II. Ecological Impacts. John Wiley and Sons.

Linzon, S. N., P. J. Temple, and R. G. Pearson. 1979. Sulfur concentrations in plant foliage and related effects. J. Air Pollut. Control Assoc. 29(5):520–526.

Lyons, W. A., J. C. Dooley, Jr., and K. T. Whitby. 1978. Satellite detection of long-range pollution transport and sulfate aerosol hazes. Atmos. Environ. 12: 621–632.

National Academy of Sciences. 1978. Sulfur Oxides. Nat. Acad. Sci. Washington, D.C.

Ohlrogge, A. J. 1960. Mineral nutrition of soybeans. Adv. Agron. 12:230–263.

PES—Pacific Environmental Services. 1978. Minnesota Oxidant Analysis. Rept. Minnesota Pollution Control Agency. Roseville, Minn.

Rennie, P. J., and R. L. Halstead. 1977. The effects of sulfur on plants in Canada. In sulfur and Its Inorganic Derivates in the Canadian Environment. NRCC/CNRS. Pub. NRCC 15015, Ottawa, Canada.

Sehmel, G. A., and W. H. Hodgson. 1974. Predicted dry deposition velocities. In Coord, R. J., Engelmann, and G. A. Sehmel (eds.), Atmosphere-Surface Exchange of Particulate and Gaseous Pollutants. Proceedings of a symposium held at Richland, Washington. 1974. Battelle, Pacific Northwest Laboratories. Pp. 339–419.

DISCUSSION

Lance Evans, Brookhaven: I think that the percentage of agricultural area that requires input of sulfur in the United States is very low. From my literature reading I think it's some areas of Tennessee and some areas of Georgia, but I think the percentage is fairly low. What do you come up with?

S. V. Krupa: I haven't come up with anything, Lance, in the sense that I've not made inventories of sulfur-deficient soils. I can understand your point. What I was only saying is that when you say there are no known areas of deficit, you should consider both processes in the sense of what's coming from the soil and what is the atmospheric input. I essentially agree with what Ellis was pointing out earlier.

L. Newman, Brookhaven: You showed a surprising lack of correlation between hydrogen ion and sulfate and nitrate in rainwater. I assume this is due to the fact that the rainwater was relatively high in pH and that the measurements of hydrogen are derived from these numbers and are not very accurate numbers.

S. V. Krupa: The rainfall was sampled with a sampler which seals the rain samples in situ, and pH varied anywhere from about 4.1 to 7.5. The study area is high in sand, and the region is being impacted from roughly east to west by alkaline dust movement. The other reason is that the power plant uses a calcium carbonate scrubber, and you do have high calcium concentrations in a lot of this rain. There are three reasons for the specific chemical basis of our rainfall: (1) the atmospheric SO_2 concentrations are rather low to account for the sulfate concentrations in rain, (2) there is a high neutralization capacity or aging occurring in the air, and (3) if I calculated excess sulfate in our rain samples, a small percentage of the high sulfate is important.

CHAPTER 34

SULFUR NUTRITION OF FORESTS

John Turner*

Marcia J. Lambert*

ABSTRACT

Sulphur is an essential major tree nutrient, and the biochemical relationship between S and N in plant proteins dictates that neither element can be adequately
assessed without reference to the other. Since trees accumulate as sulphate sulphur
any excess S beyond that required to balance the N available, the sulphate sulphur
status of the foliage provides an indicator of the S status of both the tree and the
site. Sulphur deficiency is often associated with dieback in *Pinus radiata*. It also
leads to the accumulation of particular amino acids, such as arginine, and this is
associated with high foliar disease susceptibility. It appears that S toxicity in trees is
confined to limited areas where concentrations of sulphur dioxide are high, whereas
in most situations S inputs are as sulphate, and damage symptoms are not produced. A comparison of various conifer species has indicated that species native to
coastal situations, with high maritime S inputs, were more elastic in their utilization
of S than species native to inland locations. Additions of N to forest stands result in
the incorporation of sulphate sulphur into organic forms, and thus the sulphate
cycle in forest ecosystems is regulated to a large extent by N status and N cycling.

INTRODUCTION

Sulphur is an essential major plant nutrient. However, within forest ecosystems,
limited information is available on S requirements and patterns of S cycling, and this
element has received little attention in laboratory, glasshouse, and field research in
comparison with that received by N, P, and K. Recent reports from different areas
of the world suggest that the incidence of S deficiency in soils is increasing (Beaton et
al., 1971; Blair, 1979). Possible reasons for this are the increased use of high-
analysis fertilizers with low S content, environmental protection legislation which
has restricted the input of sulphur dioxide to the atmosphere from industrial processes, and a growing awareness of the importance of S as a plant nutrient. The aim
of this paper is to discuss the role of S in trees (with particular emphasis on
Australian forests) and the relationships between S and N.

The Forestry Commission of New South Wales (N.S.W.), Australia, functions as
a forest service, with research being directed toward community needs such as
timber production, water quality, wildlife, and recreation. From a forest management viewpoint, a healthy, more highly productive forest is more desirable than one

---

*Forestry Commission of New South Wales, P.O. Box 100, Beecroft. 2119, New South Wales, Australia.

with lower productivity, since more forest products are produced and it recovers more rapidly from any particular disturbance. Within N.S.W., the emphasis on S nutrition research is related to maintaining healthy productive forests.

THE ROLE OF SULPHUR IN TREES

Sulphur is principally required by plants for the synthesis of three amino acids—cystine, cysteine, and methionine—which are essential components of protein and which contain approximately 90% of the organic sulphur found in plants (Anderson, 1975). Three possible sources of S to meet plant requirements are the soil, the atmosphere, and fertilizers. The various chemical forms of S in soils (Williams, 1975) include both organic and inorganic compounds, and their amounts vary widely both between soil types and within the soil profile itself. The various chemical forms of S also differ greatly in their availability to plants. In the surface horizons of most well-drained soils, sulphate concentrations are generally low, but considerable seasonal fluctuations in the amounts of soluble sulphate in the surface soil occur as a result of the mineralization of organic sulphur, leaching of the soluble sulphate, and its uptake by plants. The sulphate content of soils will also be affected by the application of fertilizers and the sulphate content of rain and irrigation waters. Soils vary widely in their capacity to adsorb sulphate. Many possess little or no sulphate adsorption capacity, but in others adsorption plays an important part in the retention of sulphate against leaching and in determining its distribution in the profile.

Forest trees can be extremely efficient at trapping S compounds in their foliage (Smith, 1974). Clearly, evergreen trees will act as filters all through the year, while deciduous trees, including conifers such as larch, will intercept far less sulphur dioxide during the winter and early spring. Elements reach a forest canopy not only in bulk precipitation but also in aerosols and gases that are trapped from the atmosphere by impaction or absorption onto the surfaces of the trees. This additional input can then be washed down in subsequent rain. With living vegetation, unlike other inert vertical surfaces exposed to the wind, there are the two further processes of foliar uptake and crown leaching, the first of which removes elements from the rain and the second adds to it. Thus, the element loading in rainwater beneath trees (throughfall) differs from that in bulk precipitation by an amount that is the sum of trapped input plus crown leaching minus foliar absorption, this amount being the gain of that particular element (Miller et al., 1976).

Reported studies have indicated that tree S uptake is affected by species, location, and stand characteristics. For example, S uptake by Northern Hemisphere hardwoods varied from 25 to 64 kg/ha (Ulrich et al., 1971; Likens et al., 1977; Shriner and Henderson, 1978), in contrast to 6 kg/ha for a *Quercus ilex* stand (Rapp, 1973) and from 0.3 to 2.6 kg/ha for evergreen hardwood stands in the Southern Hemisphere (Lambert and Turner, 1978; Lambert, 1979). Sulphur uptake for conifer stands varied from 1.5 to 7.0 kg/ha (Switzer and Nelson, 1972; Lambert and Turner, 1978; Bringmark, 1977; Lambert, 1979; Turner et al., 1979), but in general the studied conifer stands were located in environments with much lower S inputs.

A constant ratio of 0.030 has been found between organic sulphur and total nitrogen in the foliage of *Pinus radiata* D. Don (Kelly and Lambert, 1972), other *Pinus* species (e.g., *P. taeda* L., *P. elliottii* Engelm., *P. pinaster* Loud., and *P. ponderosa* Laws), other coniferous species such as Douglas-fir (*Pseudotsuga menziesii* (Mirb.) Franco), and *Eucalyptus* spp. (Lambert, 1979), and this is the same ratio as that in their foliar protein. Unlike certain other plants, the foliar total nitrogen has always been found to be equal to organic nitrogen, and there has been no evidence of nitrate nitrogen even when S has been limiting. Apparently N is only taken up at the rate at which S is available, and therefore protein formation is limited by the amount of S available. When the S supply is adequate, the conifer accumulates as sulphate sulphur any excess S beyond that required to balance the N available, and protein formation proceeds at the rate at which N becomes available. The sulphate sulphur status of the foliage provides an indicator of the S status of both the tree and the site and is a much more sensitive and indicative measure than foliar total sulphur (Lambert and Turner, 1977). The relationship between S and N has indicated that neither element can be adequately assessed without reference to the other and that for diagnostic work, estimates of either S or N alone are poor parameters. Since inorganic nitrogen is not accumulated in conifers when S is limiting, symptoms of S deficiency are sometimes confused with those of N deficiency.

SULPHUR RELATIONSHIPS WITHIN THE TREE

The relationship between S and N can be used for diagnostic purposes to define the health status of trees. General S ratings based on foliage sulphate sulphur levels (winter sampling—dormant growth period) have been derived for *P. radiata* (Turner et al., 1977)—trees with less than 80 ppm sulphate sulphur are S-deficient, levels between 80 and 200 ppm are marginal to adequate, between 200 and 400 ppm is adequate to high, and over 400 ppm is high and probably N-deficient. These levels indicate S storage concentrations which are readily available for immediate utilisation and growth in spring. These ratings need to be moderated when the status of other nutrients (especially N) are taken into account and when the stand is to be manipulated (for example, with fertilizer additions).

Symptoms of S deficiency are generally similar to those for N deficiency. In *P. radiata*, moderate S deficiency is associated with overall winter yellowing, especially on cold exposed sites. When severe deficiency occurs, 1 or 2 of the leading shoots, or sometimes merely the terminal buds, die back, often with spectacular brownish-red or red-orange colouration. Needle bases are frequently much yellower than the remainder of the needle (Lambert and Turner, 1977). Symptoms of S toxicity are somewhat harder to define. Damage to trees from high sulphur dioxide levels has been described as discolouration of foliage including bleaching, yellowing, and browning of needles or leaves (Sidhu and Singh, 1977; Linzon, 1965, 1966). This can be followed by irregular necrotic patches on leaf apices and margins in broad-leaved species, while in conifers, necrosis appears at the needle tip and extends to the needle base. There is also often premature leaf shedding leading to partial or complete defoliation. Although tree damage has been attributed to excessive sulphur dioxide levels, foliar sulphate sulphur levels can be accumulated to very high concentrations from soil sources before tip scorch, similar to salt damage, occurs. Relatively high concentrations of foliage total sulphur or sulphate sulphur do not necessarily cause apparent damage symptoms unless the S source is as sulphur dioxide. For coniferous species in close proximity to S-emitting sources where sulphur dioxide damage has occurred, foliar total sulphur levels as low as 1300 ppm have been reported, but typically total sulphur levels were between 1600 and 2000 ppm and as high as 2900 ppm, with corresponding sulphate sulphur levels between 700 and 1600 ppm S (Sidhu and Singh, 1977; Bjorkman, 1967). Atmospheric sulphur dioxide can enter leaf stomata directly or be absorbed in moisture on the needles, and in either case, sulphite sulphur and then sulphate sulphur are soon formed. In locations where sulphur dioxide levels are low but sulphate inputs are high, damage can potentially occur as a result of high sulphate concentration (scorching). The levels required in foliage are much higher than when the source is sulphur dioxide, however (Table 1).

Table 1. Comparison of S levels in *Pinus radiata* foliage from forests in close proximity or distant from atmospheric S sources in Australasia[a]

Form of S	Distance from S source (km)	Soil type	Foliar sulphate sulphur (ppm)	
			Mean	Range
H_2S[b]	0.1	Pumices	755	205–985
SO_2	2	Quaternary sands	1115	380–1840
SO_4	5–40	Quaternary sands	520	270–850
SO_2	5	Permian sediments	355	120–740
SO_2	18	Triassic sandstone	250	100–525
SO_2	25	Silurian	210	0–445
H_2S[b]	30	Pumices	190	80–310
SO_2	>28	Quaternary sands	125	50–250
SO_2?	>100	Tertiary basalt	45	0–120
SO_2?	>100	Diorite	30	0–110
SO_2?	>100	Basic extrusives	15	0–95

[a]Source: Lambert, Marcia J., and J. Turner. 1978. Interaction of nitrogen with phosphorus, sulphur and boron in N.S.W. *Pinus radiata* plantations. Proc. 8th Internat. Colloq. Plant Analysis and Fertilizer Problems, Auckland, New Zealand.
[b]New Zealand.

In Australian *P. radiata* plantations, S deficiency does not appear to cause a direct reduction in primary productivity. However, there appears to be a high correlation between S deficiency and foliar pathogen infection (Lambert and Turner, 1977), and associated with this are stem malformations, with subsequent losses in merchantable productivity and volume. Sulphur deficiency produces a metabolic malfunction which leads to apical bud resinosis (hence facilitates the entry of the fungus into the shoot) and also leads to an accumulation of particular amino acids, for example arginine, which would be favourable for the rapid growth of the fungus. This has been demonstrated from glasshouse work and in a field fertilizer trial with *P. radiata* which was established in an N-rich-S-deficient environment to determine the effects of N and S fertilizers on the concentrations of these nutrients in the foliage and to relate any differences to tree growth and health (Turner and Lambert, 1979). The results for the foliar S concentrations indicated that the most striking changes occurred in the plots where N fertilizer had been applied (Fig. 1). Within three months after treatment, S deficiency was induced and has persisted to the present. The foliar sulphate sulphur in the N-treated plots had been utilised, with a corresponding increase in the total concentration of free amino acids, particularly arginine. It was further noted that *Dothistroma septospora* (Dorog.) Morelet infection, a severe needle-cast fungus, was highest in the N-treated plots, with the degree of infection depending upon the intensity of S deficiency.

SULPHUR UTILIZATION BY FORESTS

Sulphur requirements for current growth can be fulfilled from three potential sources: (*a*) uptake by the roots from the soil-humus complex, (*b*) intake by the foliage, (*c*) redistribution from older tissues (older leaves, litter, wood, etc.). These

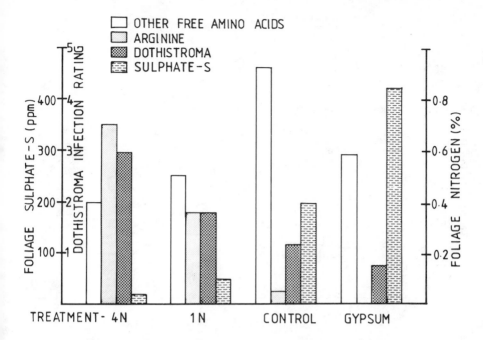

Fig. 1. Mean sulphate sulphur, *Dothistroma* infection, arginine, and free amino acids in foliage from fertilizer treatments in radiata pine trial in N.S.W., Australia. Source: Turner, J., and Marcia J. Lambert. 1979. Sulphur nutrition of conifers in relation to response to fertilizer nitrogen, to fungal infections and to soil parent materials. In Youngberg, C. T. (ed.), Proceedings Fifth North American Forest Soils Conference. Colorado State Univ., in press.

sources are interrelated, so that in a location with high S uptake and S intake, S redistribution will be low.

Foliar total sulphur and sulphate sulphur concentrations in tree species have been related to distances from sources of atmospheric S accessions in a range of studies (Malcolm and Garforth, 1977; Sidhu and Singh, 1977; Lambert and Turner, 1977). Due to the relationship between S and N (discussed earlier), foliar sulphate sulphur is a good comparative index in situations where S is adequate, but if S is deficient or the N availability is variable, foliar total sulphur provides an improved index. Lambert (1979) compared S uptake between coniferous species using a series of arboreta and trial plantings in N.S.W., Australia (Fig. 2). The sources of S available to these trees were from soil and atmospheric origins (specifically oceanic inputs). In the case of *P. radiata* (for which most data were available), foliar total sulphur concentrations decreased with increasing distance from the coast. There was an effect from soil fertility in that trees located on the most fertile soils (usually basalt-derived) had the highest S accumulations, whereas trees on the nutritionally poor sedimentary soils (Triassic, Permian, and Devonian sediments) had the lowest S accumulations.

Species other than *P. radiata* had different accumulation patterns, with the data suggesting that when species native to coastal situations were planted near the coast, higher concentrations of S were accumulated than when planted inland where the concentrations of total sulphur were lower. Species native to inland locations (e.g., *P. ponderosa*) had a pattern of foliar total sulphur accumulation which was in-

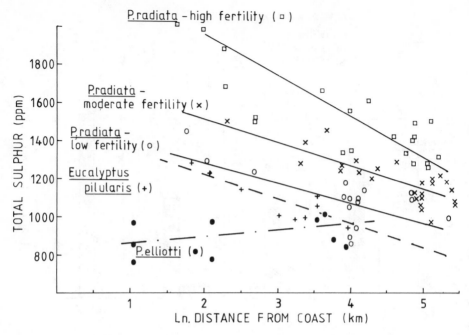

Fig. 2. The relationship between foliar total sulphur and logarithm of the distance from the coast for *Pinus radiata* with differing levels of fertility, *P. elliottii*, and *Eucalyptus pilularis* in N.S.W., Australia. Source: Lambert, Marcia J., 1979. Sulphur relationships of native and exotic tree species. M.Sc. (Hons.). Macquarie University, Sydney, Australia.

dependent of their geographical location. This tended to show that inland species were adjusted to S uptake from the soil but that in areas of higher atmospheric S inputs they could not efficiently utilize this additional source of S.

To further compare strategies of S uptake and utilization, planted and natural conifer stands from a range of sites have been compared—a planted Douglas-fir stand with moderate S availability at Bago, N.S.W. (Lambert, 1979); a natural Douglas-fir stand located on a site at Cedar River, Washington, which is subject to elevated inputs of S from a nearby copper smelter (Turner et al., 1979); and a nutritionally poor *P. radiata* site at Lidsdale, N.S.W., with comparatively recently increased S inputs following the installation of a coal-burning power station nearby (Lambert, 1979).

At the Bago site, which was located inland, the Douglas-fir had low accumulations of sulphate sulphur in the various tree components (Table 2 and Fig. 3). The S inputs to the site were extremely low, but the soil had relatively high S concentrations, and tree S uptake fulfilled about 90% of the requirement for growth. The data from the Douglas fir stand at Cedar River clearly showed that S was in excess at that site and that excess S was cycled in sulphate form (Table 2 and Fig. 3). The stand was N-deficient, and there was a high proportion of N redistribution within the tree to fulfill current growth requirements. However, S in the tree followed a different pattern, with the accent on accumulation in old foliage rather than on redistribution. The excess of S in the system accumulated as sulphate in both current and older tissue. While N uptake was less than the growth requirement, the S uptake was very much higher than the amount required for growth.

Table 2. Nutrient contents and transfers in Douglas-fir forests with low[a] and high[b] S inputs

Component	Low S inputs, Bago, N.S.W., Australia				High S inputs, Cedar River, Washington, U.S.A.			
	O.M.[c]	Total N	Total S	Sulphate sulphur	O.M.[c]	Total N	Total S	Sulphate sulphur
Nutrient content (kg/ha)								
Tree								
Foliage	16,720	186	16	1.3	9,440	98	18	11.2
Total	404,290	737	63	11.4	229,400	316	41	18.6
Understorey	0	0	0	0	3,390	29	4	1.1
Forest floor								
Woody	2,990	19	1	0	3,480	15	2	0.2
Total	35,680	369	27	1	26,670	223	26	3.0
Soil (0–15 cm)		1,680	80	5		1,300	220	40
Total ecosystem	439,970	2,786	170	19	259,460	1,868	290	63
Nutrient transfers (kg/ha/year)								
Litter fall	3,920	29.5	3.1	1.2	3,745	25.5	7.0	5.2
Total return to forest floor	3,920	35.7	6.1	4.1	3,745	27.7	12.7	10.8
Requirement for growth[d]	8,730[e]	51	3.8		7,460	45.8	5.6	
Uptake[f]		35	3.5			25.1	6.8	
Percent uptake requirement		69	92			55	120	
Percent contained in tree component	92	26	37		88	17	14	

[a]Source: Lambert, Marcia, J. 1979. Sulphur relationships of native and exotic tree species. M.Sc. (Hons.), Macquarie Univ. Sydney, Australia.
[b]Source: Turner, J., D. W. Johnson, and Marcia J. Lambert. 1979. Sulphur cycling in a Douglas-fir forest and its modification by nitrogen applications. Oecol. Plant. 14(4), in press.
[c]Organic matter.
[d]Nutrient content of current tissue.
[e]Aboveground primary productivity.
[f]Difference between requirement, internal redistribution, and leaching. Includes foliar intake.

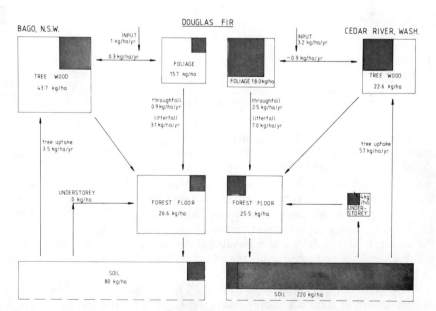

Fig. 3. Sulphur contents and transfers in Douglas-fir forests with low (Bago, N.S.W.) and high (Cedar River, Washington) S inputs (see Table 2). Shaded areas represent sulphate-sulphur.

At the Lidsdale site, which was also located inland but in a high-S environment as a result of a coal-burning power plant located 5 km away, the planted *P. radiata* (a species native to coastal environments and hence adapted to high S inputs) accumulated high levels of S (as sulphate sulphur) in the foliage and maintained relatively high growth rates (Lambert, 1979). This showed up in the per unit area estimates rather than the foliage concentrations (Table 3). The per area accumulations of sulphur were further compared with adjacent native eucalypt woodland species, and it was found that the S content of the *P. radiata* standing crop was one-third greater than that in the native woodlands and in addition an amount equivalent to the present standing biomass of the pine had been removed as thinnings. The adjacent native eucalypts had developed in a low S, P, and Ca environment, and these species appeared to be inelastic in their utilization of the elevated S (mainly as SO_4 inputs) and redistributed S to fulfill 54–70% of their S requirements. Generally, species adapted to low-S situations have low demands for S (actual lower requirements and/or lower growth), are efficient at uptake from the soil, and redistribute S efficiently within the tree.

Table 3. Biomass, S content, and transfers in a *Pinus radiata* plantation and an adjacent eucalypt woodland (*Eucalyptus rossii/E. maculosa*) in N.S.W., Australia[a]

	Eucalypt			Pine		
Component	Biomass	Total S	Sulphate S	Biomass	Total S	Sulphate S
	Nutrient distribution (kg/ha)					
Tree foliage	2,340	2.3	0.6	7,240	8.8	3.1
Total vegetation	62,355	11.6	3.3	160,500	22.3	9.1
Forest floor						
Wood	5,205	1.5	0.5	9,880	2.4	1.0
Total	10,480	4.5	1.6	13,900	6.0	2.1
Soil (0–10 cm)		5.9	16		55	14
	Nutrient transfers (kg/ha/year)					
Input (open-area precipitation)		9	9		9	9
Throughfall		11	9		19	16
Litter fall	2,910	3.1		3,790	1.3	
Logging removal				5,590	0.50	

[a]Source: Turner, J., and Marcia J. Lambert. 1979. Comparison of sulphur cycling between a conifer and a native forest in the vicinity of a coal-burning power station. Internat. Symp. Sulphur Emissions and the Environment, Socity of Chemical Industry, London. Pp. 228–230.

While species differ in their S accumulation patterns, S concentrations in different provenances (Table 4) and different families within species (Table 5) have also been shown to vary [Turner (1979) and Lambert and Turner (1977) respectively].

Communities with higher productivity and organic matter turnover appear to be buffered against environmental changes. However, in some situations where S and other nutrients are limiting to growth, productivity is generally low, and responses of species to changes in nutrient status vary from virtually nil (as in the case of S in the eucalypts at Lidsdale, where elevated S inputs were not absorbed) to species which accumulate and often utilize the added nutrients (e.g., for S in the *P. radiata* at Lidsdale).

When nutrient inputs are altered in N.S.W. forest management areas, there are several options to maintain healthy stands: (*a*) Leave the stand unaltered when it is healthy and productive. (*b*) Change either the species or the provenance of a species to one which can more efficiently utilize this changed resource. (*c*) Amend the site with other nutrients so that the stand can more fully utilize the added inputs.

Table 4. Mean foliar nutrient concentrations for provenances of Pacific Northwest Douglas-fir (age 15 years) grown on basaltic soil in N.S.W., Australia[a]

Provenance	Total N (%)	Total S (ppm)	Sulphate S (ppm)
Darrington	1.00	855	170
Ashford	1.04	930	210
Baker	1.17	930	125
Santian	1.31	1170	265
Pe-ell	1.34	1040	120
Cowlitz	1.40	1105	145
Molalla	1.50	1095	70
Vader	1.51	1285	245
L.S.D.$_{0.05}$	0.30	240	n.s.

[a]Source: Turner, John. 1979. Interactions of sulphur with nitrogen in forest stands. In Kennedy, R. (ed.), Proc. Forest Fertilization Conference, College of Forest Resources, Seattle, Washington, in press.

Table 5. Mean and range of foliar sulphate sulphur concentrations in *Pinus radiata* clones at Bago S.F., N.S.W., Australia[a]

Clone No.	Sulphate S (ppm) Mean	Range
E62	174	120–260
E64	55	0–150
E70	63	0–180
E80	8	0–35
E83	136	110–175
E87	154	65–235
L.S.D.$_{0.05}$	47	

[a]Source: Lambert, Marcia J., and J. Turner. 1977. Dieback in high site quality *Pinus radiata* stands—the role of sulphur and boron deficiencies. N.Z. J. For. Sci. 7(3):333–348.

This latter option has been studied in a western Washington Douglas-fir forest in which a N manipulation experiment was carried out in a 42-year-old poor-quality stand limited in growth by a deficiency of N (Table 6) (Turner et al., 1979). Sulphur was in excess. Treatments consisted of control, a high-N treatment (880 kg/ha N), a treatment of carbohydrate (8,000 kg/ha sucrose plus 10,000 kg/ha sawdust) to widen the C/N ratio and reduce N availability, and a carbohydrate treatment as above plus P, K, Ca, and S additions. The primary effects on the N cycle have been described elsewhere (Turner and Olson, 1976; Turner, 1977). The S cycle has also been described (Turner et al., 1979) and showed that excess S can cycle as sulphate in forest ecosystems and in that sense operate somewhat independently of the N or C cycles. The N additions resulted in the incorporation of sulphate sulphur into organic forms, and even though the stand was overstocked, a stimulus to growth was obtained. The changes (kilograms per hectare) also indicated that in the foliage, and presumably in the wood because of the growth stimulus, there were increases in

Table 6. N and S contents of specific stand components in the various fertilizer treatments of a Pacific Northwest Douglas-fir field experiment[a]

Component	880 kg/ha N			220 kg/ha N			Control			Carbohydrate			Carbohydrate plus P, K, S, Ca		
	Total N	Total S	Sul-phate S	Total N	Total S	Sul-phate S	Total N	Total S	Sul-phate S	Total N	Total S	Sul-phate S	Total N	Total S	Sul-phate S
Foliage (kg/ha)															
Current	50	3.38	0.06	39	2.97	0.27	20.7	3.26	1.86	20.02	2.71	1.28	22.6	2.86	1.41
Total	160	18.82	3.67	138	15.89	6.17	76.6	17.40	12.60	86.2	15.29	9.41	92.9	17.68	11.33
Leaf litter (kg/ha/year)	11	1.03	0.27	13	1.69	1.00	14	3.9	2.8	14	5.0	3.6	16	5.0	3.6
Foliage redistribution (kg/ha/year)	−20[b]	0.78		6	1.58		10	−0.80		16	−1.59		15.4	−2.06	
Foliage uptake (kg/ha/year)	70	2.60		33	1.97		12	4.11		4	4.37		7.2	4.29	

[a]Turner, J., D. W. Johnson, and Marcia J. Lambert. 1979. Sulphur cycling in a Douglas-fir forest and its modification by nitrogen applications. Oecol. Plant. 14(4), in press.
[b]Negative sign indicates accumulation.

N and S uptake. In the first years, the N applications reduced leaf litter fall since older needles were retained, but this was probably only a temporary phase. A further extension of this N fertilization indicated that unless in excess of 400 ppm sulphate sulphur is accumulated in tree foliage, no response to added N could be expected (Turner and Lambert, 1979; Turner et al., 1979) and that many of the soil types involved were naturally low in S and atmospheric sources provided the necessary forms of S inputs.

CONCLUSIONS

1. The biochemical relationship between S and N in plant protein dictates that one element cannot be adequately studied without reference to the other.

2. Sulphate sulphur is accumulated in trees when there is insufficient N to utilize it. When applied to stands in the sulphate sulphur form rather than as SO_2, no damage has been noted.

3. Tree species vary intrinsically in their ability to utilize S and particularly so when the S status of the site is altered.

4. On sites where there are elevated inputs, species or provenances capable of utilizing this asset should be planted and/or with suitable fertilizer regimes to improve forest productivity.

5. Coastal conifer species appear to accumulate high levels of S when in close proximity to a high atmospheric input of S, while exotic conifers whose seed sources come from inland areas (removed from strong maritime influence) tend to maintain a constant level of foliage S.

REFERENCES

Anderson, J. W. 1975. The function of sulphur in plant growth and metabolism. In McLachlan, K. D. (ed.), Sulphur in Australasian Agriculture, pp. 87–97. Sydney Univ. Press, Sydney.

Beaton, J. D., S. L. Tisdale, and J. Platou. 1971. Crop responses to sulphur in North America. The Sulphur Institute Tech. Bull. No. 18.

Bjorkman, E. 1967. Manuring and resistance to diseases. Pp. 328–331 in Colloquium on Forest Fertilization. Proc. 5th Colloquium (Jyvaskyla, Finland). Int. Potash. Inst., Berne, Switzerland.

Blair, G. 1979. Sulphur in the tropics. International Fertilizer Development Center Tech. Bull. IFDC-T12. Muscle Shoals, Alabama.

Bringmark, L. 1977. A bioelement budget of an old Scots pine forest in central Sweden. Silva Fenni. 11:201–209.

Kelly, J., and Marcia J. Lambert. 1972. The relationship between sulphur and nitrogen in the foliage of Pinus radiata. Plant Soil 37:395–408.

Lambert, Marcia J. 1979. Sulphur relationships of native and exotic tree species. M.Sc. (Hons.). Macquarie Univ., Sydney, Australia.

Lambert, Marcia J., and J. Turner. 1977. Dieback in high site quality Pinus radiata stands—the role of sulphur and boron deficiencies. N.Z. J. For. Sci. 7:333–348.

Lambert, Marcia J., and J. Turner. 1978. Interaction of nitrogen with phosphorus, sulphur and boron in N.S.W. Pinus radiata plantations. Pp. 255–262 in Ferguson, A. R., R. L. Bieleski, and I. B. Ferguson (eds.), Plant Nutrition 1978. Proceedings of the 8th International Colloquium on Plant Analysis and Fertilizer Problems, Auckland, New Zealand.

Likens, G. E., F. H. Bormann, R. S. Pierce, J. S. Eaton, and N. M. Johnson. 1977. Biogeochemistry of a Forested Ecosystem. Springer-Verlag, New York Inc.

Linzon, S. N. 1965. Sulphur dioxide injury to trees in the vicinity of petroleum refineries. For. Chron. 41:245–250.

Linzon, S. N. 1966. Damage to eastern white pine by sulphur dioxide, semimature tissue needle blight, and ozone. J. Air Pollut. Control Assoc. 16:140–144.

Malcolm, D. C., and M. F. Garforth. 1977. The sulphur:nitrogen ratio of conifer foliage in relation to atmospheric pollution with sulphur dioxide. Plant Soil 47: 89–102.

Miller, H. G., J. M. Cooper, and J. D. Miller. 1976. Effect of nitrogen supply on nutrients in litter fall and crown leaching in a stand of Corsican pine. J. Appl. Ecol. 13:233–248.

Rapp, M. 1973. Le cycle biogéochimique du soufre dans une forêt de *Quercus ilex* L. du sud de la France. Oecol. Plant. 8:325–334.

Shriner, D. S., and G. S. Henderson. 1978. Sulphur distribution and cycling in a deciduous forest watershed. J. Environ. Qual. 7:392–397.

Sidhu, S. S., and Pritam Singh. 1977. Foliar sulphur content and related damage to forest vegetation near a linerboard mill in Newfoundland. Plant Dis. Rep. 61(1):7–11.

Smith, W. H. 1974. Air pollution—effects on the structure and function of the temperate forest ecosystem. Environ. Pollut. 6:111–129.

Switzer, G. L., and L. E. Nelson. 1972. Nutrient accumulation and cycling in loblolly pine (*Pinus taeda* L.) plantation ecosystems: the first twenty years. Soil Sci. Soc. Am. Proc. 36:143–147.

Turner, J. 1977. Effect of nitrogen availability on nitrogen cycling in a Douglas-fir stand. For. Sci. 23:307–316.

Turner, J. 1979. Interactions of sulphur with nitrogen in forest stands. In Kennedy, R. (ed.), Forest Fertilization Conference. College of Forest Resources, University of Washington, Seattle, in press.

Turner, J., D. W. Johnson, and Marcia J. Lambert. 1979. Sulphur cycling in a Douglas-fir forest and its modification by nitrogen applications. Oecol. Plant. 14(4), in press.

Turner, J., and Marcia J. Lambert. 1979. Sulphur and nitrogen nutrition of conifers in relation to soil parent materials. Fifth North American Forest Soils Conference, August 1978. Colorado State Univ., in press.

Turner, J., Marcia J. Lambert, and S. P. Gessel. 1977. Use of foliage sulphate concentrations to predict response to urea application by Douglas-fir. Can. J. For. Res. 7:476–480.

Turner, J., and P. R. Olson. 1976. Nitrogen relations in a Douglas-fir plantation. Ann. Bot. 40:1185–1193.

Ulrich, B., R. Mayer, and M. Pavlov. 1971. Investigations on bioelement stores and bioelement cycling in beech and spruce stands including input-output analysis. Bull. Ecol. Res. Comm. 14:87–113.

Williams, C. H. 1975. The chemical nature of sulphur compounds in soils. Pp. 21–30 in McLachlan, K. D. (ed.), Sulphur in Australasian Agriculture. Sydney Univ. Press, Sydney.

DISCUSSION

George Sehmel, Pacific Northwest Laboratory: I was very interested in your data showing the parts per million of sulfur vs distance from the coast. Could these data be interpreted as the depletion of the airborne plume as a function of the distance from the source, so this is a data base for validating atmospheric transport models?

J. Turner: I don't think so. We find that we have relatively high sulfur inputs near the coast, the ocean being our main source of sulfur. Once we move onto our tablelands, which may be only 40 miles or so from the coast, we find a very dramatic drop-off in sulfur concentrations. There is a high degree of foliage sulfur concentration, partly as a result of soil types, but I think probably measuring inputs directly is better at this stage.

A. C. Chamberlain, Atomic Energy Research Establishment: Is the sulfate input proportional to the sodium chloride input as you go a good distance from the coast, and following up from that, is it true that the trees benefit from the sulfates despite the fact that they possibly get rain near the coast, getting excess sodium chloride?

J. Turner: We do get similar relationships to sulfate for sodium and chloride various distances from the coast. It's very difficult to progress further than this at present, as we haven't measured actual inputs over a wide range of sites for sodium and chloride. We have based sodium and chloride inputs on foliage analysis, and it's complicated by the fact that once you start looking at a range of species, some species reject sodium, others reject chloride, some reject both, and some accumulate both. Further research is required.

Stephen Wilson, New York State Energy Research and Development Authority: We are sponsoring some research that is being carried out by the College of Environmental Science and Forestry in Syracuse, a part of which is looking at the incidence of needle blights of red and Scotch pine. You mentioned tip death and some other examples when you were speaking of pathogens; did you infer that it's possible that physiology of tip death is related to some pathogens—that's the A part of the question—the B part is, are you looking at *Scleroderris* as a specific pathogen?

J. Turner: I don't know that pathogen. We have *Dotheastroma* and *Diplodea* pathogens, which are very distinctive. The tip death is certainly related to *Diplodea*.

CHAPTER 35

CONSEQUENCES OF SULFURIC ACID INPUTS TO A FOREST SOIL

Christopher S. Cronan*

ABSTRACT

One of the important concerns regarding acid precipitation and global sulfur pollution is the question of how acid rain and snowmelt interact with the soil system. Studies concerned with the ecological effects of atmospheric H_2SO_4 and HNO_3 upon soils have shown the following results: (i) atmospheric inputs of sulfuric acid may shift leaching processes from organic and carbonic acid control to sulfuric acid dominance throughout the soil zone of some forest ecosystems; (ii) in regions of noncalcareous bedrock, acid rain and acid snowmelt may alter aluminum transport, resulting in increased aluminum leaching through soils and into streams; (iii) depending upon environmental conditions, acid precipitation may cause increased leaching of essential nutrient cations from soils; and (iv) the impact of acid precipitation upon forest soils may be influenced in part by vegetation status and acid buffering processes in the forest canopy, by soil physical-chemical properties, and by geographic factors.

INTRODUCTION

One of the important concerns regarding acid precipitation and global sulfur pollution is the question of how acid rain and snowmelt interact with the soil system. This question actually raises two different considerations: (i) What are the potential ecological effects of acid precipitation upon soils, and alternatively, (ii) what are the effects of different soils upon the patterns of mineral acid neutralization during rainwater movement through watershed ecosystems? Unfortunately, there are few simple answers to either of these two questions. The particular impact of atmospheric H_2SO_4 upon soil processes and properties may vary between different sites, depending upon such factors as the rate and history of mineral acid deposition to the soil, the character of the vegetation, the characteristics of the soil parent material, the physical-chemical properties of the soil, and the like. Similarly, the degree to which atmospheric H_2SO_4 is titrated by the soils of different watersheds may vary depending upon an array of biological, chemical, and physical factors. Given the intriguing complexity of this problem area, I would like to discuss a few of the more recent research results pertaining to the consequences of sulfuric acid inputs to a forest soil.

*Department of Biological Sciences, Dartmouth College, Hanover, N.H. 03755. Research supported by Department of Energy Contract No. EVO-4498.

SOIL LEACHING PROCESSES

As a starting point, we can first consider the potential impact of acid precipitation upon general soil leaching and pedogenetic processes. In regions which are subject to intense acid precipitation, one might hypothesize that the biologically controlled production of carbonic and organic acids would not be as critical to soil leaching as the input of anthropogenic mineral acids introduced by precipitation. Under these circumstances, one would expect most of the H$^+$ for cation replacement in weathering and ion exchange reactions and mobile anions involved in cation transport to result from atmospheric additions of H$_2$SO$_4$ to the soil. Alternatively, one might predict that atmospheric H$_2$SO$_4$ has little impact upon the well-buffered forest ecosystem. In this case, one would hypothesize that the unbuffered mineral acids in precipitation are quickly neutralized by the forest canopy and upper soil horizons, so that the majority of mobile anions for cation transport are actually provided by the integrated contributions from organic acid and soil CO$_2$ reservoirs.

Results from our own studies in New Hampshire and from the literature can be used to examine these alternate hypotheses. As shown in Fig. 1, there are several

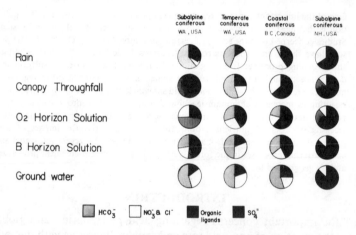

Fig. 1 Relative major anion compositions for solutions percolating through the different strata of four contrasting forest ecosystems. Each graph illustrates the percentage of the total cation charge balance at that ecosystem level which is contributed by sulfate, organic ligands, bicarbonate, and other anions. From left to right: (i) A forest with carbonic acid rain and with organic acid and carbonic acid soil leaching. From Ugolini et al., 1977, and Johnson, 1975. (ii) A forest with carbonic acid rain and with carbonic acid soil leaching. From Johnson, 1975. (iii) A forest with acid precipitation inputs and with sulfate- and bicarbonate-dominated soil leaching. From Feller, 1977. (iv) A forest in which atmospheric H$_2$SO$_4$ dominates solution chemistry from rainfall to groundwater levels. From Cronan and Schofield, 1979.

potential patterns of leaching which may result from inputs of either acid precipitation or "nonacid" precipitation. The first set of circular graphs illustrates an unpolluted subalpine ecosystem in which the pH 5.8 rainwater solution is first transformed to an organic-anion-dominated soil solution and then reverts to a bicarbonate-dominated groundwater. The next sequence of graphs shows another relatively unpolluted forest exposed to pH 4.95 rain and characterized by classic bicarbonate dominance throughout the system. The last two sequences represent two different ecosystem responses to acid precipitation inputs at pH 4.0 to 4.5. In the coastal-plain forest, the atmospheric H_2SO_4 is neutralized in the canopy and upper soil, so that carbonic acid leaching dominates the lower soil profile. In the extreme case illustrated by data from our studies in the mountains of New Hampshire, SO_4^{2-} dominates the anion chemistry in every stratum of the forest ecosystem. This implies that atmospheric H_2SO_4 provides the dominant source of both H^+ for cation replacement processes and mobile anions for cation transport throughout the soil profile and even into groundwater and lower-order streams. As indicated in Fig. 1, this pattern contrasts with soils of relatively unpolluted regions and ecosystems characterized by greater buffer intensities, where organic and carbonic acids dominate the leaching process.

SOIL ALUMINUM LEACHING RESPONSE

One of the major ecological consequences of acid precipitation in many non-calcareous regions of the United States, Canada, Scandinavia, and Germany is the increased aluminum mobilization and leaching that results from acute and chronic inputs of atmospheric H_2SO_4 and HNO_3 (Cronan and Schofield, 1979). This phenomenon appears to result from modern increases in soil aluminum leaching associated with soil solution acidification by atmospheric H_2SO_4.

The solubility of aluminum is very pH-dependent and drops dramatically in the pH range between 5 and 7. However, in the presence of organic ligands, the solubility of aluminum can be greatly enhanced by the formation of soluble organic-aluminum complexes. Table 1 shows the soil aluminum fractions that may potentially be affected by atmospheric H_2SO_4 and may contribute to dissolved aluminum in the soil solution. As indicated, each of these aluminum pools probably exhibits a separate thermodynamic and kinetic leaching response to changes in hydrogen ion activity. Thus, the aluminum transport that one observes in a given soil profile is the result of a complex of factors, including solution pH, the concentration of organic ligands, rate of water percolation, and abundances of different aluminum fractions.

In Fig. 2, the pattern of aluminum leaching for a podzolized soil in Washington is contrasted with the aluminum leaching response observed for a subalpine Spodosol in New Hampshire dominated by atmospheric H_2SO_4. For most podzolized soils, the solution acidification and organometallic complexation promoted by organic acids

Table 1. Soil aluminum fractions which may release inorganic aluminum to the soil solution in forest ecosystems exposed to acid precipitation. The fractions are ranked according to their hypothesized ability to release dissolved aluminum to a H_2SO_4-dominated soil solution.

Aluminum Fraction	Reaction Rate with H_2SO_4
Exchangeable aluminum	rapid
Amorphous aluminum hydroxide	moderately rapid
Organic-aluminum complexes	moderately rapid
$Al_x(OH)_y$ interlayers in expansible clay minerals (e.g. vermiculite)	relatively slower
Aluminum in clay lattices	relatively slower
Undecomposed aluminum silicates	relatively slower

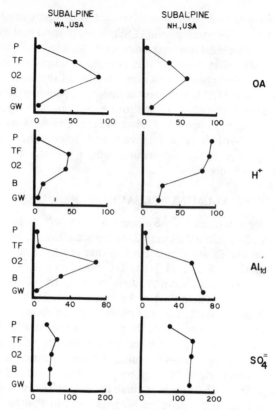

Fig. 2. Comparison of aluminum leaching processes for a podzolized soil located away from the influence of sulfur pollution (Washington subalpine) versus a podzolized soil dominated by atmospheric deposition of H_2SO_4 (New Hampshire subalpine). Graphs show changes in solution concentrations (uEq/L) during rainfall (P) percolation through the forest canopy (TF), forest floor (02), B horizon (B), and into ground water (GW). Trends of change in solution chemistry are illustrated for organic anions (OA), hydrogen ion (H^+), dissolved inorganic aluminum (Al_{td}), and sulfate. Washington subalpine podzolized soil located away from the influence of sulfur pollution, from Ugolini et al., 1977, and Johnson, 1975. New Hampshire subalpine podzolized soil dominated by atmospheric deposition of H_2SO_4, from Cronan and Schofield, 1979.

result in the transport of aluminum and other trace metals into lower horizons. During water percolation through the B horizon of such soils, aluminum and other metals leave solution in response to the removal of organic ligands by immobilization or decomposition and the associated rise in the soil solution pH. This pattern has been described for podzolized soils in the mountains of Washington and New Mexico and coastal Southeast Alaska (Ugolini et al., 1977; Johnson et al., 1977; Graustein, 1976) and is presumed to represent the historical leaching pattern for podzolized soils, including those of the Northeast.

By contrast, podzolized soils that are now being leached by atmospheric H_2SO_4 may show a very different pattern of aluminum transport. As shown in Fig. 2, soils like the higher elevation Spodosols of the Northeast may exhibit a continuous increase in dissolved aluminum concentrations during water percolation through the soil profile. This is interpreted to indicate that the historical trend of aluminum accumulation in the B2 horizon is being altered in soils that are dominated by atmospheric H_2SO_4 and HNO_3. The explanation for this modern increase in soil aluminum leaching may involve two factors: (i) atmospheric H_2SO_4 may produce a small but significant decrease in the soil solution pH throughout the soil profile, thereby increasing the solubility and hence mobility of aluminum, and (ii) whereas the acidity produced by natural organic acids may be removed from solution by both chemical precipitation and chemical neutralization reactions, the neutralization process for mineral acidity may be primarily controlled by H^+-ion exchange and weathering reactions which may be comparatively slow and thermodynamically unfavorable in an acid soil exposed to a solution pH around 4 to 4.5.

The consequences of this trend toward soil solution acidification in H_2SO_4-dominated soils may vary for different environments. For ecosystems which do exhibit increased aluminum leaching, there may be effects upon pedogenesis, clay morphology and ion exchange capacity, nutrient cycling, and biological communities.

NUTRIENT LEACHING RESPONSE

Environmental scientists have long been concerned with the potential effects of acid precipitation upon nutrient leaching rates from different soils. Experiments with field and laboratory lysimeters have already indicated that atmospheric H_2SO_4 may cause increased leaching of essential nutrient cations from some soils (Overrein, 1972; Abrahamsen et al., 1976; Wiklander, 1975; Mayer and Ulrich, 1977; Cronan, 1980). One potential implication of these findings is that there will be a long-term reduction in soil fertility and plant productivity in forest ecosystems exposed to acid rain and snowmelt. Although many investigators have addressed this problem area, we are only beginning to gather enough understanding to evaluate the potential impact of this problem on the whole forest ecosystem.

In New England, we have studied nutrient leaching responses with a number of approaches. Our initial laboratory studies were designed to examine whether cation leaching rates from forest floor microcosms changed in response to changes in the acidity of simulated canopy throughfall inputs. From experiments with undisturbed coniferous forest floor microcosms lacking plants, we found that there were significant differences in leaching of calcium, magnesium, potassium, and ammonium between lysimeters exposed to pH 4.0 vs pH 3.5 simulated throughfall. A drop in throughfall pH from 4.0 to 3.5 (a tripling in H_2SO_4 concentration) produced 63-87% higher calcium, 75% higher magnesium, 59-100% greater potassium, and 26-77% higher ammonium concentrations in microcosm percolates (Cronan, 1980). These experiments were followed up with a comparative microcosm study in which we attempted to evaluate whether leaching rates under current conditions of acid precipitation are markedly different from what might have been the historical pattern in the absence of atmospheric H_2SO_4. The experimental conditions and solution chemistry results for that study are shown in Table 2. As indicated, we found that inputs of pH 4.0 simulated acid throughfall to hardwood and coniferous forest floor microcosms caused significantly more leaching of calcium and magnesium than pH 5.7 simulated throughfall. In response to these tentative laboratory results, we are now attempting to evaluate the increased leaching rates in the context of other important ecosystem biogeochemical processes, nutrient fluxes, and nutrient pools. Toward that end, we are conducting field experiments to examine the effects of plant uptake, geochemical buffering, and rain chemistry upon the comparative leaching responses of intact ecosystems exposed to inputs of ambient acid rain vs "nonacid" rain.

Table 2. Comparison of ion leaching from forest floor microcosms exposed to pH 4.0 vs pH 5.7 simulated throughfall.

The null hypothesis was that cation leaching losses from the soil profile in response to acid precipitation are not quantitatively different from what they were historically when organic and carbonic acid leaching dominated forest soils. The hardwood microcosms were intact forest floors removed from a beech-maple-birch northern hardwood stand, while the coniferous forest floors were obtained from a balsam fir subalpine forest in New Hampshire.

								Concentration (microequivalents per liter)			
	pH	H^+	Ca^{2+}	Mg^{2+}	K^+	Na^+	NH_4^+	Al_{td}	Total cations	SO_4^{2-}	Other anions
Simulated throughfall[a]						**Inputs**						
5.7		2	25	10	20	4	5	0	66	25	41
4.0		100	25	10	20	4	5	0	164	125	39
				Percolate outputs[b]								
Hardwood microcosms												
pH 5.7 throughfall	4.60	25	76	29	68	11	171	108	488	87	399
pH 4.0 throughfall	4.43	37	123**	48**	82	18	185	116	609	179***	430
Coniferous microcosms												
pH 5.7 throughfall	4.33	47	10	12	75	5	201*	67	417	97	320
pH 4.0 throughfall	4.15	72	14*	15*	63	6	155	69	394	160***	234

[a] Experimental conditions: The experimental design consisted of treating groups of seven hardwood and coniferous microcosms with "non-acid" simulated throughfall at pH 5.7, while identical groups of microcosms were treated with simulated contemporary throughfall at pH 4.0. Microcosms each received a 3.5 cm application of simulated throughfall every 5–7 days over the course of three months. After that, lysimeter percolates were sampled for the next five simulated precipitation events.

[b] Comparison between pH 4.0 and 5.7 treatments for each microcosm type. Mann-Whitney two-tailed probabilities: * significant at $P < 0.05$, ** significant at $P < 0.01$, *** significant at $P < 0.001$.

At this point, it has become clear that in order to assess the ultimate impact of acid rain and snowmelt upon soil fertility, one must consider not only the potential acceleration of leaching that may result from atmospheric inputs of H_2SO_4 and HNO_3 but also the compensating processes which may counteract the adverse effects of increased leaching. These processes include the deposition of nutrient elements from the atmosphere (Wiklander, 1975), accelerated weathering of mineral material, anion adsorption by soil colloids (Johnson and Cole, 1977), and biological processes of nutrient release and accumulation (Abrahamsen et al., 1975).

HEAVY-METAL LEACHING RESPONSE

Atmospheric inputs of heavy-metal pollutants may accompany acid precipitation inputs to some forest ecosystems. Although these metals are typically immobilized in upper soil horizons by organic complexation, there has been some concern that acid rain might increase the currently low leaching rates of these metals. Recent studies by Tyler (1978) in Sweden suggest that H_2SO_4 inputs in rain have very little effect upon leaching rates of these metals from mor forest soils. However, additional studies may be needed to test this observation for other soil systems with varying amounts of soil organic matter.

CONCLUSIONS

In conclusion, the consequences of sulfuric acid inputs to a forest soil may vary greatly, depending upon such factors as the rates and history of mineral acid deposition to the soil, the character of the vegetation, the characteristics of the soil parent material, the physical-chemical properties of the soil, and the like. Some of the potential ecological effects of atmospheric H_2SO_4 upon soils are summarized in Table 3. Important areas for future research include the following: (i) development

Table 3. A partial summary of studies concerned with the impact of atmospheric H_2SO_4 on forest soils.

Observation	Location	Reference
Acid precipitation causes increased aluminum mobilization and leaching	Northeastern U.S. New Hampshire, U.S. Norway Germany Canada Canada Sweden	Cronan and Schofield 1979 Johnson 1979 Abrahamsen et al. 1976 Mayer and Ulrich 1977 Hutchinson and Whitby 1976 Baker et al. 1977 Wiklander 1975
Acid precipitation shifts the historic leaching regime to H_2SO_4-dominated leaching.	New Hampshire, U.S.	Cronan et al. 1978 Johnson et al. 1972
Acid rain and snow-melt may accelerate nutrient cation leaching	Norway Norway Canada Sweden New Hampshire, U.S. Germany	Overrein 1972 Abrahamsen et al. 1976 Baker et al. 1977 Wiklander 1975 Cronan 1980 Mayer and Ulrich 1977
Leaching by atmospheric H_2SO_4 may be retarded by soil sulfate adsorption	Wash., U.S. U.S. and Costa Rica Germany Sweden	Johnson and Cole 1977 Johnson et al. 1979 Khanna and Beese 1978 Wiklander 1975/76
Acid precipitation may alter the chemistry of soil humic materials	Canada	Hutchinson and Whitby 1976
Acid precipitation does not appear to accelerate heavy metal leaching in organic-rich soils	Sweden	Tyler 1978

of a predictive model for soil aluminum leaching from different types of soils and ecosystems, (ii) investigation of the effects of acid precipitation on organic matter decomposition, organic decay products, and soil humic materials, (iii) examination of the possible effects of increased inorganic aluminum mobilization on plant phosphorus nutrition, (iv) effects of soil parameters on neutralization patterns in different forested watershed systems, and (v) whole-system nutrient cycling responses to acid precipitation.

REFERENCES

Abrahamsen, G., K. Bjor, and O. Teigen. 1975. Field experiments with simulated acid rain in forest ecosystems. 1. Soil and vegetation characteristics, experimental design, and equipment. Research report no. 4, 15 pp. SNSF-project, 1432 As-NLH, Norway.

Abrahamsen, G., K. Bjor, R. Horntvedt, and B. Tveite. 1976. Effects of acid precipitation on coniferous forest. Pp. 38–63 in Braekke, F. H. (ed.), Impact of Acid precipitation on Forest and Freshwater Ecosystems in Norway. SNSF-project.

Baker, J., D. Hocking, and M. Nyborg. 1977. Acidity of open and intercepted precipitation in forests and effects on forest soils in Alberta, Canada. Water, Air, Soil Pollut. 7:449–460.

Cronan, C. S., W. A. Reiners, R. C. Reynolds, Jr., and G. E. Lang. 1978. Forest floor leaching: contributions from mineral, organic, and carbonic acids in New Hampshire subalpine forests. Science 200:309–311.

Cronan, C. S., and C. L. Schofield. 1979. Aluminum leaching response to acid precipitation: effects on high elevation watersheds in the Northeast. Science 204:304–306.

Cronan, C. S. 1980. Controls on leaching from forest floor microcosms. Plant and Soil (in press).

Feller, M. C. 1977. Nutrient movement through Western hemlock–Western red-cedar ecosystems in southwestern British Columbia. Ecology 58:1269–1283.

Graustein, W. C. 1976. Organic complexes and the mobility of iron and aluminum in soil profiles. Geol. Soc. Am., Abstr. Programs 8:891.

Hutchinson, T. C., and L. M. Whitby. 1976. The effects of acid rainfall and heavy metal particulates on a boreal forest ecosystem near the Sudbury smelting region of Canada. Pp. 745–767 in Dochinger, L.S., and Seliga, T. A. (eds.), Proc. First Internat. Symp. on Acid Precip. and Forest Ecosystem. NE For. Exp. Sta, USDA For. Serv. Gen. Tech. Rep. NE-23.

Johnson, D. W. 1975. Ph.D. thesis, University of Washington, Seattle.

Johnson, D. W., D. W. Cole, S. P. Gessel, M. J. Singer, and R. V. Minden. 1977. Carbonic acid leaching in a tropical, temperate, subalpine, and northern forest soil. Arct. Alp. Res. 9:329–343.

Johnson, D. W., and D. W. Cole. 1977. Sulfate mobility in an outwash soil in western Washington. Water, Air, Soil Pollut. 7:489–495.

Johnson, D. W., D. W. Cole, and S. P. Gessel. 1979. Acid precipitation and soil sulfate adsorption properties in a tropical and in a temperate forest soil. Biotropica 11:38–42.

Johnson, N. M., R. C. Reynolds, and G. E. Likens. 1972. Atmospheric sulfur: its effect on the chemical weathering of New England. Science 177:514–516.

Johnson, N. M. 1979. Acid rain: neutralization within the Hubbard Brook ecosystem and regional implications. Science 204:497–499.

Khanna, P. H., and F. Beese. 1978. The behavior of sulfate on salt input in podzolic brown earth. Soil Sci. 125:16–22.

Mayer, R., and B. Ulrich. 1977. Acidity of precipitation as influenced by the filtering of atmospheric sulfur and nitrogen compounds—its role in the element balance and effect on soil. Water, Air, Soil Pollut. 7:409–416.

Overrein, L. N. 1972. Sulfur pollution patterns observed; leaching of calcium in forest soil determined. Ambio 1:145–147.

Tyler, G. 1978. Leaching rates of heavy metal ions in forest soil. Water, Air, Soil Pollut. 9:137–148.

Ugolini, F. C., R. Minden, H. Dawson, and J. Zachara. 1977. An example of soil processes in the *Abies amabilis* zone of Central Cascades, Washington. Soil Sci. 124:291–302.

Wiklander, L. 1975. The role of neutral salts in ion exchange between precipitation and soil. Geoderma 14:93–105.

Wiklander, L. 1975/76. The influence of anions on adsorption and leaching of cations in soils. Grundfoerbaettring 27:125–135.

DISCUSSION

Ron McLean, Domtar Co., Montreal: I was interested to note that as far as calcium and magnesium were concerned you actually get less output than input. Were these figures for calcium in your rainfall realistic?

C. S. Cronan: Yes. I tried to pattern the simulated input after the average composition of the throughfall entering the soil profile in our field systems. There's quite a bit of abiotic capability to retain and show a net retention of calcium and magnesium over the throughfall input, and in these systems we were lacking monovalent elements like potassium and ammonia. We now need to get back to the real system and see how the plant component sorts out the net effects. You know, calcium and magnesium really have quite an ability to remain in the soil due to their divalent charge.

CHAPTER 36

PROCESSES LIMITING FISH POPULATIONS IN ACIDIFIED LAKES

Carl L. Schofield*

ABSTRACT

Acidified lakes which have lost bicarbonate buffering and maintain chronically depressed pH levels are generally devoid of fish life. Less acidic but poorly buffered waters experience episodic acidification during snowmelt and periods of increased surface runoff. Fish mortalities, which occur during these episodes as a result of stresses associated with the extreme temporal variations in pH and metallic cation concentrations, can lead to recruitment failure and alterations in population structure. Aluminum was identified as a primary toxicant present in acidified runoff entering streams and lakes during snowmelt. The mobilization of toxic species of aluminum from the edaphic to the aquatic environment is an integral part of the acidification process which impacts fish populations.

INTRODUCTION

The loss of fish populations from dilute lakes and streams exposed to acid precipitation inputs has been documented in Sweden (Almer et al., 1974), Norway (Leivestad et al., 1976), Canada (Beamish and Harvey, 1972), and the eastern United States (Schofield, 1976). Extensive lake surveys in these regions all suggest a relationship between fish survival and lake acidity (Schofield, 1976; Wright and Snekvik, 1978; Harvey, 1975). A generalized view of the lake acidification process and attendant fish population impacts, which has evolved from these observations, is analogous to a regional-scale acidimetric titration of a bicarbonate solution. The "titration curve" can be segmented into three stages, which define both the extent and nature of water quality change and the expected biological impacts. The initial stage of acidification is characterized by decreasing alkalinity but maintenance of bicarbonate buffering and pH levels above 6. No significant impacts on fish populations have been described at this level of acidification. Loss of bicarbonate buffering, with associated temporal fluctuations in pH, occurs in the second stage of acidification. Physiological stress, reproductive inhibition, and episodic mass mortalities may initiate recruitment failure and eventual extinction of fish populations during this stage of acidification. The final stage of acidification is characterized by chronically depressed pH levels (generally less than pH 5) and elevated toxic metal concentrations. Fish are usually absent from waters at this level of acidification.

---

*Department of Natural Resources, Cornell University, Ithaca, N.Y. 14853.

This simplified description of the acidification process may adequately describe the susceptibility of fish populations to acidification-induced perturbations, based on lake pH-alkalinity levels, but it does not identify the mechanisms which lead to fish mortality in acidified waters. Although hydrogen ion concentration has been identified as the primary variable involved in fish population response to acidification, other chemical species are important in determining modes of toxicity and regulating tolerance to given levels of hydrogen ion. Calcium is of importance in ameliorating osmoregulatory stress, through control of cell membrane permeability and ion transport at the level of the gill. Unacclimated fish exposed to low-calcium acid environments may experience severe sodium imbalance as a result of changes in gill permeability (McWilliams and Potts, 1978). Elevated aluminum concentrations, typically found in acidified lakes (Wright and Gjessing, 1976), are potentially toxic to fish at pH levels which are otherwise not directly harmful. Aluminum toxicity to fish is also exerted at the level of the gill and may involve damage to respiratory epithelium, in addition to osmoregulatory disturbance. Speciation of aqueous aluminum in acidified waters is controlled primarily by pH and the concentrations of potential ligands (fluoride and organic complexes). Toxicity appears to be exerted primarily by the monomeric inorganic species of aluminum, which are predominant in acidified lake and stream water during periods of increased surface runoff (Driscoll et al., 1979).

Episodic changes in water quality (decreased pH and calcium and increased aluminum concentrations) of lakes and streams associated with acid snowmelt runoff can produce mass mortalities of various age classes of fish inhabiting drainage systems sensitive to this type of perturbation (Muniz et al., 1975). It has been hypothesized that these episodic changes in water quality, associated with acid snowmelt, are primarily responsible for the demise of fish populations in acidified regions of Sweden and Norway (Hultberg, 1976; Wright and Snekvik, 1978). This hypothesis also seemed applicable to acidified headwater regions in the eastern United States. Comparable or even greater levels of acid deposition in winter precipitation occur in the mountainous areas of New York State and New England (Likens, 1976). Compared with the affected Scandinavian watersheds, however, snowmelt runoff is much less dilute due to more extensive soil development in the North American watersheds (cf. Hornbeck et al., 1976; Muniz et al., 1975). Thus, although comparable pH changes may occur in watersheds from both regions, the differences in ionic strength and cationic species present could significantly influence both the levels and mechanisms of effect on fish populations in the two areas. This paper describes the nature and magnitude of water quality changes associated with episodic acidification in the Adirondack Mountain region of New York and evaluates the potential significance of this process to fish survival.

PROCEDURES

Field and experimental studies were conducted at Little Moose Lake, near the village of Old Forge, New York, in Herkimer County. Experimental fish populations were held in water supplied to the Little Moose field laboratory by gravity flow through PVC pipe from either Acid Brook or Little Moose Lake (3-m-depth intake). Stocks of brook trout (*Salvelinus fontinalis*) eggs, yearlings, and adults were transferred to the laboratory in October–November 1976 and maintained in Little Moose Lake water throughout the winter. Trial exposures of brook trout fry in Acid Brook water were conducted from January to March 1977 in Acid Brook water transported to the Cornell fish laboratory in Ithaca, New York, for comparative bioassay with synthetic acid-aluminum solutions.

Water samples were collected on a weekly basis and more frequently during thaws from Little Moose outlet, the laboratory intakes from Acid Brook, and Little Moose Lake. Routine analyses included pH, alkalinity, conductivity, and reactive aluminum. Additional analyses for major ions (Ca, Mg, Na, K, SO_4, NO_3, Cl) and trace metals (Fe, Zn, and Cu) were obtained during thaw periods.

The toxicity of aluminum, at controlled pH levels, to brook trout fry was determined in the laboratory utilizing synthesized aluminum solutions. The experiments were conducted from January through March 1977 at the Cornell Fishery Laboratory, Ithaca, New York. The water used was a mixture of tap and deionized water and had a mean base conductivity of 27.50 μmhos/cm (range 25.35–29.46) in order to approximate the dilute waters commonly found in the Adirondack Mountains. Aluminum stock solution of 2.5 g/liter was prepared by dissolving ACS-grade aluminum wire in concentrated hydrochloric acid. Aliquots of the stock solution were added to 100-liter volumes of water to obtain the desired aluminum concentrations. The water was further acidified with 6 N sulfuric acid to obtain the necessary pH levels; it was aerated for at least two days before being placed in 76-liter plastic containers, which delivered it through capillary and Tygon tubing to 7.6-liter polyethylene test units. The turnover time of water in the units was approximately 24 hr. The test units were placed in a temperature-controlled recirculating water bath which maintained temperature between 9.5 and 11°C. Daily samples of treatment water without added aluminum were composited for analysis of major cations and anions. The pH of two replicates from each treatment was recorded daily.

Trials A, B, and C had respective nominal pH levels of 4.0, 4.4, and 4.9 with added aluminum concentrations of 0, 0.1, 0.5, and 1.0 mg/liter. Based on the results of these treatments, trial D was conducted at pH 4.9 with 0.25 and 0.50 mg/liter of aluminum added. Trial E had nominal pH 5.2 and added aluminum of 0, 0.1, 0.25, and 0.5 mg/liter. Trial F compared water mixed in the laboratory with Adirondack water. Water was obtained from Acid Brook on March 9, 1977, at the beginning of peak spring runoff. This water had a pH of 4.8 and total aluminum concentration of 0.63 mg/liter. Brook trout are occasionally seen in the brook during the summer, but it does not harbor a permanent population. Trial F treatments consisted of laboratory water with added aluminum concentrations of 0, 0.32, and 0.65 mg/liter and brook water at a nominal pH of 4.9 and aluminum concentration of 0.63 mg/liter.

Brook trout fry were obtained from the U.S. Fish and Wildlife Service Tunison Laboratory of Fish Nutrition located in Cortland, New York. Fry obtained for trial A on January 12, 1977, had just started to feed. Fry in subsequent trials were progressively larger. Trial F ended April 4. New fry were obtained from Tunison Laboratory for each trial except trial F, which used extra fish that had not been used in the previous trials and had been held at the Cornell Fishery Laboratory. Tunison Laboratory has water of pH 8.3–8.4 and a conductivity of approximately 470 μmhos/cm. The fry were placed in dilute water of pH 7.1 and a conductivity of approximately 27 μmhos/cm for 24 hr prior to the start of the experimental trials. Twenty fish were placed in each of three replicates of each treatment. Fish were fed a dry trout diet four to five times a day on weekdays and once a day on weekends. Fecal material and waste food were siphoned out twice daily on weekdays and once a day on weekends. Trials lasted until 50% mortality occurred or were terminated in 14 days if this point was not reached. Only the 0.25-mg/liter treatment of trial D was allowed to go for 23 days. Fish were judged to be dead when opercular movement had ceased and no swimming response could be elicited through stimulation of the caudal peduncle. If the time of 50% mortality was reached during the night, it was estimated by assuming a constant rate of mortality during that period. Surviving fish were placed in 10% formalin, following anesthesia in MS-222, for later histological examination. In trials C, E, and F, fish in treatments which did not experience 50% mortality were measured to the nearest millimeter and weighed on a Mettler balance to the nearest 0.0001 g before being placed in formalin. A sample of 20 fish were also weighed and measured at the start of trials E and F.

RESULTS

Stream and Lake Chemistry

Discharges of acid meltwater from the snowpack resulted in sharp drops in pH of both stream and surficial lake water (Fig. 1) in early March. Although the pH drop in Little Moose Lake was not as severe as the stream water change, alkalinity was reduced by 80% (from 100 μeq/liter prior to thawing to 20 μeq/liter after the thaw) at a depth of 3 m. Recovery to prethaw alkalinity and pH levels occurred after 12 days, just prior to the second thaw, which decreased alkalinity to a minimum of 43 μeq/liter on March 29. This level of alkalinity was sufficient to prevent any marked drop in pH of the lake water.

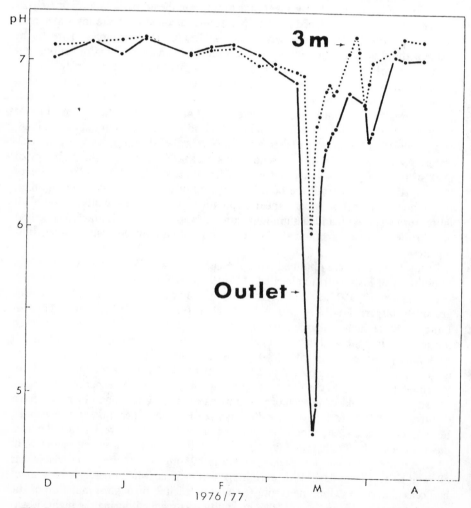

Fig. 1. Levels of pH measured in Little Moose Lake outlet and in water entering the laboratory from the 3-m intake from Little Moose Lake.

Major and minor ionic constituents of Acid Brook water collected at the beginning of the early March thaw are presented in Table 1. The high concentrations of strong-acid anions (sulfate and nitrate) are probably primarily a result of ion separation from the snowpack but may also reflect an additional contribution from the soil. The very high aluminum concentration is indicative of a significant soil influence on stream runoff composition. Precipitation aluminum concentrations were less than 10 μg/liter, and thus unmodified meltwater would not be a significant source of aluminum input. The seasonal changes in Acid Brook pH and aluminum levels are illustrated in Fig. 2. The strong correlation between brook pH and aluminum suggests that the process of exchange of meltwater hydrogen ion for soil aluminum exerts a significant control over stream runoff pH during thaws. Little Moose Lake water aluminum concentrations increased from prethaw levels of less than 20 μg/liter to 320 μg/liter during the early March thaw but only reached 170 μg/liter during the second thaw at the end of March.

Table 1. Ionic composition of treatment waters without added aluminum

Analyses of daily composites from treatments

Trial		Concentration (mg/liter)										
	pH	Ca	Mg	Na	K	Fe	Al	Cu	Zn	SO$_4$	NO$_3$	Cl
A	4.0	3.5	< 0.5	1.0	0.6	0.031	0.020	0.022	0.075	12.8	0.35	1.51
B	4.4	3.7	< 0.5	0.8	0.4	0.028	0.020	0.015	0.030	10.5	0.38	1.48
C, D, F	4.9	3.7	< 0.5	0.8	0.2	0.012	0.010	0.015	0.075	10.3	0.35	1.48
E	5.2	3.7	< 0.5	1.2	0.1	0.015	0.020	0.021	0.020	9.2	0.40	1.18
F[a]	4.9	3.7	< 0.5	1.6	0.2	0.028	0.630	0.021	0.075	4.5	6.60	0.60

[a]Obtained from Acid Brook, March 9, 1977.

Effects of Acid Snowmelt Episodes on Fish Survival

Captive populations of adult, yearling, and larval brook trout, which had been maintained over winter in Little Moose Lake water (3-m-depth intake) without incident, experienced distress and mortality during the early March episodes of water quality change described above. Eyed brook trout eggs exposed to the same water did not experience significant mortality. Yearlings and adults exhibited irregular opercular movements, lassitude, and cessation of feeding for the five-day period (March 13–17) during which mortalities were incurred. Moribund yearling trout exhibited gill damage similar to that observed in the experimental exposures described below. No stress or mortality was observed in the surviving trout populations during the second thaw, at the end of March, which produced less marked changes in lake water pH and aluminum levels.

The high mortality rates, behavioral responses, and subsequent histopathology of moribund specimens of trout exposed to the early March episodes of relatively minor lake water pH change (minimum pH 5.9) all suggested that hydrogen ion concentration was not the lethal factor in this episode. The changes in gill structure observed in moribund trout (lamellar fusion, epithelial desquamation, and general loss of structure) indicated that mortalities most likely resulted from severe gill damage and subsequent anoxia. Strong acids are not known to induce gill damage at the pH levels observed in this situation. Aluminum was suspected as the agent responsible for the pathological changes in gill structure, since aluminum concentrations during the lethal episode of water quality change did reach potentially toxic levels. Previous studies have indicated that gill damage is a primary effect of aluminum intoxication; however, these studies of aluminum toxicity either have been limited to alkaline pH levels (Freeman and Everhart, 1971) or did not consider

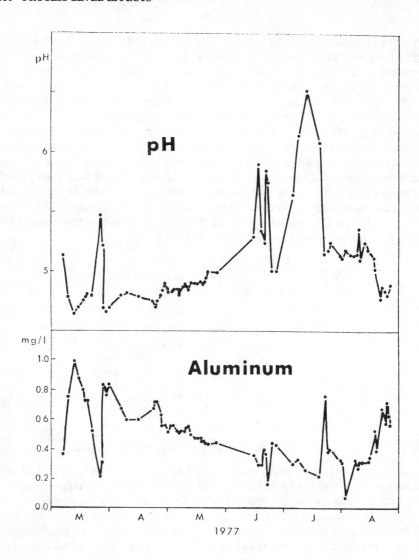

Fig. 2. Seasonal changes in pH and aluminum levels in Acid Brook during the period March–August 1977.

potential pH and speciation effects on toxicity (Burrows, 1977). In order to further evaluate the role of aluminum intoxication as a potential factor responsible for fish mortalities during acid snowmelt episodes, controlled laboratory experiments were conducted to determine the effects of aluminum at the concentrations and pH levels observed in the field.

Laboratory Experiments

When fish were first placed in the test units they swam actively for several hours in all treatments. Symptoms of stress were a darkening of skin coloration and cessation of feeding. Fish exhibiting these symptoms became inactive and rested on the bottom. These symptoms were characteristic of all fish at pH 4.0 and 4.4; however, they took longer to develop at pH 4.4 with 0 and 0.1 mg/liter aluminum and did not develop at pH 4.9 and 5.2 with 0 and 0.1 mg/liter aluminum. Darkening of the skin and cessation of feeding occurred at 0.24 mg/liter and higher concentrations of aluminum at all pH levels. Heavy accumulations of mucous and cellular debris on the gills were present on dead and surviving fish at 0.25 mg/liter aluminum and higher at pH 4.4 and above, but not at pH 4.0. Fish in this condition would occasionally flare their opercles and shake their head. At 0.25 and 0.32 mg/liter aluminum, fish exhibited these stress symptoms, but after one week they lightened in color and resumed feeding to a limited extent.

Gills removed from 15 preserved survivors of each treatment were scored on a scale of 0–3 for relative levels of gill damage. The distribution of these scores in the various treatments is given in Table 2. Scoring was based on the condition of the gill filaments and lamellae, as observed at magnification of 10–70× under a dissecting microscope. A score of zero indicated no visible damage to the gills. Fish exhibiting cell proliferation and clubbing at the distal ends of the gill filaments were assigned a score of 1, indicating "slight" damage. Extensive clubbing and lamellar fusion and edema over the entire length of the filaments was considered as "moderate" damage and assigned a score of 2. A "severe" damage score of 3 was assigned to fish exhibiting epithelial desquamation, filament collapse, and general loss of gill structure. At pH 4.0 there was no evidence of gill damage in fish exposed to aluminum concentrations of 0–1.0 mg/liter. At pH 4.4 and higher, moderate to severe gill damage was predominant at aluminum levels of 0.5 and 1.0 mg/liter, and the severity of damage appeared to increase with increasing pH. Levels of gill damage observed in survivors of the Acid Brook water treatment were intermediate to those found in the 0.32 and 0.65 mg/liter aluminum synthetic treatments (Table 2).

Table 2. Frequency of levels of macroscopic gill damage observed in surviving fish from trial F

See text for explanation. Fifteen fish from each treatment examined

Treatment	Frequency of gill damage score of —				Days exposed
	0	1	2	3	
Controls	100%				0
Brook water		47%	37%	16%	12
pH 4.8, Al 0.0	100%				14
pH 4.8, Al 0.32	20%	67%	13%		14
pH 4.8, Al 0.65				100%	6

Mean lengths and weights of fish in treatments lasting 14 days are given in Table 3. Mean weights were different between fish in 0 and 0.1 mg/liter aluminum at pH 4.9 in trial C ($t = 4.93, p = 0.001$), but weight differences were not as significant at pH 5.2 in trial E ($t = 1.87, p = 0.10$). Fish in 0.25 mg/liter aluminum at pH 5.2 experienced a loss in weight over 14 days ($t = 3.14, p = 0.01$). A decrease in weight was not as significant for fish exposed to 0.32 mg/liter aluminum at pH 4.9 in trial F ($t = 1.76, p = 0.10$), but there was certainly no significant growth in comparison with fish in the same water with no aluminum added.

Times to 50% mortality in the treatments are given in Table 4. In trial A, at pH 4.0, there were no discernible negative effects of aluminum on fish survival. In fact,

Table 3. Length and weight data for surviving fish from trials C, E, and F

Trial	Nominal pH	Aluminum (mg/liter)	Length (mm)		Weight (mg)	
			Average	Range	Average	Range
C	4.9	0.0	30.2	25–34	219.5	101–310
		0.1	28.8	21–32	173.7	56–272
E	5.2	0.0[a]	27.5	23–31	147.2	68–233
		0.0	31.1	24–32	212.3	78–444
		0.1	30.3	25–32	191.6	97–316
		0.25	27.3	21–31	111.1	44–162
F	4.9	0.0[a]	34.9	31–38	320.0	217–423
		0.0	37.3	29–42	401.5	152–595
		0.32	35.3	27–41	286.3	112–458

[a]At start of treatment.

Table 4. Times to 50% mortality for fish in trials A–F

Percent survival is given in those trials where the 50% mortality level was not reached after 14 to 23 days exposure

Trial	Nominal pH	Aluminum (mg/liter)	Time (days)			
			Replicate A	Replicate B	Replicate C	Mean
A	4.0	0.0	4.1	4.4	5.1	4.5
		0.1	5.4	4.8	5.4	5.2
		0.5	1.9	2.6	4.0	2.8
		1.0	4.8	5.4	4.9	5.0
B	4.4	0.0	9.7	10.0	11.4	10.4
		0.1	11.2	11.2	12.0	11.5
		0.5	1.5	1.8	2.9	2.1
		1.0	3.7	3.3	3.9	3.6
C	4.9	0.0	100%	100%	100%	100%
		0.1	95%	95%	95%	95%
		0.5	2.1	2.1	2.8	2.3
		1.0	1.7	1.6	2.1	1.8
D	4.9	0.25	90% (80%)[a]	90% (70%)[a]	90% (75%)[a]	90% (75%)[a]
		0.5	3.1	3.1	3.8	3.3
E	5.2	0.0	90%	90%	100%	93%
		0.1	100%	95%	90%	95%
		0.25	65%	70%	55%	63%
		0.5	1.6	1.7	1.5	1.6
F	4.9	0.0	100%	90%	100%	97%
		0.32	95%	75%	100%	90%
		0.65	4.7	4.1	5.3	4.7
		0.63[b]	11.9	5.4	4.5	7.3

[a]23-day survival.

[b]Brook water.

initial mortality rates were lowest at the 1.0-mg/liter aluminum level. Although the 50% mortality times were shorter at 0.5 mg/liter aluminum, the times for 0, 0.1, and 1.0 mg/liter were similar. In trial B, at pH 4.4, toxicity specific to aluminum is apparent. The 50% mortality times at 0.5 and 1.0 mg/liter aluminum are much lower than those at 0 and 0.1 mg/liter. In trials C, D, and E, at pH 4.9 and 5.2, aluminum levels of 0.5 mg/liter or greater yielded 50% mortality times of 1.6–3.8 days, while mortality at 0 and 0.1 mg/liter was minimal (0–10%) over 14 days. Aluminum concentrations of 0.25 and 0.32 mg/liter were not sufficient to induce 50% mortality within 14 days; however, there were indications that delayed mortality would occur at these concentrations. Fish in trials E and F were in poor condition at these concentrations after 14 days exposure, and the additional mortality after 23 days in trial D of the 0.25-mg/liter treatment supports this conclusion. In trial F, 50% mortality times were similar between the 0.65-mg/liter treatment and Acid Brook water, except for replicate A of the brook water.

DISCUSSION

Snowpack storage of strong acid from precipitation, concentration of this acid by ion separation during thaws, and subsequent release in snowmelt represent a series of processes that led to marked alteration of surface water quality in the Little Moose Lake watershed during March 1977. The increased concentration of aluminum associated with the discharge of this acid meltwater points to a soil interaction, involving exchange of meltwater hydrogen ion for soil aluminum, that has significant chemical and biological implications. Reductions in bicarbonate alkalinity to near-zero levels, due to strong-acid input, with associated increases in aluminum concentration, represent a major shift in the predominance of buffering components and the pH range of maximum buffering intensity ($pK_{H_2CO_3} = 6.3$, $pK_{Al} \cong 4.9$). This is demonstrated by the apparent control of runoff pH in Acid Brook by aluminum concentrations (Fig. 2) during the spring runoff period. The primary biological implication of this process is that toxic conditions for fish are produced by aluminum at pH levels which are otherwise not directly harmful.

Comparison of mortality rates, behavior, and gill pathology of brook trout exposed to acidic synthetic solutions and natural Adirondack waters with aluminum levels above 0.2 mg/liter indicated a specific toxic response to aluminum at pH levels down to 4.4. Significant sublethal effects of aluminum on brook trout growth in the laboratory were found at aluminum concentrations of 0.1–0.3 mg/liter. Although both hydrogen-ion and aluminum toxicity were evident at pH 4.4, none of the mortality at pH 4.0 could be attributed to aluminum at concentrations up to 1.0 mg/liter. The lag in initial mortality rate in the latter treatment suggests an antagonistic effect of Al^{3+} to hydrogen-ion toxicity, comparable with that exerted by calcium. This antagonism may not be specific to either ion but rather may reflect the role of polyvalent cations in general on limiting membrane permeability and ion transport (Burrows, 1977). At pH 4.0, approximately 95% of the aluminum would be present as Al^{3+}, yielding an equivalent $3[Al^{3+}]/[H^+]$ ratio of 1.1 for the 1.0-mg/liter treatment, where antagonism was detected. The lack of antagonism by lower Al^{3+} concentrations at lethal pH levels may result from the decrease in this ratio. Gill damage and mortality attributable to aluminum became evident in the pH range of 4.4–5.2. Increased aluminum toxicity with increasing pH suggests that changes in speciation involving hydroxy complexing in some way enhances the toxicity of the aluminum cation. If the mechanism of gill damage involves binding and transformation (polymerization and/or precipitation) of monomeric aluminum species at the gill surface, then increasing pH should enhance this process. The absence of gill damage at pH 4.0 (mutual antagonism of H^+ and Al toxicity) and increasing severity of gill damage from pH 4.4 to 5.2 at aluminum concentrations above 0.2 mg/liter suggest that this hypothesis may be applicable in solutions oversaturated with aluminum. The synthetic solutions used in the experimental trials and early spring stream runoff in the Adirondacks both represent nonequilibrium conditions with respect to aluminum solubility and species distribution. Additional studies to test this hypothesis of aluminum intoxication, together with further investigation of aluminum speciation in acidified waters, are currently in progress.

This study has demonstrated that acid snowmelt episodes can result in short-term changes in water quality which are lethal to fish. The fish kill observed in Little Moose Lake water indicates that even fish populations inhabiting lake and stream waters which are moderately well buffered at pH levels near 7 most of the year are subject to damage from these brief but highly toxic acid episodes during spring snowmelt. The extent of fish mortalities incurred by these events is unknown and would be difficult to assess on a regional basis. It is likely that these episodes occur in many other Adirondack lakes and streams, but changes in fish distribution and access to refugia (e.g., deeper water in large lakes, springs, or groundwater seeps) during periods of lethal water quality change could permit survival and population maintenance. However, many fish populations do not have the option of avoiding exposure to this toxic meltwater. For example, in small lakes which develop winter oxygen deficits in deeper waters, fish would not have the option of avoiding an influx of acid meltwater under ice cover by retreating to greater depths. The potential for development of such situations is illustrated by the changes in pH and dissolved oxygen concentration observed in a small lake in the northern Adirondacks during the winter of 1977 (Fig. 3). Larvae of species such as brook trout, lake trout

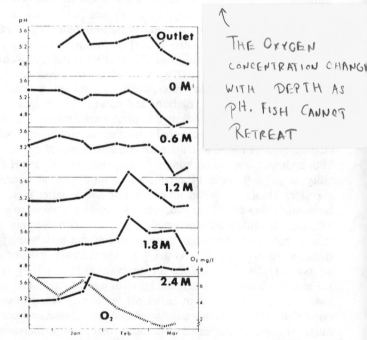

THE OXYGEN CONCENTRATION CHANGE WITH DEPTH AS pH. FISH CANNOT RETREAT

Fig. 3. Changes in pH and dissolved oxygen concentration at selected depths below ice cover in Loon Pond (Town of Brighton, Franklin County, N.Y.; surface area 4.9 ha, maximum depth 4.5 m) during the winter of 1976–1977.

(*Salvelinus namaycush*), yellow perch (*Perca flavescens*), lake chub (*Couesius plumbeus*), and white sucker (*Catostomus commersoni*), which would typically be present in the littoral areas of Adirondack lakes and in streams during acid snowmelt episodes, are at greatest risk due to relative immobility and greater sensitivity to water quality change. Repeated annual mortalities of these young fish could lead to recruitment failure and eventual population extinction.

REFERENCES

Almer, B., W. Dickson, and V. Miller. 1974. Effects of acidification on Swedish lakes. Ambio 3(1):30–36.

Beamish, R. J., and H. H. Harvey. 1972. Acidification of the LaCloche Mountain lakes, Ontario, and resulting fish mortalities. J. Fish. Res. Bd. Canada 29:1131–1143.

Burrows, W. D. 1977. Aquatic aluminum: chemistry, toxicology, and environmental prevalence. CRC Crit. Rev. Env. Control 7(2):167–216.

Driscoll, C. T., J. P. Baker, J. J. Bisogni, and C. L. Schofield. 1979. Aluminum speciation in dilute acidified waters and its effect on fish. Nature, ms. in review.

Freeman, R. A. and W. H. Everhart. 1971. Toxicity of Aluminum hydroxide complexes in neutral and basic media to rainbow trout. Trans. Am. Fish Soc. 100:644–658.

Harvey, H. 1975. Fish populations in a large group of acid-stressed lakes. Int. Ver. Theor. Angew. Limnol., Verh., vol. 19.

Hornbeck, J. W., G. E. Likens, and J. S. Eaton. 1976. Seasonal patterns in acidity of precipitation and their implications for forest stream ecosystems. Proc. 1st Int. Symp. Acid Precip. For. Ecosystems, U.S.D.A. For. Serv. Gen. Tech. NE-23, 397–407.

Hultberg, H. 1976. Thermally stratified acid water in late winter—a key factor inducing self-accelerating processes which increase acidification. Proc. 1st Int. Symp. Acid Precip. For. Ecosystem, U.S.D.A. For. Serv. Gen. Tech. Rep. NE-23, 503–517.

Leivestad, H., G. Hendry, I. Muniz, and E. Snekvik. 1976. Effects of acid precipitation on freshwater organisms. In SNSF Research Report 6/76, ed. F. H. Braekke. Pp. 87–111.

Likens, G. E. 1976. Hydrogen ion input to the Hubbard Brook Experimental Forest, New Hampshire, during the last decade. Proc. 1st Int. Symp. Acid Precip. For. Ecosystems, U.S.D.A. For. Serv. Gen. Tech. Rep. NE-23, 297–407.

McWilliams, P. G., and W. T. W. Potts. 1978. The effects of pH and calcium concentrations on gill potentials in the brown trout, *Salmo trutta*. J. Comp. Physiol. B 126:277–286.

Muniz, I. P., N. Leivestad, E. Gjessing, E. Joranger, and D. Svalastag. 1975. Fish death in connection with snow melting in the Tovdal River, spring 1975. SNSF Internal Report, IR 13/75, 60 pp., Oslo, November 1975.

Schofield, C. L. 1976. Acid precipitation: effects on fish. Ambio 5(5-6):228–230.

Wright, R. F., and E. J. Gjessing. 1976. Acid precipitation: changes in the chemical composition of lakes. Ambio 5(5-6):219–223.

Wright, R. F., and E. Snekvik. 1978. Acid precipitation: chemistry and fish population in 700 lakes in southermost Norway. Int. Ver. Theor. Angew. Limnol., Verh., vol. 20 (in press).

DISCUSSION

Questioner: What is your definition of monomeric aluminum, and what are the aluminum numbers that you have had by just estimating the toxicity?

C. L. Schofield: In answer to your first question, the fraction that we're calling monomeric aluminum consists principally of the simple hydroxy complexes, and also included in that fraction analytically would be fluoride complexes and any sulfate complexes.

Questioner: But if you are going up to, say, AlF_2, will that be more toxic?

C. L. Schofield: We haven't been able to differentiate to that degree in terms of the relative toxicity of various fluoride species that could be present. As I indicated, there was some reduction in toxicity to fish when fluoride was present in the test solution, but what the specific species are that could induce that, we're not sure.

Eric Edgerton, University of Florida: I was wondering, you mention that the presence of aluminum presumably interferes with the sodium transport actively or passively; I'm wondering about the counterions to fluoride and citrate in your aluminum amelioration experiments, whether an increase in the concentration of sodium may be in fact responsible for the increased ability of the fish to survive.

C. L. Schofield: In the experiment we tried to negate that effect by adjusting the sodium levels to a constant level in all of these conditions. Sodium levels were adjusted with sodium chloride.

CHAPTER 37

EFFECTS OF ACIDITY ON PRIMARY PRODUCTIVITY IN LAKES: PHYTOPLANKTON

George R. Hendrey*

ABSTRACT

Relationships between phytoplankton communities and lake acidity in three Adirondack Mountain lakes are being studied at Woods Lake (pH ca. 4.9), Sagamore Lake (pH ca. 5.5), and Panther Lake (pH ca. 7.0). Numbers of phytoplankton species observed as of July 31, 1979, are Woods, 27; Sagamore, 38; and Panther, 64, conforming to observations at many other sites that species numbers decrease with increasing acidity. Peak chlorophyll a and productivity values, respectively, were Woods, 6.8 mg m^{-2} and 21 mg m^{-2} hr^{-1}; Sagamore, 12.2 mg m^{-2} and 16 mg m^{-2} hr^{-1}; and Panther, 23 mg m^{-2} and 52 mg m^{-2} hr^{-1}. Patterns of increasing biomass and productivity in Woods Lake may be atypical of similar oligotrophic lakes in that they develop rather slowly to maxima six weeks after ice-out, instead of rapidly very close to ice-out. Phytoplankton productivities averaged from ice-out through July 31, 1979, were 12 mg m^{-2} hr^{-1}, 10 mg m^{-2} hr^{-1}, and 30 mg m^{-2} hr^{-1} for Woods, Sagamore, and Panther. Contributions of net plankton (net > 48 μm), nannoplankton (48 > nanno > 20 μm), and ultraplankton (20 > ultra > 0.45 μm) to productivity per square meter show that the smaller plankton are relatively more important in the more acid lakes: Woods > Sagamore > Panther ($p < 0.05$). This pattern could be determined by nutrient availability (lake acidification is suspected of leading to decreased availability of phosphorus). The amount of ^{14}C-labeled dissolved photosynthate (^{14}C-DOM), as a percent of total productivity, is ordered Woods > Sagamore > Panther. This is consistent with a hypothesis that microbial heterotrophic acitivity is reduced with increasing acidity, but the smaller phytoplankton may be more "leaky" at low pH.

INTRODUCTION

There have been few studies of the effects of acidification on phytoplankton productivity in lakes affected by acidic precipitation, and much of the work currently in progress is not generally available. This paper presents preliminary results of an ongoing investigation which is far from complete and for which much of the critical data analysis has not yet been done. The objective is to make other interested persons aware of this research effort on the effects of acidic precipitation on phytoplankton. The hypothesis behind this paper, and the work described, is that acidification alters both the structure and functioning of phytoplankton communities.

*Land and Freshwater Environmental Sciences Group, Department of Energy and Environment, Brookhaven National Laboratory, Upton, N.Y. 11973.

Certain lakes have become more acidic over the past one to three decades, as a consequence of deposition from the atmosphere of strong mineral acid (H_2SO_4 and HNO_3) and acid-forming substances (e.g., NH_4^+) which result in the formation of acids when entering terrestrial or aquatic ecosystems (Likens et al., 1977). This deposition and ensuing environmental alterations are often referred to as the acid rain problem. Decreased alkalinity and pH values have been reported for many lakes and streams in Norway (Wright and Gjessing, 1976; Wright, 1977), Sweden (Almer et al., 1974; Dickson, 1975; Grahn, 1976), Canada (Beamish and Harvey, 1972; Beamish and Van Loon, 1977; Watt et al., 1979) and the United States (Schofield, 1976; Davis et al. 1978). The comparison of data obtained one or more decades ago with current data has been criticized on the grounds that methods used earlier altered the chemistry of the sample or were lacking in accuracy. For example, Zimmerman and Harvey (1979) note that older data for alkalinity were determined by fixed-end-point potentiometric or colorimetric titrations. In systems of less than 300 microequivalents of alkalinity per liter (where species other than those of the carbonate system are significant) these fixed-end-point titrations cannot be expected to give accurate results. The error is an overestimate of about 35 microequivalents per liter, and the error is so indeterminate "that little hope exists for any mathematical salvaging operations," according to these authors. This substantial and helpful critique of temporal comparative studies may call into question much of the chemical evidence which indicates that lakes have been acidified by acidic precipitation.

There is, however, another independent line of evidence, based on the aquatic biota. Fish have disappeared from many lakes and streams in which acidification is reported and from many others where no chemical data from an earlier period are available (Statens Naturvårdsverk, 1975; Leivestad et al., 1976; Schofield, 1976). Previously the fish were present, now they are absent, because the water is too acidic and perhaps because the concentrations of materials related to watershed acidification (e.g., aluminum) are too high (Almer et al., 1978; Cronan and Schofield, 1979). In fact, many waters formerly renowned for their good fishing are now barren and so acidic that fish cannot survive at all.

INTEGRATED LAKE-WATERSHED ACIDIFICATION INVESTIGATIONS

This study of phytoplankton is a component of an integrated multidisciplinary, multiinstitutional research project being conducted in three separate lake watersheds in the Adirondack Mountains. The project, known as the Integrated Lake-Watershed Acidification Investigation (ILWAI), seeks to describe the effects of acid precipitation on lake-watershed ecosystems. The three lakes being studied are listed in Tables 1 and 2, and the major components of this project directly related to lake studies are listed in Table 3.

Studies of phytoplankton productivity are necessarily linked to both physical and chemical limnology. Data on pH, conductivity, temperature, dissolved inorganic and organic carbon (DIC, DOC), and seston ash-free dry weight (AFDW) are determined by the phytoplankton project at the ILWAI laboratory, Raquette Lake, New York. In addition, chemical samples are collected from lake profiles and from inlet and outlet streams. These are analyzed by a cooperating laboratory (J. Galloway, University of Virginia) for alkalinity (acidity), NH_4, Na, K, Ca, Mg, SiO_4, SO_4, NO_3, Cl, Al, Fe, and Mn. Samples for total P analyses are frozen and stored for analysis during the winter months at BNL. All of these data obtained by this project and a closely related project on benthic plant communities are entered in a common data base maintained by Tetra Tech, Inc.

Table 1. Lakes of the Integrated Lake-Watershed Acidification Investigation

Name	pH range	Elevation	Coordinates	USGS quadrangle
Woods Lake	4.7–5.1	615 m	43°53′ N, 74°57′ W	Big Moose
Sagamore Lake	5.0–6.4	586 m	43°46′ N, 74°38′ W	Raquette Lake
Panther Lake	5.3–7.8	562 m	43°41′ N, 74°55′ W	Old Forge

Table 2. Areal and hydrological characteristics of the ILWAI watersheds and lakes

	Panther	Sagamore	Woods
Watershed area, km^2	1.24	49.65	2.07
Lake surface area, km^2	0.18	0.72	0.23
Surface-to-watershed ratio	1:6.9	1:69	1:9.0
Volume, 10^6 m^3	0.709	7.54	0.813
Mean depth, m	3.51	11.6	4.22
Outflow, 10/77–9/78, cm/year	98.9	84.6	76.9
Outflow, 10/77–9/78, 10^6 m^3/year	1.19	42.0	1.59
Flushing time (volume divided by mean annual flow), days	230	65	180

Table 3. Major components of the ILWAI

OWRT; Office of Water Research and Technology, Department of the Interior; NYS-ERDA; New York State Energy Research and Development Authority; EPRI; Electric Power Research Institute; USGS; United States Geological Survey, Department of the Interior

Task	Principal investigator	Sponsor
Phytoplankton	G. Hendrey, Cornell University	OWRT
Benthic Plant Communities	G. Hendrey and L. Conway, Brookhaven National Laboratory	NYS-ERDA
Allochthonous Litter Decomposition	G. Hendrey and A. J. Francis, Brookhaven National Laboratory	EPRI
Lake Acidification Investigation, Chemical Studies	J. Galloway, University of Virginia	EPRI
Hydrologic Studies	J. Peters, U.S. Geological Survey	USGS
Precipitation Studies	N. Clesceri, Rensselaer Polytechnic Institute	EPRI
Data Management and Biogeochemical Modeling	C. Chen, Tetra Tech, Inc.	EPRI
Watershed Vegetation and Groundwater Chemistry	C. Cronan, Dartmouth University	EPRI

PHYTOPLANKTON METHODS

Primary production is determined by ^{14}C in situ incubation at five or six depths in each lake. Samples are returned to the laboratory for filtration and analysis of chlorophyll a and phaeophyton (fluorometry) and ^{14}C activity of particulate and dissolved phases (filtration and LSC). From August 1978 onward, chlorophyll a and ^{14}C production samples were filtered through a fractionation series to determine the contributions from phytoplankton in the following size ranges: (a) net plankton > 48 μm, (b) 48 μm > nannoplankton > 20 μm, and (c) 20 μm > ultraplankton > 0.45 μm and from the dissolved photosynthate ^{14}C-labeled dissolved organic matter (DOM). Samples of phytoplankton and of zooplankton are taken along with the primary production samples and preserved for analysis. Phytoplankton have been collected either in a van Dorn bottle or with a peristaltic pump, preserved with acid Lugol's solution, settled by the Utermöhl technique, and analyzed microscopically using a Wild inverted microscope.

PHYTOPLANKTON RESULTS AND DISCUSSION

The species of phytoplankton identified in each lake are shown in Table 4. The total numbers of species identified as of July 31, 1979, in the lakes are Woods, 27; Sagamore, 37; and Panther, 64, decreasing with increasing acidity. This is quite consistent with observations in numerous other locations (Almer et al., 1974; Kwiatkowski and Roff, 1976; Hendrey and Wright, 1976; Conroy et al., 1976; Yan, 1979).

Table 4. Phytoplankton taxa identified in Woods, Panther, and Sagamore Lakes, as of August 1979

Relative frequency of accuracy indicated: 4, dominant; 3, common/frequent; 2, occasional; 1, rare/infrequent, —, absent. Eight taxa occurring in January–April samples (winter) are tabulated separately from those occurring in May–August samples. Microscopic examinations conducted by K. Baumgartner

	Woods			Sagamore			Panther		
	Feb	Mar	Apr	Jan	Feb	Mar	Feb	Mar	Apr
Chlorophyta									
Ankistrodesmus falcatus	—	—	—	—	—	—	—	2	—
Carteria sp.	—	—	—	—	—	—	—	—	1
Chlamydomonas	—	—	2	—	—	1	—	—	2
Oocystis novae-semliae	—	—	—	—	—	—	—	—	1
Oocystis parva	—	—	—	2	2	—	—	—	—
Staurastrum megacanthum	—	—	—	—	—	—	—	—	1
Cyanophyta									
Anabaena	—	—	—	—	—	—	—	—	1
Chroococcus minimus	—	—	—	—	—	—	—	—	1
Chrysophyceae									
Chromulina (3 or less)	3	3	3	3	3	3	3	3	4
Chromulina (6)	2	2	2	3	3	3	2	3	2
Chromulina (10)	2	2	2	3	2	—	2	2	—
Diceras chodati	—	—	—	—	—	—	—	—	3
Dinobryon bavaricum	—	—	—	1	2	—	—	—	1
Dinobryon cylindricum	—	—	—	—	—	—	2	2	2
Dinobryon divergens	—	—	—	—	—	—	—	1	—
Dinobryon sertulana	—	—	—	—	—	—	—	2	—
Chrysococcocystis ovoides	—	—	—	—	—	—	—	—	2
Kephyrion sitla	—	—	—	—	—	—	—	—	2
Ochromonas (4)	2	2	2	2	2	2	2	2	2
Ochromonas (9)	1	—	—	—	—	2	1	—	2
Ochromonas nannos	1	—	—	2	2	—	2	—	2
Ochromonas scintillans	—	—	—	—	2	—	—	—	2
Pseudopedinella gallica	—	—	2	1	1	—	—	—	2

Table 4 (continued)

	Woods			Sagamore			Panther		
	Feb	Mar	Apr	Jan	Feb	Mar	Feb	Mar	Apr
Bacillariophyceae									
Diatoms (unknown)	1	—	1	1	1	—	—	—	2
Asterionella	—	—	—	—	1	—	—	—	—
Cocconeis	—	—	—	—	—	1	—	—	—
Eunotia	—	—	—	—	—	1	—	—	—
Eunotia triodon	—	—	—	—	1	—	—	—	—
Navicula	—	—	—	—	1	—	—	—	—
Nitschia sigmoidea	—	—	—	1	—	—	—	—	—
Tabellaria	—	—	—	1	—	—	—	2	—
Pyrrhophyta									
Gymnodinium	—	1	—	1	—	—	—	—	2
Gymnodinium varians	—	—	—	—	—	—	—	—	2
Peridinium inconspicuum	—	—	—	—	—	—	—	—	2
Cryptophyta									
Chroomonas acuta	—	—	—	—	—	—	1	2	2
Chroomonas pulex	—	—	—	—	—	—	—	—	1
Cryptomonas evosa	—	—	—	1	2	—	2	2	2
Rhodomonas minuta	—	—	—	—	—	—	—	—	2

	Woods				Sagamore				Panther			
	M	J	J	A	M	J	J	A	M	J	J	A
Chlorophyta (greens)												
Ankistrodesmus falcatus	—	—	—	—	—	—	—	—	—	—	1	2
Arthrodesmus incus	—	1	—	1	—	—	—	—	—	—	1	—
Asterococcus	—	—	—	—	2	—	—	—	—	—	—	—
Ankistrodesmus convolutus, v. minutus	—	—	2	2	1	1	—	—	—	—	—	—
Botryococcus Braunii	—	—	—	—	—	—	—	1	—	—	2	—
Chlamydomonas	2	2	1	2	2	2	—	—	—	2	2	2
Closteriococcus vierheimensis (Ankistrodesmus)	—	—	—	—	2	—	—	—	—	—	—	—
Chlorococcum gigas	—	—	—	—	—	—	—	—	—	1	—	—
Coelastrum microporum	—	—	—	—	—	—	—	—	—	—	1	—
Cosmarium	—	—	—	—	—	—	—	—	—	—	1	—
Elakatothrix gelatinosa	—	—	—	—	—	—	—	—	—	2	—	1
Eudorina elegans	—	—	—	—	—	—	—	—	—	1	1	—
Gloeocystis (small)	—	—	—	—	—	—	2	—	—	1	1	—
Gloeocystis major	—	—	—	—	—	—	—	—	—	—	1	2
Gloeocystis gigas	—	—	—	—	—	—	—	1	—	—	2	1
Oocystis borgei	—	—	—	2	—	2	—	—	—	2	2	—
Oocystis novae-semliae	—	—	—	—	—	—	—	—	—	3	3	2
Oocystis parva	—	—	—	—	—	—	3	3	—	2	—	—
Oocystis pusilla	—	—	—	2	1	3	3	3	—	—	2	2
Oocystis submarina	—	—	—	—	—	2	2	2	—	2	2	—
Pediastrum tetras	—	—	—	—	—	—	—	—	—	1	—	—
Planktosphaeria gelatinosa	—	—	—	—	—	—	—	—	1	—	1	—
Quadrigula chodatii	—	—	—	—	—	—	—	—	—	—	2	—
Quadrigula closterioides	—	—	—	—	—	—	—	—	—	2	2	—
Quadrigula lacustris	—	—	—	—	—	—	—	—	—	2	—	—
Scenedesmus abundans	—	—	—	—	—	—	—	—	—	1	—	—
Scendesmus denticulatus	—	—	—	—	—	—	—	—	—	2	2	1
Scenedesmus quadricauda	—	—	—	—	—	—	—	—	—	—	1	1
Sphaerocystis	—	—	—	—	2	4	2	—	—	—	—	—
Staurastrum dejectrum	—	—	—	—	—	—	—	—	—	—	1	1
Sphaerocystis Schroeteri	—	—	—	—	—	—	—	2	—	3	4	4
Staurastrum jaculiferum	—	—	—	—	—	—	—	—	1	—	—	—
Tetraedron caudatum	—	—	—	—	—	—	—	—	2	—	—	—
Xanthidium	—	—	—	—	—	—	—	—	1	—	1	—

Table 4 (continued)

	Woods				Sagamore				Panther			
	M	J	J	A	M	J	J	A	M	J	J	A
Cyanophyta (blue greens)												
Aphanotheca nidulans	—	—	—	—	—	—	—	—	—	—	3	3
Anabaena flos-aquae	—	—	—	—	—	—	—	—	—	2	—	—
Aphanocapsa elachista	—	—	—	—	—	—	—	—	—	2	2	2
Chroococcus limneticus	—	—	—	—	—	—	—	—	—	—	3	4
Chroococcus minimus	—	—	—	—	—	—	—	—	—	1	—	—
Coeosphaerium naegelianum	—	—	—	—	—	—	—	—	—	2	3	3
Dactylococcopsis smithii	—	—	—	—	—	—	—	—	—	—	—	2
Merismopedia tenuissima	—	—	—	—	—	3	4	4	—	—	—	—
Dactylococcopsis acularis	—	—	—	—	—	—	—	1	—	—	—	—
Merismopedia glauca	1	—	1	—	—	—	—	—	—	—	—	—
Oscillatoria	—	—	—	—	—	—	—	—	—	—	—	2
Chrysophyta; Chrysophyceae (golden)												
Chromulina (3)	3	3	3	4	3	3	3	3	3	3	3	3
Chromulina (6)	3	3	3	3	3	3	3	3	3	3	3	3
Chrysamoeba radians	—	—	—	—	—	—	—	—	2	—	—	—
Chrysococcus radians	—	—	—	—	1	—	—	—	—	—	—	—
Chrysoikos angulatus	2	—	—	—	—	—	—	—	—	—	—	—
Chrysosphaerella longispina	—	—	—	—	—	—	2	2	1	—	—	1
Diceras b	2	—	—	—	—	—	—	—	—	—	—	—
Diceras chodati	3	—	—	2	—	2	1	—	3	1	—	—
Dinobryon bavaricum	2	4	4	3	1	3	2	3	—	—	—	—
Dinobryon cylindricum	—	—	—	—	—	—	3	2	4	1	—	—
Dinobryon divergens	2	2	2	—	1	1	—	—	—	2	2	2
Dinobryon sertularia	2	—	—	—	1	—	—	—	—	—	—	—
Dinobryon sociale	3	3	2	3	1	2	1	—	—	1	—	—
Kephyrion valkovnovi	—	—	1	—	1	—	—	—	—	—	1	1
Kephyrion sitta	2	2	2	2	1	2	1	2	3	1	1	1
Mallomonas A	—	2	3	3	2	2	—	—	—	—	1	1
Mallomonas B	—	—	—	—	—	—	2	2	—	—	—	—
Mallomonas C	—	—	—	—	—	—	—	2	—	—	—	—
Mallomonas D	2	2	2	3	—	—	—	—	—	1	—	—
Mallomonas E	1	1	1	—	—	—	—	—	—	—	—	—
Mallomonas F	—	—	—	1	—	—	—	—	—	—	—	—
Ochromonas	2	2	2	2	3	2	2	2	—	—	1	1
Ochromonas nannos	—	—	—	—	2	2	2	2	3	3	2	1
Ochromonas scintillans	—	—	—	—	—	—	—	—	3	1	—	—
Pseudopedinella gallica	—	—	—	—	—	—	2	2	—	—	2	2
Uroglena americana	—	1	—	1	2	2	3	—	—	2	2	—
Bacillariophyceae (diatoms)												
Diatoms (general)	2	3	2	2	2	1	1	2	2	2	2	2
Melosira crotonensis	—	—	1	—	2	1	—	1	1	—	1	—
Eunotia	—	—	—	—	1	1	—	1	—	—	—	—
Navicula	2	3	2	2	—	—	—	1	2	2	1	—
Stauroneis	—	—	1	—	1	—	—	—	—	—	—	—
Frustulia	—	1	2	—	—	1	—	—	—	—	1	—
Surirella	—	—	1	—	—	—	—	—	—	—	—	—
Synedra	—	—	—	—	—	—	—	—	2	2	—	—
Asterionella formosa	—	—	—	—	—	—	—	2	—	—	—	—
Cymbella	—	—	—	—	—	—	—	—	—	2	—	—
Tabellaria	—	2	1	—	1	—	—	—	—	—	—	1
Nitzchia	—	—	—	—	—	—	—	1	—	—	1	—
Fragillaria	—	—	—	—	2	—	—	—	—	—	—	—
Pyrrhophyta (dinoflagellates)												
Gymnodinium sp.	—	—	1	2	—	—	—	—	—	—	—	—
Gymnodinium varians	3	3	3	3	2	2	2	1	2	—	1	1
Gymnodinium lantzschia	—	—	—	—	2	—	—	—	—	—	—	—
Gymnodinium paradoxum	1	—	—	—	1	—	—	—	—	—	—	—
Ceratium hirundinella	—	—	—	—	—	—	—	—	—	—	1	—

Table 4 (continued)

	Woods				Sagamore				Panther			
	M	J	J	A	M	J	J	A	M	J	J	A
Peridinium inconspicuum	2	3	2	2	—	—	1	1	1	—	1	1
Peridinium pusillum	—	—	1	1	—	—	—	—	—	—	1	—
	Cryptophyta (cryptomonads)											
Chroomonas acuta	—	—	—	1	—	1	2	2	2	—	2	1
Cryptomonas erosa	1	2	3	3	2	3	3	3	3	3	3	3
Cryptomonas ovata	1	1	1	2	2	—	—	—	4	2	2	1
Cryptomonas marssonii	—	—	—	—	—	—	—	2	—	—	—	—
Rhodomonas minuta	—	—	—	—	—	1	—	—	3	3	3	3
Sennia parvula	—	—	—	—	—	..	—	—	3	2	—	—
Cryptomonas norstedtii	—	—	—	1	—	—	—	—	—	—	—	—

Chlorophyll *a*

Chlorophyll *a* concentrations are an indication of photosynthetically active algae and are used here as a measure of biomass. Problems with estimating biomass by chlorophyll are recognized. For example, algal cells are able to adjust the content of photosynthetic pigment in accordance with their light regime (Jørgensen and Steeman Nielsen, 1965). In these lakes, the light extinction coefficients are Woods, 0.24; Sagamore, 1.13; and Panther, 0.39. Sagamore, a brown-water lake, absorbs light more rapidly than either Panther or Woods, so that the euphotic zone is shallower. The phytoplankton, mixing to the depth of the thermocline, will have less available light in Sagamore than in the other two lakes, and the concentration of chlorophyll *a* per cell may increase. This will be described quantitatively in subsequent reports, where chlorophyll *a* will be analyzed as a function of cell volume determined by microscopy.

Chlorophyll *a* concentrations integrated over the water columns of these lakes (Fig. 1) are typically low in winter under ice cover and increase in springtime as light begins to penetrate through the thinning ice-snow pack. In many oligotrophic lakes, such as these three, it is common to see a rapid increase in algal density just after ice-out; this did occur in Panther and to some extent in Sagamore. The pattern in Woods Lake, however, is slightly atypical of normal oligotrophic lakes. Both productivity and chlorophyll *a* built more slowly, if steadily, to a peak in June, six weeks after ice-out, in 1979. It is not yet possible to relate the Woods Lake observations to chemical changes, since nutrient data are not available (samples are frozen for analysis during the winter, when field work decreases).

Phytoplankton Productivity

Phytoplankton productivity (milligrams of fixed carbon per square meter per hour) integrated over the photic zone was averaged from ice-out (which occurred at the end of April 1979) through July 1979 and is shown in Fig. 2. This average areal productivity was much greater in Panther Lake than in either of the other two lakes, while Sagamore had the lowest value (Panther > Woods > Sagamore, $p < 0.05$). A Li-Cor integrating PAR (photosynthetically active radiation) meter operates at each lake from April into November. Data are collected on tape and will be entered at one time into the ILWAI data base. When these light values become available, they will be used to calculate daily and annual production values (productivity is the hourly rate of photosynthesis; production is the mass of carbon fixed over a specified time period).

The means of the maximum volumetric productivity values observed in each lake averaged from ice-out to July 31 and the corresponding chlorophyll *a* values are

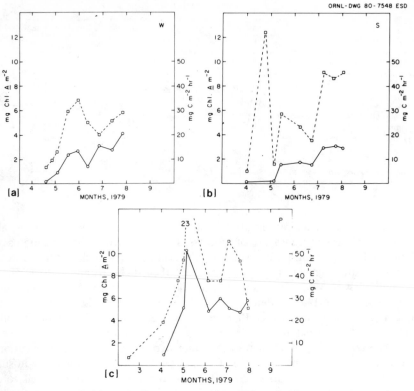

ORNL-DWG 80-7548 ESD

Fig. 1. Integrated chlorophyll *a* concentration (□) and primary productivity (O). (W) Woods Lake; (S) Sagamore Lake; (P) Panther Lake.

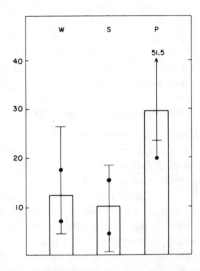

Fig. 2 Productivity (milligrams of carbon per square meter per hour) averaged over the period April 30–July 31, 1979. Range, —; standard deviation, ●.

shown in Fig. 3. These were both greater in Panther than in either of the other two lakes ($p < 0.05$). Mean specific productivity values (P/B = productivity/chlorophyll a), calculated for each of the maximum volumetric productivity observations then averaged over the corresponding time period, are Woods, 2.94 ± 1.29 hr^{-1}; Sagamore, 3.16 ± 1.19 hr^{-1}; and Panther, 3.87 ± 2.04 hr^{-1}. This may be an indication that phytoplankton photosynthesis was somewhat less effective in the two more acidic lakes for the first half of the growing season compared with phytoplankton in Panther Lake.

ORNL-DWG 80-7477   ESD

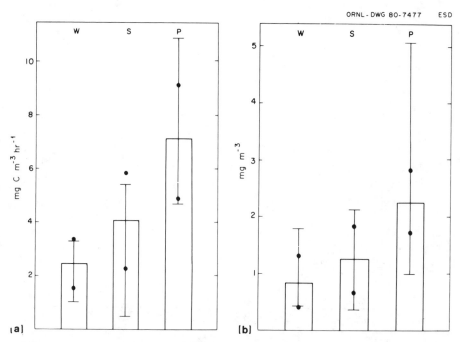

Fig. 3 Maximum productivity value and corresponding chlorophyll a value observed in each water column profile. (*a*) Maximum productivity (milligrams of carbon per cubic meter per hour); range, —; standard deviation, ● . (*b*) Chlorophyll a (milligrams per cubic meter).

In Fig. 4, all of the areal productivity values are plotted against pH (the average of the hydrogen ion concentrations over the five or six depths at which productivity was measured). The pH of Woods Lake is always low, so no relationship of individual productivity observations to pH is evident, except that the overall average is low. In Sagamore Lake, on the other hand, there does appear to be a relationship between productivity and pH. In Panther Lake, the lowest productivity value is associated with a single low pH value of 5.97.

The percent contributions of the net plankton (> 48 μm), nannoplankton (48 > nanno > 20 μm), and ultraplankton (20 > ultra > 0.45 μm) to areal phytoplankton productivity, from ice-out through July 31, 1979, and the corresponding chlorophyll a concentrations are shown in Fig. 5. There is an obvious shift in both productivity and biomass (chlorophyll a) toward smaller cell size, going from Panther to Sagamore to Woods. This certainly may be related to nutrient availability, but interpretation of these trends must be delayed until nutrient data become available.

The percent of phytoplankton productivity which appears in the filtrate passing a 0.45-μm filter, ^{14}C-labeled DOM, is presumed to have been released by the living

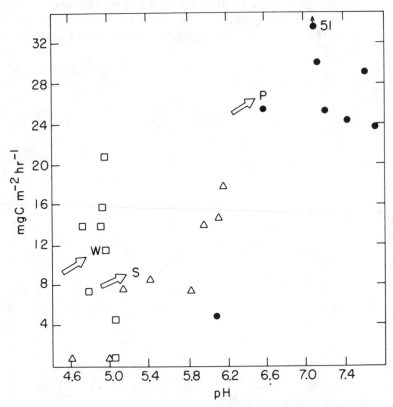

INTEGRATED PRODUCTIVITY AS A FUNCTION OF pH, THE
MEAN H$^+$ CONCENTRATION OF THE EUPHOTIC ZONE.
PANTHER = ●, SAGAMORE = △, AND WOODS = □. ARROWS
INDICATE OVER–ALL MEAN FOR EACH LAKE.

Fig. 4. Integrated productivity as a function of pH, the mean H$^+$ concentration of the euphotic zone. Panther, ● ; Sagamore, △; Woods, □. Arrows indicate overall mean for each lake.

phytoplankters during the ^{14}C incubation, but questions have been raised as to whether this release might be influenced by the filtration process, especially from the ultraplankton (Sharp, 1977; Fogg, 1977). The percent contribution of ^{14}C-labeled DOM to areal productivity averaged from April through July is significantly lower in Panther Lake ($p > 0.05$) than in the other lakes. There is considerable evidence that lake acidification inhibits microbial decomposition. This topic has been previously reviewed (Hendrey et al., 1976; Hendrey, 1979). Microheterotrophs normally are able to utilize excreted algal photosynthate (DOM) and convert it into particulate material rather rapidly if environmental conditions are suitable (Paerl, 1978). Only a small portion of this algal DOM is refractory material likely to survive longer than 24 hr (Saunders and Storch, 1971). The higher percent of ^{14}C-DOM in the two acidic lakes is consistent with the hypothesis that microbial acitivity is reduced in this more acidic environment, but questions need to be resolved concerning the effect of filtration on the amount of ^{14}C-DOM in the filtrate (Fogg, 1977). This may be caused by a greater "leakiness" of ultraplankton, which are more significant in Woods Lake than in the other two lakes.

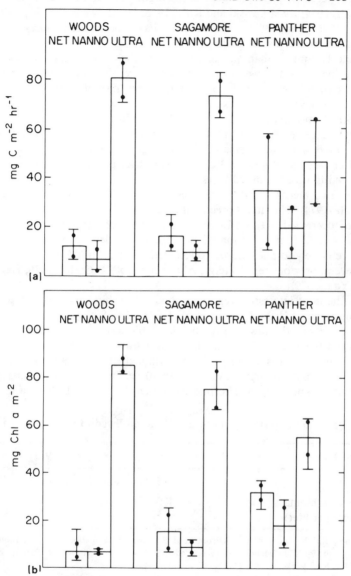

Fig. 5. Percent contributions to the particulate fraction of phytoplankton productivity and corresponding contributions to chlorophyll *a* concentration. (*a*) contributions to particulate fraction; (*b*) contributions to chlorophyll *a* concentration. Productivity in milligrams of carbon per square meter per hour; chlorophyll *a* concentrations in milligrams per square meter. Range, —; standard deviation, ●.

GENERAL DISCUSSION

Lakes and streams which are acidified exhibit marked alterations in the communities of phytoplankton and attached algae. Several factors related to acidification appear to contribute to these changes. Other investigations have found that the numbers of species of phytoplankton occurring in lakes are related to lake acidity (Almer et al., 1974; Kwiatkowski and Roff, 1976; Hendrey and Wright, 1976; Conroy et al., 1976; Yan, 1979). Some studies have found the biomass density of phytoplankton to decrease as well.

A notable feature of many acidified lakes is their remarkable clarity. This is especially evident when viewed from a hillside above the lake or from an aircraft. Often, acidic lakes have a dark, blue-green hue. Several studies have documented increases in the transparency of lakes as they become acidified. Almer et al. (1974) noted that the transparency of Lake Stora Skarsjön increased by 7 while, from 1943 to 1973, pH decreased from 6.2 to 4.5 and the amount of organic material in the water, measured as $KMnO_4$ demand (C.O.D.), decreased from 24 to 8 mg/liter in the period between 1958 and 1973. Increases in transparency by as much as 10 m were observed in other acidified lakes.

There are two mechanisms generally considered to cause this increased clarity (Almer et al., 1974). The first is precipitation of humic colored substances as a consequence of decreased pH and increased aluminum concentration (Almer et al., 1978). The second is a reduction of phytoplankton density, which seems to be related more closely to phosphorus concentrations than to pH per se, as has been noted in several previous reports (e.g., Almer et al., 1974; Grahn, 1976; Hendrey et al., 1976; Yan, 1979).

Perhaps the most interesting report on the relationship between phytoplankton, pH, and nutrients is that of Almer et al. (1978). In 58 oligotrophic lakes of the Swedish west coast, the lowest phytoplankton biomass was found in eight lakes ranging in pH from 5.1 to 5.6, while biomass density was greater in lakes of both higher and lower pH, as shown in Table 5. Although all of these Swedish lakes had low concentrations of phosphorus (less than 10 μg/liter), these authors suggested that aluminum complexing of phosphorus, which they found to have a maximum

Table 5. Variation in phytoplankton density in 58 Swedish lakes grouped by pH

Number of lakes	7	16	8	9	18
pH	<4.5	4.5–5.0	5.1–5.6	5.7–6.2	>6.2
Phytoplankton biomass (micrograms per liter)	660	400	212	287	394

Source: Almer, B., W. Dickson, C. Ekstrom, and E. Hornshröm. 1978. Sulfur pollution and the aquatic ecosystem. Pp. 271–311 in J. O. Nriagu (ed.), Sulfur in the Environment, Part II: Ecological Impacts. Wiley, New York.

near pH 5.5, might play an important role in regulating the availability of the limiting nutrient to these acidic or acidifying lakes. Transparency of six other Swedish lakes increased while pH decreased by 1.4 to 1.6 units. Similar observations have been made in Adirondack Mountain lakes (Schofield, 1973). Yan (1979) found phytoplankton biomass to be correlated to the concentration of phosphorus in an acidified lake contaminated with heavy metals, but biomass was not related to [H⁺]. The observations of Almer et al. (1978) and others that phosphorus concentration may be linked to [H⁺] provides an obvious link between decreased phytoplankton biomass and lake acidification, the direct relationship between [P] and phytoplankton biomass being well established (Kalff and Knoechel, 1978). While the maximum Al-P complexing occurs at about pH 5.5, the concentration of Al increases with watershed and lake acidification, so that more Al is available to complex phosphorus at lower pH.

Whatever the reason for the increased clarity of acidified lakes, the consequence is increased light penetration and deepening of the euphotic zone. With the resultant increased thickness of the trophogenic layer in an acidified lake, the production by phytoplankton per square meter of lake surface may be as great as, or greater than, it is in a more neutral lake. This speculation was stated by Almer et al. in 1974, and our knowledge has not yet improved much since then.

The integrated intensive study of three Adirondack Mountain lakes and their watersheds, ILWAI, will provide an opportunity for comparing the productivity of an acidified lake, including nutrient budgets and phytoplankton removal mechanisms, with similar data from nearby lakes which are less acidic.

ACKNOWLEDGMENTS

This work is funded by the Office of Water Resources Technology, through its annual allotment program at Cornell University. The Electric Power Research Institute also provides support through its Integrated Lake-Watershed Acidification Investigation (ILWAI). Additional support, in the form of cooperation in field work and in data sharing, is also obtained from the United States Geological Survey and New York State Energy Research and Development Authority. A summary of the participants and institutions is presented in the text. The authors gratefully acknowledge this cooperation and support. Special acknowledgment is due to K. Baumgartner, Cornell University, for phytoplankton identification.

REFERENCES

Almer, B., W. Dickson, C. Ekström, and E. Hornström. 1978. Sulfur pollution and the aquatic ecosystem. Pp. 271-311 in J. O. Nriagu (ed.), Sulfur in the Environment, Part II: Ecological Impacts. Wiley, New York.

Almer, B., W. Dickson, C. Ekström, E. Hornström, and U. Miller. 1974. Effects of acidification of Swedish lakes. Ambio 3:330-336.

Beamish, R. J., and H. H. Harvey. 1972. Acidification of the La Cloche Mountain lakes, Ontario, and resulting fish mortalities. J. Fish. Res. Board Can. 29:1131-1143.

Beamish, R. J., and J. C. van Loon. 1977. Precipitation loading of acid and heavy metals to a small acid lake near Sudbury, Ontario. J. Fish. Res. Board Can. 34:649-658.

Conroy, N., K. Hawley, W. Keller, and C. LaFrance. 1976. Influences of the atmosphere on lakes in the Sudbury area. J. Great Lakes Res. 2 (suppl. 1):146-165.

Cronan, C., and C. L. Schofield. 1979. Aluminum leaching response to acid precipitation: effects on high-elevation watersheds in the Northeast. Science 204:304-306.

Davis, R. B., M. O. Smith, J. H. Bailey, and S. A. Norton. 1978. Acidification of Maine (U.S.A.) lakes by acidic precipitation. Verh. Int. Ver. Limnol. 20:532-537.

Dickson, W. 1975. The acidification of Swedish lakes. Inst. Freshwater Research, Drottningsholm, Sweden, Rep. Nr. 54:8-20.

Fogg, G. E. 1977. Excretion of organic matter by phytoplankton. Limnol. Oceanogr. 22:576-577.

Grahn, O. 1976. Macrophyte succession in Swedish lakes caused by deposition of airborne acid substances. In Proc. First Intl. Symp. Acid Precipitation and the Forest Ecosystem, Ohio State Univ. USDA Forest Service Gen. Tech. Rep. NE-23, 1976:519-530.

Hendrey, G. R. 1979. Aquatics task force on environmental assessment of the Atikokan power plant: effects on aquatic organisms. BNL-50932, Brookhaven National Laboratory.

Hendrey, G. R., K. Baalsrud, T. S. Traaen, M. Laake, and G. Raddum. 1976. Acid precipitation: some hydrobiological changes. Ambio 5:224–227.

Hendrey, G. R., and R. F. Wright. 1976. Acid precipitation in Norway: effects on aquatic fauna. J. Great Lakes Res. 2 (suppl. 1):192–207.

Jørgensen, E. G., and E. Steeman Nielsen. 1965. Adaptation in plankton algae. Mem. Inst. Idrobiol. 18 (suppl.):37–46.

Kalff, J., and P. Knoechel. 1978. Phytoplankton and their dynamics in oligotrophic and eutrophic lakes. Annu. Rev. Ecol. Syst. 9:475–495.

Kwiatkowski, R. E., and J. C. Roff. 1976. Effects of acidity on the phytoplankton and primary productivity of selected northern Ontario lakes. Can. J. Bot. 54: 2546–2561.

Leivestad, H., G. Hendrey, I. P. Muniz, and E. Snekvik. 1976. Effects of acid precipitation on freshwater organisms. Pp. 87–111 in F. H. Braekke (ed.), Impact of Acid Precipitation on Forest and Freshwater Ecosystems in Norway. FR 6/76. SNSF Project. 1432 ÅspNHL, Norway.

Likens, G. F., F. H. Borman, R. S. Pierce, J. S. Eaton, and N. M. Johnson. 1977. Biogeochemistry of a Forested Ecosystem. Springer-Verlag, New York.

Paerl, H. W. 1978. Microbial organic carbon recovery in aquatic ecosystems. Limnol. Oceanogr. 23:927–935.

Saunders, G. W., and T. A. Storch. 1971. Coupled oscillatory control mechanisms in a planktonic system. Nature 230:58–60.

Schofield, C. L. 1973. The ecological significance of air-pollution-induced changes in water quality of dilute-lake districts in the Northeast. Trans. Northeast Fish and Wildlife Conf., 1972:98–112.

Schofield, C. L. 1976. Acid precipitation: effects on fish. Ambio 5:228–230.

Sharp, J. H. 1977. Excretion of organic matter by marine phytoplankton: Do healthy cells do it? Limnol. Oceanogr. 22:381– 399.

Statens Naturvårdsverk. 1975. Rodingsjoar söder om Dalälven (Char-lakes south of River Dalalven). Publication 1975:9.

Watt, W. D., D. Scott, and S. Ray. 1979. Acidification and other chemical changes in Halifax County lakes after 21 years. Limnol. Oceanogr. 24:1154–1161.

Wright, R. F. 1977. Historical changes in the pH of 128 lakes in southern Norway and 130 lakes in southern Sweden over the period 1923–1976. SNSF project TN 34/77. NISK 1432 Ås-NLH, Norway.

Wright, R. F., and E. T. Gjessing. 1976. Acid precipitation: changes in the chemical composition of lakes. Ambio 5:219–223.

Yan, N. D. 1979. Phytoplankton community of an acidified, heavy metal-contaminated lake near Sudbury, Ontario: 1973–1977. Water, Air, Soil Pollut. 11:43–55.

Zimmerman, A. P., and H. H. Harvey. 1979. Sensitivity to acidification of waters of Ontario and neighboring states: final report for Ontario Hydro. University of Toronto, Ontario.

DISCUSSION

David Schindler, Fresh Water Institute, Winnipeg: George, people who live in another world and study eutrophication have found that in order to properly predict the level of productivity they have to consider the nutrient input and the rate of water renewal, and, for example, by doubling the water renewal rate they have found that you roughly halve the productivity. If you take into account those considerations for these lakes, would that alter your results at all?

G. Hendry: I rather suspect, David, that the reason we see these differences in productivity is related to the nutrient concentration. We don't have the nutrient data; it's expensive to get it; we'd have to dedicate at least half a man-year, I think, to doing an adequate job on the phosphorus, because all of it has to be done with this extractive technique.

David Schindler: Presumably, though, if your precipitation is of the same concentration, you could simply take watershed area and water renewal times and come up with a relative index.

G. Hendry: That's a good suggestion; we have not done that, we'll certainly try that.

Ron Hall, Cornell: George, I would like to ask you about mercury level, particularly in this high-elevation lake with the increased sphagnum. Were you able to get samples below this sphagnum layer in terms of methylation of mercury?

G. Hendry: No. In response to David's question, when I said that I thought it was probably the phosphorus that's responsible for these differences we see in the lake, it's because, as was hypothesized by the Swedes in '74, the process of nutrient renewal, nutrient regeneration, and supply to the algae is probably closely linked to these problems of decomposition and mineralization, and I rather suspect that's what is happening in our lake, and the only way to find that out is to analyze the phosphorus input and to look at decomposition and nutrient recycling. David's dying to respond.

David Schindler: An alternative explanation, though, that hasn't been proved to my satisfaction, is that, in general, lakes with smaller watersheds are the ones that are most susceptible to oligotrophication, as the Swedes call it, presumably due to acid rain input, but, in general, lakes with smaller watersheds also have lower nutrient input, and there could be an autocorrelation here. It's possible that the oligotrophication, in lacking control data, which we do, is simply due to the fact that nutrient inputs were lower in the first place. I'd like to see that solved before I would draw any conclusions in my own mind.

G. Hendry: Before we can solve that, you have to find somebody that's willing to pay for it.

SECTION VII

ECOSYSTEM-LEVEL EFFECTS OF SULFUR DEPOSITION

Chairman: J. M. Kelly

CHAPTER 38

ECOSYSTEM-LEVEL EFFECTS OF SULFUR DEPOSITION

Chairman: **J. M. Kelly**
Tennessee Valley Authority

OVERVIEW

J. M. Kelly*

In recent decades, human activities have greatly increased total emissions to and deposition of substances from the atmosphere. This is due mainly to increases in combustion of fossil fuels, use of fertilizers and other chemicals in intensive agriculture and other forms of land management, and disposal of industrial, urban, and agricultural wastes. Atmospheric pollutants are imported, transported, exported, accumulated, and often changed prior to deposition. The result of their deposition may be beneficial or detrimental; their effects may be immediate and visible, or they may be low-level and accumulative with a more lasting effect.

State and federal agencies concerned with environmental affairs need to have the necessary information available that will allow them to cope with environmental crises as they occur and, more important, to prevent the occurrence of such economically and ecologically costly crises. General as well as specific information about the state, structure, and function of the terrestrial and aquatic environment is essential for determining the extent of changes that are man-induced and, therefore, controllable.

The most appropriate way to solve these problems and prevent their future occurrence is to thoroughly understand ecosystem processes and have the ability to integrate ecological information with other factors of the environment, including economic and sociological parameters. Thus research on the effects of impacts of atmospheric emissions should be approached on three levels—species/process, community/subsystem, and landscape/system. Each level provides unique opportunities to increase our understanding of environmental response on that level and, when integrated into the next level of complexity, adds to our understanding of response at that level as well.

Studies conducted at the species/process level generally deal with one or two species or processes, they are conducted under highly controlled conditions where

---

*Tennessee Valley Authority, Oak Ridge National Laboratory, P.O. Box X, Oak Ridge, Tennessee 37830.

only the condition being studied is varied, and instantaneous or short-term responses are generally measured. Studies on this level generally provide insight into the physical or chemical response of individuals or processes and aid in the determination of the actual response mechanism.

While research conducted at the species/process level provides excellent opportunities to study the mechanisms of response, it is not reasonable to assume that the magnitude or the expression of the response observed would be the same under less controlled conditions. Since a community/subsystem represents an aggregation of living organisms or processes having mutual relationships among themselves and to their environment, studies at the community/subsystem level tend to approximate real-world response to a much greater degree than studies at the species/process level. Many of the parameters evaluated in studies conducted at the species/process level can also be done at the community/subsystem level except that response is evaluated more in terms of the collective response of the community or subsystem. The degree of control obtainable with studies on the community/subsystem level is generally much less because of the increased importance of interacting factors, the addition of competition, and, in many cases, the introduction and utilization of the naturally occurring physical environment.

In an ecological frame of reference, community studies are generally of longer duration and can deal with mixed stands or monocultures; the communities may be relatively small aggregations of plants, such as field plots or microcosms, or they may be rather extensive, such as a forest stand or an agricultural crop. The papers presented in the preceding section on process-level effects provide some excellent examples of both species/process and community/subsystem approaches.

CHAPTER 39

IMPACT OF ATMOSPHERIC SULFUR DEPOSITION ON AGROECOSYSTEMS

U. S. Jones*

E. L. Suarez†

ABSTRACT

In South Carolina, S supplied in precipitation increased from an annual average of 6.3 kg/ha in 1953–1955 to 11.3 kg/ha in 1973–1975. Annual S fallout in precipitation in Iowa, Tennessee, and Wisconsin rural areas during the 1970s was about twice that in South Carolina. Twenty years ago, S in the atmosphere made a minor contribution to the S required by agroecosystems in South Carolina, but current data indicate that it now makes a major contribution.

Should S in precipitation approach the 200-kg/ha/year quantity added by S fungicides and be deposited on a peach ecosystem growing on poorly buffered sandy soils, the health of the trees would decline. South Carolina sandy soil data show that, at 0 to 30 cm deep, soil S, soil pH, and soil Al are significantly correlated with each other and with peach tissue Al concentration and that the higher the Al the greater the mortality of trees. The health of cotton and soybean ecosystems also declines when planted on peach orchard sites which had been formerly sprayed with 200 to 250 kg/ha/year of S fungicide.

At the present rural levels of air SO_2 and subsoil SO_4, the possibility that plant health of eastern U.S. agroecosystems is being influenced by too much or too little atmospheric S is slight. In South Carolina, one crop (corn) out of eight crops and one soil (Norfolk loamy sand) out of five indicated response to S at the 10% probability level. At none of the 15 locations was there an indication of too much atmospheric S for healthy agroecosystems.

INTRODUCTION

The purpose of this paper is to present recent and varied aspects of research on atmospheric deposition of S and its influence on soil acidity and plant growth.

Significant quantities of S are being added to soil in dry deposition and precipitation. In some ecosystems this input may be beneficial in the maintenance of soil fertility, and in others it may not be.

Amendments used on cultivated soils have a large effect on soil acidity and may overshadow acid-precipitation influences. However, there are large areas that have poorly buffered soils and are susceptible, especially in uncultivated forest areas.

Time and space variations of atmospheric S deposition are important considerations. The temporal environmental changes considered in this paper are those having

*Professor, Department of Agronomy and Soils, Clemson University, Clemson, SC 29631.

†Graduate student, Department of Agronomy and Soils, Clemson University, Clemson, SC 29631.

taken place in the southeastern United States since 1950. Spatial variations considered are those that have taken place during the 1970s in the Midwest as well as in the southeastern United States.

Selected data from Iowa, Wisconsin, Tennessee, Louisiana, Georgia, and North and South Carolina are presented as evidence supporting the conclusions drawn.

That atmospheric depositions are changing is clearly documented in the discussion by Cogbill (1974) on the history and character of acid precipitation in eastern North America.

Sulfur is the major contributor to acid rain, which, upon infiltration into soils, increases the hydrogen ion concentration by leaching Ca and other bases from the soil profile. On the other hand, S in the atmosphere can be used to meet the requirements for this essential plant growth element in areas where soils have limited supplies.

This paper reports results obtained from S soil fertility research carried on concurrently with atmospheric S monitoring programs and compares results obtained in South Carolina with those reported for other areas.

MATERIALS AND METHODS

Air, precipitation, and soil samples were collected at 15 South Carolina locations (Fig. 1). Air and precipitation were analyzed for S every 30 days, and soil S was determined annually in June. Data were recorded for all locations from 1973 through 1977. The collection sites were near Blackville, Cameron, Darlington, Clemson, Columbia, Wedgefield, Wateree, Fort Motte, Sumter, Saint Matthews, and Charleston, South Carolina.

During 1953–1955, Jordan (1964) collected precipitation for S analysis from sites at Clemson, Columbia, and Charleston that were close to the sites used in 1973–1977.

Fig. 1. Location of atmospheric sulfur collection sites in South Carolina.

Collection sites in Iowa, Wisconsin, and Tennessee are described in papers by Tabatabai and Leflin (1976a), Hoeft et al. (1972), and Shriner and Henderson (1978), respectively. Materials and methods used by these authors were generally the same as those described here and are included in the publications.

Sulfur Soil Fertility Tests

Uniform field experiments were conducted during 1973 to 1978 with barley (*Hordeum vulgare* L.) on Cecil sandy loam near Clemson; with turnip (*Brassica rapa* L.) and snapbean (*Phaseolus vulgaris* L.) on Lakeland sand near Columbia; with tomato (*Lycopersicum esculentum* Mill.) on a Wagram loamy sand near Columbia; with sweet corn (*Zea mays* L.), cucumber (*Cucumis sativus* L.), and soybean (*Glycine max* L.) on Marlboro loamy sand near Blackville; and with field corn (*Zea mays* L.) on Norfolk loamy sand near Darlington. The basic fertilizer contained sulfur-free materials and comprised one treatment—the control. Other treatments consisted of 9, 18, and 36 kg/ha of S broadcast as $CaSO_4 \cdot 2H_2O$, 18 kg/ha of S and 14 kg/ha of Mg as $MgSO_4 \cdot 7H_2O$, and 14 kg/ha of Mg as $CaMg(CO_3)_2$. The pH and calcium concentration were maintained at the optimum level for the crop to be grown by the use of $Ca(OH)_2$. There were six treatments with five replicates in a randomized complete block design. The treatments were repeated on the same plots during successive years. The plot size was 41 m² for sweet corn, cucumber, and soybean; 49 m² for field corn; 5.5 m² for turnip, tomato, and snapbean; and 18.6 m² for barley.

Cecil sandy loam is a clayey, kaolinitic thermic Typic Hapludult. Wagram loamy sand is a loamy, siliceous thermic Arenic Paleudult. Lakeland sand is a thermic coated Typic Quartizipsamment. Marlboro loamy sand is a clayey, kaolinitic thermic Typic Paleudult, and Norfolk loamy sand is a fine-loamy, siliceous thermic Typic Paleudult.

Air S Determination

Sulfur in the air was determined in a representative sample collected in a standard lead peroxide sampler obtained from Research Appliance Company, Gibsonia, Pennsylvania. The sampler consists of a cylinder around which a cotton fabric with a surface area of 100 cm² is wrapped. The fabric was coated with a paste containing lead peroxide. The cylinders were exposed freely to the air but were protected from rain by a cowl. The amount of SO_2 absorbed depends not only upon the concentration in the air but also upon the column of air coming into contact with the peroxide. After a 30-day exposure, the fabric and coating were removed from the cylinder, and the amount of sulfur was determined by the turbidimetric barium sulfate method (Jordan et al., 1959).

Precipitation S Determination

Sulfur in precipitation was determined in a representative sample collected in a standard 3-liter plastic bucket 20 cm in diameter. The bucket was supported in a metal canister about 180 cm above ground. A protective metal ring was installed as an extension of the canister to prevent pollution of the rainwater by birds. After one-month exposure periods, corrected to 30-day intervals, the entire quantity of water was taken from the bucket, screened through a 20-mesh sieve, and evaporated to dryness in an S-free hood; the residue was taken up in a solution of 0.5 N $NH_4C_2H_3O_2$ in 0.25 N CH_3COOH, filtered, and washed. Sulfur was precipitated as barium sulfate and determined tubidimetrically (Jordan et al., 1959).

Soil Analysis

Samples were taken from holes dug with an ordinary hand posthole digger. They were protected from contamination and removed at 0–15, 15–30, and 30–45 cm-depth. The peach (*Prunus persica* L) orchards were sampled to depths of 180 cm. Each soil sample consisted of a mixture of three samples in the immediate area at the appropriate depth.

Sulfur

Five grams of soil were extracted with 20 ml of 0.5 N $NH_4C_2H_3O_2$ in 0.25 N CH_3COOH by shaking for 5 min; about 0.2 g of charcoal was added, and the extract was filtered. Sulfate was precipitated as $BaSO_4$ by adding $BaCl_2$ solution. After 5 min the $BaSO_4$ concentration was estimated employing the standard turbidimetric method by reading on a spectrophotometer at 420 nm.

Aluminum

Soil was extracted with 1 N KCl at a 1:10 soil-solution ratio, and exchangeable Al was determined in the extract by atomic absorption spectroscopy.

pH

Twenty grams of sieved soil were placed in a 100-ml beaker, and 20 ml of distilled water was added. After stirring and allowing to settle for 1 hr, pH was measured with a glass electrode pH meter.

Plant Analysis

Field crop, vegetable, and orchard leaf samples were taken from the part of the plant and at a time suggested by Jones (1974). Leaf samples were finely ground and digested with a mixture of perchloric and nitric acids. Magnesium was determined by atomic absorption spectroscopy. Sulfur was precipitated as $BaSO_4$ and determined turbidimetrically as described above. Tissue from two replicates of each treatment was analyzed. The data were adequate for the purpose: to determine sufficiency or deficiency of S and Mg.

Aluminum concentrations in leaves and twigs were determined with an arc spectrograph using an ARL Quantometer.

RESULTS AND DISCUSSION

Precipitation S Concentration Trends

It was possible to make comparisons of the 1973–1975 collections with 1953–1955 collections of S in precipitation from sites near Columbia, Clemson, and Charleston as noted in Table 1.

The data indicate an increase in S supplied through precipitation from an annual average of 6.3 kg of S per hectare in the 1953–1955 period to 11.3 kg of S fallout per hectare 20 years later in 1973–1975. A comparable study (Hoeft et al., 1972) in Wisconsin reported an annual average sulfur addition from precipitation in rural and urban areas of about 30 kg/ha/year. The S in precipitation at experiment stations at Watkinsville, Georgia, and Waynesville, North Carolina, for the 1953–1955 period is given for comparison in Table 1. Values for these two locations, 6.4 and 6.2 kg of S per hectare, respectively, are not significantly different from 6.3 kg of S per hectare, the mean for three South Carolina towns during the same period.

Table 1. Sulfur collected in precipitation, 1953–1955 vs 1973–1975

Location	Average annual amount of S collected (kg/ha)	
	1953–1955[a]	1973–1975
Clemson, S.C.	8.9[b]	10.1
Columbia, S.C.	4.2	11.8
Charleston, S.C.	6.0	12.1
Mean	6.3[c]	11.3
Watkinsville, Ga.	6.4	
Waynesville, N.C.	6.2	

[a]Source: Jordan, H. V., C. E. Bardsley, L. E. Ensminger,and J. A. Lutz. 1959. Sulfur content of rainwater and atmosphere in Southern states as related to crop needs.USDA Tech. Bull. 1196.

[b]Average for years 1953 and 1955. Data for 1954 reported.

[c]The 1953–1975 mean is significantly different from the 1973–1975 mean at the 1% level.

Effect of locations

The lowest 1973–1975 average annual amount of precipitation S was 8.2 kg/ha found in rural areas of South Carolina. Precipitation in urban and industrial areas deposited 11.3 and 11.2 kg/ha, respectively. These data and comparable data from Iowa (Tabatabai and Laflen, 1976a), from Wisconsin (Hoeft et al., 1972), and from Tennessee (Shriner and Henderson, 1978) are summarized in Table 2. An industrial area in Wisconsin accumulated an amount of S in precipitation that was significantly greater than that from urban areas, but this was not so in South Carolina. The reason for this is not readily apparent except that the same amount (11.8 kg of S per hectare) was collected at an industrial location, about 3 km from a coal-burning steam electric plant which began operation in 1972, and at the Sandhill Experiment Station in the suburbs of Columbia, South Carolina. Comparable atmospheric S concentrations, determined by other methods, are reported by the South Carolina Department of Health and Environmental Control (1978) for urban and industrial areas. Their data point to the same conclusion: the concentration of SO_2 in the air in heavily populated areas is about the same as that in industrial centers.

Air S Concentration Trends

At the same 15 locations in South Carolina, average annual collections of S in the air for the four-year period 1973–1976, were 12.7, 18.6, 42.7, and 58.3 kg of S per hectare respectively (Fig. 2). The highest values recorded were in the January–March quarter and the October–December quarter of each year. The lowest values were April–June and July–September quarters. Each year, seasonal variations were significant, and each year the total air S was greater than the year before.

Table 2. Sulfur collected in precipitation at rural, urban, and industrial locations in southeastern and midwestern United States

State	Years	Sulfur collected (kg/ha/year)			Total number of locations
		Rural	Urban	Industrial	
Iowa[a]	1971–1973	16.0	16.8[b]		6
South Carolina	1973–1975	8.2	11.3	11.2	15
Wisconsin[c]	1969–1971	16.0	42.0	168.0[d]	20
Tennessee[e]	1973–1976			18.1[f]	5

[a]Source: Tabatabai, M. A., and J. M. Laflen. 1976a. Nutrient content of precipitation over Iowa. Water, Air, Soil Pollut. 6:361–373.

[b]Ames, Iowa.

[c]Source: Hoeft, R. G., D. R. Keeney, and L. M. Walsh. 1972. Nitrogen and sulfur in precipitation and sulfur dioxide in the atmosphere in Wisconsin. J. Environ. Qual. 1(2):203–208.

[d]Alma, Wis.

[e]Source: Shriner, D. S., and G. S. Henderson. 1978. Sulfur distribution and cycling in a deciduous forest watershed. J. Environ. Qual. 7(3):392–397.

[f]Oak Ridge, Tenn.

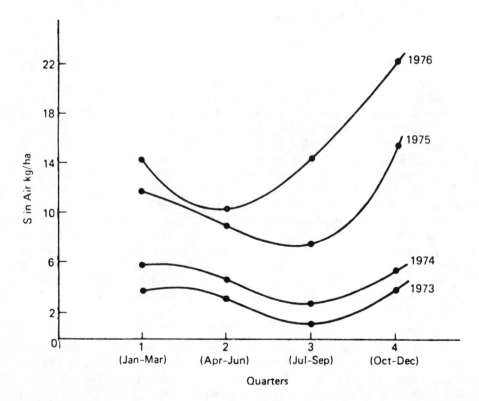

Fig. 2. Sulfur collections in air averaged by quarters for the 15 locations, 1973–1976.

Air and Precipitation S Combined

According to Alway (1937) about 22% of the S absorbed by PbO_2 candles will be absorbed on the soil surface. Air S values were adjusted, using the Alway factor, to estimate S absorbed by the soil (Table 3). The adjusted values were then combined with the appropriate precipitation S data to generate approximate values for total S added to soil during 1973–1975.

Hoeft et al. (1972) reported that, by using Alway's estimate of 22%, soil absorption from air in Wisconsin would be 3.1 and 6.3 kg S/ha/year for rural and urban sites, respectively. Adding these figures to the 1969–1971 values of 16 kg of S per hectare from rural and 42 kg of S per hectare from urban sources (Table 2) would give a total atmospheric deposition to land and surface waters in Wisconsin of 19.1 and 48.3 kg of S per hectare per year respectively.

Data for 1974 (Jones et al., 1979) indicate that 4.1 kg of S per hectare per year was absorbed by soil from air (Table 3), and this amount added to 10.6 kg of S per hectare per year from precipitation would amount to 14.7 kg of S per hectare per year.

These observations suggest that S deposition at rates of 15 to 20 kg of S per hectare per year were not uncommon in rural locations of the southeastern and midwestern United States in the early 1970s. Subsequent research in South Carolina indicates that beginning at about the time of the 1972 oil embargo and continuing until 1977 there has been a substantial increase in the amount of S as SO_2 in the air (Table 3). The South Carolina Department of Health and Environmental Control (1978) data point to the same conclusion.

No attempt is made in this paper to evaluate the absorption of atmospheric S by plants, but several workers (Hoeft, 1972; Olsen, 1957) reported that the amount is considerable, varying from 44 to 73% of the total S needed, the higher percentages being utilized by plants growing on low-S soils.

Since 1975 the South Carolina data show that about 20 kg of S per hectare is being added to the land and surface waters from both air and precipitation.

Average annual precipitation amounts for the period 1975–1977 were 139, 136, and 116 cm, respectively, and the quantities of air S absorbed by PbO_2 candles were 42.7, 58.3, and 59.0 kg of S per hectare per year, respectively, suggesting an inverse relation between average annual precipitation and S in air. Tabatabai and Laflen

Table 3. Total sulfur added to the soil from air and precipitation in South Carolina, mean of 15 locations
(In kilograms per hectare per year)

Year	Air S × % S absorbed by soil			S in precipitation		Total S added to the soil from air and precipitation
1973	12.7 × 22% =	2.8	+	8.4	=	11.2
1974	18.6 × 22% =	4.1	+	10.6	=	14.7
1975	42.7 × 22% =	9.4	+	9.8	=	19.2
1976	58.3 × 22% =	12.8	+	7.5	=	20.3
1977	59.0 × 22% =	13.0	+	6.8	=	19.8

(1976b) reported that S concentration in precipitation at three locations in Iowa was inversely related to the amount of monthly precipitation. As the precipitation increases, it washes more S from the air, resulting in greater amounts of S being added to the soil and a lesser amount of S remaining in the atmosphere.

Several workers [Tabatabi and Laflen (1976b), Shriner and Henderson (1978)] have indicated that S concentration in precipitation is correlated not only with the total amount of annual precipitation but more directly with the frequency of rains. A moderate rainfall, say 1.5 cm, removes practically all sulfur from the atmosphere. If the rain continues to a total of 6 cm, the last 4.5 cm serves only to dilute the concentration of sulfur. Thus, four intermittent rains of 1.5 cm each may bring down more sulfur than a single rain of 6 cm. A more complete report of the relationship between sulfur in rainfall and quantity and frequency of rains must await more data-gathering.

Contribution of Atmospheric S to Soil Acidity

The contribution of 20 kg of S per hectare to soil acidity can be calculated in terms of $CaCO_3$ equivalent by the following stoichiometric neutralization reaction:

$$H_2SO_4 + CaCO_3 \leftrightarrows CaSO_4 + H_2O + CO_2 \uparrow .$$

Utilizing the reaction above it can be calculated that 1 kg of S after oxidation and hydrolysis will neutralize about 3 kg of $CaCO_3$. Therefore, 20 kg of S per hectare per year fallout in South Carolina would neutralize about 60 kg of $CaCO_3$ per hectare per year. To lower sandy loam soil pH from 7.5 to 6.5 would require about 600 kg of S per hectare. Considering atmospheric deposition of S as the only soil-acidifying agent in nature, which it is not, it would require 30 years of 20 kg S per hectare per year to reach the level of 600 kg/ha required to lower the pH from 7.5 to 6.5.

When adding lime to reduce soil acidity or S to reduce soil alkalinity, quantities of S or lime are added in amounts of the order of 1000 to 3000 kg per hectare, making the 20 to 60 kg/ha of S or $CaCO_3$ a small consideration in terms of the total needs of the soil. However, it should not be discounted entirely, because on poorly buffered acid sands and loamy sands such as those found in humid sandhill regions of the world, impact of atmospheric S deposition can be significant, as will be explained later.

Soil sulfate sulfur and pH

In a study to determine if there is correlation between soil pH and soil sulfate sulfur in sandy soils, sulfate sulfur and pH were determined on 29 Norfolk loamy sands at three depths: 0 to 15, 15 to 30, and 30 to 45 cm. Soil sulfate sulfur was inversely correlated with soil pH at the 15-to-30- and 30-to-45-cm depths in 1972 and 1974 and at the 0-to-15-cm depth in 1974 (Table 4).

The supply of S in surface soils is largely in organic matter. Since cultivated Ultisols and Entisols are relatively low in organic matter, their surface S supply is usually low. Most of these soils do have accumulations of larger amounts of extractable S in subsoils. Low pH has been correlated with sulfate retention in subsoils of several soil types (Kilmer and Nearpass, 1960).

Soil Al and pH

It has been demonstrated that S additions to soil increase H ion concentrations. Hydrogen ion concentration in acid mineral soils is, for the most part, the result of

Table 4. Soil sulfate sulfur as related to soil pH of 29 Norfolk loamy sand soils at three depths

Soil depth (cm)	Sulfate sulfur (ppm)		pH		
	Range	Average	Range	Average	r
1972					
0–15	6–42	13.96	5.0–6.6	5.87	0.11
15–30	10–97	41.30	4.6–6.6	5.47	−0.58[a]
30–45	27–172	110.93	4.7–6.0	5.10	−0. 59[a]
1974					
0–15	13–32	17.48	5.4–6.6	5.98	−0.30[b]
15–30	14–44	20.90	5.3–6.7	5.91	−0.33[b]
30–45	14–216	56.13	5.2–6.7	5.66	−0.49[a]

[a]Highly significant correlation.

[b]Significant correlation.

exchangeable Al which hydrolyzes to produce H ions. Where there is a sufficient concentration of sulfate sulfur, the reaction in soil follows:

$$2Al(OH)_3 + 3H_2SO_4 \rightarrow Al_2(SO_4)_3 + 6H_2O .$$

Aluminum enters the soil solution in very high concentrations as the pH drops below 5.0 in mineral soils (Black, 1957).

Cotton leaf S as influenced by soil sulfate sulfur

Leaf and petiole samples were taken from the youngest fully mature leaves on the main stem of cotton at the time of first square (bud). At this position and stage of growth, S deficiencies do not generally occur until the S level is less than 0.18% in the leaves (Jones, 1974). Sulfur concentrations of cotton leaves and petioles sampled in this manner were significantly correlated to soil sulfate sulfur in the 15–30 and 30–45 cm depth of 29 Norfolk loamy sands. The S concentration mean of the tissue was 0.85% (Table 5).

Table 5. Soil sulfate sulfur as related to total sulfur concentration of mature cotton leaves and petioles sampled before bud formation (squaring) on 29 Norfolk loamy sands

Soil depth (cm)	Soil sulfate sulfur (ppm)		Tissue sulfur concentration (%)		
	Range	Average	Range	Average	r
0–15	13–32	17.48	0.29–1.5	0.85	0.15
15–30	14–44	20.90	0.29–1.5	0.85	0.44[a]
30–45	14–216	56.13	0.29–1.5	0.85	0.44[a]

[a]Significant correlation.

Sufficiency level of sulfate sulfur in soils is 14 ppm at the 0–30 cm depth (Bardsley and Kilmer, 1963). The range of soil sulfate sulfur concentration of Norfolk soils at the 0–15 cm depth was 13–32 ppm and at the 15–30 cm depth was 14–44 ppm (Table 5).

Based on plant tissue analyses, none of the 29 loamy sands was deficient in available S for cotton growth, and based on soil analysis, only one of them was.

Effects of S and Soil Acidity on Agroecosystems

Data on temporal changes in atmospheric sulfur indicate that there have been substantial increases of S from 6.3 kg/ha in 1953–1955 to 11.3 kg/ha in 1973–1975 in South Carolina (Table 1). Also, deposition of 16 kg/ha in 1969–1973 in Iowa and Wisconsin farming areas was common (Table 2). The question one might ask is, How much can the S concentration of rainfall be increased before atmospheric S depositions will be detrimental to soils in soybean ecosystems in the Midwest or peach ecosystems in the Southeast?

There are no pat answers to this question because the effects of S on ecosystems will depend on temporal conditions and on many other factors. One of the other factors is the proximity of the ecosystem to urban and industrial areas. In Wisconsin, for example, it has been shown that there is ten times as much S in industrial as in rural areas (Table 2). Another factor influencing the effects of S and acid rain on agroecosystems is soil texture. Poorly buffered sandy soil can tolerate much less sulfur than well-buffered fine-textured silt and clay loam soil. Still another factor influencing the effect of sulfur is the plant species. Grain sorghum can tolerate more S and resulting acidity than soybeans (Brupbacker et al., 1974).

There is not a large background of scientific data on the effects of levels of atmospheric S on agroecosystems, but there is some S soil fertility research from which inferences of effects can be made for a given set of climatic conditions. Two of these agroecosystems, for which some data are available (Jones and Jones 1975; Brupbacker et al., 1974), are peaches growing on poorly buffered loamy sand soils in South Carolina and a soybean–grain-sorghum cropping sequence on a well-buffered silt loam soil in Louisiana. These are both agroecosystems of considerable economic value in their respective regions, and they will be briefly discussed.

Peach tree short life

Early senescence of peach trees has been prevalent on poorly buffered sandy soils of the southeastern United States. Crops or trees planted following these peach orchards do not grow normally. The influence of S on soil conditions and its effect on longevity of peach trees on poorly buffered soils has been recently defined (Jones and Jones, 1975).

For many years, trees have been sprayed with 150 to 250 kg/ha of elemental sulfur to control brown rot (*Sclerotinia fructicola* Wint. Rehn). After being washed into the soil by precipitation, S is oxidized by *thiobacilli* to SO_3 and converted to H_2SO_4 by hydrolysis.

Orchards which have been sprayed with high levels of S annually offer leads as to the amount of S that the sandy soil can tolerate and the magnitude of damage that can be created by large amounts of SO_2 dissolved in precipitation and washed into the soil.

To determine the persistence of S accumulation in orchards, samples were taken from Lakeland sand on which peaches are now growing, where peaches formerly were grown, and where peaches had never been grown (Table 6).

Table 6. Effects of peach spraying program at Sandhill Experiment Station on soil S concentration and pH

Soil[a] sample depth (cm)	Never an orchard		Orchard 7 years ago		Orchard at present	
	S (ppm)	pH	S (ppm)	pH	S (ppm)	pH
0–14	4	6.7	9	6.5	22	6.2
60	5	6.7	14	5.9	50	4.7
120	5	5.4	14	5.3	48	4.6
180	7	5.2	5	5.2	11	4.7

[a]Soil type is Lakeland sand; good soil sulfur content for peaches is 14 ppm, and suggested soil pH range is 6.0–6.5.

Loamy sand on which there was never a peach orchard was very low in S to a depth of 180 cm. Soil S concentration of the growing orchard site at 60 cm deep was 10 times the soil S concentration of the site where peaches had never been grown and more than 3 times the sulfur content of the site where peaches were grown seven years before. This contrast prevailed to a depth of 120 cm. At 180 cm deep in all three sites, the S content was very low, probably because of percolation, a fluctuating water table, and absence of ionic exchange colloids in the underlying pale yellow, loose sandy parent material.

In the area where peaches had never been grown, the sulfate S content varied from 4 ppm on the surface to 5 ppm at 120 cm deep. Considering 14 ppm extractable S as a minimum S content for normal peach tree growth, it appears that an untreated sandy soil at this location would be deficient in S. On the other hand, soil S in the growing orchard varied from 22 ppm on the surface to 48 ppm at 120 cm deep, and this amount of S is much more than necessary for good peach tree growth.

Soil pH in all three sites at the 0-to-15-cm depth was in the range desirable for peaches because the topsoil had been limed. At the 60-cm depth, in the growing orchard, the H ion concentration was about 10 times as great as in soil where an orchard was growing seven years ago, and it was 100 times as great as in soil that was never planted to an orchard. The most striking drop in pH was in the growing orchard: 6.2 at the surface, 4.7 at 60 cm, and 4.6 at 120 cm deep (Table 6). This is a drastic change in the root zone, and such high soil acidity means greatly increased Al solubility and uptake by trees (Tables 7 and 8). Also, a greater solubility and uptake of the metals Zn and Mn by cotton following peaches was reported by Lee and Page (1967).

Like soil S concentration, the H ion concentration at 60 and 120 cm on both orchard sites was much greater than desirable for peach tree growth. As pH decreased, the sulfate sulfur concentration increased (Table 6) as did the Al concentration (Table 9).

Results indicate that S fungicide sprays were responsible for the high concentration of S in subsoils to a depth of 120 cm and that sulfate retention in the subsoil was responsible for high active H and Al ion concentration there.

The Al concentration of peach leaves taken in June 1973 was related to the condition of trees (Table 10). A total of 45 trees found in six different orchards were classified as healthy, weak, or dead. Tissue analyses indicated that as Al concentration increased, the health of trees declined significantly.

Table 7. Soil pH as related to total Al concentration of plant tissue samples taken from peach orchards growing on 12 Troup and Ruston loamy sand sites

Soil depth (cm)	Soil pH		Al in plant tissue (ppm)		r
	Range	Average	Range	Average	
0–15	5.1–6.5	5.8	55–490	213	−0.46[a]
30	4.9–6.3	5.6	55–490	213	−0.37[a]
60	4.8–6.4	5.3	55–490	213	−0.28

[a]Significant correlation.

Table 8. Soil Al as related to total Al concentration of plant tissue samples taken from peach orchards growing on Troup and Ruston loamy sand sites

Soil depth (cm	Al in soil (ppm)		Al in plant tissue (ppm)		r
	Range	Average	Range	Average	
0–15	1–20	6.75	55–490	213	0.69[a]
30	1–36	10.83	55–490	213	0.61[a]
60	1–32	11.10	55–490	213	0.15

[a]Highly significant correlation.

Table 9. Soil pH as related to soil Al concentration in samples taken from peach orchards growing on 12 Troup and Ruston loamy sand sites

Soil depth (cm)	Soil pH		Al in soil (ppm)		r
	Range	Average	Range	Average	
0–15	5.1–6.5	5.8	1–20	6.75	−0.30[a]
30	4.9–6.3	5.6	1–36	10.83	−0.33[a]
60	4.8–6.4	5.3	1–32	11.10	−0.49[a]

[a]Significant correlation.

It was concluded that should the S in precipitation increase to the 200-kg/ha level added by S fungicides and be deposited over a period of 10–15 years on poorly buffered sandy soils that are planted to peaches, serious economic effects would be experienced by growers.

At a rate increase of 5 kg/ha (6.3 to 11.3 kg/ha) in 20 years (Table 1) it will require 760 years to reach an S precipitation deposition level of 200 kg of S per hectare per year.

Table 10. Condition of peach trees as related to tissue Al concentration in six orchards on sandy Coastal Plain soils

Condition of trees	Number of trees	Plant tissue Al means[a] (ppm)
Healthy	18	96 a
Weak	15	233 b
Dead	12	329 c

[a]Means not followed by a common letter are different, $P < 0.05$.

Soybeans and grain sorghum

An investigation was initiated in 1971 and continued through 1974 at Alexandria, Louisiana (Brupbacker et al., 1974) to determine the influence of large applications of S on soil pH and on the yield of Funk BR-79 hybrid grain sorghum and Bragg soybeans grown on Norwood silt loam.

Prior to the application of S, the soil pH was 7.4. A preliminary laboratory investigation indicated that Norwood soil was well buffered and that relatively large applications of S would be required to adjust the soil reaction to pH values of 7.0, 6.5, 6.0, 5.5, and 5.0.

Sulfur rates varying from 0.1 to 33.3 metric tons per hectare were broadcast on the surface of the soil on May 18, 1971, and incorporated into the top 10 cm by disking.

Three months after the application of 11.8 metric tons of S per hectare, the soil reaction was reduced from pH 7.4 to 6.9. Immediately after the sorghum was harvested in 1971 the soil reaction on plots that received 33.3 metric tons of S per hectare was pH 6.3. Sulfur treatments did not result in a further decrease in pH in 1972 and 1973. In 1974 the soil reaction on plots that received the higher rates of S was pH 7.3.

Sulfur applications had no significant effect on the yield of grain sorghum in 1971. Sulfur apparently had not reacted sufficiently to influence the yield of sorghum during the first year.

In the second year, 1972, the application of S at rates exceeding 23.5 metric tons of S per hectare resulted in statistically significant reductions in yield of Bragg soybeans. This was attributed to a decreased stand noted on all plots that received the higher rate of S.

In 1973 the 33.3 metric tons per hectare rate of S resulted in a significant decrease in the yield of sorghum. During the 1973 growing season, a salt incrustation appeared on the surface of the plots that received higher rates of S.

The application of 0.11 metric ton of $CaSO_4$ per hectare in 1971 resulted in a yield increase of 484 kg/ha of grain sorghum in 1974. The other S treatments had no significant influence on the yield of grain sorghum in 1974.

It is apparent from this experiment on Norwood silt loam that S fallout in precipitation would have to reach a level of more than 23.5 metric tons per hectare before detrimental effects to soils growing soybeans would occur and to a level of 33.3 metric tons per hectare before the yield of grain sorghum would be adversely affected by soil conditions at this location. Sulfur in precipitation could never, as a practical matter, attain a high enough S concentration to deposit 23 to 33 metric tons per hectare on land.

Vegetable and Field Crop Responses to S on Sandy South Carolina Soils

Sulfur status of sandy soils, crops, and fertilizers used at 15 locations (Fig. 1) in 1976 is shown in Table 11. The first ten sandy soil types listed in the table are from farmers' fields, and the last five are from the no-S treatment of the experimental units at Clemson, Blackville, Darlington, and Columbia experiment stations.

The ten cooperating farmers were representative of farmers in the area. As may be seen by the grower-reported fertilizer use (Table 11), they recognize the need of plants for S. All ten growers reported good yields in 1976, a year of good rainfall distribution amounting to a total of about 137 cm at Columbia.

Table 11. Sulfur status of sandy soils, crops, and fertilizers in 1976 at five experiment stations and at ten farm locations where atmospheric sulfur was measured in South Carolina

Soil type	Sulfate sulfur[a] (ppm), top 30 cm	Crop[b]	Leaf sulfur[c] Concn. (%)	Sufficiency (%)	Fert. grade	Grower-reported S in fertilizer
Lakeland s	22	Z	0.14	0.15	0-9-27	Sul. of pot. mag. 20% superphos.
Faceville sl	58	C	0.61	0.18	14-7-14	Sul. of pot. mag. Sul. of ammonia
Faceville sl	34	C	0.17	0.18	0-17-34	S, B, Mg added
Faceville sl	50	C	0.29	0.18	4-12-12	Part of P from 20% superphos.
Norfolk sl	20	G	0.79	0.30	4-12-12	Ammoniated 20% superphosphate
Marlboro fsl	30	Z	0.30	0.15	5-15-30	Premium grade
Norfolk sl	14	Z	0.35	0.15	0-12-36	Premium blend
Orangeburg sl	12	C	0.33	0.18	6-12-12	Ammoniated 20% superphosphate
Orangeburg sl	15	C	0.57	0.18	5-15-30	Premium, 2% S
Norfolk ls	12	Z	0.32	0.15	N-P-K	P from 20% superphosphate
Dothan fls[d]	17	P	0.32	0.16	None	Satisfactory soil test
Cecil sl[d]	32	H	0.14	0.14	27-9-18	Sulfur-free[e]
Marlboro sl[d]	17	G	0.32	0.30	27-9-18	Sulfur-free[e]
Norfolk ls[d]	9	Z	0.18	0.15	27-9-18	Sulfur-free[e]
Lakeland ls[d]	4	P	0.22	0.16	27-9-18	Sulfur-free[e]
Wagram ls[d]	4	T	0.50	0.20	27-9-18	Sulfur-free[e]

[a]Soil sulfur sufficiency; 14 ppm according to Bardsley and Kilmer (1963).

[b]C, cotton; G, soybean; Z, corn; H, barley; P, green bean; T, tomato.

[c]Leaf S content sampled according to Jones (1974).

[d]Experiment station fields.

[e]All data on these rows are from no-S plots at Simpson (Clemson), Edisto (Blackville), Pee Dee (Darlington), and Sandhill (Columbia) Experiment Stations.

Only two of the ten farmers' fields, Orangeburg sandy loam and Norfolk loamy sand, contained less than 14 ppm of sulfate sulfur in the top 30 cm of soil. The cotton leaf S concentration in samples taken on the Orangeburg sandy loam field was 0.33%, and the corn leaf S concentration in samples from the Norfolk loamy sand field was 0.32%, both concentrations being about twice the S leaf concentration considered sufficient for cotton and corn.

Crop response 1973–1978

Long-term uniform field experiments were conducted with commercially important agronomic and horticultural field crops at four experiment stations using materials and methods described earlier. Complete data obtained in the uniform field experiment with corn at Darlington, South Carolina, on a Norfolk loamy sand are given in Table 12. Increased yield ($P < 0.10$) of corn grain and silage were obtained with 9 and 18 kg/ha of S at Darlington after the first year of cropping, but there were no indicated differences among rates or sources of S (Table 12).

Complete data for the other crops at the other locations are available, but they are not shown because at this time, after five years of cropping, there is no response to rates or sources of S by the other crops. The only treatment effects shown in this paper for the other crops and locations are the control and the 18 kg of S per hectare rate (Table 13).

The acetate-extractable soil sulfate sulfur level of the 0-to-30-cm depth was 9 ppm in the no-treatment control at Darlington. At the other experiment station locations the sulfate sulfur level of the 0-to-30-cm soil depth varied from 4 to 32 ppm. And at these locations, there was no significant crop response to rates or sources of fertilizer S during 1974–1978 because S was supplied to the crops from either the subsoil or atmosphere.

Table 13 summarizes the response to 18 kg of S per hectare of field corn at Darlington and seven other crops to fertilizer S at Clemson, Blackville, and Columbia. There has been no response during 1974–1978 to fertilizer S, even at the 10% probability level, by sweet corn, barley, cucumber, snapbean, soybean, tomato, or turnip at the latter three locations.

Research data (Jones and Jones, 1978) show response to Mg but the absence of response to S or irrigated mulched and unmulched Walter tomatoes on a Wagram loamy sand containing 7.4 ppm of sulfate sulfur in the 0–30 cm depth. During this experiment (1976–1977), about 20 kg/ha of S (Table 3) were added from air and precipitation to the soil each year. The authors reported a higher level of S in soil from the uncovered than the covered (polyethylene plastic mulch) plots. Atmospheric deposition of S accounts for the difference in S content of the covered and uncovered soil.

Crop response 1953–1959

Response to S (Jordan, 1964) was obtained with Ladino clover and carpet grass at Summerville, South Carolina, on Goldsboro fine sandy loam. The sulfate sulfur level of the 0-to-30-cm soil depth was 9.5 ppm. There was no response to S with eight cuttings of clover-grass mixtures during 1956–1959 at Clemson, South Carolina, on an alluvial Chewacla silt loam with 9.5 ppm S at the 0–30 cm depth. At Blackville, South Carolina, there was a cotton response to S the third and fifth year of a 1953–1959 continuous cotton cropping sequence on a Norfolk sandy loam that contained 10 ppm S at the 0–30 cm depth. While these experiments were being conducted, about 6.3 kg/ha of S was being deposited in precipitation (Table 1).

Table 12. Sulfur concentration of ear leaf and yield of grain as affected by rates and sources of broadcast applications of sulfur on Norfolk loamy sand

Rates and sources of sulfur broadcast each year (kg/ha)		Corn (kg/ha) produced in:				Leaf S[a] and Mg concentration (%)							
S	Mg	1974	1975	1976	1978	1974 S	Mg	1975 S	Mg	1976 S	Mg	1978 S	Mg
0	0 (control)[b]	4,039	3,804	5,349	2,878	0.20		0.21	0.21	0.18	0.14	0.18	0.16
9	0 (Gyp)[c]	3,931	4,825	6,115	2,947	0.20		0.21	0.19	0.18	0.14	0.17	0.14
18	0 (Gyp)	4,160	4,845	6,108	3,322	0.20		0.22	0.19	0.18	0.15	0.16	0.16
36	0 (Gyp)	4,736	4,549	6,014	2,871	0.20		0.19	0.17	0.25	0.16	0.19	0.16
18	16 (Eps)[c]		4,012	5,934	3,131			0.23	0.24	0.20	0.14	0.22	0.17
0	16 (Dol)[c]		5,208	6,310	3,294			0.24	0.25	0.20	0.17	0.21	0.18
LSD.10		NS	1,020	720	423								

[a]Sufficiency level at tasseling, 0.13% (Allan, 1976).

[b]Control soil at 0–30 cm contains 9 ppm sulfate sulfur.

[c]Gyp, gypsum; Eps, Epsom salt; Dol, dolomite.

Table 13. Crop response to 18 kg of S per hectare in South Carolina, 1974–1978

Location	Soil symbol	Crop	Yield unit	No. tests[a]	Yield per hectare Without sulfur	Yield per hectare With sulfur
Clemson	CIC2	Barley	Metric ton	3	4.10	3.84
Darlington	NoB	Corn grain	Metric ton	4	4.02	4.61[b]
Darlington	NoB	Corn silage	Metric ton	1	8.05	8.86[b]
Blackville	MbA	Sweet corn	Crate[c]	1	16.22	16.72
Blackville	MbA	Cucumber	$100	1	16.13	15.34
Blackville	MbA	Soybean	Metric ton	2	2.55	2.51
Columbia	LaB	Snapbean	Metric ton	3	2.70	2.51
Columbia	LaB	Turnip	Metric ton	3	23.40	26.30
Columbia	WgB	Tomato	Metric ton	3	13.21	13.44
Columbia	WgB	Tomato (mulched)	Metric ton	3	16.37	17.99

[a]Defined as one experiment at one location for one year.

[b]$P < 0.10$.

[c]60 Marketable ears.

On the basis of research in South Carolina and elsewhere in the southeastern United States during the decade from 1950 to 1960 (Jordan, 1964), crops commonly grown in South Carolina, e.g., clover (*Trifolium repens* L.), cotton (*Gossypium hirsutum* L.), corn (*Zea mays* L.), wheat (*Triticum aestivum* L.), and vegetables, were fertilized with sulfur-bearing fertilizers. It was suggested, based on research in the 1950s, that cotton fertilizers be used at a rate to supply not less than 9 kg of S per hectare. Soils having less than 14 ppm of acetate-extractable S in the top 30 cm were found to be responsive to applied S (Bardsley and Kilmer, 1963).

The crop-response research reported for the 1973–1977 period supports, in general, the critical level of 14 ppm reported earlier. However, the general overall recommendation of 9 kg of S per hectare for all cotton fertilizer was made when the average amount of sulfur coming down in precipitation was about 6 kg/ha (Table 1). It is evident from research during 1973 to 1977 that S in the air is increasing. It is also evident, by comparing S in precipitation in 1953–1955 with the fallout in 1973–1975, that S in precipitation is increasing. For these reasons the findings and general recommendations for S fertilizer should be reevaluated from time to time with carefully designed long-term experiments such as those reported here.

Twenty years ago, S in the atmosphere made a minor contribution to the S required by crops in South Carolina (Jordan, 1964). But current data indicate that atmospheric S is making a major contribution to the agronomic and horticultural crop needs for sulfur as a plant nutrient.

Sulfur Trends in Air, Precipitation, and Soil at Five Experiment Stations

From 1973 to 1978, temporal trends in the S supply of air, precipitation, and soil were determined at five Experiment Station fields (Table 11). The fields were cropped continuously for six years. The crops were fertilized with a 27-9-18 sulfur-free fertilizer made from urea, ammonium phosphate, and potassium chloride.

Air and precipitation were monitored monthly, but soil samples were taken annually in June (Fig. 3). The differences in deposition of S in precipitation over the six-year period were small but significant. For the same period the differences in air S absorbed on PbO_2 candles were highly significant, and from 1976 to 1978 they are correlated with the S in precipitation (Table 14).

The S concentration in soil at the 0–15 cm depths was quite variable from year to year, probably because concentrations represented sampling only in the month of June each year. Rainfall data and field observations indicate that sulfate sulfur in soil fluctuates with the amount and intensity of rainfall. Following wet springs, e.g. 1975 and 1978, the soil level of sulfate sulfur was lower than after the drier springs of 1974 and 1976. Other workers (Shriner and Henderson, 1978) have reported annual soil S losses due to leaching of as much as 4.3 kg/ha from forest soils. More frequent monthly sampling for several years will probably minimize variability between soil S concentrations from one year to the next.

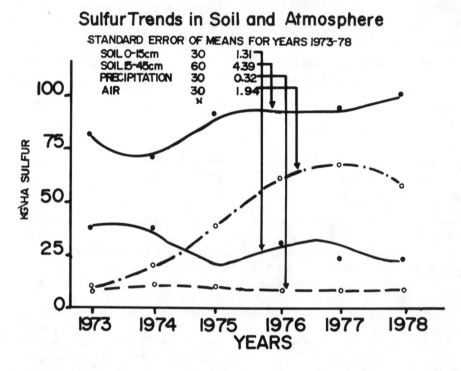

Fig. 3. Temporal trends in sulfur supply of air, precipitation, and soil at five South Carolina Experiment Stations where soils were cropped for six years with sulfur-free fertilizers made from urea ammonium phosphate and potassium chloride. Air and precipitation were monitored monthly and soil on a yearly schedule.

Field crop observations over a period of many years have indicated brief periods of S deficiency in the spring on sandy topsoils by young growing plants whose meager root system has not yet penetrated below the sand to the S-rich subsoil. When the root system has developed, plants usually grow to a vigorous high-yielding maturity with no significant loss in yield. Such ecosystems are the most difficult to evaluate in quantitative terms as regards S needs of the crop. Growers who see these

Table 14. Sulfur in precipitation as correlated with the S absorbed from the air by the PbO_2 candle.

Years	Sulfur in precipitation (kg/ha)		Sulfur in air (kg/ha)		r
	Range	Mean	Range	Mean	
1972	3.6–16.8	8.6	5.4–23.6	12.6	N.s.[a]
1974	5.6–16.2	10.6	10.0–28.0	18.6	N.s.[a]
1976	3.2–14.2	7.4	20.8–101.8	58.2	0.51[b]
1978	5.4–11.0	7.9	34.5–87.2	57.0	0.73[c]

[a]Not significant at the 95% probability level.

[b]Significant at the 95% probability level.

[c]Highly significant (99% probability level).

temporary deficiencies are concerned and are willing to take immediate corrective action to improve the appearance of the crop whether or not there is promise of an increase in crop yield.

The subsoil 15–45 cm deep (Fig. 3) seldom contains a sulfate sulfur concentration of less than 14 ppm (32 kg/ha) except on soils containing 70% or more sand. Fourteen parts per million is the concentration of sulfate sulfur, in the 0–30 cm depth, which is considered to be sufficient for economic crops (Bardsley and Kilmer, 1963).

At the 15–45 cm depth, soil S data in Fig. 3 indicate ample sulfur for plant nutrition in the subsoil of these five sandy soils.

These results following continuous cropping from 1973 to 1978 indicate small S reductions in the 0–15 cm depth of soil. Sulfur in precipitation is also slightly reduced over the same period because of reduced rainfall. On the other side of the S balance sheet, the concentration of sulfate sulfur in the 15–45 cm depth and concentration of sulfate sulfur in the air are significantly increasing.

Considering air SO_2 and available subsoil SO_4 to be usable by crops in agroecosystems described here, the possibility of plant health being influenced by too much or too little soil or air S is slim indeed. In South Carolina, one crop (corn) out of eight crops and one soil (Norfolk loamy sand) out of five indicated response to sulfur at the 10% probability level (Table 13). At none of the 15 locations (Table 4) was there indication of too much atmospheric S for healthy plant growth.

REFERENCES

Air Pollution Measurements of the South Carolina Air Quality Surveillance Network. Columbia, S.C.

Allan, A. Y. 1976. The effects of sulfur on maize yields in Western Kenya. East Afr. Agric. For. J. 41:312–322.

Alway, F. J., A. W. Marsh, and W. J. Methley. 1937. Sufficiency of atmospheric sulfur for maximum crop yields. Soil Sci. Soc. Am., Proc. 2:229–238.

Bardsley, C. E., and V. J. Kilmer. 1963. Sulfur supply of soils and crop yields in the southeastern United States. Soil Sci. Soc. Am., Proc. 27:197–199.

Black, C. A. 1957. Soil-Plant Relationships. John Wiley and Sons, Inc., New York. 1st ed.; p. 142.

Brupbacker, R. H., et al. 1974. Influence of applications of sulfur on soil reaction and yield of grain sorghum and soybeans. Louisiana State Univ. Agronomy Dept., Report of projects.

Cogbill, C. V., and G. E. Likens. 1974. Acid precipitation in the northeastern United States. Water Resour. Res. 10:1133–1137.

Hoeft, R. G., D. R. Keeney, and L. M. Walsh. 1972. Nitrogen and sulfur in precipitation and sulfur dioxide in the atmosphere in Wisconsin. J. Environ. Qual. 1(2): 203–208.

Jones, J. B., Jr. 1974. Plant analysis handbook for Georgia. Univ. of Georgia Coop. Ext. Serv. Bull. 735.

Jones, T. L., and U. S. Jones. 1975. Influence of soil pH, aluminum and sulfur on short life of peach trees growing on loamy sands in southeastern United States. Proc. Fla. Soc. Hort. Sci. 87:367–371.

Jones, U. S. 1978. Sulfur content of rainwater in South Carolina, pp. 394–402 in D. C. Adriano and I. L. Brisbin, Jr. (eds.), Environmental Chemistry and Cycling Processes. DOE Symposium Series 45 (CONF-760429).

Jones, U. S., M. G. Hamilton, and J. B. Pitner. 1979. Atmospheric sulfur as related to fertility of Ultisols and Entisols in South Carolina. Soil Sci. Soc. Am., J., in press.

Jones, U. S., and T. L. Jones. 1978. Influence of polyethylene mulch and magnesium salts on tomatoes growing on loamy sand. Soil Sci. Soc. Am., J. 42:918–922.

Jordan, H. V. 1964. Sulfur as a plant nutrient in the southern United States. USDA Tech. Bull. 1297.

Jordan, H. V., C. E. Bardsley, L. E. Ensminger, and J. A. Lutz. 1959. Sulfur content of rainwater and atmosphere in Southern states as related to crop needs. USDA Tech. Bull. 1196.

Kilmer, V. J., and D. C. Nearpass. 1960. The determination of available sulfur in soils. Soil Sci. Soc. Am., Proc. 24:337–340.

Lee, C. R., and N. R. Page. 1967. Soil factors influencing the growth of cotton following peach orchards. Agron. J. 59:237–240.

Olsen, R. A. 1957. Absorption of sulfur dioxide from the atmosphere by cotton plants. Soil Sci. 84:107–111.

Shriner, D. S., and G. S. Henderson. 1978. Sulfur distribution and cycling in a deciduous forest watershed. J. Environ. Qual. 7(3):392–397.

South Carolina Department of Health and Environmental Control. 1978.

Tabatabai, M. A., and J. M. Laflen. 1976a. Nutrient content of precipitation over Iowa. Water, Air, Soil Pollut. 6:361–373.

Tabatabai, M. A., and J. M. Laflen. 1976b. Nitrogen and sulfur content and pH of precipitation in Iowa. J. Environ. Qual. 5(1):108–112.

CHAPTER 40

IMPACT OF ATMOSPHERIC SULPHUR DEPOSITION ON FOREST ECOSYSTEMS*

Gunnar Abrahamsen†

†Norwegian Forest Research Institute, 1432 Aas-NLH.

ABSTRACT

Results from experiments in coniferous forests supplied with artificial rain of varying acidity are discussed. The acidity of the "rain" was adjusted using H_2SO_4.

The enrichment of K, Ca, and Mg in throughfall beneath spruce crowns is increased by decreased pH of the artificial rain. This has not significantly influenced the nutrient concentrations in the needles. The content of NH_4 and NO_3 in throughfall is less than that of incident rain, and it is not influenced by rain acidity.

The soil chemical properties have been studied in field plots and lysimeters. Artificial rain with pH at 3 and below has significantly increased the net leaching of K, Ca, and Mg. As a consequence, the base saturation and soil pH have decreased. There are strong indications that the present acidity of precipitation has also reduced the concentration of exchangeable metal cations in the soil. A significant net leaching of NH_4 and NO_3 has been found in the most acidified plots.

Variations in soil chemical properties have not significantly influenced the concentrations of most nutrient elements in the needles. In one experiment the foliar concentration of N was temporarily enhanced. This was most likely the reason for a slight increase in the tree growth.

INTRODUCTION

During the last decade, much attention has been paid to the ecological consequences of increased deposition of acidic air pollutants. In previous times the emission of SO_2 produced effects on vegetation which were mainly of local significance. Today, however, improved cleaning of emissions and the use of taller stacks have reduced the local effects and increased the dispersal of air pollutants into large, previously uncontaminated, rural areas.

In such areas the pH of the precipitation has been reduced from the natural value of 5.6 to annual averages between 4.0 and 5.0 (OECD, 1977). The acidity of the precipitation is often very episodic. In southernmost Norway, where the annual average pH of precipitation is 4.3, almost 10% of the total precipitation has pH values equal to, or less than, 4.0 (Dovland et al., 1976). The wet deposition of excess sulphate in Europe appears to be of the order of 5 to 25 kg of S per hectare per year.

*SNSF-contribution FA[51]/79.

The total deposition, both dry and wet, is of the order of 10 to 100 kg of S per hectare per year (OECD, 1977). In addition to the hydrogen ion content, sulphate, nitrate, and ammonium ions have similarly increased in precipitation.

The deposition of nitrogen has not been studied within the OECD project. However, estimates from Norwegian data indicate a wet deposition in the order of 2 to 20 kg of N per hectare per year (Dovland et al., 1976; Wright et al., 1978). The dry deposition of nitrogen is not known. Deposited nitrogen is composed equally of nitrate and ammonium; as with sulphur, part of the atmospheric burden originates from natural sources.

It is difficult to predict the impact of the increased deposition of sulphur and nitrogen on various forest ecosystems. In some forests, sulphur or nitrogen or both might be limiting elements for plant growth, and increased deposition would most likely result in increased productivity. In other forests, other nutrient elements might be limiting, and increased deposition of sulphur and nitrogen might not influence the productivity or even increase the deficiency of limiting elements.

The aim of this paper is to give results from studies with artificial application of sulphuric acid in coniferous forests on entisols in Norway. Results concerning movement of plant nutrients in the ecosystem are stressed. Ecological effects such as changes in plant and animal communities are not discussed, as they are most likely functions of changes in nutrient availability. Preliminary results from these experiments have been previously published by Abrahamsen et al. (1976a, 1977), Horntvedt (1979), and Tveite and Abrahamsen (1979).

DESCRIPTION OF THE FIELD EXPERIMENTS

The studies include a total of five field plot experiments and one lysimeter experiment. Three of the field plot experiments (A-1, A-2, A-3) and the lysimeter experiment are located in an area (area A) 40 km north of Oslo. The two other field plot experiments (B-1 and B-2) are situated in an area (area B) 180 km southwest of Oslo.

The normal temperature and precipitation for area A are given in Table 1. Vegetation and soil characteristics of field plot and lysimeter experiments are given in Table 2. The experiments are located on flat plains of glacifluvial deposits in area A and alluvial deposits in area B. The deposits are dominated by sand and are exceedingly deep, in area A approximately 60 m.

All experiments, including the lysimeters, are in the open and receive natural precipitation in addition to artificial rain with different concentrations of sulphuric acid. The treatments in experiment A-1 include the application of two quantities of

Table 1. The monthly normal temperature and precipitation of the experimental area measured at Gardermoen weather station.

	January	February	March	April	May	June
Temp., °C	− 6.9	− 6.3	− 2.3	3.2	9.4	13.6
Precip., mm	59	43	32	48	51	72

	July	August	September	October	November	December	Annual
	16.0	14.6	10.0	4.5	− 0.6	− 3.9	4.3
	94	105	84	86	82	76	832

Table 2. Vegetation and soil characteristics for the field experiments with artificial acid rain. CEC was determined by NH_4OAc at pH 7; the figures are from the upper 10 cm of the Bs horizon

Experiment	Tree species	Tree height in 1975 (m)	Dominating ground-cover species	Soil profile	Soil horizon	Silt (%)	Clay (%)	Loss on ignition (%)	CEC (meq per 100 g)
A-1	Pinus contorta	3	Deschampsia flexuosa	Typic Udipsamment (USDA) Cambic Arenosol (FAO) Semipodsol (Norway)	0 Bs	14	5	28 4	39 11
A-2	Picea abies	4	D. flexuosa Pleurozium schreberi	Typic Udipsamment Cambic Arenosol Iron podsol	0 Bs 1	15	6	63 2	73 3
Lysimeter in Area A			D. flexuosa V. myrtillus	Typic Udipsamment Cambic Arenosol Iron podsol	0 Bs 1	15	6	73 2.5	28 9
B-1	Pinus sylvestris	1.5	V. vitis-idaea D. flexuosa Calluna vulgaris Iron podsol	Typic Udipsamment Cambic Arenosol	0 Bs	30	1	87 5	135 6

artificial rain, viz., 25 and 50 mm per month (Table 4). In the other experiments, including the lysimeters, 50 mm of artificial rain are applied each month in the frost-free period. Usually, five waterings have been carried out a year. The artificial rain is produced by mixing sulphuric acid with groundwater. The chemical characteristics of the groundwater and natural rain are given in Table 3.

Table 3. Average ion concentration in the natural rain during the period 1973 to 1978 and in the artificial rain applied to the lysimeters and plots of experiment A-2

In milligrams per liter

	pH	Cl	Na	K	Ca	Mg	SO$_4$-S	NH$_4$-N	NO$_3$-N
Rainwater	4.36	0.64	0.35	0.17	0.29	0.06	1.01	0.32	0.40
Artificial rain	6.1	2.9	2.32	0.61	3.88	0.65	1.29	0.01	0.10

Table 4. Plot size, commencement of artificial watering, and treatments of the field experiments

The number of replications is 3 in the field plot experiments and 4 in the lysimeter experiment; all experiments are freely randomized

Experiment	Plot size	Commencement of watering	No artificial watering	Supply of water with pH of—				
				6.1–5.6	4	3	2.5	2
A-1[a]	3 × 5 m	Aug. 11, 1972	x	x	x	x		
A-2[b]	150 m^2	June 28, 1973	x	x	x	x	x	
B-1[b]	75 m^2	Aug. 14, 1974	x	x	x	x	x	x
Lysimeter[b]	Diam: 29.5 cm, height: 40 cm	July 1974	x	x	x	x		x

[a]Include plots supplied with both 25 and 50 mm of water per month.
[b]Supplied with 50 mm of water per month.

The experimental design, treatments, number of replications, plot sizes, and commencement of rain application are given in Table 4. The experiments also include lime treatment; however, results of this will not be considered in the present paper.

The lysimeters are fibre-glass cylinders filled with undisturbed soil monoliths, approximately 35 cm deep, and intact ground vegetation (Teigen et al., 1976).

The artificial rain has been applied above the canopy in experiments A-2 and B-1. In A-1 the "rain" was applied above the canopy during the first three years, but thereon from approximately 1 m above ground level. The intensity of the artificial rain is high. The 50 mm in experiments A-1 and B-1 were applied during a period of 20–30 min, whereas in experiment A-2 the application time is approximately 2½ hr. In the lysimeters the 50 mm were applied in three equal amounts during one day.

More details concerning vegetation, soil, and methods used in the experiments are given by Abrahamsen et al. (1976b) and Stuanes and Sveistrup (1979).

The problems arising when using groundwater with high concentrations of elements, such as Ca and Mg in particular, to simulate the effect of rain with low concentrations of these elements need some extra comments. Groundwater acidified to, e.g., pH 4 with high concentrations of Ca and Mg has to be less acidifying in the soil than natural rainwater with the same pH. However, to estimate the joint effect of certain concentrations of H$^+$, Ca, and Mg in the water, calculation of the lime potential of the solutions would be valuable. The lime potential (LP) is given by the function LP = pH − ½ p[Ca + Mg] (Schofield et al., 1955). Solutions with the same lime potential would be expected to have the same acidifying effect on the soil

(Bache, 1979). Using the figures in Table 3 it will be found that the lime potential of the natural rain and artificial rain of various pH values is as follows:

	pH	LP
Natural rain	4.3	1.8
Artificial rain	6.1	4.2
	4.0	2.1
	3.0	1.1
	2.5	0.6
	2.0	0.05

Compared with natural rain, the acidifying effect of artificial rain at pH 3 and below is obvious. However, it is seen that the artificial rain at pH 4 is less acidifying than the natural rain with a pH of 4.3. In evaluating the total effect of both artificial and natural rain, the bulk LP must be considered.

ANALYTICAL METHODS AND MEASUREMENTS

Throughfall studies were carried out in experiment A-2 in 1976 (Horntvedt, 1979). In connection with the application of the artificial rain, throughfall water was sampled from each plot by means of five gauges beneath the spruce crowns. "Rain" was also sampled above and between the trees. The summer of 1976 was unusually dry, and there was almost no rainfall during the irrigation periods. In the autumn of 1976, following the irrigation of the summer, natural rain was sampled on the same plots as the artificial rain sampling.

Soil samples from experiment A-2 were collected in October 1975 and 1978, more than one month after the previous watering. During this period there were more than 100 mm of natural rain in 1975 and approximately 60 mm in 1978. The soil was therefore naturally leached of free acid prior to sampling. Twenty subsamples from the O, E, Bs1 (3–5 cm), and Bs2 (18–23 cm) horizons were taken from each plot. The subsamples from the separate horizons were pooled. Soil samples have also been collected from the other experiments, but only those from experiment A-2 have been finalized.

Water samples have been collected weekly from the lysimeters. Samples used for measurements of pH, Cl, and conductivity were analysed as soon as was possible. Samples for other ions were preserved with HCl, stored under cool conditions, and analysed monthly.

The nutrient status of the trees has been examined by needle analyses in the three sampling experiments. Pooled samples of needles from five to eight trees per plot were taken in October–December every year. In this paper, only results from the analyses of the current year's needles are considered. The samples were collected from the third or fourth branch whorl at the top of the tree.

The chemical analyses of soil, needles, and water were as described by Ogner et al. (1975, 1977). According to these procedures, CEC was determined by NH_4OAc extraction at pH 7.

In the three sapling stands (A-1, A-2, and B-1), the height growth of all trees was measured at the end of each growth period. Height growth before commencement of the experimentation was also measured for use in covariance analyses (Tveite and Abrahamsen, 1979).

RESULTS

Throughfall Experiments

Experiment A-2 has been used to estimate the total, average annual deposition of chemical elements from the natural rain and irrigation water and also the adsorption or leaching of elements from the spruce crowns. Estimates of the amount of natural precipitation from 1973 to 1978 have been obtained from Gardermoen weather station, situated at the same altitude and approximately 5 km south of the experimental field. Analyses of concentrations in the rainwater on precipitation samples collected at the experimental field have been given by the Norwegian Institute for Air Research.

The plots in experiment A-2 were supposed to receive 50 mm of irrigation water at each watering. As the water is supplied through a rotating boom above the tree crowns, large amounts of water are obviously lost by drift. Thus only approximately 41 mm appear to be deposited on the plots. Approximately 61% of these 41 mm reach the ground below the trees; the remainder is lost by interception.

On this evidence, rough estimates of the average annual deposition of water and various elements in open terrain and beneath the tree crowns have been calculated (Table 5). The estimates are based on the assumptions that the annual variation in

Table 5. Annual average deposition of water and various elements in precipitation and irrigation water above and beneath spruce crowns in plots exposed to artificial rain of different acidity, Experiment A-2

Average for the period 1973 to 1978

Position and treatment	Precipitation (mm)	Average pH	Deposition (meq per square meter)					Lime potential
			K	Ca	Mg	SO_4	$NH_4 + NO_3$	
Natural precipitation	742	4.36	3	11	4	47	38	1.85
pH 6								
Above crown	913	4.46	6	44	13	61	39	2.20
Beneath crown	553	4.79	53	58	21	80	17	2.72
Net leached	−360		47	14	8	19	−22	
pH 4								
Above crown	913	4.34	6	44	13	107	39	2.08
Beneath crown	553	4.12	54	77	24	159	21	2.10
Net leached	−360		48	33	11	52	−18	
pH 3								
Above crown	913	3.38	6	44	13	255	39	1.57
Beneath crown	553	3.57	57	92	25	305	18	1.58
Net leached	−360		51	48	12	50	−21	
pH 2.5								
Above crown	913	3.40	6	44	13	685	39	1.14
Beneath crown	553	3.24	62	92	26	575	18	1.25
Net leached	−360		56	48	13	−110	−21	

throughfall chemistry is negligible and that there is little interaction between the snow and the canopy during the winter. The latter assumptions are probably incorrect (Eaton et al., 1973; Henderson et al., 1977; Fahey, 1979), but we have no observation on the importance of these factors.

Decreasing the pH of the irrigation water increased the net leaching of Ca and, to some extent, K, Mg, and SO_4. As throughfall water in the acidified plots contained more H^+, Ca, and Mg, the acidifying effect of throughfall does not appear to be greater than that in the corresponding incident rain. This is evident when looking at the lime potential above and beneath trees. The absorption of N in the canopy was not influenced by the acidity of the artificial rain.

Leaching from Soil

Annual leaching losses of some elements from the lysimeters are shown in Figs. 1 and 2. Figure 1 also gives the amount of leachate together with the amount of precipitation divided into natural and artificial. The figures for the natural precipitation only include 50 mm of the total snowfall during the winter periods. During snowmelt, the lysimeter soil was frozen, and most of the meltwater ran off the surface. However, approximately 50 mm of the meltwater seeped into the lysimeters due to the edge of the fibre-glass cylinder being approximately 50 mm above the surface of the soil of the lysimeters.

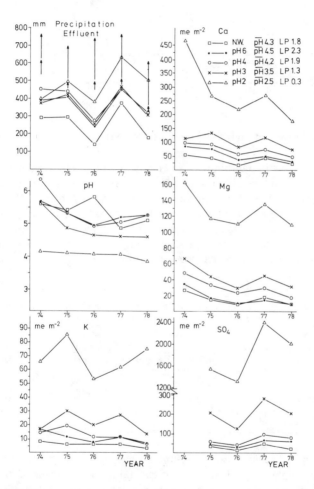

Fig. 1. Annual average losses of water, K, Ca, Mg, and SO₄ from the lysimeters. The annual average pH is calculated by weighing the H⁺ concentration with the amount of water. The annual input of water (arrows) is divided into natural (lower arrow) and artificial rain (upper arrow). Symbols for treatments give the pH of artificial rain as well as bulk pH and lime potential for the natural and artificial precipitation. N.W., nonartificially watered.

The amount of leachate varies considerably between years and reflects the amount of precipitation. Evapotranspiration does not differ significantly between the

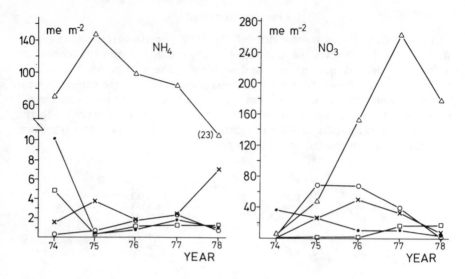

Fig. 2. Annual losses of NH₄ and NO₃ from the lysimeters. Symbols as in Fig. 1.

lysimeters receiving artificial rain with a pH of 6, 4, and 3. On average for all years the evapotranspiration was 335 mm for these lysimeters. For the unwatered and the pH 2 treated lysimeters, evapotranspiration amounts were 236 and 229 mm respectively. As a percentage of the input of water the annual average evapotranspiration was 48% for both the unwatered and watered lysimeters, except those receiving water of pH 2, where the evapotranspiration is only 32%. The small evapotranspiration in the last-mentioned lysimeters was caused by the reduction in the plant cover (Table 6).

Table 6. Percentage cover of the field layer, moss layer, and litter layer in lysimeters exposed to artificial rain of various acidity

Vegetation layer	No artificial watering	Artificial rain			
		pH 6	pH 4	pH 3	pH2
Field layer	71	78	58	59	0
Moss layer	23	28	38	36	0
Litter layer	31	13	15	30	100

The pH of the leachate from the lysimeters has been significantly influenced by the amount of sulphuric acid in the artificial rain. There are no noticeable differences between the unwatered and the two least acidified treatments. This, however, was hardly to be expected when considering the lime potential of the artificial and natural rain together. The leachate from the lysimeters watered with dilute sulphuric acid at pH 3 and especially pH 2 have been greatly acidified.

The leaching of K, Ca, Mg, and SO₄ follows, to some extent, the same pattern as the amount of water percolating through the lysimeters and also the amount of sulphuric acid in the artificial rain. The smallest leaching is found from unwatered lysimeters, the greatest leaching from the most acidified lysimeters.

The leaching of NH₄ and NO₃ does not exhibit any clear relationship with the amount of leachate or pH of the artificial rain when the pH varies between 6 and 3.

At pH 2, however, NH₄ and especially NO₃ are leached in large amounts. The maximum leaching of NO₃ comes after the maximum leaching of NH₄.

Figure 3 shows the annual average input-output budgets for the same elements as in Figs. 1 and 2 including Al. Al has been treated as Al^{3+} in the calculations. This

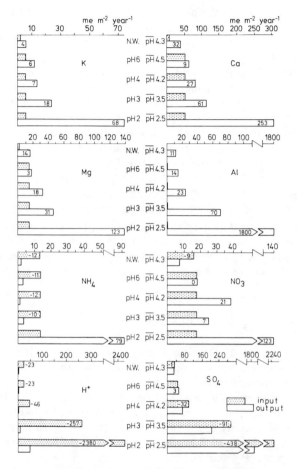

Fig. 3. Annual average input and output budgets for K, Ca, Mg, Al, NH₄, NO₃, H, and SO₄ in the lysimeters. The figures give the difference between input and output, i.e., the net leaching. Symbols for treatments give the pH of the artificial rain as well as the bulk pH for natural and artificial precipitation.

most likely overestimates the Al contribution. The reason is partly that hydroxy complexes may be important (Dalal, 1975).

The leaching of all metal cations increases with decreasing pH of the artificial rain. In general, there are small differences in net leaching between the lysimeters receiving only natural rain and those watered with "rain" with a pH of 4. "Rain" at pH 6 and an LP of 2.3 has given a smaller net leaching of Ca and Mg than the two former treatments. When the pH of the artificial rain decreases to 3 (LP 1.3), the net leaching of K, Mg, Ca, and Al increases 2–3 times compared with the less acidic treatments. "Rain" with a pH of 2 has further increased the net leaching of K, Mg, and Ca approximately 4 times compared with the leaching at pH 3.

The increased leaching of the metal cations provides considerable buffering to the effluent. This buffering appears from the H^+ budget, which shows that the output of H^+ is approximately 9% of the input in the lysimeters watered with "rain" at pH 6, decreasing to 2% in the lysimeters watered with "rain" at pH 2. In the three least acidic treatments the buffering is obviously mainly caused by ion exchange between H^+ and K, Mg, and Ca. In the pH 3 treatment the same ions appear to account for approximately 40% of the buffering, and Al accounts for approximately 30%. In the pH 2 treatment, breakdown of hydroxy complexes of Al is obviously the main cause for the buffering.

In the nonwatered lysimeters and in those supplied with water at pH 6 the output of SO_4 equals the input. When increasing the input of SO_4, the adsorption increased, producing a large net adsorption in the lysimeters supplied with water at pH 3 and 2. In spite of the adsorption, SO_4 appears to be an important anion in the leachate from this soil. In the nonwatered lysimeters, approximately 30 meq of SO_4 are leached annually. This amounts to approximately 50% of the total leaching of K, Ca, and Mg. At pH 3, approximately 200 meq of SO_4 are leached annually whereas the total leaching of K, Ca, Mg, and Al amounts to 260 meq.

The leaching of NH_4 and NO_3 appears not to be influenced by "rain" acidity when varying from pH 6 to 3. Within this field, 10 to 12 meq m^{-2} (1.4 to 1.7 kg ha^{-1}) of NH_4-N is net absorbed by the soil annually. NO_3, however, appears to be leached in the same amount as that which is added or even higher. In the lysimeters supplied with water at pH 2 a considerable net leaching of both species has occurred. On average, for all years the net annual leaching of N amounts to 202 meq m^{-2}, or 28 kg ha^{-1}.

Soil Chemical Properties

Chemical properties of the soil in experiment A-2 are given in Fig. 4. The histograms for soil pH, base saturation, and partly SO_4 show results from both 1975 and 1978. Otherwise only results for 1978 are given.

The amount of water-soluble plus exchangeable K, Ca, and Mg is greater in the topsoil (O and E layers) than in the B layer. The content of K only appears to be influenced by "rain" acidity in the raw-humus layer, and even here the influence is relatively small. In the lower B horizon there appears to be increased K content in most acidified plots ($P < 0.058$).

Mg and Ca have been influenced in the same direction by increasing the acidity of the percolating solution. Application of groundwater with a pH of 6 and 4 has increased the content of exchangeable Mg and Ca in the O and E layers compared with that of the nonwatered plots. The increase is equivalent to 2.9 g m^{-2} of Ca for the O and E layers jointly. The total annual input of Ca and Mg is equivalent to 1.14 g m^{-2} of Ca per year. The decrease in Ca and Mg concentration in the plots supplied with pH 3 "rain," compared with those receiving pH 4 "rain," is equivalent to approximately 4 g m^{-2} of Ca. Further down into the soil there are no statistically significant effects of the treatments on the Ca and Mg content.

The decreased concentration of exchangeable K, Ca, and Mg is also displayed by the variation in the base saturation. As for Ca and Mg, the base saturation is influenced by "rain" acidity in the O and E layers but not in the B layer. The decrease in base saturation from the sampling in 1975 to 1978 should also be noticed. The changes in base saturation are reflected by a corresponding change in the pH of the soil.

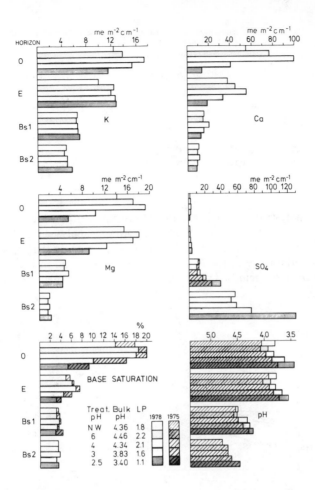

Fig. 4. Some chemical properties of different soil layers from plots exposed to natural and artificial rain of various acidity N.W., nonartificially watered. The amounts of water-soluble plus exchangeable K, Ca, Mg, and SO₄ are given per square meter per centimeter of soil depth as average for each horizon.

Figure 4 also gives the amount of water-soluble and extractable SO_4 in the soil. Very small quantities of SO_4 were found in the O and E layers. However, the amounts of SO_4 increase considerably with soil depth; thus in the B horizon large amounts of SO_4 are adsorbed. The influence of SO_4 input is obvious.

Nutrient Content of Conifer Needles

The content of K, Ca, Mg, and Al in the spruce needles of 1978 for experiment A-2 is given in Table 7. The variations between treatments are not significant for K, Ca, and Mg. It should be noticed, however, that all three elements are less concentrated in the needles from the plots supplied with artificial rain at pH 2.5. The Al concentration in the needles from the same plots (pH 2.5) is significantly enhanced when compared with the other treatments.

Table 7. Concentrations of K, Ca, Mg, and Al in the current year's spruce needles
in the plots of experiment A-1

Measurements from 1978

	Not artificially watered	Supply of water with pH of—				Prob. level of F
		6	4	3	2.5	
Bulk pH	4.4	4.5	4.3	3.8	3.4	
K	0.48%	0.53%	0.55%	0.53%	0.50%	0.11
Ca	0.43%	0.42%	0.44%	0.44%	0.38%	0.63
Mg	0.11%	0.11%	0.11%	0.11%	0.10%	0.12
Al	0.09%	0.10%	0.07%	0.09%	0.11%	0.001

So far, the analyses of K, Ca, Mg, and Al have not been concluded for the needles of the other field plot experiments. We have, however, analysed the N and S content of the needles from experiments A-1, A-2, and B-1 (Fig. 5). In the Scots pine experiment (B-1) a certain rise in the N concentration was found in the current year's needles from the most acidified plots, the autumn the experiment began. This effect

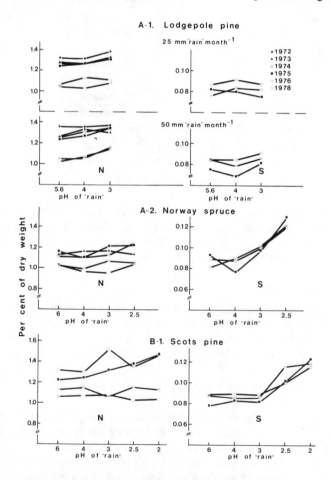

Fig. 5. N and S concentrations (percent) in the current year's needles from field experiments with application of artificial rain of various acidity.

was accentuated the following year (1975) when a definite increase in the N concentration was found with increased "rain" acidity. In the years 1976 and 1978, however, no such effects were found. In the Norway spruce experiment (A-2) a rise in the N level of the needles in the plots exposed to pH 2.5 "rain" was found in the autumn of 1974. Later on, similar effects were not found. Nor in the lodgepole pine experiment (A-1) have significant effects on the N concentration in the needles been observed.

The S concentration of the current year's needles was analysed in 1975, 1976, and 1978. In experiments A-2 and B-1 the trees have responded to increased S supply by increasing the foliar concentration. In experiment A-1, there is no significant effect of the increased S deposition.

Effects on Plants

Our main concern, with regard to effects on plants, has been on tree growth. However, the experiments have also enabled us to look for effects on plant tissue and ground-cover vegetation.

Results from the tree growth measurements have revealed small effects of the treatments (Fig. 6). In experiment A-2, no effect on height growth or girth increment

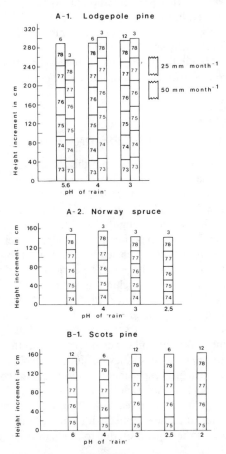

Fig. 6. Height increment in the different field experiments with application of artificial rain of varying acidity. Numbers within columns: growth year. Numbers at top of the column: number of replications.

has been observed. The lodgepole pine experiment (A-1) is more difficult to interpret. The plots receiving 50 mm of pH 4 and pH 3 "rain" have apparently improved in growth compared with those receiving pH 5.6 "rain." However, the differences between the plots supplied with 25 mm of pH 5.6 "rain" and those receiving 50 mm of pH 4 and pH 3 "rain" are not statistically significant. We therefore at present ascribe the low growth rate of the plots supplied with 50 mm "rain" at pH 5.6 to random variation. The height growth of the Scots pine in experiment B-1 has been significantly increased by increased application of sulphuric acid. The increased height growth was first observed in 1976, two years after commencing the watering. Also, in 1977 the most acidified plots had the highest growth rate. In 1978, however, no differences between the treatments could be found.

Detailed studies of effects of dilute H_2SO_4 on plant tissue have not been carried out. Sporadic examinations of leaves and needles have, however, revealed necrotic spots on pine needles and leaves of birch (*Betula pubescens*), rose-bay willow herb (*Chamaenerion augustifolium*), chickweed wintergreen (*Trientalis europaea*), and may lily (*Maianthemum bifolium*) in plots exposed to water at pH 2.5 and 2. In general, plant species of the herbacous layer appear to tolerate very acid water, and significant changes in the plant cover and species diversity have only been observed on plots exposed to water with a pH of 2 and 2.5. Mosses, however, are obviously more sensitive to dilute sulphuric acid. Water with a pH of 3 has reduced the moss layer. In plots exposed to water at pH 2.5 and 2 the mosses are almost completely extinguished. The mosses in the lysimeters appear to have a greater tolerance of the pH 3 water than those growing in the field plots (Table 6). The dominating moss species in these lysimeters are *Polytrichum commune* and *Dicranum* spp.

DISCUSSION

The increasing leaching of substances from the foliage with increasing acidity of rain is a phenomenon that has been discussed with great interest during the latter years (e.g., Eaton et al., 1973; Bjor et al., 1974; Wood and Bormann, 1975; Abrahamsen et al., 1976; Tukey, 1979; Horntvedt, 1979). The present study enables us to compare the enrichment in throughfall with the leaching from the soil. This comparison is valid, as the soil of experiment A-2 is almost identical to the lysimeter soil. Table 5 and Fig. 3 show that the enrichment of K in throughfall appears to be greater than the net leaching from the soil. The enrichment of Ca and Mg is of the same order of magnitude as the leaching from the soil. Part of the gain in throughfall derives from dry deposition (Mayer and Ulrich, 1974; Miller, 1976). However, when also considering the amount of nutrients accumulated in the tissue of growing plants and that returned to the soil by litter fall, it is obvious that the total uptake of nutrients from the soil is much greater than the leaching losses from the soil. This is in agreement with studies on nutrient cycling (e.g., Likens et al., 1977). It also shows that leaching experiments without a plant cover are likely to give estimates of leaching losses from soil that are much too great. On this evidence it appears likely that the large increase in leaching from lysimeters supplied with pH 2 water compared with that of those supplied with water at pH 3 is, to a large degree, caused by the extinction of the vegetation due to the high concentration of sulphuric acid.

The gain in chemical elements in throughfall which is caused by leaching of the plant tissue is known to be affected by plant species and soil fertility (Eaton et al., 1973; Päivänen, 1974; Miller et al., 1976; Abrahamsen et al., 1976a). Conifers are often believed to acidify the soil partly because throughfall appears to be more acidic than incident rain (Nihlgard, 1970). However, as the increasing acidity of throughfall appears to be accompanied by increased concentrations of Ca and Mg, the throughfall might not acidify the soil more than incident rain. This result appears when comparing the lime potentials of throughfall and incident rain (Table 5).

The soil chemical studies have clearly demonstrated increased leaching of Ca and Mg, followed by increased acidity and decreased base saturation of the soil, when applying water at pH 3 and below. The quantity of H_2SO_4 added in these treatments corresponds to 20–200 years wet deposition regarding the present average concentrations in precipitation. From equilibrium considerations it appears obvious that small amounts of "rain" at high concentrations will give greater effects than large amount of "rain" at low concentrations. Also, the injury of the vegetation implies that the short-term effects of high concentrations of H_2SO_4 will be greater than long-term effects of low concentrations. However, there are effects of less acidic treatments that indicate that the present acidity of precipitation has increased the leaching of Ca and Mg especially and reduced soil pH and base saturation. When ranging the net leaching according to the lime potential of the "rain" it is seen that the net leaching increases with decreasing lime potential. The Ca and Mg content of the soil is highest for the plots supplied with water at pH 4 and an LP of 2.1. The content is slightly less for the plots supplied with water at pH 6 and LP of 2.2. The base saturation for these two treatments is, however, not different. In the plots which have only received natural rain with an average pH of 4.3 and an LP of 1.8, the base saturation is somewhat less than for the two former treatments. Therefore, the conclusion is that the base saturation appears to be positively correlated with the lime potential of the "rain." It should be remembered that the effects of the acid treatments should be compared with the effects of unpolluted rain. If it is assumed that unpolluted rain with a pH of 5.7 contains 0.009 millimols per liter of Ca plus Mg, the lime potential would be 3.2.

It is not inconceivable that a slight reduction in the exchangeable amount of Ca and Mg in the soil influences the tree growth or other properties of the vegetation. The analyses of the needles have shown that substantial reductions in the content of Ca and Mg in the soil (pH 3 and pH 2.5 treatments) have not significantly influenced the concentrations in the needles. This also means that the increased foliar leaching of K, Ca, and Mg observed with decreasing pH of the "rain" has not significantly reduced the concentrations in the needles (Abrahamsen and Dollard, 1979).

Many experiments have demonstrated that the increase in leaching of cations with increased input of H_2SO_4 might be reduced because of high SO_4 adsorption in the soil (Johnson and Cole, 1977; Tamm, 1977; Johnson et al., 1979; Abrahamsen and Dollard, 1979). The present paper shows that considerable sulphate adsorption can be demonstrated when increasing the concentration of SO_4. However, in the non-artificially watered lysimeters and in those supplied water with a pH of 6 where the annual input of SO_4 equals 35–55 meq m^{-2}, SO_4 output equals the input. Similar results have been reported by Tamm (1977). This indicates that high sulphate adsorption observed experimentally might be a concentration effect; therefore, after a long exposure to elevated input of SO_4, the output will gradually approach the input. Therefore, the actual leaching of metal cations may not be reduced in the long term in spite of considerable experimentally observed SO_4 adsorption. In this connection, it is interesting that watershed studies indicate small net adsorption in areas where soil formation has taken place after the last glaciation (Likens et al., 1977; Gjessing et al., 1976), whereas studies from older soils with a higher sesquioxide content show high SO_4 adsorption (Shriner and Henderson, 1978; Johnson et al.. 1979).

Fertilizer experiments in boreal forests have shown that N is the main element limiting the growth of trees. The NH_4 and NO_3 deposited with precipitation would therefore most likely be used by plant and soil microorganisms and retained in the system. As shown in Table 5, there is a net adsorption of these ions from the throughfall. In the lysimeters supplied with water at pH 3 and above, approximately 0.15 g of ammonium nitrogen per square meter is retained annually. NO_3 is retained in the nonartificially watered lysimeters, whereas the supply of water with a pH of 3 and above has increased the leaching to amounts equal to the input. Studies of watersheds in the boreal zone also often show a net adsorption of mineral nitrogen (Likens et al., 1977; Gjessing et al., 1976). In lysimeters supplied with water at pH 2 a net leaching of NH_4 and NO_3 amounting to 2.8 g of N per square meter per year has taken place. Since large amounts of N are taken up by vegetation (Likens et al., 1977), the extinction of the plants in these lysimeters is probably the cause of the high leaching. It is, however, also possible that the mineralization of N in the soil is influenced by the acidification. It is surprising that so much NO_3 is produced in the most acidified soil. The pH of the soil probably varies from much below 4 in the top to approximately 4.5 in the bottom of the lysimeters. According to Alexander (1979) the nitrification rate is often undetectable much below pH 4.5.

Increased availability of mineral N to the trees appears to have influenced the N concentration in the needles of the Scots pine experiment (B-1). However, this effect was only observed in 1975. In 1976 and 1977 there was a slight but statistically significant increase in the height growth. When considering the results of the needle analysis it appears that the increased supply of N is the only explanation for the increased growth. These results clearly demonstrate that nutrient elements such as K, Ca, Mg, and S are not limiting factors for tree growth in our experiments. This is in agreement with general experience from studies on fertilizers in Scandinavia (e.g., Tamm, 1965; Brantseg et al., 1970).

In our experiments, which have been restricted to studying the effect of sulphuric acid, no decline in the tree growth has been observed. Acid precipitation, however, is enriched in NH_4 and NO_3. As N is the main element limiting forest growth in Scandinavia, the most likely effect of acid precipitation in this area and other similar areas of the boreal zone is increased tree growth. A similar effect would be expected in areas suffering from sulphur deficiency. In other areas of the world where Mg, and possibly Ca and K, are limiting factors, acid precipitation is likely to reduce the productivity of forests.

REFERENCES

Abrahamsen, G., K. Bjor, R. Horntvedt, and B. Tveite. 1976a. Effects of acid precipitation on coniferous forest. In Braekke, F. H. (ed.), Impact of acid precipitation on forest and freshwater ecosystems in Norway. SNSF-project, Norway, FR 6/76: 36–63.

Abrahamsen, G., K. Bjor, and O. Teigen. 1976b. Field experiments with simulated acid rain in forest ecosystems. I. Soil and vegetation characteristics, experimental design and equipment. SNSF-project, Norway, FR 4/76.

Abrahamsen, G., R. Horntvedt, and B. Tveite. 1977. Impacts of acid precipitation on coniferous forest ecosystems. Water, Air, Soil Pollut. 8:57–73.

Abrahamsen, G., and G. J. Dollard. 1979. Effects of acid precipitation on forest vegetation and soil. Proc., Workshop on Ecological effects of acid precipitation, Gatehouse-of-Fleet, Galloway, September 4–7, 1978. EPRI SOA77-403.

Alexander, M. 1980. Effects of acidity on microorganisms and microbial processes in soil. Proc. NATO Advanced Research Institute, Effects of Acid Precipitation on Terrestrial Ecosystems, Toronto, May 22–26, 1978.

Bache, B. W. 1980. The acidification of soils. Proc., NATO Advanced Research Institute, Effects of Acid Precipitation on Terrestrial Ecosystems, Toronto, May 22–26, 1978.

Bjor, K., R. Horntvedt, and E. Joranger. 1974. Nedbørens fordeling og kjemisk innhold i et skogbestand på Sørlandet (Juli–desember 1972). [Distribution and chemical enrichment of precipitation in a southern Norway forest stand (July–December 1972).] SNSF-project, Norway, FR 1/74.

Brantseg, A., A. Brekka, and H. Braastad. 1970. Gjødslingsforsøk i gran- og furuskog. [Fertilizer experiments in stands of *Picea abies* and *Pinus silvestris*.] Medd. Nor. Inst. Skogforsk. 27:537–607.

Dalal, R. C. 1975. Hydrolysis products of solution and exchangeable aluminium in acidic soils. Soil Sci. 119:127–131.

Dovland, H., E. Joranger, and A. Semb. 1976. Deposition of air pollutants in Norway. In Braekke, F. H. (ed.), Impact of acid precipitation on forest and freshwater ecosystems in Norway. SNSF-project, Norway, FR 6/76:14–35.

Eaton, J. S., G. E. Likens, and F. H. Bormann. 1973. Throughfall and stem-flow chemistry in a Northern hardwood forest. J. Ecol. 61:495–508.

Fahey, T. J. 1979. Changes in nutrient content of snow water during outflow from Rocky Mountain coniferous forest. Oikos 32:422–428.

Gjessing, E. T., A. Henriksen, M. Johannessen, and R. F. Wright. 1976. Effects of acid precipitation on freshwater chemistry. In Braekke, F. H. (ed.), Impact of acid precipitation on forest and freshwater ecosystems in Norway. SNSF-project, Norway, FR 6/76:64–85.

Henderson, G. S., W. F. Harris, D. E. Todd, Jr., and T. Grizzard. 1977. Quantitiy and chemistry of throughfall as influenced by forest-type and season. J. Ecol. 65: 365–374.

Horntvedt, R. 1979. Leaching of chemical substances from tree crowns by artificial acid rain. Proc., IUFRO Tagung 2.09 "Luftverunreinigung—Rauchschäden," Ljubljana, Yugoslavia, September 18–23, 1978. In press.

Johnson, D. W., and D. W. Cole. 1977. Sulfate mobility in an outwash soil in western Washington. Water, Air, Soil Pollut. 7:4879–495.

Johnson, D. W., D. W. Cole, and S. P. Gessel. 1979. Acid precipitation and sulfate adsorption properties in a tropical and a temperate forest soil. Biotropica 11:38–42.

Likens, G. E., F. H. Bormann, R. S. Pierce, J. S. Eaton and N. M. Johnson. 1977. Biogeochemistry of a forested ecosystem. Springer-Verlag, New York, Heidelberg, Berlin.

Mayer, R., and B. Ulrich. 1974. Conclusions on the filtering action of forests from ecosystem analysis. Ecol. Plant. 9:157–168.

Miller, H. G., J. M. Cooper, and J. D. Miller. 1976. Effect of nitrogen supply on nutrients in litter fall and crown leaching in a stand of Corsican pine. J. Appl. Ecol. 13:233–248.

Nihlgård, B. 1970. Precipitation, its chemical composition and effect on soil water in a beech and a spruce forest in south Sweden. Oikos 21:208–217.

OECD. 1977. The OECD programme on long range transport of air pollutants. Measurements and findings. Organisation for Economic Co-operation and Development, Paris.

Ogner, G., A. Haugen, M. Opem, G. Sjøtveit, and B. Sørlie. 1975. Kjemisk analyse-program ved Norsk institutt for skogforskning. [The chemical analysis program at the Norwegian Forest Research Institute.] Medd. Nor. Inst. Skogforsk. 32: 207–232.

Ogner, G., A. Haugen, M. Opem, G. Sjøtveit, and B. Sørlie. 1977. Kjemisk analyse-program ved Norsk institutt for skogforskning. Supplement I. [The chemical analysis program at The Norwegian Forest Research Institute. Supplement I.] Medd. Nor. Inst. Skogforsk. 33:85–101.

Päivänen, J. 1974. Nutrient removal from Scots pine canopy on drained peatland by rain. Acta For. Fenn. 139:1–19.

Schofield, R. K., and A. W. Taylor. 1955. The measurement of soil pH. Soil Sci. Soc. Am., Proc. 19:164–167.

Shriner, D. S., and G. S. Henderson. 1978. Sulfur distribution and cycling in a deciduous forest watershed. J. Environ. Qual. 7:392–397.

Stuanes, A., and T. E. Sveistrup. 1979. Field experiments with simulated acid rain in forest ecosystems. 2. Description and classification of the soils used in field, lysimeter and laboratory experiments. SNSF-project, Norway, FR 15/79.

Tamm, C. O. 1965. Some experiences from forest fertilization trials in Sweden. Silva Fenn. 117:1–24.

Tamm, C. O. 1977. Skogsmarkens försurning, orsaker och motåtgärder. Sver. Skogsvardsforb. Tidskr. 75:189–200.

Teigen, O., G. Abrahamsen, and O. Haugbotn. 1976. Eksperimentelle forsur-ingsforsøk i skog. 2. Lysimeterundersøkelser. [Acidification experiments in conifer forest. 2. Lysimeter investigations.] SNSF-project, Norway, IR 26/76.

Tukey, H. B., Jr. 1980. Some effects of rain and mist on plants, with implications for acid precipitation. Proc., NATO Advanced Research Institute, Effects of Acid Precipitation on Terrestrial Ecosystems, Toronto, May 22–26, 1978.

Tveite, B., and G. Abrahamsen. 1980. Effects of artificial acid rain on the growth and nutrient status of trees. Proc., NATO Advanced Research Institute, Effects of Acid Precipitation on Terrestrial Ecosystems, Toronto, May 22–26, 1978.

Wright, R. F., H. Dovland, C. Lysholm, and B. Wingård. 1978. Inputs and outputs of water and major ions at 9 catchments in southern Norway, July 1974–June 1975. SNSF-project, Norway, TN 39/78.

Wood, T., and F. H. Bormann. 1975. Increases in foliar leaching caused by acidifi-cation of an artificial mist. Ambio 4:169–171.

DISCUSSION

Dale Johnson, Oak Ridge National Laboratory: In the most acid treatment, where you had increased nitrate leaching, have you done any incubations to find out whether this is due to enhanced populations of nitrifying bacteria? It obviously isn't due to reduced plant uptake.

G. Abrahamsen: No.

Ron Hall, Cornell Univ.: I would like to know about the acid that you add with the increase in calcium and magnesium. Did you try to do experiments to sort out what happened to just the leaves of the plants when you added your simulated rain without this additional amount of calcium and magnesium? Was there a visual dif-ference that you could sort out? In other words, was the physiological response of the plant different with the addition of the increased amount of calcium and magnesium relative to what would actually be in the rain?

G. Abrahamsen: Well, I think it's quite a difficult question. We have studies, which I have not presented here, in which the observed effects are different from what is normally found; namely we get very high calcium and magnesium leaching from the trees, but this has not influenced needle concentration. It seems to us that increased leaching may be due simply to increased uptake. We don't really know what's happening. I think that the main problem in this connection is not the direct effect on the trees, at least at the concentrations we have in the rainfall in Norway, but rather the indirect effects mediated by the soil, and by using the lime potential I think you could answer your question.

Danny Rambo, EPA Labs, Corvallis: I'm wondering if you have looked at the mycorrhizal populations in the soil and if that might have affected the fact that the needles seemed to maintain good nutrition.

G. Abrahamsen: I think the soil mycorrhiza is well developed. There are many studies showing that mycorrhizal species are very tolerant to acidification and that acidification caused by acid rain is very small compared to the acidification produced by plant roots.

CHAPTER 41

IMPACT OF ATMOSPHERIC SULFUR DEPOSITION ON GRASSLAND ECOSYSTEMS

W. K. Lauenroth*
J. E. Heasley*

ABSTRACT

We examine impacts of sulfur deposition on grassland ecosystems, focusing upon carbon flow and nutrient cycling. Results from a five-year field experiment with controlled sulfur dioxide (SO_2) concentrations are presented and discussed in terms of their system-level implications. The most important system-level results of this experiment are decreased decomposition rates and increased sulfur concentrations in all system components as a result of exposure to low SO_2 concentrations. A projection of these results to a hypothetical equilibrium state under conditions of substantial sulfur deposition from the atmosphere is presented.

INTRODUCTION

Discussions of coal development in the western United States must address impacts of that development upon native grasslands. Approximately 37% of the mineable coal reserves in the United States are located beneath the grasslands of the northern Great Plains. The low sulfur content and access by strip mining of these deposits assure that this region will be a focal point for coal development and associated environmental impacts for the next several decades. Data compiled by Durran et al. (1979) from permit applications projected an increase in electric generating capacity for the northern Great Plains from approximately 3,000 MW in 1976 to 17,000 MW in 1986. Associated with this, they predicted a 350% increase in sulfur oxide emissions.

In order to make intelligent decisions about the environmental impacts of this projected development, one must understand the potential impacts of atmospheric sulfur compounds upon the living systems within this region. Because native grasslands comprise approximately 70% of the land area of the northern Great Plains and are a substantial component of the agriculturally based economy of the region, they are an extremely valuable resource. The perennial nature of these

*Natural Resource Ecology Laboratory, Colorado State University, Fort Collins, Colorado 80523.

grasslands makes them more vulnerable to cumulative impacts from low-concentration sulfur deposition than annual agricultural crops, such as wheat, or short-lived perennial crops, such as alfalfa.

Our comments regarding the impacts of atmospheric sulfur deposition on grassland ecosystems will focus upon the way it affects the distribution and movement of carbon and mineral nutrients through the system. First, we will present a view of the distribution of carbon and mineral nutrients under current background concentrations of atmospheric sulfur. Then we will discuss responses we have observed during a five-year field experiment with controlled SO_2 concentrations. We will summarize by presenting an interpretation of long-term consequences of increased atmospheric inputs of sulfur to a grassland ecosystem.

DESCRIPTION OF FIELD EXPERIMENT

The field experiment was conducted in southeastern Montana in cooperation with the Environmental Protection Agency. Four 0.5-ha plots were exposed to controlled levels of SO_2 during the growing season beginning in 1975 and continuing for five years (Heitschmidt et al., 1978). Treatments were begun on four additional 0.5-ha plots in 1976 and continued for four years. The SO_2 was delivered through a network of pipes supported approximately 75 cm above the soil surface (Lee and Lewis, 1978). Flow rates through the system were constant, and canopy-level concentrations varied with meteorological conditions, principally wind speed and temperature. Thirty-day median concentrations over the plots were less than 26 $\mu g \cdot m^{-3}$ (less than 0.01 ppm) for the control, 52 $\mu g \cdot m^{-3}$ (0.02 ppm) for the low-concentration treatment, 105 $\mu g \cdot m^{-3}$ (0.04 ppm) for the medium-concentration treatment, and 183 $\mu g \cdot m^{-3}$ (0.07 ppm) for the high-concentration treatment (Lee et al., 1979). The control plot received ambient air through the gas delivery system, and the greater-than-background concentrations recorded were the result of occasional drifting of SO_2 from the treatment plots (Preston and Gullett, 1979). Frequencies of SO_2 concentrations were distributed log-normally, as predicted, with the highest concentrations occurring between 8:00 PM and 7:00 AM. The highest concentrations measured over an 8-min averaging time for one of the sites during 1976 and 1977 were 0.7 mg $\cdot$ m^{-3} (0.28 ppm) for the control, 1.7 mg $\cdot$ m^{-3} (0.66 ppm) for the low-, 2.6 mg $\cdot$ m^{-3} (1.0 ppm) for the medium-, and 3.8 mg $\cdot$ m^{-3} (1.4 ppm) for the high-concentration treatment.

SUBSYSTEM RESPONSES

Carbon in grassland ecosystems is concentrated in two compartments, primary producers and soil organic matter (Fig. 1). An average upland site in the northern Great Plains contains approximately 5000 g $\cdot$ m^{-2} of organic carbon. Primary producers account for 23% of the organic carbon; soil organic matter, 75%; decomposers, 2%; and consumers, less than 1%. Annual net primary production is approximately 250 g of carbon per square meter.

Mineral nutrients (viz., nitrogen, sulfur, phosphorus) that are actively cycling or are tied up in organic compounds are distributed much the same as carbon. The largest portions are located in soil organic matter and primary producers. The soil nutrient pool represents the labile fraction of the mineral forms.

The potential impacts of atmospheric sulfur deposition on an ecosystem can be expected to affect carbon flow by altering the standing states of carbon, the rates of transfer between states, or both (Fig. 1). After five years of exposure of a grassland to median concentrations of SO_2 up to 183 $\mu g \cdot m^{-3}$, the distribution of carbon among

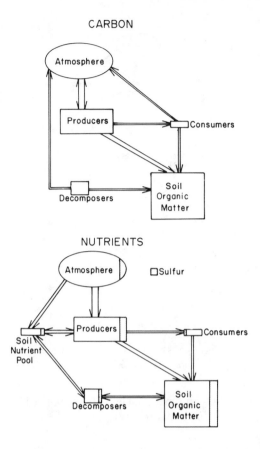

Fig. 1. Simplified diagram of carbon flow and nutrient cycling in grasslands.

compartments is only slightly different from the nominal condition. The short-term major impacts largely impinge upon transfer rates.

Field measurements of photosynthesis on the dominant species, Western wheatgrass (*Agropyron smithii* Rydb.), using ^{14}C, did not result in a clear indication of an impact of SO_2 exposure. Leaf-to-leaf and tiller-to-tiller variabilities in $^{14}CO_2$ uptake were so large that no conclusive statements can be made about these data. We did find indirect evidence of an alteration in the transfer of carbon from the atmosphere to primary producers in terms of alterations in leaf area and chlorophyll content. Figure 2 illustrates the impact of SO_2 on the live and dead area of a single age class of Western wheatgrass leaves. Compared with the control, exposure to 52 $\mu g \cdot m^{-3}$ of SO_2 prolonged the duration of maximum live leaf area and increased the functional life by several weeks. This apparent stimulatory effect of low-concentration SO_2 exposure on plant processes is common throughout many of our experimental results. In contrast to this, exposure to both 105 and 183 $\mu g \cdot m^{-3}$ of SO_2 resulted in the peak live leaf area occurring two weeks earlier than the control and a three-week decrease in the functional life of the leaf.

Chlorophyll concentration data for Western wheatgrass support these results (Fig. 3a). Average chlorophyll concentrations were increased by approximately 10% on the 52-$\mu g \cdot m^{-3}$ SO_2 treatment, unchanged at 105 $\mu g \cdot m^{-3}$ of SO_2, and decreased by 20% at 183 $\mu g \cdot m^{-3}$ of SO_2. Chlorophyll data for a variety of species variously agree

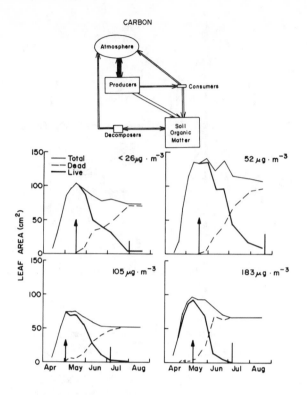

Fig. 2. Impact of SO₂ on functional leaf life. Peak live leaf area is identified by an arrow and the end of functional life by a bar.

and disagree with the results for Western wheatgrass. Figure 3*b* includes data for species in which chlorophyll concentrations were increased by SO₂ treatment. Figure 3*a* includes species for which chlorophyll content was decreased. The three species with observed increases in chlorophyll are rather minor components of the producer community, although silverleaf scurfpea (*Psoralea tenuiflora* Pursh.) is a legume and therefore a potential nitrogen-fixing species. Of the three species for which a decrease in chlorophyll concentration was observed, two of them, Western wheatgrass and prairie junegrass [*Koeleria cristata* (L.) Pers.], comprise approximately 60% of annual aboveground net primary production. The remaining species, scarlet globemallow [*Sphaeralcea coccinea* (Pursh.) Rydb.], is the most constant component of the forb group and is an important item in the diets of the pronghorn. It is difficult to assess the subsystem- and system-level impacts of the change in photosynthetic capacity by SO₂. One would expect that if the functional lives of individual leaves are reduced and their chlorophyll concentrations during their functional lives are reduced, the result would be measureable in terms of net primary production. Over the five years of measurements of net primary production from harvested samples, we did not detect a significant treatment response.

The single significant change in the standing crop of carbon in the producer compartment observed over the five years of exposure to SO₂ was a decrease in the amount of carbon stored in the rhizomes of Western wheatgrass (Fig. 4). Western wheatgrass plants consist of an indeterminate number of tillers connected by a rhizome. Growth initiation in the spring and after defoliation by grazing is accomplished by drawing upon carbohydrates stored in rhizomes and roots. Before the

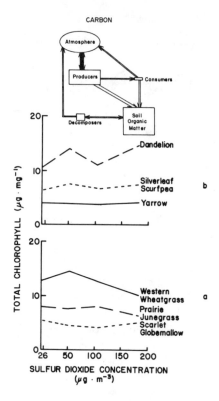

Fig. 3. Impact of SO₂ on chlorophyll concentration of six grassland plants. (*a*) Species whose chlorophyll content was decreased by SO₂; (*b*) species whose chlorophyll content was increased or unchanged by SO₂.

experiment was begun, in 1975, the area had been grazed by cattle, and the increase in rhizome biomass observed on the control after fencing is a commonly reported recovery from grazing stress. The lack of recovery on the 183-$\mu g \cdot m^{-3}$ SO₂ treatment suggests that SO₂ exposure is having an effect similar to that of grazing. The density of Western wheatgrass tillers was the same for the two treatments.

Although consumers constitute a minor component of the distribution of total system carbon, they may perform important regulatory functions for the primary-producer compartment (McNaughton, 1979) and the entire system (Mattson and Addy, 1975). Our experimental areas were much too small to observe responses of the large vertebrate consumers to SO₂; consequently, our efforts were largely concentrated on invertebrates. Grasshoppers are an important invertebrate herbivore group in grasslands (Mitchell and Pfadt, 1974; Detling et al., 1979) and are present in sufficient numbers to provide statistically valid data. Figure 5*a* demonstrates the observed changes in grasshopper density as a function of SO₂ exposure. These results are based upon two years of data. In addition to reducing their numbers by 25%, when grasshoppers were presented with forages grown on each of the SO₂ treatments, they preferred plants grown on the control (Fig. 5*b*). It is not clear at this point which constituent of the exposed plants the grasshoppers were finding undesirable. We have not measured any consistent differences in forage quality among the various treatments except increased concentrations of sulfur. Cattle are known to reduce their voluntary intake of forages high in sulfur (Rumsey, 1978). The grasshoppers may have been responding negatively to increased sulfur content.

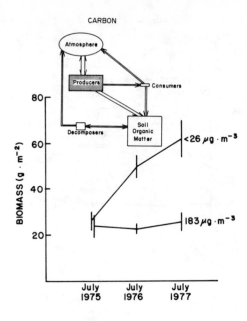

Fig. 4. Impact of SO₂ on the biomass of rhizomes of Western wheatgrass.

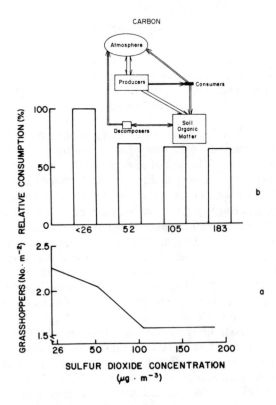

Fig. 5. Impact of SO₂ on grasshoppers. (*a*) density; (*b*) relative forage consumption.

We conducted two field experiments to determine the effects of SO_2 exposure on decomposition. In the first we placed Western wheatgrass plants in nylon bags on the control and high-concentration treatment. Bags located on the control contained either plants grown on the control plot or plants grown on the high-concentration treatment. Bags on the high-concentration treatment contained only plants grown on that treatment. We were testing two hypotheses in this experiment. The first was that exposure to SO_2 during decomposition would alter the processes involved. The second was that the sulfur content of the decomposing material would influence the constituent processes. Figure 6a represents the relative amounts decomposed over the growing season. The two dashed lines represent the plants placed on the control plot. The solid line represents plants decomposed under the high-concentration treatment. These results convinced us that SO_2 was directly affecting decomposition processes. We found no influence of plant sulfur concentration on decomposition. The following year, we placed plants collected only from the control plot on each SO_2 treatment. The results indicated that even at the lowest-SO_2 treatment, there was a reduction in decomposition (Fig. 6b). Preliminary results from a laboratory experiment with similar plant materials support these results. Although decomposer population levels, in general, have not been monitored, densities of microarthropod populations (tardigrades) inhabiting the surface layers of the soil have exhibited a marked decline in response to SO_2 exposure (Fig. 7). However, at this time very little is known regarding the role of tardigrades in the system, particularly with respect to decomposition.

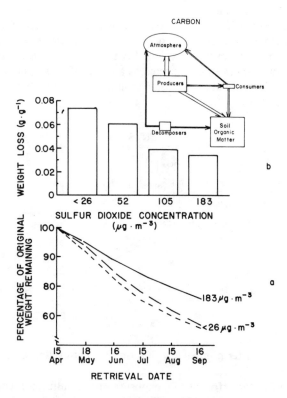

Fig. 6. Impact of SO_2 on decomposition of Western wheatgrass.

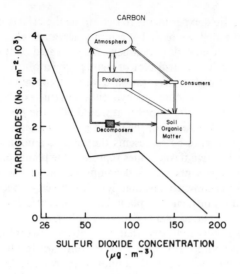

Fig. 7. Impact of SO_2 on the density of tardigrades.

Overall, the impacts of SO_2 we observed during our experiment indicated that carbon flow in grasslands is not drastically altered in the short term. Any estimation of long-term changes in carbon distribution in response to SO_2 exposure must be somewhat speculative at the present time.

We have collected less information to date concerning impacts of SO_2 on nutrient cycles than on carbon flow. Much of our experimental information concerning the impact of SO_2 on sulfur cycling has not been completed. Figure 8 indicates the general short-term changes to be expected as a result of atmospheric sulfur inputs. One should see an increase in the standing crop and flow of sulfur in and among all compartments.

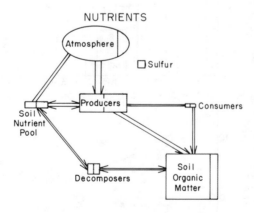

Fig. 8. Short-term impact of SO_2 on nutrient cycling in grasslands.

Figure 9 illustrates the effect of SO_2 exposure on the sulfur content of the dominant species (Lauenroth et al., 1979). These are regression lines fitted to the data from different years. Sulfur contents averaged over the entire growing season were

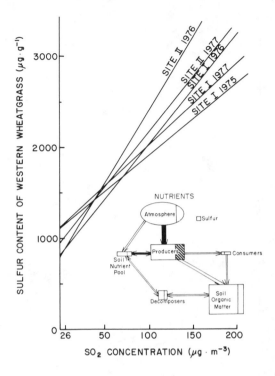

Fig. 9. Impact of SO₂ on the sulfur content of Western wheatgrass.

roughly three times as high in plants exposed to 183 $\mu g \cdot m^{-3}$ of SO_2 as in the control. This increase was the result of increased transfer from the atmosphere. It is possible that because of negative feedback between leaf sulfur content and root sulfur up- take, the transfer from the soil nutrient pool to producers was reduced by exposure to SO_2. Differences in sulfur concentrations among years are largely explained by differences in rainfall and its influence on plant growth.

Although grasslands in the northern Great Plains are not commonly thought of as sulfur-deficient, observed nitrogen-to-sulfur ratios (Fig. 10) together with the results of fertilization experiments (Fig. 11) indicate that exposure to SO_2 may have a beneficial effect, particularly early in the growing season (Fig. 10). These results most likely reflect differences in the mechanisms by which grassland plants take up and utilize nitrogen and sulfur, as well as the interaction of those mechanisms. Nitrogen is handled very conservatively. A large portion of the nitrogen in any year's live standing crop is translocated to storage organs as senescence proceeds (Clark, 1977). Sulfur, in contrast, has been found to be relatively immobile in plants and is not subjected to the same conservation mechanisms during leaf senescence as nitrogen (Bouma, 1975). The amount of sulfur required for growth in any one year apparently must be supplied largely from the soil nutrient pool during the growing season. Early in the growth period, nitrogen is readily available from storage loca- tions in the plant (Clark, 1977). Nitrogen-to-sulfur ratios are high, as sulfur must be taken up from the soil. Atmospheric sulfur augments that taken up from the soil, thus resulting in lower nitrogen-to-sulfur ratios than those observed under background sulfur conditions.

Figure 11 is a further illustration of the important interactions that occur between nitrogen and sulfur. Here we applied sulfur, nitrogen, and sulfur-plus-nitrogen fer-

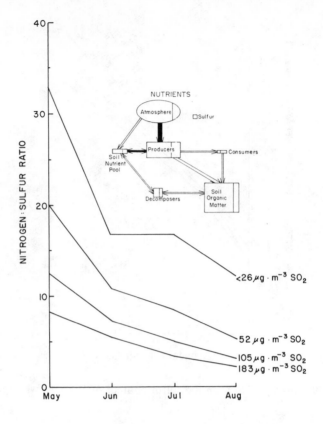

Fig. 10. Impact of SO₂ on nitrogen-to-sulfur ratios of Western wheatgrass.

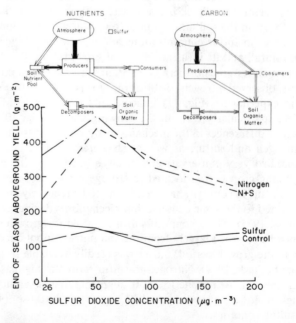

Fig. 11. Impact of nitrogen and sulfur fertilizers and SO₂ on the end-of-the-growing-season aboveground yield.

tilizers to each of our SO₂ treatments. The end-of-the-growing-season aboveground yield was the same on the fertilizer control and sulfur-only fertilizer treatment, regardless of SO₂ concentration. The addition of nitrogen resulted in a significant increase in aboveground yield, both with and without sulfur, on all SO₂ treatments. The interaction of nitrogen fertilizer with the 52 -μg·m^{-3} SO₂ produced a significant increase in yield, compared with the control and other SO₂ treatments. These data clearly indicate that addition of mineral nitrogen increases the ability of grassland plants to utilize sulfur either from the soil nutrient pool or from the atmosphere. When atmospheric sulfur inputs exceed an optimum value, yield decreases can be expected.

SYSTEM-LEVEL RESPONSES

In summary, we are putting forth a speculative view of a grassland in equilibrium with substantial inputs of sulfur from the atmosphere (Fig. 12). We are in the process of constructing a simulation model of a grassland subjected to atmospheric sulfur inputs, and within a year we will be in a much stronger position to anticipate long-term effects and equilibrium conditions.

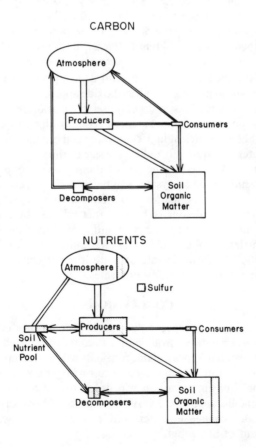

Fig. 12. Hypothetical carbon flow and nutrient cycling in a grassland equilibrated with atmospheric sulfur inputs.

The results of our field and laboratory experiments indicate that excess sulfur entering the system as sulfur oxides results in the reduction in chlorophyll in the dominant primary producer species. This decrease in chlorophyll will result in reduced energy capture by individual leaves. Exposure to SO_2 also results in phenological changes, producing a shift toward younger leaf-age classes. Because of the higher efficiency of energy capture of younger plant tissue, these changes will tend to offset the effects of chlorophyll destruction in terms of total biomass production. The energetic cost of young tissue growth is high relative to maintenance of older leaf tissue. Shifts in the population age structure toward younger age classes will likely place greater demands upon energy reserves as well as utilize energy that would normally be stored. Measurements of rhizome biomass over the five-year experiment support this hypothesis (Fig. 5). Reductions in stored carbon levels may result in greater susceptibility to perturbations such as fire, drought, or grazing, thus reducing ecosystem stability. Phenological changes in primary producers may also disrupt consumer life cycles, resulting in a reduction in consumer population levels or a shift in species composition.

The most immediate impact of exposure to low concentrations of SO_2 is the increase in sulfur content in all of the ecosystem components. Changes in primary-producer sulfur content may reduce the palatability of certain species for herbivores, thus reducing herbivore numbers or causing a shift in grazing pressure to more palatable species. Such was observed for grasshoppers in this study. Changes in grazing pressure together with differential tolerance for excess sulfur could result in a shift in species composition in all biotic components of the grassland ecosystem.

Perhaps the most significant impact of sulfur deposition in the long term is the impact upon decomposition processes. Sulfur dioxide appears to reduce decomposition rates directly through reductions in decomposer population levels and decomposer activity or through shifts in decomposer species composition to species that are more tolerant of sulfur but less efficient in decomposing soil organic matter. A decrease in nutrient cycling rates will gradually reduce the size of the soil nutrient pool available for uptake by plants. As a result, the size of the primary-producer and associated consumer components of the ecosystem will be reduced. Carbon and nutrients will accumulate in the soil organic matter pool. This accumulation will continue until nutrient inputs from various sources balance that released by decomposition. The sulfur content of all of the components should come into equilibrium at a higher value. The soil nutrient pool should have a large sulfur component compared with nonsulfur nutrients. The turnover rate of nonsulfur nutrients should be rapid but low in volume due to lower decomposition rates.

CONCLUSIONS

In the long term, we would expect a decrease in primary-producer carbon, largely as a result of decreased nutrient availability. Because of reduced decomposition, a larger portion of the potentially available nutrients will be tied up in organic matter. This will not apply to sulfur, which will constitute a larger portion of the nutrients in each compartment. Consumer carbon will be decreased as a result of decreased primary production. Because of the very large size of the soil organic matter compartment, it will not change in a perceptible way. Decomposer carbon will be decreased as a result of SO_2 input.

ACKNOWLEDGMENTS

This paper reports on work supported in part by U.S. Environmental Protection Agency Grant R805320-02-0.

REFERENCES

Bouma, D. 1975. The uptake and translocation of sulfur in plants. Pp. 79–86 in McLachken, K. D. (ed.), Sulfur in Australasian Agriculture. Sydney Univ. Press, Sydney.

Clark, F. E. 1977. Internal cycling of ¹⁵nitrogen in snortgrass prairie. Ecology 58: 1322–1333.

Detling, J. K., M. I. Dyer, and D. T. Winn. 1979. Effect of simulated grasshopper grazing on CO_2 exchange rates of Western wheatgrass leaves. J. Econ. Entomol. 72:403–406.

Durran, D. R., M. J. Meldgin, M. K. Liu, T. Thoen, and D. Henderson. 1979. A study of long range air pollution problems related to coal development in the northern Great Plains. Atmos. Environ. 13:1021–1037.

Heitschmidt, R. K., W. K. Lauenroth, and J. L. Dodd. 1978. Effects of controlled levels of sulfur dioxide on Western wheatgrass in a southeastern Montana grassland. J. Appl. Ecol. 14:859–868.

Lauenroth, W. K., C. Bicak, and J. L. Dodd. 1979. Sulfur accumulation in Western wheatgrass exposed to three controlled SO_2 concentrations. Plant Soil 53:131–136.

Lee, J. J., and R. A. Lewis. 1978. Zonal air pollution system: Design and performance. Pp. 322–344 in Preston, E. M., and R. A. Lewis (eds.), The Bioenvironmental Impact of a Coal-Fired Power Plant. Third Interim Report, Ecological Research Series EPA600/3-78-021, Corvallis, Oregon.

Lee, J. J., E. M. Preston, and D. Weber. 1979. Temporal variation in SO_2 concentration on ZAPS. Pp. 284–305 in Preston, E. M., and T. L. Gullett (eds.), The Bioenvironmental Impact of a Coal-Fired Power Plant. Fourth Interim Report, Environmental Protection Agency, Corvallis, Oregon.

Mattson, W. J., and N. D. Addy. 1975. Phytophagous insects as regulators of forest primary production. Science 190:515–522.

Mitchell, J. E., and R. E. Pfadt. 1974. A role of grasshoppers in a shortgrass prairie ecosystem. Environ. Entomol. 3:358–360.

McNaughton, S. J. 1979. Grazing as an optimization process: Grass-ungulate relationships in the Serengeti. Am. Nat. 113:691–703.

Preston, E. M., and T. L. Gullett. 1979. Spatial variation of sulfur dioxide concentrations on ZAPS during the 1977 field season. Pp. 306–331 in Preston, E. M., and T. L. Gullett (eds.), The Bioenvironmental Impact of a Coal-Fired Power Plant. Fourth Interim Report, Environmental Protection Agency, Corvallis, Oregon.

Rumsey, T. S. 1978. Effects of dietary sulfur addition and synovex-S ear implants on feedlot steers fed an all-concentrate finishing diet. J. Anim. Sci. 46:463–477.

DISCUSSION

Danny Rambo, EPA Labs, Corvallis: Did you notice any change in populations of pollinating insects or any decrease in fertilization of plants that are insect-pollinated?

W. K. Lauenroth: To answer the first part of your question, no, we did not, but sampling problems with insects are almost insurmountable, and we didn't focus upon any particular pollinators, so no, we did not measure any differences. In terms of sexual reproduction, most of the species in this grassland are perennial, and in the short term, asexual reproduction is the most common way of reproducing. There are some people who are working on the project that are looking at interference with sexual reproduction processes.

Ellis Cowling, North Carolina State University: Could you speculate in the same way that you did at the end of your talk about the consequences of having nitrogen added together with sulfur by the same atmospheric dispersal mechanisms that apply to sulfur—in other words, a deliberate attempt either by injection of ammonia into the flue gases or by other means to increase emissions of oxides of nitrogen? What would be the consequences in the ecosystem if that were done on a planned regime?

W. K. Lauenroth: I think that our fertilizer experiment showed clearly that primary production could be increased if both sulfur and nitrogen deposition were increased. What we don't know is the long-term consequences of altering competitive relationships for that nitrogen at the primary-producer level as well as consumer and decomposer levels.

Ellis Cowling: Will your future experiments take that into account and attempt to characterize the consequences of concomitant nitrogen and sulfur fertilization?

W. K. Lauenroth: Yes, they will be limited to simulation modeling experiments.

Charles Goodman, Southern Company Services: If your simulation models are applied to a new power plant, will you be able to deal with the real world situation in which SO_2 levels decay as you go away from the source, so that you will be able to get an overall assessment of impact?

W. K. Lauenroth: Yes, we are currently constructing a series of most probable scenarios for this area that we feel that our models should be able to respond to.

Lance Evans, Brookhaven National Laboratory: Have you considered the effect of grazing?

W. K. Lauenroth: We have not explicitly. We have not been able to exclude the invertebrate grazing. We did conduct an experiment this summer simulating grazing by clipping, but those results are not available at this time.

Sandy McLaughlin, Oak Ridge National Laboratory: Could you comment on how well your distribution system approximates the distribution of concentrations that would occur with normal meteorological dispersion, and also are you getting stomatal closure during the nighttime, when your concentrations are highest?

W. K. Lauenroth: We don't know about all night long; we have done some measurements using the porometer predawn, and we find that their resistances are extremely high, but we're fairly convinced that even if the stomates are not closed at night—are not entirely closed at night—they're closed enough that the model resistances are very high. I don't think I can comment about the first part of your question, in terms of the distribution; in fact, maybe I don't understand it; can you rephrase it?

Sandy McLaughlin: I wonder if the ratio of your highest concentrations to your 30-day averages that you're using in calculating your total dose represent what you feel are realistic ratios for a point-source situation.

W. K. Lauenroth: No, they don't. We have a higher number of high-concentration measurements, most of which we are considering to be not physiologically significant, since they occur between, say, 6 PM and 7 AM.

CHAPTER 42

IMPACT OF SULFUR DEPOSITION ON THE QUALITY OF WATER FROM FORESTED WATERSHEDS*

Gray S. Henderson†

Wayne T. Swank‡

James W. Hornbeck§

ABSTRACT

Nutrient discharge in streamflow resulting from five manipulative treatments to experimental watersheds was used to evaluate the potential impact of sulfur and associated acid deposition on cation-leaching and content of water. Observed stream discharge of nitrate nitrogen was used to calculate H^+ production due to accelerated nitrification in response to forest harvest. This value or a calculated value for H^+ production following urea fertilization was combined with observed cation discharge, which had been corrected for release from organic matter decomposition, to estimate cation release ratios from the soil exchange complex due to H^+. These ratios were then used to estimate potential increases in watershed discharge of cations if precipitation acidity were to change from pH 4.3 to pH 4.0, 3.5, or 3.0. The potential increases in annual Ca, Mg, K, and Na discharge were calculated to be less than 0.5 kg/ha at pH 4.0 and less than 2.5 kg/ha at pH 3.5, increases which would be difficult to detect among natural variations in stream water chemistry. Calculations for precipitation at pH 3.0 suggest a potential increase in cation discharge which could be as great as 8.5 kg/ha for individual elements, a change which would be more easily detected.

INTRODUCTION

We have a difficult task addressing the title of this paper in the emphatic nature in which it is stated. The path of any ion through a forested ecosystem is tortuous at best, and interactions with biotic and abiotic components are numerous and complex. As such it is difficult to relate quality of streamflow to atmospheric inputs of sulfur, hydrogen, or any other element.

As added evidence of the difficulty of our task, we know of no studies which have quantified the impact of precipitation chemistry on quality of streams draining forested watersheds. The reported impacts of atmospheric deposition on chemistry of lakes and streams in Scandinavia and northeastern North America have not been

---

*Contribution from the Missouri Agricultural Experiment Station. Journal Series Number 8443.

†School of Forestry, Fisheries, and Wildlife, University of Missouri, Columbia, Missouri.

‡U.S.D.A., Forest Service, Franklin, North Carolina.

§U.S.D.A., Forest Service, Durham, New Hampshire.

accompanied by information about the relative contributions of direct atmospheric inputs vs outputs from terrestrial watersheds. Gorham and McFee (1979) emphasize this lack of quantitative understanding when they state, "There is an urgent need for mass-balance studies to determine the degree to which the acids, heavy metals, nutrients and organic molecules reaching aquatic ecosystems have originated from direct precipitation upon the lake surface, transfer from atmospheric deposition upon the land, or—in some cases—enhanced soil leaching due to acid precipitation."

Gorham and McFee's proposal of enhanced soil leaching is of special interest to our paper since sulfur occurring as sulfate in precipitation is closely associated with hydrogen ion content (Hornbeck et al., 1977). Any increase in sulfate is accompanied by an increase in hydrogen ions, which may in turn accelerate the leaching of basic cations from soil exchange sites and into streams. In this paper we will use results from some experimental watershed studies to indirectly assess the potential impact of increased sulfate and its accompanying acidity upon chemistry of streams draining forested headwaters.

The complexity of sources and pathways for streamflow is a major reason why there have been no direct studies of impacts of atmospheric deposition on stream chemistry. Streamflow from forested watersheds is a combination of four possible sources: base flow, direct channel input, surface storm flow, and subsurface storm flow. These flow components are generated by water moving through different paths, and thus chemical properties are influenced by different factors. Base flow commonly has contact with bedrock and soil for a relatively long period of time, so the chemistry of base flow reflects the chemical composition and weathering characteristics of the geology. Acid precipitation could be expected to increase bedrock weathering rates, but its influence will be diminished as acidity is neutralized on passage through the vegetation canopy and soil profile (Eaton et al., 1973; Hornbeck et al., 1977).

Water entering the stream system through direct channel input is only a small proportion (less than 1%) of the total streamflow; however, its chemistry can be altered by the vegetation canopy. Concentration increases of some nutrients in streamflow during the initial periods of storms have been attributed to leaching from vegetation (e.g., Henderson et al., 1977; Johnson and Henderson, 1979). If acid precipitation accelerates foliar leaching through cuticular erosion and/or altering concentration gradients between internal and external leaf surfaces (Tukey, 1970), then stream chemistry will likewise be changed.

Surface storm flow (also termed overland flow) is rainfall which fails to infiltrate the soil. It rarely occurs in undisturbed forested watersheds in the eastern United States except when headwater streams expand during extreme events. At such times, much of the surface flow is actually exfiltrating subsurface storm flow. Subsurface storm flow is made up of precipitation which infiltrates and then moves vertically and laterally downslope to become streamflow. The chemistry of this large proportion of annual streamflow is largely determined by interaction with the cation and anion exchange reactions of the soil.

Results from the gaged watershed studies we will be discussing are an integration of all the above sources of streamflow. Obviously, much research lies ahead if we are to fully understand the impacts of precipitation chemistry upon streamflow.

WHAT CAN BE LEARNED FROM PREVIOUS MANIPULATIVE WATERSHED EXPERIMENTS?

Forest cutting and fertilization experiments conducted on experimental catchments may provide a way to approximate the potential impact of acid precipitation on leaching of cations from soils and into streams. Harvesting and fertilization generally result in increases in stream nutrient discharge. The increases in cation discharge have been attributed to acidity generated during the process of nitrification of ammonium (Likens et al., 1970). Given the cation discharges, if the amount of acidity generated in the studies can be determined, the quantity of cations released can be expressed as a function of the acidity produced. In turn, projections can be made of increased cation leaching which might be expected if sulfate content and precipitation acidity were to increase.

Harvesting Effects on Acidity

When forests are cut, both soil moisture and soil temperature are affected. Soil moisture increases during summer months because transpiration and interception have been reduced. Soil temperature is elevated due to increased radiation reaching the forest floor in the absence of a shading forest canopy. Increases in both soil moisture and soil temperature are conducive to accelerating the rate of organic matter decomposition. One product of decomposition is the release of the ammonium ion. The ammonium ion then undergoes a two-step oxidation to nitrate (Alexander, 1967):

$$NH_4^+ + 1\frac{1}{2}O_2 \rightarrow NO_2^- + 2H^+ + H_2O, \tag{1}$$

$$NO_2^- + H_2O \rightarrow NO_3^- + 2H^+ . \tag{2}$$

The first stage is mediated by bacteria in the genus *Nitrosomonas,* while the second reaction is controlled by the genus *Nitrobacter.* Thus, in the conversion of one equivalent of ammonium to one equivalent of nitrate, four equivalents of hydrogen ion are released. It is this increase in acidity which has been thought to partly cause the increase in cation discharge observed after forest cutting (Likens et al., 1970).

Calculations from Harvesting Studies

Hubbard Brook deforestation

In 1965–1966, all vegetation on watershed 2 of the Hubbard Brook Experimental Forest in New Hampshire was felled, and subsequent regrowth was inhibited with herbicides for three successive years (Likens et al., 1970). Streamflow amounts and element concentrations were monitored at weekly intervals and compared with similar values from a control watershed. The changes in element discharge from the deforested watershed are shown in Table 1. An increase of 245.9 kg/ha of nitrogen was discharged over a two-year period due to the deforestation and herbicide application. If this increased nitrogen discharge was the result of accelerated organic matter decomposition in the soil, which had an N concentration of 2.34% (Bormann et al., 1977), then decomposition of organic matter increased by 10,500 kg/ha. Cations are also released during decomposition, and part of the observed increase in nutrient discharge can be attributed to this source, the remainder to displacement by hydrogen ions on the soil exchange. The cation release due to decomposition was calculated by multiplying the amount of decomposition (10,500 kg/ha) by the concentration of the nutrient in soil organic matter. These concentrations were 0.79, 0.08, 0.14, and 0.008% for Ca, Mg, K, and Na, respectively (calculated from Likens et al., 1977, Table 9). Cation release due to decomposition was subtracted from stream discharge values to estimate the amount of each cation lost from the soil exchange (Table 1).

Table 1. Changes in element discharge due to deforestation at Hubbard Brook
Experimental Forest and the estimation of the proportions arising from
organic matter decomposition and the soil exchange complex

Data for the period from June 1, 1966, to May 31, 1968, and based on
Likens et al. (1970)

	Increase in streamflow loss (kg/ha)	Release from organic matter (kg/ha)	Loss from soil exchange	
			(kg/ha)	(eq/ha)
NH$_4^+$-N	1.0	Assumed all		
NO$_3^-$-N	244.9	Assumed all		
Ca	147.9	83.1	64.8	3240
Mg	28.5	8.5	20.0	1650
K	55.6	14.7	40.9	1050
Na	21.7	0.8	20.9	910
Al	36.4	?	36.4	4040

The amount of hydrogen ion produced was estimated from the nitrate ion discharge. Increased nitrate-nitrogen discharge was 244.9 kg/ha, or 17,490 equivalents per hectare. If four equivalents of H$^+$ are produced during the transformation of one equivalent of NH$_4^+$ to one equivalent of NO$_3^-$, then 69,960 equivalents per hectare of H$^+$ were produced. This is the amount of acidity which presumably produced the loss of cations from the soil exchange complex shown in Table 1.

Hubbard Brook strip cut

In October 1970, one-third of the vegetation on watershed 4 (89 ha) of the Hubbard Brook Experimental Forest was cut and harvested. The harvest was accomplished by removing vegetation from 25-m-wide strips running perpendicular to the stream and leaving 50-m uncut strips between those that were cut (additional 25-m strips from the uncut area were harvested in 1972 and 1974). Details of the experiment are given by Hornbeck et al. (1975). Changes in nutrient discharge due to the strip cut during the first two years are presented in Table 2 together with the amount of cation release from organic matter decomposition and estimates of loss from the soil exchange complex. Organic matter decomposition was calculated to be 500 kg/ha. Accelerated nitrate-nitrogen discharge amounted to 11.9 kg/ha (850 equivalents per hectare), which translates to 3400 equivalents per hectare of H$^+$ produced during the N mineralization and nitrification process.

Fernow clear cut

In 1969–1970 a 34-ha watershed on the Fernow Experimental Forest near Parsons, West Virginia, was commercially clear-cut, and water chemistry before and after the cut was compared with that of an adjacent control watershed. Data for the period from the 1970 growing season through the 1971–1972 dormant season were obtained

Table 2. Changes in element discharge due to strip-cutting at Hubbard Brook
Experimental Forest

Estimates of the amounts arising from organic matter decomposition and the soil
exchange complex are included for the period from the 1970–1971 dormant
season through the 1972 growing season and are derived from Hornbeck et al (1975)

	Increase in streamflow loss (kg/ha)	Release from organic matter (kg/ha)	Loss from soil exchange	
			(kg/ha	(eq/ha)
NO_3^--N	11.9	Assumed all		
Ca	9.8	4.0	4.8	240
Mg	1.4	0.4	1.0	80
K	1.7	0.7	1.0	30
Na	1.1	0.1	1.0	40

from Aubertin and Patric (1974) and are presented in Table 3. Calculations for
organic matter decomposition release were similar to those for the Hubbard Brook
studies, but on-site concentrations of nutrients in soil organic horizons were not
available. Values used were 1.23, 1.42, 0.07, 0.13, and 0.002% for N, Ca, Mg, K,
and Na, respectively, and were average values for similar vegetation types in the Appalachian region (Henderson et al., 1978). Organic matter decomposition was
calculated to be 195 kg/ha based on an increase in N loss of 2.4 kg/ha (increase in
nitrate-nitrogen loss of 3.4 kg/ha and a decrease in ammonium-nitrogen loss of 1.0
kg/ha). Nitrate-nitrogen discharge was 3.4 kg/ha, or 240 equivalents per hectare,
which means 960 equivalents per hectare of H^+ was generated during nitrification.

Table 3. Changes in element discharge due to a commercial clear cut on the Fernow
Experimental Forest

Data represent the period from the 1970 growing season through the 1971–1972
dormant season (Aubertin and Patric, 1974), and estimates of release from
organic matter decomposition and loss from the soil exchange complex are
also included

	Change in streamflow loss (kg/ha)	Release from organic matter (kg/ha)	Loss from soil exchange	
			(kg/ha)	(eq/ha)
NH_4^+-N	− 1.0			
NO_3^--N	3.4	2.4		
Ca	3.0	2.8	0.2	10
Mg	1.1	0.1	1.0	80
K	2.2	0.3	1.9	50
Na	0.7	Negligible	0.7	30

Coweeta clear cut and cable logging

In 1976–1977, roads were built and stabilized on watershed 7 at the Coweeta Hydrologic Station near Franklin, North Carolina, and it was clear-cut and harvested using cable logging. Changes in nutrient discharge the second year after cutting (June 1, 1978, to May 31, 1979) and not including the period of road building are presented in Table 4. Organic matter decomposition and element release were calculated using soil organic horizon nutrient concentrations of 1.25, 1.16, 0.07, 0.17, and 0.002% for N, Ca, Mg, K, and Na, respectively (Henderson et al., 1978). Accelerated organic matter decomposition was calculated to be 100 kg/ha. H^+ production was estimated to be 360 equivalents per hectare based on the observed increase in nitrate-nitrogen discharge of 1.26 kg/ha (90 equivalents per hectare).

Table 4. Changes in element discharge due to clear-cutting and cable logging at the Coweeta Hydrologic Station for the period June 1, 1978, to May 31, 1979

Estimates of release from organic matter decomposition and loss from the soil exchange complex are also included

	Change in streamflow loss (kg/ha)	Release from organic matter (kg/ha)	Loss from soil exchange (kg/ha)	(eq/ha)
NH_4^+-N	0.0			
NO_3^--N	1.26	Assumed all		
Ca	a	1.16	a	a
Mg	0.35	0.07	0.28	25
K	0.98	0.17	0.81	20
Na	0.10	< 0.01	0.10	5

[a]Calcium loss in streamflow from the clear-cut and cable-logged watershed was 0.25 kg/ha less than that from the control catchment.

Fernow urea fertilization

When urea is added to soils, it has a pronounced influence on soil reaction. Initially, urease present in the soil breaks down urea as follows (Alexander, 1967):

$$CO(NH_2)_2 + H_2O \xrightarrow{\text{urease}} 2NH_3 + CO_2 . \tag{3}$$

The ammonia produced then reacts with water:

$$NH_3 + H_2O \longrightarrow NH_4^+ + OH^- . \tag{4}$$

The result of this reaction causes soil pH to rise shortly after urea is added to soils. Thereafter, the NH_4^+ undergoes nitrification as shown in Eqs. (1) and (2), and H^+ is released, causing soil pH to drop. The net result is that for each equivalent of urea nitrogen applied, an equivalent of OH^- and four equivalents of H^+ will be generated, and the net acidity produced will be three equivalents. The above analysis assumes that no NH_3 is lost through volatilization, but, in fact, at least a small volatilization loss can be expected.

In May 1971 an experiment was conducted at the Fernow Experimental Forest in which 260 kg/ha of urea nitrogen was applied to a 30-ha forested watershed. Following the application, streamflow chemistry was monitored closely for a year (Aubertin et al., 1973). The increases in nutrient discharge for this period are shown in Table 5. The 260 kg/ha of urea nitrogen applied is equal to 18,750 equivalents per hectare, yielding a net increase in H^+ of 55,710 equivalents per hectare.

Table 5. Changes in element discharge due to urea fertilization
at the Fernow Experimental Forest for the period
May 1, 1971, to April 30, 1972

Data are from Aubertin et al., 1973.

	Increase in streamflow loss	
	(kg/ha)	(eq/ha)
Ca	33.5	1680
Mg	6.7	550
K	3.3	80
Na	2.5	110

Cations Leached as a Function of H+ Produced

Results from the five watershed studies discussed above are summarized in Table 6 with respect to the amount of total cations (Ca, Mg, K, and Na) released in response to the amount of H^+ produced. The cation release associated with an

Table 6. Summary of total cation release, hydrogen production, and the cation release ratio for five manipulated-watershed studies

	H^+ produced (eq/ha)	Total cation release (eq/ha)	Cation release ratio (eq/eq H^+)
Hubbard Brook			
Deforested	69,960	6,850	0.10
Strip-cut	8,400	390	0.05
Fernow			
Clear-cut	960	170	0.18
Fertilization	55,710	2,420	0.04
Coweeta			
Clear-cut and cable-logged	360	50	0.14

equivalent of H^+ ranges from a low of 0.04 for the urea fertilization to a high of 0.14 for the Fernow clear cut. The value for the urea fertilization may be slightly low due to overestimation of the amount of H^+ actually produced. This overestimation could be due to volatilization of NH_3 and/or uptake and incorporation of NH_4^+ by vegetation and microorganisms prior to nitrification. It is interesting that these diverse treatments on a variety of sites produce ratios which are so similar. Additional data from this and other sites would help to further establish the variability of this ratio. The release of individual cations (Ca, Mg, K, and Na) as a function of H^+ produced for each of the five studies is given in Table 7.

Table 7. Release ratios for individual cations observed in five manipulated- watershed studies

	Cation release ratio (eq/eq H^+)			
	Ca	Mg	K	Na
Hubbard Brook				
Deforested	0.046	0.024	0.015	0.013
Strip-cut	0.029	0.010	0.004	0.005
Fernow				
Clear-cut	0.010	0.083	0.052	0.031
Fertilization	0.030	0.010	0.001	0.002
Coweeta				
Clear-cut and logged	0.0	0.069	0.056	0.014

CATION LEACHING POTENTIAL UNDER VARIOUS PRECIPITATION ACIDITY AND SULFUR CONTENTS

The ratios of total and individual cation release to H^+ produced were used to assess leaching potential under different precipitation acidity regimes. To do this it was assumed that annual precipitation amounts to 150 cm and has a weighted average pH of 4.3, which would correspond to a sulfate-sulfur input of 12.0 kg/ha if the H^+ were associated only with sulfate. These conditions are very similar to those found in much of the eastern United States. These baseline precipitation characteristics were used to calculate potential changes in leaching for conditions where precipitation acidity changed to pH values of 4.0, 3.5, and 3.0. Additional H^+ inputs of 750, 3990, and 14,250 equivalents per hectare would be experienced for 150 cm of precipitation at pH 4.0, 3.5, and 3.0, respectively (Table 8). If all of the H^+ were associated with sulfate, these conditions would be associated with sulfate-sulfur inputs of approximately 25, 75, and 240 kg/ha.

Three values for increased leaching are given in Tables 8 and 9—low, high, and median—which were determined using the low, high, and median values for total and individual cation release presented for the study locations in Tables 6 and 7.

Based on the median value (Table 9), cation discharge could be expected to increase by less than 0.5 kg/ha for any nutrient for precipitation at pH 4.0 and by less

Table 8. Potential increases in cation leaching expected under three precipitation acidity regimes

Low, high, and median values are based on corresponding cation release ratios in Table 6

pH	Additional H$^+$ (eq/ha)	Cation leaching potential (eq/ha)		
		Low	High	Median
4.0	750	30	135	75
3.5	3,990	160	720	400
3.0	14,250	570	2,565	1,425

Table 9. Potential increases in annual discharge of four cations under three precipitation acidity regimes

Low, high, and median values are based on corresponding values in Table 7

	Leaching potential for individual cations (kg/ha)			
	Ca	Mg	K	Na
At pH 4.0:				
Low	0.0	0.1	<0.1	<0.1
High	0.7	0.8	1.6	0.5
Median	0.4	0.2	0.4	0.2
At pH 3.5:				
Low	0.0	0.5	0.2	0.2
High	3.7	4.0	8.7	2.8
Median	2.3	1.2	2.3	1.2
At pH 3.0:				
Low	0.0	1.7	0.6	0.7
High	13.1	14.4	31.2	10.2
Median	8.3	4.2	8.4	4.3

than 2.5 kg/ha for rain at pH 3.5. For some watersheds, changes of this magnitude would be difficult to detect among the natural variations in water chemistry. The increases for pH 3.0 are larger and would be more easily detected. Leaching increases based on the high value are much more pronounced and could more easily be detected, especially for the pH 3.5 and 3.0 situations.

In Table 1, data on aluminum discharge were included. In the Adirondacks of New York, Schofield (1976) has suggested that high Al concentrations in lakes are at least partly responsible for increased fish mortality. It is postulated that Al is leached from soils by infiltrating snowmelt which is very acid. The data from the Hubbard Brook deforestation study show that 0.06 equivalent of Al per equivalent of H$^+$

produced was released to streamflow. Using the same assumptions as above, increased Al leaching of 0.4, 2.2, and 7.7 kg/ha could be expected at pH values of 4.0, 3.5, and 3.0, respectively.

In this paper we have attempted to reanalyze manipulative watershed experiments in order to arrive at some possible estimates of the potential effects of acid precipitation on cation leaching. Our calculations have assumed that the H^+ produced as the result of the various treatments interacted exclusively with cations on the soil exchange with subsequent discharge of cations in streamflow. In fact, some of the H^+ has reacted with primary and secondary minerals, causing increased weathering rates and associated cation discharge. While the net effect on water quality is the same, we do not really know the proportions of discharge coming from these different sources.

The harvesting studies reveal the influence of organic matter decomposition on cation discharge. Sulfur deposition and acid precipitation may retard decomposition and thus in part counteract the increases due to leaching and weathering.

REFERENCES

Alexander, M. 1967. Introduction to Soil Microbiology. John Wiley and Sons, Inc., New York.

Aubertin, G. M., and J. H. Patric. 1974. Water quality after clearcutting a small watershed in West Virginia. J. Environ. Qual. 3:243–249.

Aubertin, G. M., D. W. Smith, and J. H. Patric. 1973. Quantity and quality of streamflow after urea fertilization on a forested watershed: first-year results. Pp. 88–100 in Forest Fertilization, Symposium Proceedings. U.S.D.A. Forest Service Gen. Tech. Rep. NE-3.

Bormann, F. H., G. E. Likens, and J. M. Melillo. 1977. Nitrogen budget for an agrading Northern hardwood forest ecosystem. Science 196:981–983.

Eaton, J. S., G. E. Likens, and F. H. Bormann. 1973. Throughfall and stem-flow chemistry in a Northern hardwood forest. J. Ecol. 61:495–508.

Gorham, E., and W. W. McFee. 1979. Effects of acid deposition upon outputs from terrestrial to aquatic ecosystems. In NATO Adv. Res. Institute, Ecological Effects of Acid Rain. University of Toronto.

Henderson, G. S., W. F. Harris, D. E. Todd, Jr., and T. Grizzard. 1977. Quantity and chemistry of throughfall as influenced by forest-type and season. J. Ecol. 65:365–374.

Henderson, G. S., W. T. Swank, J. B. Waide, and C. C. Grier. 1978. Nutrient budgets of Appalachian and Cascade region watershed: a comparison. For. Sci. 24:385–397.

Hornbeck, James W., Gene E. Likens, and John S. Eaton. 1977. Seasonal patterns of acidity of precipitation and their implications for forest stream ecosystems. Water, Air, Soil Pollut. 7:355–365.

Hornbeck, J. W., G. E. Likens, R. S. Pierce, and F. H. Bormann. 1975. Strip cutting as a means of protecting site and streamflow quality when clear-cutting Northern hardwoods. Pp. 208–229 in Bernier, B., and C. H. Winget (eds.), Forest Soils and Forest Land Management. Les Presses de l'Universite' Laval, Quebec, Quebec, Canada.

Johnson, D. W., and G. S. Henderson. 1979. Sulfate absorption and sulfur fractions in a highly weathered soil under a mixed deciduous forest. Soil Sci. 128: 34–40.

Likens, G. E., F. H. Bormann, N. M. Johnson, D. W. Fisher, and R. S. Pierce. 1970. Effects of forest cutting and herbicide treatment on nutrient budgets in the Hubbard Brook watershed-ecosystem. Ecol. Monogr. 40:23–47.

Likens, G. E., G. H. Bormann, R. S. Pierce, J. S. Eaton, and N. M. Johnson. 1977. Biogeochemistry of a Forested Ecosystem. Springer-Verlag, New York.

Schofield, C. L. 1976. Acid precipitation: effects on fish. Ambio 5:228–230.

Tukey, H. B., Jr. 1970. The leaching of substances from plants. Annu. Rev. Plant Physiol. 21:305–323.

DISCUSSION

Bill Graustein, Yale University: I'd like to comment on your last point, on aluminum. One thing that makes it very difficult to make such predictions about aluminum is the great difference between one watershed and another. I think that perhaps if we contrast Hubbard Brook and Coweeta we might be able to make my point fairly clearly. Hubbard Brook currently shows a fairly low rate of weathering in the horizon, but I think this is partially because of the unusual hydrology of the area, where the B_2 horizon is essentially impermeable to water. The Hubbard Brook watershed must have been a site of very rapid weathering in former times, because the A_2 horizon contained nothing but quartz; all the other minerals had been removed. Over a long period of time the only thing that's available for neutralizing acid precipitation is rock weathering. Changes in biomass are virtually nil, so an acid runoff essentially means a weathering rate that's slower than the rate of acid input. In a place like Coweeta, however, where there is much deeper percolation and development of saprolite, there are abundant mineral surfaces available for weathering. These weathering reactions remove hydrogen or increase bicarbonate in solution as great as the pH. Aluminum runoff is a problem from Hubbard Brook only because the pH is so low that aluminum can become soluble. If weathering reactions proceed, they'll raise the pH of the solution, and aluminum will precipitate. Extrapolating aluminum leaching measurements from a watershed that's already so acidic that aluminum will remain in solution to one which does not have a large buffering capacity but has a large reaction capacity to raise pH and therefore reduce aluminum mobility, I think, is a little bit hazardous.

G. S. Henderson: Oh, I agree, this whole thing is a little bit hazardous. I should point out, though, that the aluminum data that I used here came just from the Hubbard Brook study. Your points are still well taken, though, and obviously the whole picture of aluminum is tremendously complex, with the change in solubility as a function of pH.

George Sehmel, Pacific Northwest Laboratory: Can you determine how much each of the cations is leached as a function of depth?

G. S. Henderson: Good question. I don't think that on the short term that we've been studying this type of thing. We can detect in soils these changes due to natural changes in acidity. Certainly over long periods of time, you can analyze soil and find differing zones of cation movement, either loss or addition. One of the best examples of this sort of thing I've found is in many of the mollisols; in that you have secondary calcium carbonate accumulations in the B horizon, and this material has come from the overlying horizons. You can see these zones of differential movement of cations, but I have never looked at differential cation concentrations in different soil profiles and tried to relate them to different levels of acidification or leaching rates at this point.

CHAPTER 43

ECOLOGICAL EFFECTS OF WHOLE-STREAM ACIDIFICATION

Ronald J. Hall*
Gene E. Likens*

ABSTRACT

A natural mountain stream was experimentally acidified to determine the effects on abiotic and biotic parameters. Sulfuric acid was added in dilute concentrations to Norris Brook within the Hubbard Brook Experimental Forest. Stream water pH was lowered to 4.0 for five months in 1977.

Increased acidity resulted in significant mobilization of Al and Ca in the stream water. Mg, K, and Na concentrations were not significantly greater in the acidified sections relative to the reference section of the stream. Macroinvertebrate collectors, scrapers, and predators showed a change in drift activity for the first week of acid addition; thereafter, drift behavior was not different for reference and treatment areas. Drift density of species within the shredder functional group was not altered by pH 4.0 stream water.

INTRODUCTION

Man, by combusting fossil fuels, has released and is releasing large quantities of sulfur and nitrogen oxides to the atmosphere. These atmospheric substances are oxidized to acids (sulfuric and nitric) at various rates depending upon environmental conditions (e.g., Likens, 1976; Likens et al., 1979). Some of these acids are not neutralized by alkaline substances, also frequently present in the atmosphere (Gorham, 1976), and ultimately fall on terrestrial and aquatic ecosystems via precipitation. The extent to which elevated hydrogen ion presently found in rain may alter the ecological interactions of plants and animals in natural aquatic ecosystems is poorly known, at least quantitatively.

Diverse communities in streams (Sutcliffe and Carrick, 1973; Leivestad and Muniz, 1976; Hendrey et al., 1977, among others) and lakes (Almer et al., 1974; Grahn et al., 1974; Hendrey and Wright, 1976; Sprules, 1975) are affected by this acid deposition, and as a result the water is acidified. The impact of increased acidity on fisheries is the most evident result (Beamish and Harvey, 1972; Leivestad et al., 1976; Schofield, 1976). Most of the field investigations, however, have consisted of qualitative field surveys in acidic surface waters with few detailed studies on the dynamic changes of biotic and abiotic components at one location (cf. Hall et al., 1980).

---

*Section of Ecology and Systematics, Division of Biological Sciences, Cornell University, Ithaca, New York 14850.

The purpose of this research was to quantitatively determine the impact of increased hydrogen ion stress on an otherwise natural stream ecosystem. We experimentally manipulated the pH of a mountain stream within the Hubbard Brook Experimental Forest. Dilute concentrations of sulfuric acid were added to Norris Brook to lower the pH to levels found in incident precipitation (pH 4.0). We present here some ecological effects of increased acidity on biotic and abiotic compartments.

THE STUDY SITE

Norris Brook watershed (WS) is located within the Hubbard Brook Experimental Forest (Fig. 1). The WS area is 87.2 ha and drains first through third-order streams. Total stream length from the origin (427 m above sea level) to the confluence with

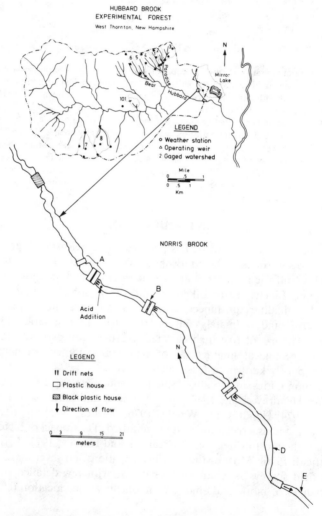

Fig. 1. Norris Brook study area within the White Mountains, New Hampshire. Reference area was located 5 m above the sulfuric acid addition. Treatment area sampling locations were 15 m (site B), 50 m (site C), 75 m (site D), and 100 m (site E) below the point of acid addition. From Hall et al., 1980.

Hubbard Brook (230 m) is 1.73 km. The elevation at the study site is 245 m. Norris Brook was desirable for our study because biotic diversity and population densities were higher relative to streams of comparable size in the Hubbard Brook Experimental Forest (Fiance, 1977). For example, Norris Brook contains brook trout (*Salvelinus fontinalis*) and slimy sculpins (*Cottus cognatus*) from the mouth to the second-order tributaries.

The 200-m study area was located on a third-order (Strahler, 1957) stream. Mean width in this area was 1.8 m. Organic debris accumulations in the stream were few. Water temperature ranged from 11 to 20°C in July and from 0 to 4.5°C in March. The dominant riparian vegetation consisted of red spruce (*Picea rubens*), hemlock (*Tsuga canadensis*), paper birch (*Betula papyrifera*), and white pine (*Pinus strobus*).

METHODS AND MATERIALS

Acidification Experiment

Stream water pH near 4.0 was maintained by manually adding dilute concentrations of sulfuric acid (0.05–1 N) from a carboy. The acid was continuously added for approximately five months (18 April–22 September 1977). The study area was divided into a reference area (site A), 5 m above the acid addition, and a treatment section 15 m (site B), 50 m (site C), 75 m (site D), and 100 m (site E) below the acid dripping location (Fig. 1). The pH of the turbulent stream water was monitored regularly 12 m below the acid addition.

Water Sampling and Chemical Analyses

Detailed chemical analyses and water sampling procedures are described elsewhere (Hall et al., 1980). Only a brief outline will be given here.

Water samples for chemical analyses were collected at frequent intervals (every 6 hr for two days, every 12 hr for the next two days, and every 24 hr for the remaining three days) during the first week of the experiment. Then, water samples were collected weekly for the duration of the experiment. Two water samples were collected at each station, A through E, in acid-washed 500-ml containers. pH and conductivity were immediately determined for one sample; cations and anions were measured promptly in the other sample, or the sample was frozen. Stream water pH was measured at all stations with a Sargent-Welch PBL pH meter equipped with an A. H. Thomas glass electrode and a Fisher Scientific calomel electrode.

Invertebrate Drift

Drift nets were positioned above and below the acid addition (Fig. 1) to estimate changes in drift behavior (see Waters, 1962) of vertebrate and invertebrate communities. Two nets with 253-μm mesh filtered the stream water at sites A, B, and C. Drift samples for the first week were collected at 12-hr intervals to determine day-night (0700–1900 and 1900–0700, respectively) differences in biotic drift behavior. Thereafter, day and night drift studies were done for 24-hr periods every week until 31 May. Only 24-hr drift collections were made from June through September because numbers of organisms collected were low. Samples were preserved in 70% alcohol and manually sorted in the laboratory. Discharge was measured at the mouth of the nets at frequent intervals with either a Marsh-McBirney 201 or a Gurley Pigmy current meter. Invertebrates were identified and assigned to invertebrate functional groups (see Merritt and Cummins, 1978).

RESULTS

Stream Discharge and Water Chemistry

Stream water discharge and water chemistry for site A (reference) and site E (100 m below acid addition) are depicted in Fig. 2. Mobilization of cations in the stream water was similar at sites B, C, and D and was generally similar to the results shown for site E, but stream water concentrations were lower in the former.

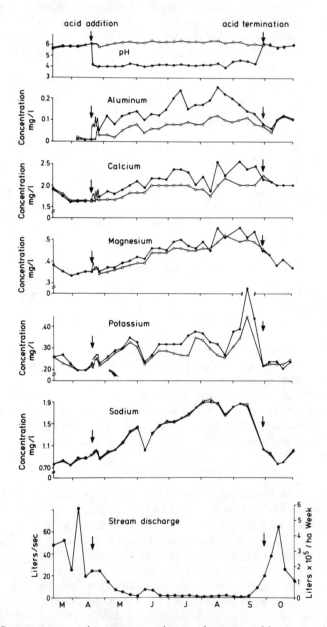

Fig. 2. Stream water cation concentrations and mean weekly stream discharge for site A (reference) and site E (treatment) from mid-March to end of October 1977. From Hall et al., 1980.

The pH in the reference section varied from 5.9 to 6.4 for the duration of our experiment, while the mean pH (range 3.9–4.5) at site E was 4.0. In general an inverse relationship existed between the cation concentrations (Al, Ca, Mg, K, and Na) and stream water discharge in both the reference and treatment areas. Al and Ca were significantly higher ($p < 0.05$, Student's t test, paired samples) in the stream water in the treatment section relative to the reference area (Table 1). Mg ($p = 0.2$), K ($p = 0.3$), and Na ($p = 0.5$) showed no significant difference between the two stations, but the former two cations were mobilized somewhat into the water from the stream bottom during low discharge (June through mid-September). Stream cation concentrations in the treatment area returned quickly (within one day) to reference values after termination of acid addition.

Table 1. Pairwise comparison (Student t test) for stream
water cation concentrations for reference (site A)
and treatment areas (site E) from 18 April to
20 September 1977

$N = 41$ or 42

Element	Site A		Site E	
	$\bar{X}$	S.D.	$\bar{X}$	S.D.
Al	0.095	± 0.15	0.15 ± 0.09[a]	
Ca	1.90	± 0.29	2.1 ± 0.33[b]	
Mg	0.42	± 0.07	0.44 ± 0.08[c]	
K	0.30	± 0.10	0.32 ± 0.11[c]	
Na	1.27	± 0.36	1.28 ± 0.35[c]	

[a] $p < 0.01$.
[b] $p < 0.05$.
[c] Not significant.

Invertebrate Functional Groups

Drift densities of invertebrates (total numbers) assigned to functional groups (collectors, scrapers, predators, and shredders; see Merritt and Cummins, 1978) before acid addition (days 1 and 2) were not significantly different ($p > 0.1$, Mann-Whitney U test), both within and between sites A, B, and C (Fig. 3). Likewise, drift density of macroinvertebrates on days 3 and 4 at site A was not significantly different ($p > 0.1$) from days 1 and 2. However, total numbers in drift increased significantly ($p < 0.05$) on days 3 and 4 (sites B and C) with elevated acidity (Fig. 3). Macroinvertebrate collectors increased the most, followed by scrapers and predators. Shredders collected in the drift samples were not significantly affected at pH 4.0.

Species within the invertebrate community were affected differentially by increased acidity. For example, drifting mayflies (Ephemeroptera), true flies (Diptera), and some stoneflies (Plepoptera) increased in drift density during the first

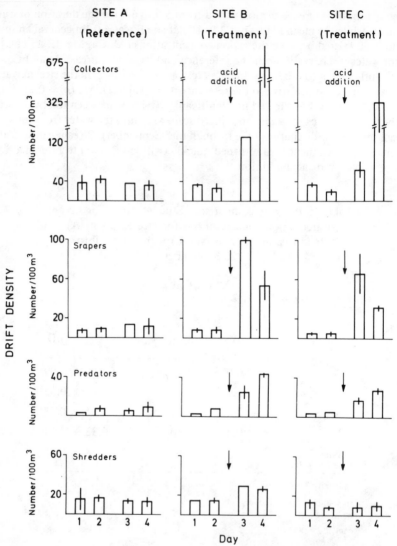

Fig. 3. Total numbers of invertebrates collected in drift nets (24-hr) apportioned to different functional groups. Bars represent range for two nets; vertical lines represent range. Days 1 and 2 (16, 17 April) were estimates of drift density before experiment; days 3 and 4 (18, 19 April) were after acid addition. Sites B and C were located 15 m and 50 m below acid addition. From Hall et al., 1980.

24 hr of acid addition (Hall et al., 1980), while the drift activity of some species of stoneflies, caddisflies (Trichoptera), an alderfly (Neuroptera), and a mayfly (Fiance, 1978) was not altered.

In contrast to the first two-day macroinvertebrate response to increased acidity, a different drift activity pattern occurred during the following month of continuous acid addition (Fig. 4). A significant increase in drift density at site B relative to A ($p < 0.05$) was discerned during the first week. However, no difference ($p > 0.1$) occurred in total numbers of invertebrates collected at sites A and B from the second to the fourth week of elevated acidity.

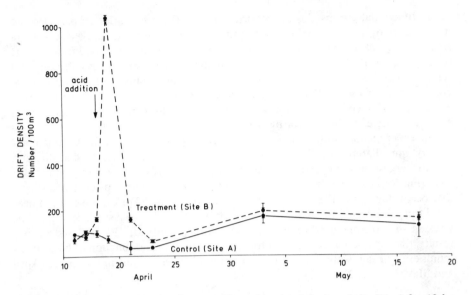

Fig. 4. Total number of invertebrates collected at site A (—) and site B(---) for 12-hr (1900–0700) intervals from 16 April to 17 May 1977. Closed circles represent mean for two drift nets; vertical lines represent range. From Hall et al., 1980.

DISCUSSION

Stream Water Chemistry

Increased acidity changed the biogeochemistry of the stream ecosystem. Al, Ca, and probably Mg and K were mobilized into the stream water from the bottom sediments. The mobilization of the metal cations was due to either replacement on organic or inorganic exchange complexes (Cronon et al., 1978; Johnson, 1979) or dissolution reactions with silicates or other minerals (Kramer, 1976; Johnson, 1979) in the stream sediments. Net losses of Ca, Mg, and Al from small watersheds on granitic rock in Norway also were directly related to the amount of hydrogen ion input (Gjessing et al., 1976).

The chemistry of headwater streams within the Hubbard Brook Experimental Forest is similar to the chemistry of incident precipitation; it is characterized by strong acids, and pH values as low as 4.0 have been observed. The higher acidity at the headwater stream reaches at Hubbard Brook reflects more closely the meteorological inputs of strong acids (Johnson, 1979) because there is hard granitic bedrock with very little surficial soil. The thin soils and hard crystalline bedrock are characteristic at higher elevations in similar areas within the Adirondack Mountains, New York (Schofield, 1977) and within the mountains in southern Scandinavian countries (Gjessing et al., 1976). The soil mantle becomes thicker downstream, and thus the stream water chemistry is more strongly influenced by the soils, resulting in the neutralization of excess strong acids from rain. Mobilization of Al plays a major role in this neutralization process (Johnson, 1979).

Invertebrate Functional Groups

The reduction in species that function as collectors, scrapers, and predators with increased acidity may have an effect on the quality and quantity of organic substrates in the stream ecosystem. All of the macroinvertebrates within the different functional groups play an important role in the degradation of particles and distribution of organic matter of terrestrial origin (predominantly leaves from riparian vegetation) and attached plants (periphyton) on the stream sediments (Cummins, 1974). The decreased number of species that function as collectors and scrapers may result in an increase of in-stream plant biomass. Quantitatively increased accumulations of attached algae have been observed in Norris Brook (Hall et al., 1980). Hendrey (1976) also noted a biomass increase in periphyton in seminatural stream channels at pH 4.0 in Norway.

Our experimental results show that sensitive organisms within the benthic community immediately modified their drift activity in response to elevated acidity and that the largest movement (total numbers) out of the study area occurred within two days (Fig. 4). The two-day period is similar to the short-term episodes of low pH that occur during the first part of snowmelt in the spring. For example, a pH near 4.0 during the beginning of spring snowmelt was recorded to last for less than one week in the Adirondack Mountains, New York (Schofield, 1977). Based on our results, this time period is sufficient to alter the structure and function of biotic communities in stream water with low dissolved ion concentrations in the spring (April and May; see Fig. 2).

ACKNOWLEDGMENTS

This is a contribution to the Hubbard Brook Study. Financial support for the field study at Hubbard Brook was obtained through the National Science Foundation and the International Paper Company Foundation. Facilities were provided by the Northeastern Forest Experiment Station, U.S. Department of Agriculture, Forest Service, Broomall, Pennsylvania. We acknowledge the technical assistance of J. Eaton, D. Buso, M. Hall, W. Martin, R. Moore, E. Soffey, N. Caraco, and R. Wesley.

REFERENCES

Almer, B., W. Dickson, C. Ekström, E. Hörnström, and U. Miller. 1974. Effects of acidification on Swedish lakes. Ambio 3:30–36.

Beamish, R. J., and H. H. Harvey. 1972. Acidification of the La Cloche Mountain Lakes, Ontario, and resulting fish mortalities. J. Fish. Res. Board Can. 29:1131–1143.

Cronan, C. S., W. A. Reiners, R. C. Reynolds, Jr., and G. E. Lang. 1978. Forest Floor leaching: contributions from mineral, organic, and carbonic acids in New Hampshire subalpine forests. Science 200:309–311.

Cummins, K. W. 1974. Structure and function of stream ecosystems. BioScience 24:631–641.

Fiance, S. B. 1977. Distribution and biology of mayflies and stoneflies of Hubbard Brook, New Hampshire. Master's thesis. Cornell University, Ithaca, New York.

Fiance, S. B. 1978. Effects of pH on the biology and distribution of *Ephemerella* (Eurylophella) *funeralis* (Ephemeroptera: Ephemerellidae). Oikos, 31:332–339.

Gjessing, E. T., A. Henriksen, M. Johannessen, and R. F. Wright. 1976. Effects of acid precipitation on freshwater chemistry. Pp. 64–85 in Braekke, F. W. (ed.), Impact of Acid Precipitation on Forest and Freshwater Ecosystems in Norway. Research Report 6/76, SNSF (Sur Nedbørs Virkning På Skog Og Fisk) Project, Oslo, Norway.

Gorham, E. 1976. Acid precipitation and its influence upon aquatic ecosystems—an overview. Water, Air, Soil Pollut. 6:457–481.

Grahn, O., H. Hultberg, and L. Lander. 1974. Oligotrophication—a self-accelerating process in lakes subjected to excessive supply of acid substances. Ambio 3(2):93–94.

Hall, R. J., G. E. Likens, S. B. Fiance, and G. Hendrey. 1980. Experimental acidification of a stream in the Hubbard Brook Experimental Forest, New Hampshire. Ecology, in press.

Hendrey, G. R. 1976. Effects of pH on the growth of periphytic algae in artificial stream channels. SNSF Project, Research Report 25/76, Oslo, Norway.

Hendrey, G. R., K. Baalsrud, T. Traaen, M. Laake, and G. Raddum. 1977. Acid precipitation: some hydrobiological changes. Ambio 5:224–227.

Hendrey, G. R., and R. F. Wright. 1976. Acid precipitation in Norway: effects on aquatic fauna. Pp. 192–207 in Proc., First Specialty Symposium on Atmospheric Contributions to the Chemistry of Lake Waters, Orillia, Ontario, Canada.

Johnson, N. M. 1979. Acid rain: neutralization within the Hubbard Brook ecosystem and regional implications. Science 204:497–499.

Kramer, J. R. 1976. Geochemical and lithological factors in acid precipitation. U.S.D.A. Forest Service Gen. Tech. Rep. NE-23, pp. 611–618.

Leivestad, H., G. Hendrey, I. P. Muniz, and E. Snekvik. 1976. Effects of acid precipitation on freshwater organisms. Pp. 87–111 in Braekke, F. W. (ed.), Impact of Acid Precipitation on Forest and Freshwater Ecosystems in Norway. Research Report 6/76, SNSF Project, Oslo, Norway.

Leivestad, H., and I. P. Muniz. 1976. Fish kill at low pH in a Norwegian river. Nature 259:391–392.

Likens, G. E. 1976. Acid precipitation. Chem. Eng. News 54:29–44.

Likens, G. E., R. F. Wright, J. Galloway, and T. Butler. 1979. Acid rain. Sci. Am. 241(4):39–47.

Merritt, R. W., and K. W. Cummins. 1978. An Introduction to the Aquatic Insects of North America. Kendall/Hunt, Dubuque, Iowa.

Schofield, C. L. 1976. Acid precipitation: effects on fish. Ambio 5:228–230.

Schofield, C. L. 1977. Acid snow-melt effects on water quality and fish survival in the Adirondack Mountains of New York State. Research Program Technical Comprehensive Report No. A-072-NY. Office of Water Research Technology, U.S. Department of the Interior.

Sprules, G. W. 1975. Midsummer crustacean zooplankton communities in acid-stressed lakes. J. Fish. Res. Board. Can. 32:389–395.

Strahler, A. N. 1957. Quantitative analysis of watershed geomorphology. Trans., Am. Geophys. Union 38:913–920.

Sutcliffe, D. W., and T. R. Carrick. 1973. Studies on mountain streams in the English lake district I. pH, calcium and the distribution of invertebrates in the River Dudden. Freshwater Biol. 3:437–462.

Waters, T. F. 1962. Diurnal periodicity in the drift of stream invertebrates. Ecology 43:316–320.

CHAPTER 44

ECOLOGICAL EFFECTS OF EXPERIMENTAL WHOLE-LAKE ACIDIFICATION

D. W. Schindler*

ABSTRACT

Experimental acidification of an oligotrophic lake in the Canadian Precambrian Shield has been attempted in order to examine details of the acidification process which are impossible to study in areas with a long previous history of acid precipitation inputs. It is apparent that a number of dramatic changes in the aquatic ecosystem occur before acidification has progressed to pH levels which are toxic to fishes. More attention must be given to the effects of acidification on organisms other than fish if ecosystems are to be maintained in something like their present state.

INTRODUCTION

This paper presents a brief summary of the experimental acidification of an oligotrophic lake in the Canadian Precambrian Shield. More detailed work on various aspects of the study will appear in the *Canadian Journal of Fisheries and Aquatic Science* early in 1980.

Lake 223 is a small lake in our Experimental Lakes Area. Its alkalinity when we began the experiment was about 100 μeq per liter, with a pH of about 6.5 to 6.8. The 150-ha watershed is untouched, containing stands of virgin black spruce and jack pine. It's several miles from the nearest road and several hundred miles from the nearest large SO_2 source. It has fish populations of about 300 lake trout, about 600 suckers, and a few other species of small cyprinids. Precipitation in the area currently averages about pH 4.9 to 5.0, as high as the pH of any precipitation being collected in the Precambrian Shield of North America today.

The experiment was designed to examine some of the details of the acidification process which are impossible to study in areas which have been subjected to acid precipitation for some time. by acidifying the lake directly, we can separate the effects of acid precipitation on lakes from secondary effects, such as the leaching of heavy metals and nutrients from terrestrial soils. We are able to examine changes which occur early in the acidification process, when buffering reserves are rapidly depleted but pH values are still high. We are also able to examine the physiological and embryological effects of acid on organisms which are sustained at controlled pH values on a year-round basis. Finally, the acid inputs to the lake are known precisely, facilitating the calculation of accurate chemical budgets.

*Department of Fisheries and Oceans, Freshwater Institute, 501 University Cresent, Winnipeg, Manitoba R3T 2N6, Canada.

Our acidification regime is shown in Fig. 1. We studied the lake for two years, 1974 and 1975, under nonacid conditions. In 1976, 1977, and 1978, we added enough sulfuric acid to lower the pH by about 0.25 unit a year. The lake is acidified heavily after ice-out in May to attain the desired pH value, then enough acid is added two or three times per week to keep the pH at the desired value. We have

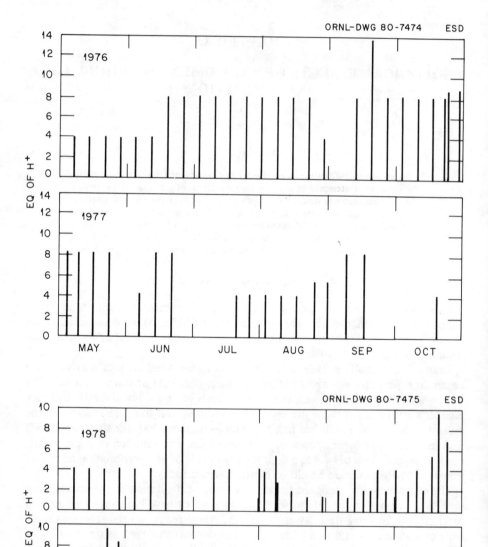

Fig. 1. Acid additions to the epilimnion of Lake 223, Experimental Lakes Area, from 1976 to 1979.

chosen this rather than a prescribed acidification regime so that people doing physiological and toxicological experiments can work with animals "conditioned" for long periods at constant pH. The regime is a reasonably natural one in that about half our precipitation falls as snow during the winter months and is flushed in during May. The rest is as periodic rains through the rest of the summer and fall. In 1977 we held the pH at 6.08, in 1978 at 5.84, and in 1979 at 5.60. Few of the data in the literature are for pH changes this small.

RESULTS

Most of the effects of acidification on water chemistry were as we expected. By late summer of the first year we had decreased the epilimnetic dissolved inorganic carbon (DIC) concentration to about 30% of what it had been in previous years (Fig. 2). During that period the pH decreased only about 2/10 of a unit. While this

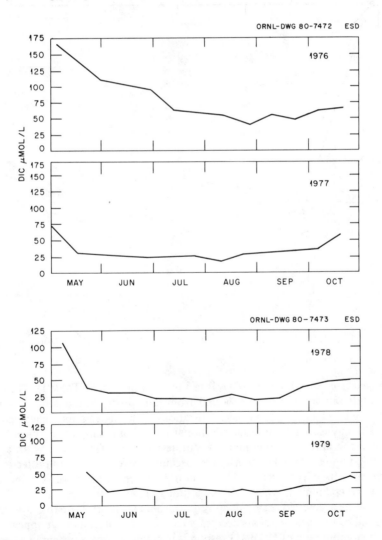

Fig. 2. Epilimnion concentrations of dissolved inorganic carbon (DIC), Lake 223, from 1976 to 1979.

would be expected by anyone familiar with typical bicarbonate titration curves, I think that the message for people who are trying to detect effects of acidification in the field is to measure alkalinity, not pH. Unfortunately, alkalinity is probably one of the least precise measurements that most limnologists make in soft waters. The Gran titration procedure (Stumm and Morgan, 1970) needs to be widely adopted.

The lost bicarbonate is, of course, transformed to CO_2 gas. The pCO_2 of surface water does not increase as a result, because even in such small lakes the gas exchange is high enough to discharge excess CO_2 quickly. If anything, surface water pCO_2's after acidification were lower than in previous years (Fig. 3). When all of the inputs and outputs for DIC are balanced, it is also clear that most of the carbon that's be-

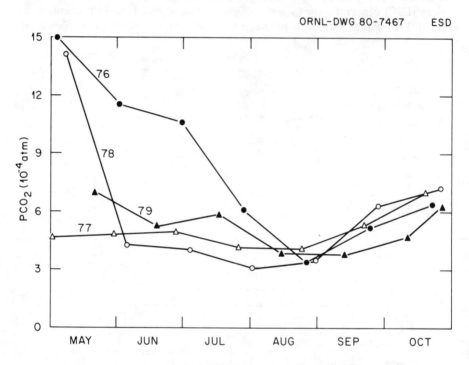

ORNL-DWG 80-7467   ESD

Fig. 3. pCO_2 in Lake 223 surface waters after acidification.

ing changed from bicarbonate to CO_2 is lost to the atmosphere (Table 1). The bicarbonate depletion and pH drop expected from laboratory titrations of lake water were not realized in the whole lake "titration." Through various chemical and biological analyses, we found that additional buffering is supplied by bicarbonate released in the anoxic hypolimnion by sulfate reducers. This lake has an anoxic hypolimnion under normal circumstances because it's very small and protected and tends to carry its oxygen deficit over from one season to the other. On the other hand, in the epilimnion, where oxic conditions prevail, the fraction of buffering which can't be accounted for by dissolved bicarbonate is very close to our limits of error for all of the measurements necessary to make the calculation. It appears as if the summer epilimnion of Lake 223 responds like a bicarbonate system in a beaker. In other words, most of the unaccounted for buffering capacity in the lake is in the anoxic hypolimnion.

Table 1. Dissolved inorganic carbon budgets for Lake 223, 1976 and 1977
In kilograms

	1976	1977
1. Inflow stream	45	204
2. Direct drainage[a]	319	439
3. Precipitation[b]	0	0
4. Total input	364	643
5. Outflow	344	149
6. Total input minus outflow	20	494
7. Observed ΔM[c]	− 134	− 1118
8. Presumed lost to atmosphere (6 − 7)	154	1612

[a]Assumed to be equal in yield/area to the northwest subbasin of Rawson Lake (Schindler et al., 1976).

[b]Atmospheric saturation with CO_2 gas is ignored. Bicarbonate content was negligible.

[c]M = mass in lake. Data used are for 8 January 1976, 27 January 1977, and 11 January 1978.

So far, acidification does not seem to have caused the decreased decomposition which has been hypothesized from Scandinavian studies, at least not in the hypolimnion (Grahn et al., 1974; Hendrey et al., 1976) (Fig. 4). What has happened is that sulfate-reducing bacteria have increased their activities as more and more sulfate is

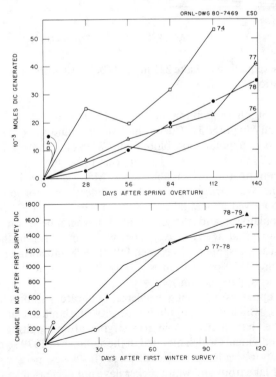

Fig. 4. Dissolved inorganic carbon generated by decomposition in Lake 223. (a) Summer hypolimnion, (b) winter under ice.

added to the lake (Fig. 5). It will be of considerable interest to see whether these decomposers become saturated as sulfate increases or whether decomposition does decrease at lower pH values.

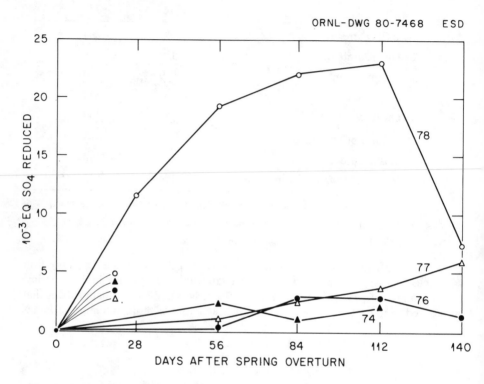

Fig. 5. Sulfate reduction in Lake 223 in 1974 (before acidification) and 1976–1978 (during acidification).

The response of phytoplankton in Lake 223 is also not consistent with the "oligotrophication" hypothesis. The production and standing crop of algae actually increased, due to a hypolimnetic bloom of *Chlorella* (Table 2). The lake has, however, increased in transparency. This appears to be due to changes in dissolved color rather than in phytoplankton abundance. Major changes have taken place in the epilithiphyton, with *Mougeotia* reaching epidemic proportions in 1979. This genus was also observed in acid lakes in Sweden (Grahn et al., 1974).

Changes in zooplankton and emerging benthos have not been completely analyzed so far. We do know that *Mysis relicta*, a prominent organism in the diets of Lake 223 lake trout, nearly disappeared between fall of 1978 and spring of 1979, at a pH of 5.55. The spring 1979 population was only about 3% of the previous one (I. Davies and W. Nero, pers. comm.).

Crayfish, too, appear to be having problems. Laboratory studies have shown that *Orconectes virilis*, which has a population density of about one animal per square meter in the lake, has difficulties in recalcifying after moulting at pH 5.5–5.6 (Malley, 1980). Individuals which moulted in the summer of 1979 remained soft, but so far the population size has not been affected (I. Davies, pers. comm.).

Populations of lake trout and white sucker have not decreased so far. The former species began experiencing reproductive problems in 1978. Eggs laid that fall were

Table 2. Phytoplankton volume and chlorophyll *a* in Lake 223, before and after acidification

Year	Phytoplankton volume (mg/m³)		Chlorophyll (mg/m³)[a]	
	$\bar{x}$ epi[a]	Whole lake[b]	$\bar{x}$ epi[a]	Whole lake[b]
1974[c]	775	991	1.9	3.9
1975	605	949		3.6
1976	852	1168	1.7	2.7
1977	948	2070	2.6	3.7
1978	1079	1501	2.2	3.5
1979	877	1662	2.5	5.4

[a]Mean concentration in the epilimnion in the ice-free season.
[b]Mean annual value for all strata.
[c]Euphotic zone.

much smaller than in control lakes, and high mortalities were observed among developing embryos. The incidence of embryonic deformities was also much higher than normal (Kennedy, 1980). It may be no accident that only a few percent of lake trout lakes in Ontario have pH values below 6.0 (Martin and Olver, 1976).

DISCUSSION

A number of dramatic changes in the aquatic ecosystem take place before acidification has progressed to pH values which are toxic to fishes. We must pay much more attention to the effects of acidification on organisms other than fish if we wish to maintain ecosystems in something like their present state.

It is strange that no evidence of "oligotrophication" has appeared in Lake 223 so far. While the possibility remains that symptoms will develop rapidly as the pH is lowered below 5.5, it is possible that there are other explanations.

Due to the lack of preacidification standing crop, respiration, and productivity data for the Scandinavian lakes which are thought to have become oligotrophic, it was necessary to deduce their condition either from short-term experimental studies or by comparison with less acid lakes (Hendrey et al., 1976).

It seems possible that the lower standing crops of invertebrates found in acid lakes could be due to a low nutrient income, rather than more acid conditions. As far as I am aware, no author has examined lakes for this possibility. Lakes which are susceptible to acidification tend to be in small headwater basins with hardrock geology, conditions which also produce low nutrient incomes. It should be possible to test this hypothesis by comparing phytoplankton standing crops in the acid lakes with recent eutrophication models based on phosphorus loading and water renewal, such as those of Vollenweider (1976), Dillon and Rigler (1974), and Schindler et al. (1978).

Current plans are to continue to acidify Lake 223 to at least pH 5.0, reaching that value in the summer of 1981. It is obvious that such experiments can help to interpret changes in lakes which have been acidified due to atmospheric pollutants, under conditions which are difficult to quantify.

REFERENCES

Dillon, P. J., and F. H. Rigler. 1974. A test of a simple nutrient budget model predicting the phosphorus concentration in lake water. J. Fish. Res. Board Can. 31:1771–1778.

Grahn, O., H. Hultberg, and L. Landner. 1974. Oligotrophication—a self-accelerating process in lakes subjected to excessive supply of acid substances. Ambio 3(2):93–94.

Hendrey, G. R., K. Baalsrud, T. Traaen, M Laake, and G. Raddum. 1976. Acid precipitation: some hydrobiological changes. Ambio 5:224–227.

Kennedy, L. A. 1980. Teratogenesis in lake trout (*Salvelinus namaycush*) in an experimentally acidified lake. Can. J. Fish. Aquat. Sci., under review.

Malley, D. F. 1980. Decreased survival and calcium uptake in the crayfish, *Orconectes virilis* in low pH. Can. J. Fish. Aquat. Sci. 37(3): (in press).

Martin, N. V., and C. H. Olver. 1976. The distribution and characteristics of Ontario lake trout lakes. Ontario Min. Nat. Res., Fish Wildlife Res. Br., Res. Rep. 97.

Schindler, D. W., E. J. Fee, and T. Ruszczynski. 1978. Phosphorus input and its consequences for phytoplankton standing crop and production in the Experimental Lakes Area and in similar lakes. J. Fish. Res. Board Can. 35(2):190–196.

Schindler, D. W., R. W. Newbury, K. G. Beaty, and P. Campbell. 1976. Natural water and chemical budgets for a small Precambrian lake basin in central Canada. J. Fish. Res. Board Can. 33:2526–2543.

Schindler, D. W., R. Wagemann, R. Cook, T. Ruszczynski, and J. Prokopowich. 1979. Experimental acidification of Lake 223, Experimental Lakes Area: I. Background data and the first three years of acidification. Can. J. Fish. Aquat. Sci. 37(3):000–000.

Stumm, G. M. W., and J. J. Morgan. 1970. Aquatic Chemistry: An Introduction Emphasizing Chemical Equilibria in Natural Waters. Wiley & Sons, New York.

Vollenweider, R. A. 1976. Advances in defining critical loading levels for phosphorus in lake eutrophication. Mem. Ist. Ital. Idrobiol. Dott Marco de Marchi 33:53–83.

DISCUSSION

Ron Hall, Cornell University: Do you think that the reduction of the humic materials may be related to any aluminum increase? I'd like to know, did the aluminum increase to levels that may account for the decrease in humic material, one, and then I'd like to know how you added acid—in concentrated form or diluted form? The reason I ask that is that it strikes me that the organisms that were affected, organisms such as the crayfish, are found in the littoral areas of the lake, and so it's important in terms of adding this high-density acid in relation to mixing and how it affects the organisms.

D. W. Schindler: Okay, firstly, I said that the change in clarity was due to the change in color of the humic acid; we had no change in quantity that you could detect as dissolved organic carbon measurement. If you look at what the brown color does with respect to pH, in general, if you go more acid, starting at about pH 7 you lose color from humic acids. The aluminum concentration has maybe doubled or tripled, which means that it's 10 micrograms per liter, perhaps, right now, so it probably isn't enough to have accounted for any decrease in humic material. With respect to the acidification, what we do is to dump concentrated sulfuric acid into the prop wash from a moving boat. Now this lake is nearly 10^6 cubic meters, and a typical addition at any one time is something about 50 liters. Going out immediately after and doing spot sampling around the lake, we don't detect any spotty heterogeneous concentration pattern, so I don't think we're doing too badly in that respect.

George Hendry, Brookhaven: David, one of the things you mentioned is, you don't have the complication of aluminum added from the watershed because you're applying it directly to the lake. Dickson has shown that aluminum effectively complexes phosphorus, especially at about the pH that you're at now, about 5.5, and as you go downward from there the complexing is less effective, so one thought that occurs to me is that if you have in natural lakes two processes going on, one is a tendency to bind phosphorus at intermediate ranges, another is perhaps releasing that binding as you go to more acid ranges, but you're also producing a great deal more aluminum from the watershed as you increase the acidity of lakes, in naturally stressed lakes.

D. W. Schindler: I think that's right, but my reply to how I'd prefer to approach that would be, I'd like to acidify the terrestrial watershed and look at the bottom of that. It's like trying to look at two ecosystems by analyzing what happens at the bottom of one; it's sort of absurd, like analyzing human populations by looking at what happens in the sewer.

George Hendry: A second question, or comment, is that if indeed you're changing the color of the humus, that is, making it invisible, your light penetration going deeper into the lake ought to really increase your phytoplankton productivity.

D. W. Schindler: I think that's exactly what's happened.

Carl Schofield, Cornell University: I wondered if you had any further insight in terms of the mechanisms that were stressing the lake trout population that caused the failure in embryogenesis. Is there any indication that the fish changed their distribution within the lake following acidification relative to the changes in oxygen pattern and sulfate reduction?

D. W. Schindler: Not so far. The bottom of the hypolimnion went anoxic even before we started, as I mentioned, and the total oxygen consumption appears to be a function of the completeness of spring overturn rather than what we're doing; I glossed over that rather superficially, pointing out that you should look at that sulfate reduction bar and ignore the oxygen component. I don't think we're affecting O_2-related decomposition at all. The spawning takes place in the same areas, and we're catching fish with nets for mark and recapture in the same areas that we were at the start; that's really all that I know on that. Various physiological parameters that have been looked at, such as serum calcium and so on, don't seem to show any effect. The only thing we've seen is the smaller egg diameter, and of course, if you look way back in the classical literature, when that's about all people could measure, and they found poorer performance of smaller eggs and embryos in general, but we really can't link that chemically.

Carl Schofield: I was thinking more in terms of the summer distribution of the trout prior to spawning, which would be the period when the developing embryos—

D. W. Schindler: I see. We normally don't sample these fish too heavily during the warm part of the summer, because they're pretty vulnerable. We sample them at spring overturn and then beginning in early September for about a month before spawning, and there's been no indication during that period of any changes in distribution; at least we're catching them with equal success, which might be a pretty rough indicator.

R. A. N. McLean, Domtar, Inc.: First of all, David, for my first question, I'm sorry I didn't meet you four years ago to ask you this, but have you done any measurements of mercury in these fish, because, as you know, there is a considerable controversy about the effects of acidification on the uptake of methylmercury into fish; the second question has to do with your humate change. In changing the composition of the humate, you would change drastically the levels of heavy metals in the lake; some of them one would expect to go into solution, others you would expect to precipitate out; have you done anything with either of these?

D. W. Schindler: Okay, first with respect to the mercury, we didn't think of that four years ago; we did about two years ago, and luckily we get a few mortalities from our mark and recaptures; we save these in sort of a frozen museum, and the plan is this winter to get out the first comparable set of specimens and see whether we have any increase. Secondly, with respect to the heavy metals, we have also done some pilot experiments in tubes that go to the sediments in the area at pH's down to 4, and so far the lake seems to be mimicking that pretty well; we're getting slight increases in aluminum, manganese, iron, and zinc, two- to threefold increases. I would predict from what we saw in the tubes that, as we approach pH 5 and lower, these will go up much more drastically. We're just beginning to see effects now; for some things like iron and manganese, these are difficult to dissect from the effects of anoxia, which, as I mentioned in response to Carl, are partly determined by seasonal phenomena rather than acidification.

SECTION VIII

REGIONAL-SCALE STUDIES OF ATMOSPHERIC DEPOSITION EFFECTS

Chairman: J. N. Galloway

CHAPTER 45

REGIONAL-SCALE STUDIES OF ATMOSPHERIC DEPOSITION EFFECTS

Chairman: **James N. Galloway**
University of Virginia

OVERVIEW

James N. Galloway*

Historically, environmental contamination has been viewed as a local problem, with effects limited to the areas surrounding the source of the contaminant. In the past decade, however, the phenomenon of regional and global environmental contamination has been realized. The reason for this increased awareness is simply one of better observation of the changes in quality of our environment.

Common examples of regional contamination are increases in the concentration of CO_2 in the global atmosphere and increases in the concentration of acids, nutrients, metals, and organics in atmospheric deposition on a regional and perhaps global scale.

Whenever anthropogenic activity causes a change in the environment, the next appropriate action is to assess the effects of that change on human health and on ecosystem welfare. This session of the symposium was designed to evaluate that welfare assessment. Specifically, on a regional basis, what are the environmental effects of increased levels of sulfur in the atmosphere?

Clearly the first step in the assessment process is to define the extent of the perturbation. The paper "National Atmospheric Deposition Program: Analysis of Data from the First Year," by Miller, discusses the national Atmospheric Deposition Program (NADP), which has been established on a long-term basis.

Since the program has just finished its first year, there is not an extensive data base. But preliminary analysis indicates that (1) the program has to be long-term to successfully assess trends in the composition of atmospheric deposition, (2) quality control on sampling and analysis has to be enthusiastic and rigorous to create a data base that is both usable and integratable with other data bases, (3) programs on the *measurement* of the composition of atmospheric deposition and programs on the *effects* of atmospheric deposition have to be tightly coupled to insure that data analysis is both complementary and integrated.

---

*Department of Environmental Sciences, University of Virginia, Charlottesville, Virginia 22903.

Sulfur in atmospheric deposition interacts with the environment in a series of steps that can be conceptualized as layers. First, the deposition interacts with the forest or grassland canopy, followed by soil, the bedrock, and then streams and lakes. The remaining papers of the session address these layers.

The paper "Atmospheric Canopy Interaction of Sulfur in the Southeastern United States" by Parker, Lindberg, and Kelly assesses the contribution of atmospheric sulfur compounds to forested ecosystems by internal cycling of sulfur from the soil through the tree and out of the leaf. The problem addressed in the paper is a difficult one, for the sulfur found in throughfall has three potential sources: wet deposition, dry deposition, and leaching from the leaf (internal cycling); however, only the former can be simply and accurately measured. By looking at the seasonal trends, volume and concentration relationships, and location of forests relative to sulfur sources, the authors address the dilemma of one equation and two unknowns. Their conclusion is that, depending on the forest type and proximity to sulfur sources, dry deposition and internal cycling of sulfur are of comparable magnitude. This, coupled with the sulfur in wet deposition, means that more than 50% of the sulfur received by the forest floor is anthropogenic in origin. Since some Southeastern forests are deficient in sulfur, this increased deposition may be beneficial. However, most of the increased deposition of sulfur is accompanied by an equivalent amount of hydrogen and ammonium ions, both of which could potentially acidify forest soils. At this time it is not known whether the beneficial effects of fertilization are greater or less than the harmful effects of acidification in these Southeastern forests.

In the final analysis, most environmental problems are presented and evaluated on the landscape/system level. By strict definition a community, in most instances, represents an ecosystem since it includes living organisms and nonliving substances interacting to produce an exchange of materials between the living and nonliving parts, while a more widely held interpretation of the ecosystem concept would be a group of communities providing a much greater variety of biological expression and physical features. The ecosystem concept is somewhat artificial in that it separates overlapping units into distinct bodies for the convenience of man's understanding. Nevertheless, response at the ecosystem or landscape level is probably the least understood of all levels and the one in greatest need of quantification. The response observed at the ecosystem level is an integrated one because of the differences in response within and between species and communities and the interaction of system processes, mechanisms, and transfers. The increased complexity and associated variability of work under essentially uncontrolled conditions require several years of constant observation as well as comparative studies at several locations in order to obtain a realistic evaluation of the many facets of system response.

The papers presented in this section deal with the impacts of atmospheric sulfur deposition at the landscape/system level. Although in many cases individual species and/or processes are evaluated, an attempt is made to relate the observed response to the other environmental factors which mediate the observed response and are themselves impacted by the mediated response. While considerable progress—as evidenced by the work of Abrahamson and his Scandinavian colleagues—has been made in evaluating ecological impacts on coniferous forests, much work is yet to be done; especially needed are studies of temperate deciduous and mixed deciduous-coniferous forests on a variety of geologic bases. The holistic approach of Lauenroth and associates to the evaluation and prediction of SO_2 effects on shortgrass ecosystems is an excellent example of a well-integrated study of impacts at the ecosystem level. Again, as in the case of the coniferous forest studies, additional work needs to be undertaken in a variety of grassland systems in both the eastern and western United States.

Studies of the impact of atmospheric sulfur on agroecosystems has generally been limited to the evaluation of negative effects of acute SO_2 injury on crop yield. It is now recognized that atmosphere-derived sulfur may play a significant role in satisfying the sulfur requirements of agricultural crops. As Jones's paper points out, there is a definite need for studies to evaluate the role of atmospheric sulfur in both crop nutrition and production as well as potential impacts on the general nutrient status of agricultural soils.

The importance of the soil solution interface between terrestrial and aquatic ecosystems is pointed out by the paper of Henderson. The terrestrial-aquatic interface is probably the least quantified of the important interecosystem transfers and certainly deserves additional evaluation in terms of the impact of sulfur additions on the quality of water entering streams and lakes. Special emphasis should be placed on anion-cation leaching and the relative roles of natural and anthropogenic sources of acidity as related to altered aluminum solubility. An increase in aluminum solubility, aside from its direct effects in aquatic systems, could alter the stability of the terrestrial ecosystem through impacts on aluminum-sensitive macro- and microfloral and faunal species.

The papers of Hall and Schindler summarize the growing body of knowledge relating to the impact of changing water quality on both lake and stream ecosystems. In most cases the impacts have been attributed primarily to sulfur through its impact on precipitation chemistry and subsequent increases in aluminum inputs. Again, the importance of the terrestrial-aquatic interface becomes critical in determining what trophic levels within these aquatic systems will be impacted either directly or indirectly.

In summary, based on the information presented here, there is a strong need for additional work at the ecosystem level if we are to fully understand the true environmental impact of atmospheric sulfur addition. Holistic evaluations of agroecosystems, deciduous forest ecosystems, and the interface between the terrestrial and aquatic environments are definitely needed.

Just as the forest canopy acts as a dynamic layer in the forested ecosystems, so do soils. The next three papers address various aspects of the soil layer in its response to the increased atmospheric deposition of sulfur. The paper "Regional Pattern of Soil Sulfate Accumulation: Relevance to Ecosystems Sulfur Budgets," by Johnson, Hornbeck, Kelly, Swank, and Todd, addresses the questions, Once sulfur enters the soil, where does it go? If it is retained in the soil, what is the form? These questions are important for our understanding of how forests as systems respond to sulfur deposition. The results of the authors support the hypothesis that sulfur does accumulate in some watersheds due to inorganic sulfate adsorption in the soils. Watersheds that accumulated little or no sulfate have smaller amounts of free iron and/or higher amounts of organic material. The study was based on detailed analysis of soils from three watersheds in Tennessee, New Hampshire, and North Carolina.

The next two studies used less detailed data from many more sites. The two papers, "A Regional Ecological Assessment Approach to Atmospheric Deposition: Effect on Soil Systems," by Klopatek, Harris, and Olson, and "Sensitivity of Regions to Long-Term Acid Precipitation," by McFee, both used extensive computer data bases on soil pH, cation exchange capacity, base saturation, and soil moisture to determine the sensitivity of soils to acidification by acid precipitation. The two papers were different in both type of data base and scope but came to similar conclusions: (1) Sensitivity is dependent on a wide variety of parameters, some well known, others not. (2) Soils sensitive to acid precipitation exist in the eastern United States. (3) The exact area containing sensitive soils is unknown.

The next layer in the ecosystem to contact precipitation is the bedrock. In actuality, acid deposition does not affect bedrock but rather the opposite. Bedrock with large amounts of buffering capacity—limestones, marbles, glassy volcanic rocks, ultramafic rocks, etc.—will weather to produce soils with such high buffering capacity as to be unaffected by acid precipitation. The paper "Geologic Factors Controlling the Sensitivity of Aquatic Systems to Acid Precipitation," by Norton, divides bedrock into four classes that vary from bedrock with little buffering capacity (type 1) to very high amounts (type 4). These types are plotted on the scale of the eastern United States to act as a guide to the location of freshwaters that will be sensitive to acidification by virtue of their association with bedrock and soils of low buffering capacity. Field checks of the analysis verify that acidification of surface waters and impact on biological systems due to acid precipitation are occurring in areas classified as types 1 and 2.

The final layers in the movement of rainwater through the environment are the freshwater systems. The composition of lakes and streams is controlled by atmospheric deposition, characteristics of watershed (type of vegetation, soil and bedrock geology), and biological processes occurring in the water and watershed. The paper "Implications of Regional Scale Lake Acidification," by Schindler, attempts to assimilate the information existing on lake acidification and assess its implications. As is stated in the paper, "The assessments . . . are miserably scanty. While we now have relatively good programs on this continent for measuring deposition and trajectories, and a good start on mapping of sensitive areas . . . , we still have no idea of how rapidly waters are being acidified under different precipitation regimes." In other words, we know it's happening, we know where it is and could be happening, but we don't know how much and how fast.

Taken together, all of the papers in this session point to a similar conclusion. Increased concentration of sulfur in the atmosphere has effects on all levels of an ecosystem. It's up to us to be clever enough to complete the assessment of their magnitude.

CHAPTER 46

NATIONAL ATMOSPHERIC DEPOSITION PROGRAM: ANALYSIS OF DATA FROM THE FIRST YEAR

John M. Miller*

ABSTRACT

The National Atmospheric Deposition Program (NADP) is an interagency project sponsored by the state agricultural experiment stations, U.S. Department of Agriculture, the Environmental Protection Agency, and others. Originally designated NC-141, one of its major goals is to determine spatial and temporal trends in the deposition of chemicals, both harmful and beneficial, on the land and surface waters. The first precipitation samples for chemical analysis were taken in July 1978. A year later, over 30 sites were part of the program, and their weekly collected rain samples were shipped to a central laboratory for the analysis of all the major ions.

This paper presents data from the first year (July 1978–June 1979), which were taken during a period of atypical meteorological conditions: a very dry period in the fall of 1978 and an extremely wet period in the winter of 1979.

Selected sulfate deposition data that were available from the network were compared with values from other networks, both active and inactive. The comparison showed that precipitation amounts and sulfate concentrations collected simultaneously at the same site by two different networks varied enough to cause disagreement in sulfate depositions. Though there are not enough data to make comprehensive evaluation, this preliminary review points out the need for long-term commitment for continuing the network, for a continuous data review process, and for a vigorous quality assurance program.

INTRODUCTION

"The eventual consequences from increasingly acid rainfall are obviously immense—both in dollar costs and in possibly irreversible damage to various forms of life" (*Washington Post* editorial, August 1979).

Recent publicity of the acid rain problem has brought to the fore an area of environmental research that not too many years ago was grossly undersupported, particularly in the United States. Europeans, who in the 1950s established a long-term precipitation chemistry network, were the first to recognize that a possible problem existed (Oden, 1968). Even when first indications of high precipitation acidity in North America were published (Likens et al., 1972), there was little support to establish a national cohesive program of measurement and research. Now that the

---

*Air Resources Laboratories, National Oceanic and Atmospheric Administration, Silver Spring, Maryland 20910.

problem has been identified, it is time to pursue an aggressive national and international program to investigate this environmental problem.

Scientists have recognized that acid rain is only a part of the larger field of atmospheric deposition. Materials of all sorts are emitted into the atmosphere, where they may be transported various distances (1 km to 1000 km) from their source. For example, sulfur dioxide and nitrogen dioxide, precursors of the acidity in rain, are emitted in large quantities into the environment. Along with these well-known substances, there are other materials, such as trace metals, organic substances, and natural substances such as sea salt, ammonia, etc. All these chemicals are then deposited on the earth's surface either by wet deposition (rain, snow) or by dry deposition. Therefore any program to measure and evaluate acid deposition should include measurements of all pathways to the earth's surface. We still do not understand the relative importance of the various paths materials follow to the land and to surface waters.

To investigate this problem, the National Atmospheric Deposition Program (NADP) was conceived. Though there had been earlier attempts to establish a national precipitation chemistry network (usually ignoring the dry-deposition aspect), all efforts were (1) only for short duration, (2) so poorly funded that the quality of the data is questionable, and (3) regional rather than national in scope (Wisniewski and Miller, 1977). In 1975, Prof. Ellis Cowling, of North Carolina State University, suggested a fresh approach using the concept of regional research developed by the Department of Agriculture and its cooperating agricultural experiment stations. A network of core sites would be established at agricultural experiment stations. These sites have two major advantages; they are located far from large local contamination sources and would have interested scientists to supervise the collection. Other federal, state, and private agencies were welcomed to join the project under the umbrella of the Department of Agriculture's North Central region. Through Prof. Cowling's leadership and help from a community of over 100 scientific colleagues, a viable program was organized which included an executive committee and numerous technical committees to guide the founding and direction of the network. The first samples were collected on July 11, 1978.

NETWORK STATUS AND PROCEDURE

After one year of operation through July 1979, the NADP network has over 30 active sites (Fig. 1). Since that time a number of applications have been accepted to include new sites in the network. The number of locations may eventually include as many as a hundred sites. A summary of the stations to date is presented in Table 1 and Fig. 2.

Details on site selection, sampling procedures, laboratory analysis, and data handling are all documented in internal reports available from the project director, Dr. Jim Gibson, Colorado State University. This information will eventually be published in the project's annual report. Basically, the network includes sites that are not influenced by local contamination and can be serviced on a weekly basis. All rainfall collections are made in the Aerochem Metric collector, which opens only during a precipitation event. Dry deposition collections are also made but will not be discussed in this paper. The weekly precipitation collection is sent to the Central Analytical Laboratory at the Illinois State Water Survey. These samples are analyzed for pH, conductivity, and the major ions (Table 2). The data are stored and then published on a quarterly basis and are thus available to everyone in the scientific community.

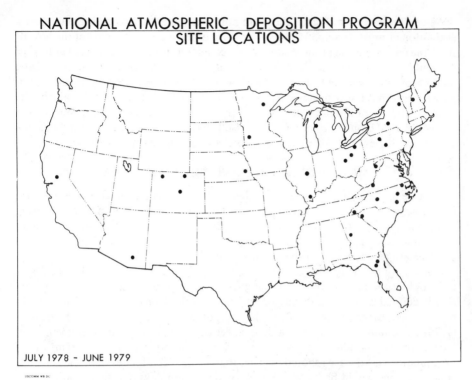

NATIONAL ATMOSPHERIC DEPOSITION PROGRAM
SITE LOCATIONS

JULY 1978 - JUNE 1979

Fig. 1. Location of active NADP sites as of July 1979.

Table 1. List of active NADP sites as of July 1979

State	Town	Station Number	Number of samples through June 19, 1979
Arizona	Tombstone	AZ01	4
California	Davis	CA88	20
Colorado	Manitou	CO21	18
Colorado	Sand Spring	CO15	10
Colorado	Pawnee	CO22	3
Florida	Bradford	FL03	26
Florida	Auston-Cary	FL00	23
Georgia	Georgia St.	GA41	30
Illinois	Bondville	IL11	13
Illinois	Dixon	IL63	17
Michigan	Wellston	MI53	25
Minnesota	Marcell	MN16	39
Minnesota	Lamberton	MN27	17
Nebraska	Mead	NE15	25
New Hampshire	Hubbard	NH02	41
New York	Huntington	NY20	31
New York	Aurora	NY08	7
N. Carolina	Lewiston	NC03	30
N. Carolina	Coweeta	NC25	33
N. Carolina	Piedmont	NC34	30
N. Carolina	Clinton	NC35	31
N. Carolina	Finley A	NC41	34
N. Carolina	Finley B	NC42	34
Ohio	Delaware	OH17	33
Ohio	Caldwell	OH49	32
Ohio	Wooster	OH71	33
Pennsylvania	Kane	PA29	22
Pennsylvania	Leading Ridge	PA92	7
S. Carolina	Clemson	SC18	9
Virginia	Horton St.	VA13	33
W. Virginia	Parsons	WV18	46

As can be seen from Fig. 2, there are not yet enough data to make a full year's evaluation at most stations. However, because of the location of some of the NADP sites, the data collected at these sites can be compared with those collected at past or presently existing sites. Table 2 shows the major United States networks and the number of sites that could be compared with NADP sites. The criterion for comparison was that the two sites must be within 100 km of each other. Details of the location and operation of the networks listed in Table 2 can be found in Wisniewski and Miller (1977) and Niemann (1979). Also listed in Table 2 are the ions that can be compared. An example of such a comparison is given later in this paper.

METEOROLOGY DURING THE FIRST YEAR

Because of the wide variation of weather of the continental United States it is hard to characterize a whole year, especially in terms of precipitation patterns. During any given month, some areas are under drought conditions while others have an overabundance of rain. Because the goal of NADP is to describe deposition, rain patterns are as important as the composition of the rain.

One way of reporting the general meteorological conditions over the United States is to look at the number of days of stagnation for a given month. A day of stagnation is defined as a period during which an anticyclone or high pressure dominates the weather. Air movement is sluggish, and pollutants tend to build up. Korshover (1976) has quantified this using a defined set of meteorological measurements that include pressure-gradient values. A stagnation can imply two things in terms of the precipitation: (1) the area should be drier than usual; (2) if it does rain, the chemical concentrations would be higher. A lack of stagnation days during a month would mean a wetter than usual period. Plotted in Fig. 3 is the average number of stagnation days over a 40-year record compared with the 1978–1979 period. From the published monthly precipitation values of the United States, there is indeed a very dry period in the fall of 1978 and an extremely wet period in winter 1979 (Fig. 4).

Though it is always difficult to characterize a given period as "typical," meteorological conditions during the first year of NADP operation were to some degree atypical. Because of these anomalies, measurements over a number of years must be made before a clear picture of deposition can be drawn.

NETWORK DATA

From Fig. 2 it can be seen that the data record is too short to make a definitive comparison. Seasonal variations cannot be shown at least until several more months of data become available. However, some preliminary comparisons can be made that might be useful.

These include

1. internal consistencies such as plotting time sequences, rain gage vs precipitation chemistry volume, laboratory vs field pH, etc.
2. comparisons with past or present precipitation collection data from other networks to determine if the deposition levels are consistent.

An example of the first type of comparison is shown in Fig. 5. Laboratory and field pH are plotted for Parsons, West Virginia. The general trend of higher acidity in the field can be seen from the data. This same phenomenon has been reported by other networks (Hakkarinen, 1979). Possible explanations for neutralization include poor field or laboratory measurements, contamination of the sample, slowly dissolving aerosols, degassing, and possibly others. This change should be investigated, because it may mean that an important chemical change takes place between collection at the site and analysis at the laboratory.

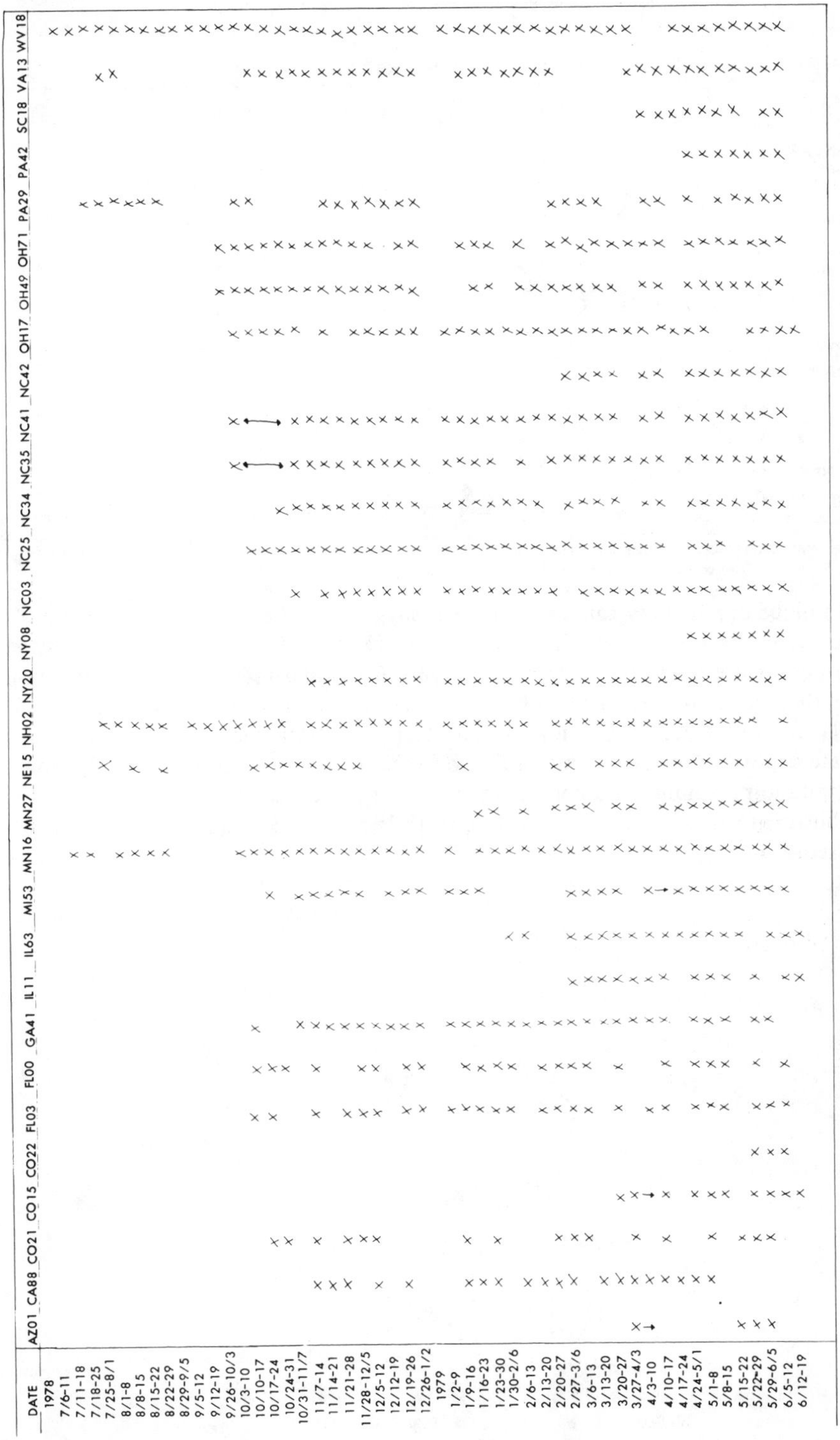

Fig. 2. Number of valid analyzed precipitation samples during the July 1978–June 1979 period.

Table 2. Major precipitation chemistry sites, past and present

Network	pH	Cond	Ca	Mg	K	Na	NH₄	NO₃	Cl	SO₄	PO₄	Type of collector	Comparable sites	Period
Junge 1955–56			X		X	X	X	X	X	X		Wet only[a]	8	Monthly
PHS-NCAR 1959–66	X		X	X	X	X	X	X	X	X		Wet only	3	Monthly
USGS 1962–64	X		X	X	X	X	X	X	X	X		Bulk[b]	4	Monthly
WMO/EPA/NOAA 1972–present	X	X	X	X	X	X	X	X	X	X		Wet only	5	Monthly
DOE 1976–present	X	X	X	X	X	X	X	X	X	X	X	Wet only	0	Monthly
MAP3S 1976–present	X	X	X	X	X	X	X	X	X	X	X	Wet only	4	Event
TVA 1976–present	X	X	X	X	X	X	X	X	X	X	X	Wet only	1	Weekly
EPRI 1978–present	X	X	X	X	X	X	X	X	X	X	X	Wet only	4	Event
NADP 1978–present	X	X	X	X	X	X	X	X	X	X	X	Wet only		Weekly

[a]A "wet only" collector is open only during precipitation events.
[b]A "bulk" collector is open constantly and thus collects both wet and dry deposition.

A unique opportunity for comparison as suggested in item 2 above presents itself at Raleigh, North Carolina. Three other networks (EPRI, WMO, and USGS) have operated or are operating at this site. Figure 6 summarizes the sulfate deposition from the four networks. The USGS and WMO data are monthly, so that NADP and EPRI data were averaged for a month. The latter data show a somewhat lower sulfate deposition value. Is this caused by variations in the sulfate concentrations, precipitation amounts, or a combination of both? Because the NADP network was established to measure deposition, this is an important question that must be addressed.

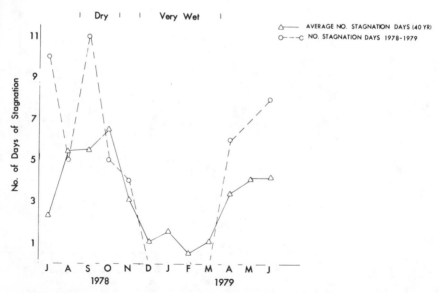

Fig. 3. Plot of number of stagnation days during a given month for a 40-year average and the period July 1978 to June 1979.

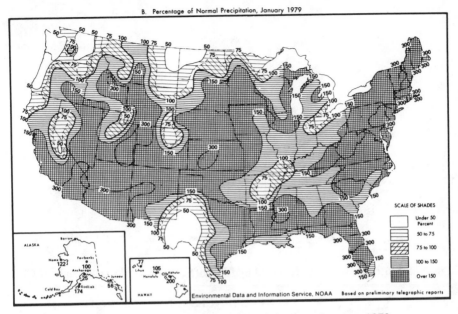

B. Percentage of Normal Precipitation, January 1979

Fig. 4. Percentage of normal precipitation, January 1979.

CONCLUSIONS

Though only a preliminary evaluation can be made of the NADP data, the following suggestions can be made concerning the network.

1. A careful study should be made of the use of precipitation amounts measured in the collector or the rain gage. These two vary enough to make a significant difference in deposition calculations.

2. A comparison of the other network data should be an integral part of the data evaluation.

3. The minimum longevity of the NADP network should be at least ten years in order to establish a climatology of atmospheric deposition.

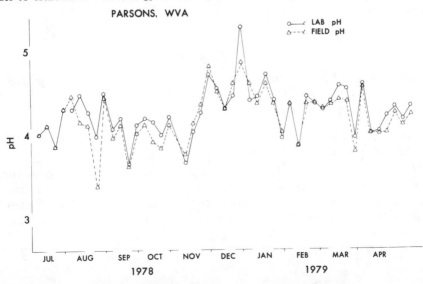

Fig. 5. Field vs laboratory pH at Parsons, W.V.

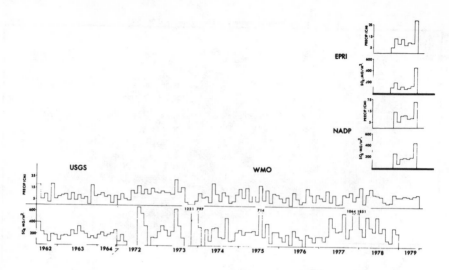

Fig. 6. Sulfate deposition as reported by four different networks at Raleigh, N.C.

REFERENCES

Hakkarinen, C. 1979. Personal communication.

Korshover, J. 1976. Climatology of stagnating anticyclones east of the Rocky Mountains, 1936–1975. NOAA Technical Memorandum ERL-ARL-55.

Likens, G. E., F. H. Bormann, and N. M. Johnson. 1972. Acid rain. Environment 14(2):33–40.

Niemann, B. L. 1979. A future acid rain monitoring network. Report developed by Teknekron for the U.S. Environmental Protection Agency, in press.

Oden, S. 1968. The acidification of air and precipitation and its consequences on the natural environment (in Swedish). Ekologikcommitten, Bull 1, Statens Naturvelenskopliga Forskningsrad, Stockholm.

Wisniewski, J., and J. M. Miller. 1977. A critical review of precipitation chemistry studies—North American and adjacent areas. WMO Special Environment Report No. 10, WMO No. 460, Geneva, Switzerland, pp. 63–69.

DISCUSSION

Sam Stephens, UTS, Toronto: Could you please tell me the difference in time between the field measurement of pH and the lab measurement?

J. M. Miller: Two to five days.

Sam Stephens: Do you have any on-site laboratories to have quality control of any of the other data other than the pH, at any of the stations?

J. M. Miller: I think that there are some special studies done at site to determine concentrations, particularly, I know, in some of the other networks, where they actually do on-site measurements of sulfate and other ions, but what I was suggesting is possibly having an ion chromatograph on site and having the samples going right into the real-time measurements.

CHAPTER 47

ATMOSPHERE-CANOPY INTERACTIONS OF SULFUR IN THE SOUTHEASTERN UNITED STATES

Geoffrey G. Parker*
Steven E. Lindberg†
J. Michael Kelly‡

ABSTRACT

Throughfall sulfate-sulfur enrichment in five Southeastern forests was found to be related to the proximity of sulfur sources, though not all of the differences between stands could be ascribed to dry-deposition inputs. Both net throughfall deposition and throughfall concentrations of sulfate sulfur exhibited maxima in the warmer months, suggesting rapid cycling as well as enhanced dry deposition. Plots of the relationship between throughfall sulfate concentrations and precipitation amounts are suggested as possible indicators of net throughfall sources. From indications in the literature and observations in these studies, it is suggested that, depending on the forest type and proximity to sulfur sources, on the order of 50% of net throughfall sulfate is due to precipitation-dissolved dry deposition.

INTRODUCTION

Precipitation becomes significantly enriched in concentrations of sulfate sulfur following its interaction with tree canopies to the extent that sulfate is often the major anion in forest throughfall as well as in incident precipitation. The literature suggests that the increase in throughfall sulfate is commonly twofold, still higher in industrialized regions. The annual flux of this additional sulfate is large relative to both the sulfur content of the canopy and, consequently, to the annual sulfur transfer in litter fall.

The enrichment originates both internally and externally to the forest system. Part of the sulfur removed during precipitation arises from some mobile fraction within the plant itself. This portion, which reflects the net effect of leaching from and uptake by plant surfaces, provides a transfer between the plant and soil reservoirs within the system. Some of the increase, however, derives from the rainfall dissolu-

*Dept. of Environmental Sciences, University of Virginia, Charlottesville, V.A. 22903.

†Environmental Sciences Division, Oak Ridge National Laboratory, Oak Ridge, Tenn. 37830. ESD Publication number 1460. Research sponsored by the Office of Health and Environ. Res., U.S. Dept. of Energy, under contract W-7405-eng-26 with Union Carbide Corp.

‡Air Quality Research Section, Tennessee Valley Authority, Muscle Shoals, Ala. 35660.

tion of sulfur gases and aerosols captured by canopy surfaces since the previous precipitation event. This component represents an externally derived input to the system.

Recent observations at four separate research sites in Tennessee and Virginia are presented in order to examine the sources and processes of throughfall sulfate enrichment. We examine the precipitation chemistry of sulfur in several Southeastern forests, including factors affecting the flux, relationships between sulfate levels in throughfall and precipitation amounts, seasonality in inputs, and other behavioral clues to the sources of the enrichment. We review several methods for estimating the inputs deriving from impaction and gaseous absorption and discuss factors possibly useful in separating these components.

METHODS

Definitions

Mass concentrations in throughfall multiplied by the respective water flux per unit area (depth of precipitation) yield the rate of mass transfer per unit area, or deposition. When such depositions in incident precipitation are subtracted from those in throughfall, the remainder is termed the net throughfall deposition or throughfall enrichment. The internal and external fractions of the net throughfall deposition will be termed leachate and wash-off, respectively, though the former is the net effect of both leaching and foliar absorption of surface-deposited material.

Sampling and Analysis

Throughfall and incident precipitation were collected in five separate plots in three locations in the southeastern United States (Fig. 1). Sampling intensity and frequency differed due to differences in study objectives.

Lindberg et al. (1979) sampled precipitation and throughfall on an event basis using HASL-type wetfall-only collectors at the Walker Branch Watershed (WB) throughout the period from April 1976 to December 1977. Walker Branch Watershed lies within 20 km of three major sulfur emission sources (Fig. 1). Incident-rain collectors were situated above the canopy (on a 43-m meteorological tower) and in a forest floor clearing, while three additional collectors were located beneath a chestnut oak canopy (*Quercus prinus*). The samples were analyzed for pH, SO_4^{2-}, Pb, Zn, Cd, and Mn. Some samples were also analyzed for strong and weak acidity. The study was directed at determining mechanisms and rates of atmospheric deposition to a forest in the vicinity of coal-fired emission sources.

The Camp Branch (CB) and Cross Creek (CC, Fig. 1) experimental watersheds in middle southern Tennessee have a similar mixed-oak cover but differ in their proximity to major emission sources, being 95 and 19 km, respectively, from a large coal-fired plant (Kelly, 1979). In each watershed, up to 24 stations containing two precipitation collectors each have been sampled monthly since November 1977 and analyzed for pH, NH_4^+, total N, total P, K^+, SO_4^{2-}, Ca^{2+}, and Mg^{2+}. Both watersheds have open plots for collection of incident precipitation. The study was designed to quantify differences in annual inputs and losses.

In the Virginia study (Va., Fig. 1), throughfall and incident precipitation were sampled on an event basis during April–August 1978 in two forest stands in the vicinity of Charlottesville, one a 44-year-old white pine (*Pinus strobus*) plantation and the other a mature oak-poplar (*Quercus-Liriodendron*) stand. There are no appreciable emission sources influencing the Charlottesville area. Up to 12 throughfall and three incident-precipitation collectors were employed in each site. Samples were

ORNL-DWG 79-19088

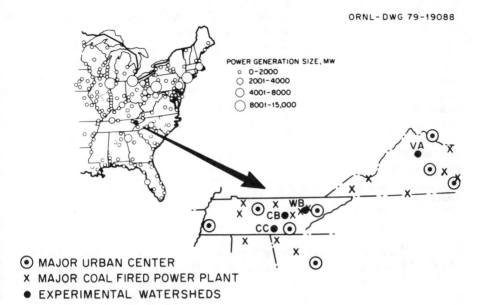

Fig. 1. Location of sampling sites and large nearby sulfur emission sources (coal-fired power plants). WB represents the location of the Walker Branch Watershed, CB the Camp Branch Watershed, and CC the Cross Creek Watershed, all in eastern Tennessee. VA represents the location of the Virginia study watersheds in western Virginia.

analyzed for SO_4^{2-}, Cl^-, NO_3^-, NO_2^-, SiO_4^{4-}, NH_4^+, PO_4^{3-}, Ca^{2+}, Mg^{2+}, K^+, Na^+, and pH. This study was directed at detailing throughfall behavior over small temporal and spatial scales in an attempt to identify sources of sulfate in throughfall.

RESULTS

Deposition Enrichment

Incident precipitation supplied between 9 and 13 kg of S per hectare per year (sulfate expressed as S) to the Tennessee watersheds, while throughfall fluxes were between 14 and 32 kg of S per hectare per year. (see Table 1). During the growing season (April–August), throughfall and incident precipitation averaged 9.1 (± 4.6, S.D.) and 5.9 (± 1.7) kg of S per hectare per year, respectively, over all five stands. Growing-season ratios of throughfall to rainwater deposition, averaging 1.5 (± 0.5), were higher in the Tennessee stands (1.4 to 2.1) than in the Virginia stands (0.9 to 1.2).

Seasonal Trends

Deposition

Both concentrations and deposition rates of incident-rain sulfate have been reported to exhibit a marked warm-season peak (Likens et al., 1977), reflecting increased rates of emission and gas-to-particle conversion, poorer atmospheric dispersion, and, consequently, higher concentrations in ground-level air. On the whole, throughfall and net throughfall sulfate depositions exhibit similar seasonal trends in

Table 1. Throughfall and incident rain sulfate deposition rates for several southeastern stands
In kilograms of S per hectare

Stand	Location	Annual deposition		Summer months (April–August)		
		Throughfall	Incident	Throughfall	Incident	Net
Quercus prinus	Tenn.	32.0	13.2	17.1	7.4	9.7
Mixed oak	Tenn.[a]	15.0	8.7	7.3	3.8	3.5
Mixed oak	Tenn.[b]	14.0	11.3	6.3	4.4	1.9
Pinus strobus	Va.[c]			8.1	6.9	1.1
Quercus-Liriodendron	Va.[c]			6.5	6.9	−0.4

[a]Camp Branch Watershed, outside power plant airshed.
[b]Cross Creek Watershed, within power plant airshed.
[c]Two forest stands at the one Virginia study site.

the studies summarized here. Seasonal trends cannot be inferred from the five months of data available for the Virginia stands. However, net throughfall deposition in the Tennessee watersheds is markedly higher in the summer months (Fig. 2a) than in the Virginia watershed. In the dormant season, net throughfall can occasionally be negative, suggesting net removal of incident sulfate by the nonfoliated canopy. Net throughfall deposition in the oak–tulip poplar forest at the Virginia site was negative throughout the growing season except for the month of July.

Concentration

The seasonality of throughfall enrichment can alternatively be indicated by volume-weighted concentrations, since deposition is directly affected by the amount of precipitation. Throughfall concentrations of sulfate (monthly deposition divided by monthly throughfall precipitation) also exhibit distinct growing-season maxima (see Fig. 2b). The autumnal peaks for the Cross Creek and Camp Branch watersheds might be attributed to sulfate leaching accompanying leaf decomposition.

Neither deposition nor concentration trends alone fully describe canopy responses to precipitation. Both are strongly influenced by precipitation amounts. Deposition is directly related to precipitation volume, while concentrations are generally decreased by dilution. In the studies summarized here the warm-season maxima in both net throughfall sulfur deposition and throughfall sulfate concentrations probably reflect the presence of materials derived from both increased metabolic recycling and higher impaction and absorption. The canopy is both cycling more internal and trapping more external material.

Volume vs Concentration Relationships

Sulfate concentrations in both throughfall and incident precipitation decrease with increasing precipitation amount. Such behavior has been reported in leaching studies (e.g., Tukey, 1970; Clements et al., 1972) but has been described for sulfate in field situations only recently (Lindberg et al., 1979). Figures 3 and 4 are plots of these relationships in samples taken at Walker Branch and the two Virginia stands respectively.

For the event-basis samples, asymptotic sulfate concentrations as plotted are estimated to be about 2 mg of SO_4^{2-} per liter for incident precipitation (Figs. 3 and 4). Throughfall concentrations level off at nearly 4 mg of SO_4^{2-} per liter under chestnut oak (Walker Branch, Fig. 3b) and white pine (Virginia, Fig. 4b) and at 3 mg

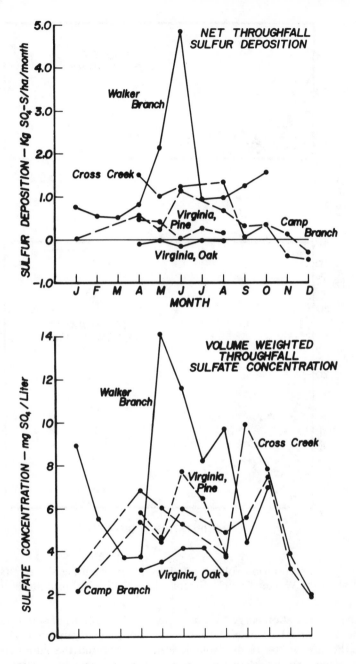

Fig. 2 Monthly net throughfall sulfur deposition and volume-weighted throughfall sulfate concentration under five Southeastern forests (Fig. 2a, top; 2b, bottom).

of $SO_4{}^{2-}$ per liter under the Oak–tulip-poplar canopy (Virginia, Fig. 4c). Asymptotic concentrations in monthly samples from the Camp Branch and Cross Creek experimental watersheds (not shown) are difficult to obtain because such concentrations are essentially volume-weighted means of several storms and cannot detail canopy response to precipitation events.

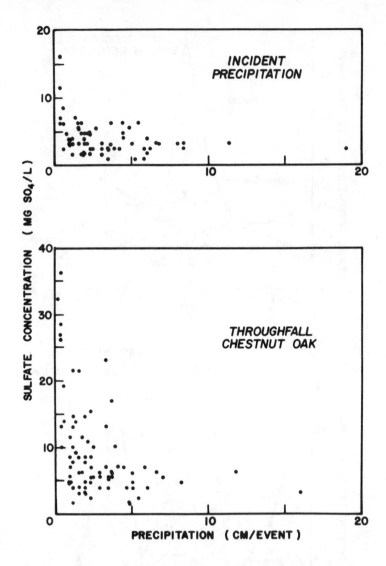

Fig.3 Sulfate concentration-volume relationships in event-basis precipitation and throughfall at the Walker Branch Watershed, 1977 (Fig. 3a, top; 3b, bottom).

Volume-concentration curves can be important descriptions of stand response to precipitation. Curve parameters such as initial slope and asymptote might possibly indicate the size of the readily soluble element pool and the minimum rate of leaching, suggesting the relative importance of various sources of elements in net throughfall. Furthermore, they detail the response of sulfate concentrations to both rainfall quantity and intensity. Small-volume events individually make negligible contributions to annual budgets but may have significant episodic effects on plant surfaces. For example, Lindberg et al. (1979) also reported similar inverse relationships between rainfall amount and concentrations of H^+ and several trace elements in rain and throughfall. Thus storms of minor hydrologic importance often result in the highest exposures of leaf surfaces to concentrations of trace metals, sulfate, and its associated acidity.

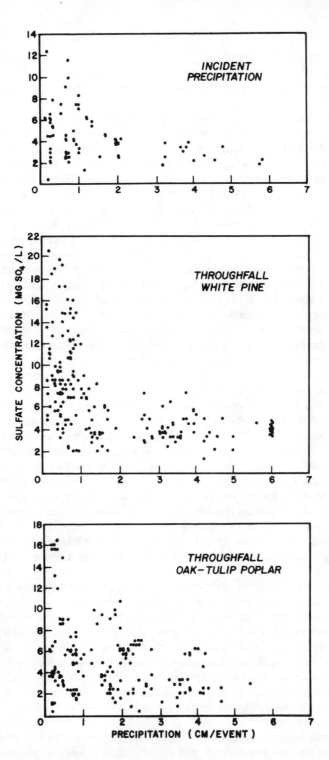

Fig. 4 Sulfate concentration-volume relationships in event-basis precipitation and throughfall in two forests near Charlottesville, Va., April–August 1978 (Fig. 4a, top; 4b, middle, 4c, bottom).

DISCUSSION

Throughfall sulfate deposition rates at the five sites are within the range of reported global values. Table 2 presents incident precipitation and throughfall deposition rates reported in the literature and calculated from our recent studies in the Southeast.

The literature suggests that source proximity can significantly affect sulfate enhancement in throughfall (Baker et al., 1976; Haughbotn, 1973; Mayer and Ulrich, 1974). While some fraction of the sulfate removal might reflect increased leaching caused by acid damage to plant surfaces, some is undoubtedly accounted for by increased dry deposition near emission sources. This is suggested by the studies summarized here (Table 1). Proximity to major emission sources appears to influence the magnitude of net throughfall sulfur deposition. The chestnut oaks of Walker Branch Watershed, which is located within ∽22 km of two major sulfur emission sources (Fig. 1), have a net throughfall flux of 9.7 kg of S per hectare during the five summer months. Cross Creek watershed, 19 km downwind of a coal-fired power plant, exhibits a net throughfall flux of 1.9 kg of S per hectare, while the flux of the Camp Branch watershed, 95 km from the same plant, is 3.5 kg of S per hectare over the same months. The Virginia white pine and oak–tulip-poplar sites, not directly influenced by a major emission source, transfer 1.2 and −0.4 kg of S per hectare, respectively, in net throughfall deposition. Dry-deposition rates are not necessarily proportional to source proximities but also depend on the emission strengths, transport routes, and their variability. The distance from sources can, however, influence the proportion of dry deposition due to gaseous or particulate sulfur, which may affect the total deposition.

Much of this trend is also due to site- and species-specific factors; the Virginia oak–tulip poplar canopy retains incident sulfate and the "impacted" Cross Creek watershed transfers less net throughfall sulfate than the "background" Camp Branch site. While there is some effect of source proximity in enhancing net throughfall, the quantitative relationship cannot be ascertained from these studies.

At the Virginia sites the larger summertime throughfall deposition beneath the white pine plantation than beneath the nearby oak–tulip poplar stand may be attributable to the differences in receptor geometry and total area of the leaves. The small-sized foliar elements of conifers should be more successful in trapping (and possibly retaining) the sulfur aerosol particles and gases (Chamberlain, 1975). Additionally, the larger total canopy area for conifers provides a greater surface for absorption.

Few data are available on the seasonal variation of the sulfate enrichment in net throughfall. Our studies in the Southeast indicate a warm-weather maximum for these deposition fluxes. This trend suggests the enriching process to be influenced by the plant-soil recycling rate, the atmospheric supply of aerosols and gases, or both. It is extremely difficult to distinguish which process is most important in deciduous forests based on these observations alone. Continuous measurements of throughfall sulfur in a coniferous stand on an annual basis could shed more light on the sources of enrichment, particularly during the dormant season, when the large and elaborate canopy surface remains, though the rate of replenishment of the pool of leachates is expected to be reduced.

Curves describing the decrease of throughfall concentrations of sulfate with increasing precipitation are perhaps more useful than the above seasonal patterns alone (Figs. 3 and 4). The fact that throughfall asymptotes are higher than those of incident rainwater suggests that the canopy interaction can result in a net increase in the throughfall sulfate regardless of how much rainwater has passed over the surfaces (Lindberg et al., 1979).

Table 2. Reported values for sulfate-sulfur deposition rates for throughfall and incident precipitation in world forests

Forest system	Reference	Sampling period (months)	Deposition[a] Incident	Throughfall	Precipitation amount (cm)
Spruce, southern Norway	Abrahamsen et al., 1976	6	4.9	9.9	47
Pine, southern Norway	Abrahamsen et al., 1976	6	4.9	9.5	47
Birch, southern Norway	Abrahamsen et al., 1976	6	4.9	6.5	47
Spruce, arctic Norway	Abrahamsen et al., 1976	4	0.9	1.1	15
Pine, arctic Norway	Abrahamsen et al., 1976	4	0.9	1.1	15
Birch, arctic Norway	Abrahamsen et al., 1976	4	0.9	0.5	15
Scots pine, England	Bache, 1977	5	5.8	19.6	12
Spruce, Alberta	Baker et al., 1976	3	0.3	0.8	
Spruce, Alberta	Baker et al., 1976	3[b]	2.4	8.6	
Subalpine balsam fir, New Hampshire	Cronan, 1978	12[c]	24.4	46.4	203[d]
Hardwoods, New Hampshire	Eaton et al., 1973	5	5.1	26.1	56
Hemlock, British Columbia	Feller, 1977	12[c]	11.0	40.0	245[d]
Conifers, southern Norway	Haughbotn, 1973	12[b]	32.3	111.2	77
Conifers, southern Norway	Haughbotn, 1973	12	17.7	69.1	77
Conifers, southern Norway	Haughbotn, 1973	12	10.0	21.1	77
Beech, central Germany	Heinrichs and Mayer, 1977	12[e]	24.1	47.6	106
Spruce, central Germany	Heinrichs and Mayer, 1977	12	24.1	80.0	106
Birch, southern Sweden	Horntvedt and Joranger, 1974	4.5	4.4	5.7	39
Spruce, southern Sweden	Horntvedt and Jorander, 1974	4.5	4.4	8.6	39
Pine, southern Sweden	Horntvedt and Joranger, 1974	4.5	4.4	7.8	39
Hemlock-spruce, southeastern Alaska	Johnson, 1975	12[c]	0	16.4	270
Tropical rain forest, Costa Rica	Johnson, 1975	12	12.5	22.3	390
Douglas fir, Washington	Johnson, 1975	12	4.0	5.2	165
Subalpine silver fir, Washington	Johnson, 1975	12	16.8	5.3	300
Hardwoods, Amazonian Venezuela	Jordan et al., 1980	12	44.5	16.7	391
Hardwoods, Amazonian Venezuela	Jordan et al., 1980	12	46.6	19.6	412
Hard beech, New Zealand	Miller, 1963	12	8.4	10.4	135

Table 2. (continued)

Forest system	Reference	Sampling period (months)	Deposition[a] Incident	Throughfall	Precipitation amount (cm)
Beech, southern Sweden	Nihlgard, 1970	12*	7.9	18.5	95
Spruce, southern Sweden	Nihlgard, 1970	12*	7.9	54.2	95
Oak, southern France	Rapp, 1973	12	16.4	22.6	
Wheat England	Raybould et al., 1977	2	2.06	4.42	6
Scots pine, southern Sweden	Richter and Granat, 1978	3[c]	7.0	21.9	73
Scots pine, southern Sweden	Richter and Granat, 1978	3[c]	6.2	13.5	38
Douglas fir, Oregon	Sollins et al., 1979	12	4.7	2.4	237
Spruce, Russia	Sokolov, 1972	6	0.5	2.0	46
Birch, Russia	Sokolov, 1972	6	0.5	1.0	46
Loblolly pine, North Carolina	Wells et al., 1975	12[c]	7.9	9.9	
Chestnut oak, Tennessee	Lindberg et al., 1979	12[b,f]	13.2	32.0	143
Mixed oak, Tennessee	Kelly, 1979	12[c]	8.7	15.0	154
Mixed oak, Tennessee	Kelly, 1979	12[b,c]	11.3	14.0	75
White pine, Virginia	Parker, 1980	5	6.9	8.1	61
Oak–Tulip poplar, Virginia	Parker, 1980	5	6.9	6.5	61

[a]Kilograms of S per hectare per sampling period.
[b]In vicinity of factory or power plant.
[c]Scaled up from a subannual estimate.
[d]Mean of extreme estimates.
[e]Includes stem flow.
[f]Several years data.

Throughfall sulfate concentrations in sequentially sampled events are consistently higher than in the corresponding incident precipitation (Richter and Granat, 1978; Lindberg, unpublished data) even during large-volume storms (Parker, unpublished data). The source of much of the sulfate in these latter samples is certainly leachate, since the desorbable surface-associated material is expected to be removed quickly. However, throughfall samples representing the initial segment of an event also will have some internal leaching component, since internally derived soluble material is leached quickly with initial wetting (Tukey, 1970; Clements et al., 1972).

To produce meaningful precipitation-concentration curves requires observations of single events. Concentrations in samples combining several storms represent volume-weighted means, thus confounding estimation of the asymptotic concentration. Even event samples can obscure the effects of the separate within-storm processes. The instantaneous response of the canopy can aid not only in studying enrichment sources but in evaluating the effect of incident rainwater quality. Thus, more attention must be paid to sequential sampling of rain and throughfall for large storms.

To date, three patterns have been observed in sequential throughfall sulfate: (1) concentrations declining regularly with successive precipitation, resembling a negative-exponential curve, (2) concentrations oscillating about a mean without temporal trend, especially in finely sampled small storms, and (3) concentration trends in throughfall similar to those in incident precipitation, though of higher value and delayed in time, (Lindberg and Parker, unpublished data; Richter and Granat, 1978).

While net throughfall sulfate concentrations can decrease with increasing precipitation volume, the corresponding deposition to the forest floor can increase or remain constant. This suggests that the source in the canopy is not readily exhausted and that it depends strongly on the amount of water passing over canopy surfaces. Such behavior suggests a source tied to a mobile and replenishable internal supply rather than a finite (at least during the event) external pool. The possibility that some of this internal SO_4^{2-} might derive from SO_2 absorbed by foliage since the previous storm further complicates the internal vs external source question.

These observations on the behavior of sulfate in precipitation and throughfall in the southeastern United States are indirect arguments for an important internal leaching source of net throughfall sulfur. Direct evidence is lacking for several reasons: (1) dry deposition is extremely difficult to estimate, being under the control of a multitude of factors and thus highly variable, (2) trapped particles and absorbed gases are not necessarily uniformly mobilized by precipitation (this is particularly true for absorbed SO_2, which can be converted to sulfate and possibly translocated away from sites of leaching), and (3) the leaching dynamics of sulfur in forest trees are poorly understood, particularly the temporal response of both attached and excised foliage to rinsing in water. Also largely unknown is the magnitude of the sulfur leaching pool and its chemical form, though Turner et al. (1977) have recently developed methods for the estimation of both the organic and inorganic sulfur pool in Douglas fir foliage.

Literature Estimates of Throughfall Sources

Several indirect methods have been applied to the determination of net throughfall sources (Table 3). However, most of them are limited in their applicability to other forest systems. Eaton et al. (1978) calculated the dry-deposition input to a Northern hardwood forest to be 6.1 kg of S per hectare per year. The value was estimated as the difference in system inputs and outputs, assuming negligible sulfur storage. This represented 29.1% of net throughfall and stem flow observed during five summer months. The annual contribution must be less, however, since positive values of net throughfall sulfur can also occur under winter hardwood canopies (Mayer and Ulrich, 1974; Lindberg et al., 1979).

From throughfall measurements and flux-gradient estimates of SO_2 deposition velocities, Bache (1977) asserted that the "enrichment (net throughfall) approximately balances the contribution arising from dry deposition." However, he noted that the correspondence between these measurements was confusing given that throughfall sulfate enhancements were observed to increase with increasing amounts of incident precipitation, even during very rainy periods, and that SO_2 could be absorbed and translocated by foliage as sulfate. He calculated a high efficiency of wash-off of 89% for his mixed-pine stand in England and of between 15 and 44% using data for a similar stand in Norway. He also found that the total enrichment per volume of throughfall (milligrams of sulfate per square meter per millimeter) normalized by the ambient sulfur dioxide loads (micrograms of SO_2 per cubic meter) yields a value close to 0.8 in both systems. The nature of this potential relationship between SO_2 concentrations and sulfate in net throughfall is worthy of further investigation.

Table 3. Incident, throughfall and net throughfall sulfate-sulfur deposition in several forested systems where attempts were made to partition net throughfall sources

| Reference | Sampling period | Deposition (kg/ha) | | | Percent of net throughfall due to deposition |
		Incident	Throughfall	Net throughfall	
Eaton et al., 1978	Incident, many years; throughfall, 5 months	12.7	33.7	21.0	21.9
Bache, 1977	153 days	5.8	19.6	13.8	89.1
	127 days[a]			4.5	14.8–44.4
Miller, 1963	1 year	8.4	10.4	2.0	~100
Mayer and Ulrich, 1974	3 years	22.2	46.1	23.9	59.4
Lindberg et al., 1979	Several years	13.0	32.0	19.0	26.3
Raybould et al., 1977[b]	66 days	2.06	4.42	2.36	13

[a]Using data of Bjor et al., 1974.
[b]Wheat canopy.

Raybould et al. (1977) also suggested that gaseous deposition, as estimated from deposition velocities, could account for all net throughfall sulfur. They concluded, however, that the low mobility of attached radio-labeled SO_2 (averaging 13% mobilized of the total attached) indicated an origin of sulfate in throughfall largely of internal leachates. As with Bache (1977), they failed to include particulate deposition, leading to an overestimate of the leaching contribution.

Mayer and Ulrich (1974) hypothesized that dry deposition to the Solling beech forest would be equivalent to net throughfall during the dormant season, assuming that the leafless forest would neither differ in filtering ability from the extensive summer canopy nor be appreciably leached. However, to the extent that growing-season leaching will outweigh the dormant-season filtering, the trapping capacity has probably been overestimated.

Lindberg et al. (1979), using a combination of artificial collection surfaces, leaf washing, and ambient aerosol measurement, determined that 26% of the 19 kg of S per hectare per year net throughfall sulfate deposition at the Walker Branch Watershed was due to deposition of particulate sulfate to foliage and stems. Since gaseous sulfur inputs were not measured during this period, the dry-deposition fraction of net throughfall has been underestimated.*

Miller (1963) concluded that net throughfall sulfate in a maritime New Zealand stand of hard beech was entirely due to dry deposition of sea salts. Ratios of throughfall to incident precipitation deposition were very similar for major seawater ions, suggesting that canopy-filtered deposits of sea salts were quantitatively transferred into throughfall. However, ratios of sulfur to major sea salt elements were not preserved in net throughfall. Some of the sulfur is apparently derived from sources other than sea salt deposits.

All of the above estimates were obtained under widely differing experimental situations with diverse hypotheses about the throughfall process (as summarized in Table 3). A distinguishing feature of these estimates is their variation. This is due, in part, to unique atmosphere or canopy situations, varying sampling and analytical

---

*More recent evaluation of the potential role of SO_2 scavenging by vegetation in this system resulted in the conclusion that an additional ~8 kg S/ha yr could be deposited to the canopy as SO_2 (Lindberg, unpublished). Taking this into account increases the relative contribution of external sources to the net throughfall sulfur flux to ~70%.

methods, and differing assumptions. Alternatively they imply that the partitioning of net throughfall sulfate can truly vary from primarily atmospheric input to wholly recycling, depending on a number of factors discussed above.

It appears that for temperate hardwoods in industrialized regions, 40%–60% of annual net throughfall sulfate is due to wash-off of dry deposition. For conifers in the same regions the values are on the order of 30–50%. For hardwoods and conifers in areas with generally background levels of sulfur aerosols and gases, possibly 0–20% of net throughfall is due to externally derived sources. The degree to which measurement of incident precipitation alone underestimates total sulfate inputs to a forest depends on the proximity of local sources as well as on the nature of the canopy and the amount of precipitation.

CONCLUSIONS

Enriched sulfate in net throughfall deposition suggests both rapid sulfur recycling by the plant and significant inputs from outside the plant-soil system. This is suggested by high net throughfall sulfate deposition and throughfall sulfate concentrations in Southeastern forests during the summer months.

The response of throughfall sulfate concentrations to increasing precipitation amounts might indicate the sources of the enrichment, particularly where samples are taken on an event or within-event basis.

Canopies in the vicinity of major emission sources generally show greater enrichment in sulfate deposition than those which are far removed. However, it is impossible to ascribe all the differences in net throughfall between such sites to dry deposition.

Our own observations on the behavior of net throughfall in Southeastern forests and indications in the literature suggest that on the order of 50% or more of the enrichment in the southeastern United States is due to recycling of sulfur by the plant itself. This estimate is derived primarily from indirect evidence and indicates a fundamental lack of knowledge and data concerning the processes whereby forest canopies can remove and utilize atmospheric material. More attention must be focused on the following basic processes:

1. The storage and circulation of sulfur forms within the plant, with particular attention to the amounts available to be leached, the maximum rates of leaching, and the factors influencing these rates. A tracer study on a forest tree would be most useful.

2. The rate of accumulation and retranslocation of SO_2-derived sulfate. How available is this material to be leached?

3. The mobility of the canopy-trapped aerosol particles. What fraction of the retained load can be washed off, and how is this fraction dependent on the composition of incident precipitation?

In particular we recommend that studies directed at quantifying throughfall sources and processes include the following:

1. routine event-basis collection, with occasional sequential sampling within large-volume storms,
2. concomitant measurements of gaseous and particulate sulfur in ground-level air, with
3. estimations or measurements of sulfur deposition velocities or trapping efficiencies.

REFERENCES

Abrahamsen, G., K. Bjor, R. Horntvedt, and B. Tveite. 1976. Effects of acid precipitation on coniferous forest. Pp. 36–63 in Braekke, F. H. (ed.), Impact of Acid Precipitation on Forest and Freshwater Ecosystems in Norway, Oslo-Ås, Norway.

Bache, D. H. 1977. Sulphur dioxide uptake and the leaching of sulphates from a pine forest. J. Appl. Ecol. 14:881–895.

Baker, J., D. Hocking, and M. Nyborg. 1976. Acidity of open and intercepted precipitation in forests and effects on forest soils in Alberta, Canada. Pp. 779–790 in Proc. First Intl. Symp. Acid Precip. Forest Ecosys. USDA Forest Service Tech. Rept. NE-23.

Bjor, K., R. Horntvedt, and E. Joranger. 1974. Distribution and chemical composition of precipitation in a southern Norway forest stand (in Norwegian). Norwegian Council for Scientific and Industrial Research. Report 1/74. Oslo-Ås (referenced by Bache, 1977).

Chamberlain, A. C. 1975. The movement of particles in plant communities. Pp. 155–203 in Monteith, H. C. (ed.), Vegetation and the Atmosphere, Vol. I, Principles. Academic Press, New York.

Clements, C. R., L. P. H. Jones, and M. J. Hopper. 1972. The leaching of some elements from herbage plants by simulated rain. J. Appl. Ecol. 9:249–260.

Cronan, C. S. 1978. Solution chemistry of a New Hampshire subalpine ecosystem: biogeochemical patterns and processes. Ph.D. thesis, Dartmouth College.

Eaton, J. S., G. E. Likens, and F. H. Bormann. 1973. Throughfall and stemflow chemistry in a northern hardwood forest. J. Ecol. 61:495–508.

Eaton, J. S., G. E. Likens, and F. H. Bormann. 1978. The input of gaseous and particulate sulfur to a forest ecosystem. Tellus 30:546–551.

Feller, M. C. 1977. Nutrient movement through western hemlock–western redceder ecosystems in southwestern British Columbia. Ecology 58:1269–1283.

Haughbotn, O. 1973. Nebørundersøkelsor i Sarpsborgdistriktet og undersøkelser over virkninger av forsurende nedfall på jordas kjemiske egenskaper. Ås-NLH, 1973. 151 pp. (referenced in Horntvedt, 1975).

Heinrichs, H., and R. Mayer. 1977. Distribution and cycling of major and trace elements in two central European forest ecosystems. J. Envir. Qual. 6:402–407.

Horntvedt, R., and E. Joranger. 1974. Nedbørens fordeling og kjemiske innhold under traer: Juli-november 1973 Teknisk notat TN 3/74. SNSF-projektet, Oslo-Ås. 29 pp. (referenced in Horntedvedt, 1975).

Horntvedt, R. 1975. Kjemisk Innhold i nedbør under traer. Et litteratursammendrag. (Chemical content of precipitation under trees, a literature review). TN 18/75. SNSF project, Ås-NLH, Norway.

Johnson, D. W. 1975. Processes of elemental transfer in some tropical, temperate, alpine and northern forest soils: factors influencing the availability and mobility of major leaching agents. Ph.D. thesis, University of Washington, Seattle. 169 p.

Jordan, C.F., F. Golley, J. Hall and J. Hall. 1980. Nutrient scavenging of rainfall by the canopy of an Amazonian rain forest. Biotropica, in press.

Kelly, J. M. 1979. Camp Branch and Cross Creek experimental watershed projects: objectives, facilities and ecological characteristics. EPA-600/7-79-053, Washington, D.C.

Likens, G. E., F. H. Bormann, R. W. Pierce, J. S. Eaton, and N. M. Johnson. 1977. Biogeochemistry of a Forested Ecosystem. Springer-Verlag, New York. 146 pp.

Lindberg, S. E., R. C. Harriss, R. R. Turner, D. S. Shriner, and D. D. Huff. 1979. Mechanisms and rates of atmospheric deposition of selected trace elements and to a deciduous forest watershed. Environmental Sciences Division Publ. no. 1299, ORNL-TM/6674, Oak Ridge National Laboratory, Oak Ridge, Tennessee.

Mayer, R., and B. Ulrich. 1974. Conclusions on the filtering action of forests from ecosystem analysis. Oecol. Plant. 9:157–168.

Miller, H. G., J. M. Cooper, and J. D. Miller. 1976. Effect of nitrogen supply on nutrients in litterfall and crown leaching in a stand of Corsican pine. J. Appl. Ecol. 13:233–248.

Miller, R. B. 1963. Plant nutrients in hard beech III. The cycle of nutrients. N.Z. J. Sci. 6:388–413.

Nihlgard, B. 1970. Precipitation, its chemical composition and effect on soil water in a beech and a spruce forest in South Sweden. Oikos 21:208–217.

Parker, G. G. 1980. Master's thesis, University of Virginia, Charlottesville, Va.

Rapp, M. 1973. Le cycle biogeochimique du soufre dans une foret de *Quercus ilex* L. du sud de la France. Oecol. Plant. 8:325–334.

Raybould, C. C., M. H. Unsworth, and P. J. Gregory. 1977. Sources of sulphur in rain collected below a wheat canopy. Nature 267:146–147.

Richter, A., and L. Granat. 1978. Pine Forest Canopy Throughfall Measurements. Report AL-43, Department of Meteorology, University of Stockholm.

Sokolov, A. A. 1972. (Chemical composition of rainfall passed through the birch and spruce canopies). Lesovedenie 3:103–105 (Russian with English summary) (referenced in Horntvedt, 1975).

Sollins, P., L. L. Grier, F. M. McCorison, K. Cromack, Jr., A. T. Brown, and R. Fogel. 1979. The internal nutrient cycles of an old growth Douglas fir stand in western Oregon. Ecological Monographs, in press.

Tukey, H. B., Jr. 1970. The leaching of substances from plants. Ann. Rev. Plant Physiol. 21:305–329.

Turner, J. M., M. J. Lambert, and S. P. Gessel. 1977. Use of foliage sulphate concentrations to predict response to area application by Douglas fir. Can. J. For. Res. 7:476–480.

Wells, C. G., and J. R. Jorgensen. 1974. Nutrient cycling in loblolly pine plantations. Pp. 137–158. in Proc. 4th North Am. Forestry Soils Conference.

Wells, C. G., A. K. Nicholas, and S. W. Buol. 1975. Some Effects of Fertilization on Mineral Cycling in Loblolly Pine. Pp. 754–764 in Howell, F. G., J. B. Gentry, and M. H. Smith (eds.) Mineral Cycling in Southeastern Ecosystems. U.S. Research and Development Administration.

DISCUSSION

John Skelly, Virginia Tech: Could you look at the length of time between precipitation events and correlate it with the comment that you made with the last scanning electron microgram? In other words, was there a difference in the amount of time for accumulation of particulates on the leaves with your different events?

G. Parker: We've looked at those data only in the case of Walker Branch. There is poor correlation between the amount of net throughfall sulfate and the time since the last storm. I think there are a number of reasons for this. The number of days since the last storm is only a rough indicator of the loading that the canopy has received. It might be a better idea to get some sort of integrated total loading, but we haven't done that. I plan to do that for the Virginia data.

Steve Lindberg: Let me add something in response to John Skelly's question, and that is, one of the reasons that it's been so difficult to establish a relationship between antecedent conditions and concentrations of net sulfate in throughfall is that we believe the volume–sulfate-concentration relationship is so strong that what you really have to look at in terms of the amount of sulfate removed in net throughfall would be constant-volume sampling of the initial fraction of a throughfall event. I believe then we'd be able to see some very strong relationships with antecedent conditions and perhaps meteorological parameters in terms of air stagnation, winds from the direction of power plants, and so on, and that's something that we're looking at now too.

A. C. Chamberlain, Atomic Energy Research Establishment: I'd like to take the opportunity to add something which I should have said at the end of my talk. It's a little late for this, but if you believe the measurements that were made of SO_2 uptake to pine shoots and then the extrapolation from that in terms of transpiration, then what you come up with roughly speaking is, if you have 1 μg per cubic meter of sulfur as SO_2 in the atmosphere, then on our data you would express the deposition as, say, 1 to 1.5 kg per hectare per year. I think that, assuming that the SO_2 concentrations in this area are possibly around 10 μg per meter cubed, if that is correct. SO_2 might well, on our data, account for somewhere between 5 and 10 kg of sulfur per hectare per year.

G. Parker: That's certainly a large proportion of the net throughfall sulfate. I think, however, that there are some assumptions that have to be made there, the main one being that the attached or absorbed SO_2 is available to be leached or removed from the canopy when the precipitation arrives. The feeling I have is that it's translocated as sulfate to other parts of the plant and is probably not available for washing.

Moderator Galloway: Geoff, would you or your other two speakers care to put some boundary conditions on what percentage of the increase in concentrations of sulfur in a throughfall could be from dry deposition or leaching?

G. Parker: Yes, I believe we could. I would have been much more willing to do so two days ago, but having heard the problems that we're all experiencing estimating deposition velocity and total input to the canopy, I believe I would revise this to about 50%. Before coming to the meeting I was convinced that probably 70% of the net throughfall sulfur deposition was due to leaching from the plants, but I'm not quite that sure anymore.

Bill Graustein, Yale: We're doing a throughfall study on adjacent aspen and spruce watersheds in New Mexico. There is no net throughfall sulfate under aspen but about a factor of 2, or 8 kg per hectare, out of our spruce canopy. Several other lines of evidence suggest there's no dust impaction of dry fallout under the aspen canopy but a large amount in spruce. I'm not sure how much these results can be generalized, but it does seem that a large fraction of the 8 kg per hectare out of the spruce is due to dust interaction.

Steve Lindberg: A comment on the moderator's question with respect to putting limits on these numbers for the amount of material derived internally. In the Walker Branch, where we've measured the total input to the canopy of particulate-associated sulfate, we can say that something on the order of 50% of the material coming through, net throughfall, can be derived from that source, that is, dry deposition of particulate-associated sulfate, so 50% may be internally leached. However, the question I'd like to raise is, when are we going to be able to accurately determine the input of SO_2 to these same forest canopies, because if that is indeed important and if our estimates are right, it could mean that as much as 80 to 90% or more of the material can be atmospherically derived, so there's some real problems involved, obviously.

G. Parker: I think it may be a good time to suggest a slightly different tack on that question as well. A lot of energy has been expended in estimating how much dry deposition the canopy receives. If we can find out how much sulfate a plant loses, this would be more tractable and possibly more useful.

CHAPTER 48

SENSITIVITY OF SOIL REGIONS TO LONG-TERM ACID PRECIPITATION

William W. McFee*

ABSTRACT

Criteria for ranking soil sensitivity to the effects of acid precipitation are discussed. A ranking scheme based on cation exchange capacity and presence or absence of carbonate in the top 25 cm of soil and presence or absence of flooding was devised. Five map units varying in potential sensitivity and percentage of the area considered sensitive were used to map the eastern United States. Maps of New York, North Carolina, West Virginia, Pennsylvania, and Indiana are presented. It is recognized that other factors could be considered to improve the ranking scheme and that land use, which is ignored in these maps, has an effect overshadowing that of acid precipitation. The maps should be useful in research planning and in selecting areas for intensive study. The ranking of sensitivities is not intended to predict severity of effects but to guide the selection of terrestrial and aquatic sites to areas where the potential for adverse impacts of long-term atmospheric deposition are the greatest. The need for field testing of the system is obvious.

INTRODUCTION

Several networks of atmospheric deposition stations are gathering conclusive data showing that eastern United States and Canada receive acid precipitation. There is a potential for deleterious effects of this acid on both terrestrial and aquatic ecosystems. Only limited information on the effects on aquatic systems is available, and essentially none is published on effects on soils and plants in naturally occurring (field) terrestrial ecosystems. Since acid precipitation is so widespread and the potential harm is subtle, there is a need to concentrate our research efforts in areas most likely to be susceptible.

Soils are continually changing, yet they remain one of the most stable components of terrestrial ecosystems. The formation of soil horizons of significant depth by weathering processes starting from rock or even unconsolidated deposits requires thousands or even millions of years. On the other hand, significant changes in root length occur in a few weeks and changes in bacterial numbers within days or even hours. In general, however, soils are very resilient. They tolerate large inputs of sundry materials and retain, or within a short time revert to, their original condition. This apparent stability sometimes lulls us into a false sense that soil productivity is indestructible. We are dependent upon the soil's ability to support cycle after cycle

---

*Agronomy Department, Purdue University, West Lafayette, Ind. 47907.

of vegetation in both natural and cultivated systems which produce the vast majority of our food and fiber. Lessons that date to antiquity in some parts of the world and at least into the 1930s in the United States taught that soils are vulnerable to erosion. Present knowledge is not adequate to predict the degree of adverse effects of acid deposition on soils, but we should not assume that soils will not be damaged by agents other than wind and water.

Sulfur deposition to soils is not all bad. It is, in fact, frequently needed, as pointed out by Noggle in this volume. If the deposition were all $CaSO_4$, it is doubtful if harmful effects would be noticed anywhere in terrestrial ecosystems, but since much of the sulfate is balanced by hydrogen ions, the result is a dilute strong acid, which has considerable potential for affecting the system. Atmospheric deposition of acid and other materials may pose a significant threat to some soils and the ecosystems they support. In recent reports, some of the known and expected effects of acid precipitation have been summarized as follows (Wood, 1979; Hutchinson, 1979; Galloway et al., 1978):

1. Natural soil-formation processes in humid regions lead to acidification, and acid precipitation accelerates this process.
2. The sensitivity of soils to acidification by acid precipitation depends on the soil buffer capacity and pH. Noncalcareous sandy soils with pH above 5 are the most sensitive to acidification.
3. Very acid soils are less sensitive to acidification. However, even a slight loss of nutrients by leaching may be detrimental.
4. Acid inputs increase leaching of exchangeable plant nutrients.
5. Acidification slows many soil microbiological processes, such as N fixation, decay of plant residues, and nitrification.
6. Acid precipitation increases the leaching losses of Al, which may be detrimental to associated aquatic environments.
7. Amendments used in field crop production have a large effect on soil acidity and overshadow acid precipitation influences on most cultivated soils.
8. There are large uncultivated areas that have soils that are poorly buffered and are potentially sensitive to acid precipitation.

There are other known and suspected effects of acid precipitation in the cited references; however, this report is concerned with the regions where most of the sensitive soils (item 8 above) occur.

The sensitivity of soils to the effects of acid precipitation has been discussed by several authors. Wiklander (1973/74, 1979) and Bache (1979) have presented excellent descriptions of cation exchange reactions involved in soil acidification and described the relationship of buffering capacity, base saturation, and "lime potential" to soil susceptibility to acidification. Reuss (1975) and McFee and Kelly (1977) have pointed out the importance of N and S transformations in influencing soil acidity and the slowness of acid precipitation in changing soil. In all of these discussions it is apparent that the four parameters which follow are important in estimating soil sensitivity to acid precipitation:

1. the total buffering capacity or cation exchange capacity, provided primarily by the clay and soil organic matter,
2. the base saturation of that exchange capacity, which can be estimated by the pH of the soil,
3. the management system imposed on the soil—whether it is cultivated and

amended with fertilizers or lime or is renewed by flooding or other additions,

4. the presence or absence of carbonates in the soil profile.

The objective of this project was to develop maps of the eastern United States that delineate soil areas according to their expected sensitivity to the effects of acid precipitation.

METHODS

The first step was to define "sensitivity" as it is used here in relation to soils and acid precipitation. Sensitivity of soils to acidification, alone, seemed too narrow even though that may be the most important long-term effect. Soils influence the quality of waters in associated streams and lakes and may be changed in ways other than simple pH-base-saturation relationships, e.g., microbiological populations of the surface layers, accelerated loss of Al by leaching. Therefore, criteria were sought that would relate to soil "sensitivity" to any change brought about by acid precipitation that would be important in the local ecosystem.

Rationale for Sensitivity Classes

Factors affecting sensitivity to acid inputs in the list above were evaluated for their wide applicability and ready availability for use in broad-scale mapping. As pointed out earlier, management that includes amendments will mask effects of acid precipitation and should be considered in the final analysis. We chose to ignore land use in this effort with the intent of eventually overlaying land-use maps on the soil maps. Most of the serious soil effects are tied to changes in pH or changes in leaching. The soil's susceptibility to changes in either of these is most closely tied to its cation exchange capacity (CEC). If the soil has a low CEC and a circumneutral pH, then acid inputs are likely to reduce pH rapidly. A soil with a high CEC is strongly buffered against changes in pH or leachate composition. Acid soils, pH near that of the acid precipitation, on the other hand, will not change in pH rapidly in response to acid inputs but are likely to release Al^{3+} ions in the leachate. Further, soils with a low CEC have a low reserve of plant nutrients, and slight losses may be significant to their productivity. Therefore, even though it is apparent that CEC, or buffering capacity, does not completely define the soil sensitivity to acid inputs, it was chosen as the primary criterion in this attempt at regional mapping of soil sensitivity to acid precipitation. Soils that contain free carbonates which will react with acid inputs are generally believed to be little affected by acid precipitation and were therefore classed nonsensitive. Likewise, any soil subject to frequent deposition, such as those in floodplains, is unlikely to be strongly affected by acid precipitation.

In much of the literature it is frequently stated that soils with low CEC or sandy soils with low organic matter are likely to be most susceptible to effects of acid precipitation. However, these "low CEC" values have not been quantified. In order to develop a working set of classes, certain assumptions and "worst case" calculations were made. Since soils are, in general, rather resistant to change by acid inputs, a fairly high acid input was assumed and the question asked, What is the maximum effect that could have on soil, and how high would the CEC have to be to resist that effect?

Arbitrarily, a 25-year span was chosen. If the maximum probable acid input (100 cm of precipitation at pH 3.7 per annum) in that period is equal to 10–25% of the cation exchange capacity in the top 25 cm of soil, a significant effect might occur.

This is the case when the top 25 cm of soil has an average CEC of 6.2–15.4 meq per 100 g (also assumess a bulk density of 1.3 g/cc), and these soils were considered slightly sensitive. If that same acid input exceeds 25% of the CEC in the top 25 cm, i.e., when the CEC is less than 6.2 meq per 100 g, the soils are considered sensitive.

I am certainly aware that these criteria are arbitrary in many ways, that arguments can be made for different assumptions about inputs, for the use of shallower or deeper depths, for the use of percent base saturation, and many other adjustments. However, both our limited knowledge of the effects of acid precipitation on soils and the generalized information available on the distribution of soil associations and their properties make this a reasonable first approach to mapping sensitive soil regions.

Sensitivity Classes

The following classes were established:

1. *Nonsensitive areas* (NS) include (*a*) all soils that are calcareous (i.e., contain free carbonates) in the surface layers (within 25 cm), (*b*) all soils that are subjected to frequent flooding (it is assumed that the renewal effects of fresh deposition would mask acid precipitation effects), and (*c*) all soils with an average CEC greater than 15.4 meq per 100 g in the top 25 cm.

2. *Slightly sensitive areas* (SS) include areas not in class 1*a* or 1*b* that have average CEC in the range 15.4 to 6.2 meq per 100 g in the top 25 cm.

3. *Sensitive areas* (S) include those areas not in class 1*a* or 1*b* that have average CEC less than 6.2 meq per 100 g in the top 25 cm.

Mapping Units

The sensitive areas (S and SS) are frequently intermingled in the landscape with nonsensitive areas; therefore, an attempt was made to indicate the proportion of the area included within each mapping unit that is a sensitive soil class. The sensitive areas were subdivided, and the following five mapping units were used at the most detailed level mapped:

NS	The area contains mostly nonsensitive areas
S1	Sensitive soils dominate the area
SS1	Slightly sensitive soils dominate the area
S2	Sensitive soils are significant but cover less than 50% of the area
SS2	Slightly sensitive soils are significant but cover less than 50% of the area

Sources of Information

The distribution of soil associations either was taken directly from state soil association maps or regional soil association maps or, in a few cases, was generalized from county maps. The level of detail, scale, and age of maps varied widely, and we used our best judgment in smoothing boundaries that were interrupted at state lines. Whatever data we could find concerning cation exchange capacity and other chemical or physical properties of the soils within the associations were utilized to estimate the CEC of the dominant soils of the association. The most frequent source was the chemical data in the "Soil Survey Laboratory Data and

Descriptions'' series published by the Soil Conservation Service, USDA. In the absence of CEC determinations we estimated it based on the soil classification, location and parent material, information given on organic matter and texture, or personal knowledge of the soils. For example, the great group of Quartzipsaments was assumed to have rather low CEC and was placed in the sensitive group when more detailed information was missing. Likewise, sandstone-derived soils on sloping areas in rough terrain or sandy outwash terraces would be considered sensitive, whereas a lacustrine deposit or medium- to fine-textured glacial till plains would be considered nonsensitive if CEC data were unavailable.

Map Preparation

In each state or region a list of mapping units on the source map or maps was compiled. Then the available literature was searched for information, and each mapping unit was assigned a sensitivity rating based on the CEC, or estimated CEC, presence of carbonates, and susceptibility to flooding. The dominant soils within an association frequently differed in their rating. If two of the dominant soils where two or three were listed were of the same sensitivity, the numeral "1" was added to the symbol (e.g., SS1 or S1). If only the second of a list of two or any one of a list of three was sensitive, an occurrence rating of 2 was assigned (SS2 or S2, indicating a significant but less than 50% occurrence of the sensitive soils in the mapping unit).

The mapping units were drawn on overlays at the same scale as the original soil map used. They were then converted to a common 1:2,500,000 scale and compiled into an eastern United States map not included in this report. The five in this report were generalized to allow reduction. Copies of the more detailed state maps are available from the author.

RESULTS

The amount of detail shown in the maps of the states depends upon the detail available in soil association maps and the limits of scale. The smallest areas mapped in most states are on the order of 20 square miles. In using such maps, it must be remembered that every area will have inclusion of other soil types. Most of the mapping units designated nonsensitive (NS) will include small areas of slightly sensitive (SS) or sensitive (S) soils, and the reverse is obviously true. Land use is ignored in making these maps. They are based only on soil features, primarily CEC and presence of carbonates. Land-use maps should be used as overlays on these to determine sensitive areas not affected by agricultural practices.

The north-central region including Wisconsin, Michigan, Illinois, Indiana, and Ohio is dominated by soils formed in glacial deposits of relatively young age or wind and water deposits of the same era. Where the soils are medium or fine textured, the soils are well buffered. Therefore, the region as a whole is nonsensitive to slightly sensitive. The soils potentially slightly sensitive are concentrated in the northern parts of Wisconsin and Michigan on coarse glacial deposits and to a lesser degree in southern Indiana (Fig. 1) and Ohio. In Indiana (Fig. 1) only two map classes were used, nonsensitive (NS) and areas containing less than 50% of soils potentially slightly sensitive (SS2). There are areas much too small to show on these maps where the soils are very poorly buffered (e.g., dune sands), but overall one must conclude that the soils in Indiana are unlikely to reflect the influences of acid precipitation. The areas mapped SS2 are mostly medium-textured sloping soils which were in the upper portion of the CEC range chosen for potentially "slightly sensitive."

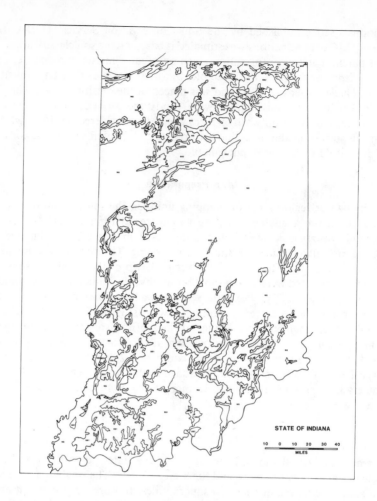

Fig. 1. Soils of Indiana potentially sensitive to long-term effects of acid precipitation. Nonsensitive, NS; slightly sensitive, SS; sensitive, S. Numeral following S or SS indicates occurrence of class within the area; 2 indicates significant occurrence but less than 50% of area, 1 indicates dominance of that class.

In Kentucky, West Virginia, and Tennessee, most of the soils are highly weathered from a variety of residual parent materials. The long weathering period in relatively warm, wet climates has produced highly leached soils, most of which have a moderate or low CEC. Therefore, by the criteria used here, large portions of these states are mapped in the slightly sensitive category with varying degrees of inclusions (SS1 or SS2). The Cumberland Mountains and Plateau area of Tennessee and Kentucky, where most of the soils are shallow and derived from sandstone, is an extensive area where nearly all the soils are sensitive or slightly sensitive. In West Virginia (Fig. 2), eastern Kentucky, and eastern Tennessee, the shallow soils of the steep slopes are usually lower in buffering capacity than the valley soils between the ridges. West Virginia (Fig. 2) contains only a small percentage of area that was sufficiently poorly buffered to be classed potentially sensitive; however, small sensitive areas are included in the other map units, especially within the SS areas.

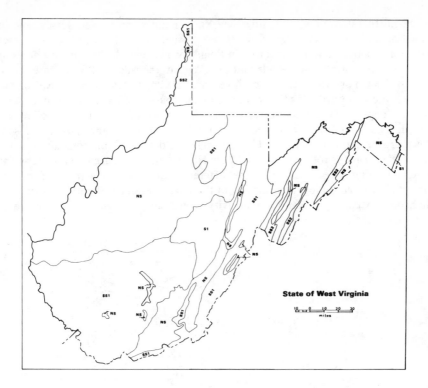

Fig. 2. Soils of West Virginia potentially sensitive to long-term effects of acid precipitation. Nonsensitive, NS; slightly sensitive, SS; sensitive, S. Numeral following S or SS indicates occurrence of class within the area; 2 indicates significant occurrence but less than 50% of area, 1 indicates dominance of that class.

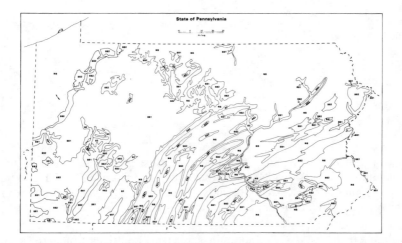

Fig. 3. Soils of Pennsylvania potentially sensitive to long-term effects of acid precipitation. Nonsensitive, NS; slightly sensitive, SS; sensitive, S. Numeral following S or SS indicates occurrence of class within the area; 2 indicates significant occurrence but less than 50% of area, 1 indicates dominance of that class.

The northeastern area, including New York, Pennsylvania, New Jersey, and the New England states, includes large regions of slightly sensitive and sensitive areas with nonsensitive area interspersed. In Pennsylvania (Fig. 3) the sensitive areas are related to the Alleghany uplands and the ridges in the ridge and valley area to the east. In New York (Fig. 4) the sensitive areas are concentrated in the Adirondack region, where most of the soils are derived from coarse nonbasic glacial till or sandy outwash deposits. Significant areas are also found east of the Hudson and on Long Island. The New England states contain a large area of slightly sensitive soils; however, the soil map detail and cation exchange data available for that area were variable. Our maps for Vermont and New Hampshire were based on a sketchy data base and do not adequately show the differences in sensitivity within, nor do they mesh well with adjoining, states. Most of the state of Maine is slightly sensitive, and that complex pattern of varying sensitivity should carry westward into New Hampshire and Vermont.

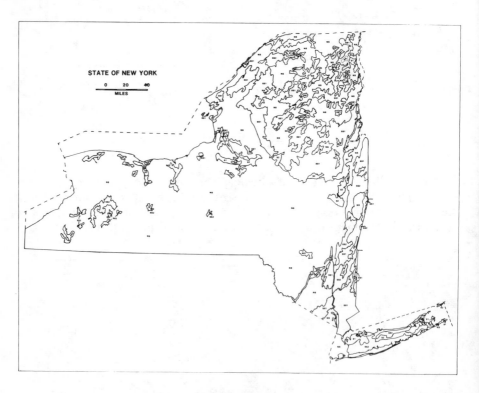

Fig. 4. Soils of New York potentially sensitive to long-term effects of acid precipitation. Nonsensitive, NS; slightly sensitive, SS; sensitive, S. Numeral following S or SS indicates occurrence of class within the area; 2 indicates significant occurrence but less than 50% of area, 1 indicates dominance of that class.

The states that form the east and south coasts from Louisiana around to Delaware and Maryland have sandy coastal plains deposits in common. The North Carolina map depicted in Fig. 5 illustrates the pattern of sensitivity common to that region. Where coastal plains deposits are well drained and have low organic matter content, as they usually do in the "sand hills" and rolling lands, they are considered sensitive or slightly sensitive. The Piedmont region contained both slightly sensitive and

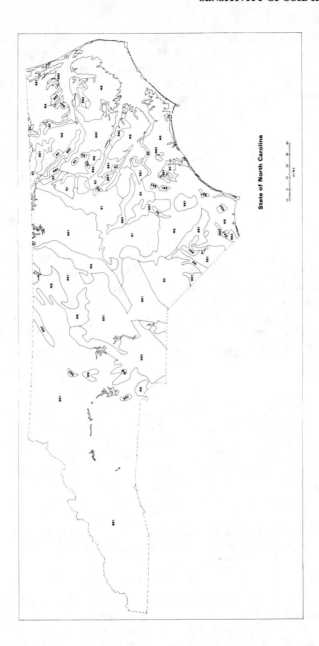

Fig. 5. Soils of North Carolina potentially sensitive to long-term effects of acid precipitation. Nonsensitive, NS; slightly sensitive, SS; sensitive, S. Numeral following S or SS indicates occurrence of class within the area; 2 indicates significant occurrence but less than 50% of area, 1 indicates dominance of that class.

nonsensitive regions, and most of the Appalachian region was in one of the sensitive categories. Where soil maps were in sufficient detail, as they were in Georgia and some other states, most of the stream valleys are mapped NS.

Using the rather tentative criteria established in this report, large portions of the eastern United States, perhaps one-half of the soils, may be slightly sensitive to long-term effects of acid precipitation. These areas are concentrated in the southeastern United States on highly weathered soils, on the shallow and sloping soils of the Appalachian Highland regions, in the Adirondack mountains, and in the outwash deposits and coarse tills of the New England states. In much of this area, the southeastern United States in particular, a high percentage of these soils are in cultivation, and any effects of acid precipitation will be overshadowed by effects of soil amendments such as lime and fertilizers.

Only rather limited soil areas are so poorly buffered that they were classed sensitive. These would appear to be the areas which should be given first consideration for investigation of effects on soils.

CONCLUSION

This exercise produced maps that appear to be useful in research planning and obtaining an overview of soil resources related to the acid precipitation phenomena. Most of the state maps appear to be reasonable based on current soil and geologic knowledge. A few (not included in this report) are based on skimpy data bases and have little value. It is apparent that additional soil data should be obtained, the criteria refined, and the maps improved at these scales or more detailed levels. Obviously, field testing is needed before new criteria are defined. Until better information is available, if land-use maps are overlaid on these maps and natural or unmanaged areas are located that also contain a high percentage of potentially "sensitive" soils, one can locate regions where research has a higher probability of detecting acid precipitation effects.

We should not ignore the possibility of serious effects accruing to soils from atmospheric deposition, but until we have a firm data base we should heed the advice of B. T. Mason (1977) of the U.K. Meteorological Office. Where he was speaking of the atmosphere, we could substitute the word "soils." "The atmosphere is a robust system with a built-in capacity to counteract any perturbation . . . . Sensational warnings of imminent catastrophe, unsupported by firm facts or figures, not only are irresponsible, but likely to prove counterproductive. The atmosphere is wont to make fools of those who do not show proper respect for its complexity and resilience."

Defining and locating soil areas most likely to be sensitive to acid precipitation is a logical step toward gathering "firm facts and figures."

REFERENCES

Bache, Byron. 1979. The acidification of soils. In Hutchinson and Havas (eds.). See Hutchinson (1979).

Galloway, J. N., E. B. Cowling, E. Gorham, and W. W. McFee. 1978. A national program for assessing the problem of atmospheric deposition (acid rain). A report to C.E.Q. National Atmospheric Deposition Program, NC-141.

Hutchinson, T. and M. Havas (eds.). 1979. Effects of Acid Precipitation on the Terrestrial Ecosystem. Proceedings of NATO Adv. Res. Institute, Toronto, Canada, in press.

McFee, W. W., and J. M. Kelly. 1977. Air pollution: impact on soils. In New Directions in Century Three. Strategies for Land Uses. Proceedings of the Soil Conservation Soc. of America, SCSA, Ankeny, Iowa.

Noggle, J. C. 1980. The role of gaseous sulfur in crop nutrition. This volume.

Reuss, John D. 1975. Chemical and biological relationships relevant to the effect of acid rainfall on the soil-plant system. In Proceedings of First International

Symposium on Acid Precipitation and the Forest Ecosystem. U.S.D.A.F.S. Tech. Report NE-23, pp. 791–813.

Wiklander, Lambert. 1973/74. The acidification of soil by acid precipitation. Grundfoerbaettring 26(4):155–164.

Wiklander, Lambert. 1979. Leaching and acidification of soils. In Wood, M. J. (ed.). See Wood (1979).

Wood, M. J. (ed.). 1979. Ecological Effects of Acid Precipitation. Report of workshop held at Cally Hotel, Gatehouse-of-Fleet, Galloway, U.K. 4–7 Sept. 1978. EPRI SOA77-403, Electric Power Res. Institute, 3412 Hillview Ave., Palo Alto, California 94303.

DISCUSSION

David Lang, Minnesota Pollution Control Agency: I would be interested in your opinion of whether you could take a map produced by Dr. Norton and overlay it with one of your maps from the same area and get a composite picture of both the bedrock and the soil sensitivity.

W. W. McFee: We certainly hope so, and we intend to do that. We think that by overlaying them, we can certainly improve the prediction of sensitivity.

Bob Goldstein, EPRI: How sensitive are your predictions to your choice of a precipitation pH of 3.8 or do the percentages of areas in the different classes change much if you use a pH 4 or pH 3.4?

W. W. McFee: No, it would not change significantly. If we simply expanded or contracted the time a little bit, it would offset the change in pH assumption. There's nothing ironclad about my assumption that you need 25% of the cation exchange capacity in acid input, so I really don't think it would make an important difference if you changed those assumptions. It might change our boundaries, but I think we'd end up with the same general areas as being most sensitive. Obviously, we're not trying to predict what's going to happen, and if we were, then the amount of acid would become very important. We're simply trying to pick those areas where, if we have a lot of acid, the effects are most likely to appear.

Bob Goldstein: I would like to ask you, what is your comment with regard to the possible acidification of lakes in relation to your sensitive soils?

W. W. McFee: We've always been faced with a quandary when we try to talk about sensitivities of soils, because we are not just concerned with whether the soil would be acidified but also with it's influence on aquatic systems. I think that the soils with low cation exchange capacity that are already acid are to be included in our sensitive zones and are the ones that will offer the least resistance to aquatic acidification.

CHAPTER 49

REGIONAL PATTERNS OF SOIL SULFATE ACCUMULATION: RELEVANCE TO ECOSYSTEM SULFUR BUDGETS*

Dale W. Johnson†

J. W. Hornbeck‡

J. M. Kelly§

W. T. Swank||

D. E. Todd†

ABSTRACT

Analyses of soils from Walker Branch, Camp Branch, and Cross Creek, Tennessee; Coweeta, North Carolina; and Hubbard Brook, New Hampshire, support the hypothesis that watershed sulfur accumulation is due to inorganic sulfate adsorption in soils. Analyses of soils from lysimeter study sites at La Selva, Costa Rica, and Thompson site, Washington, produced similar results. In laboratory adsorption studies, only soils from Coweeta retained substantial (50 to 100%) additional amounts of sulfate in insoluble forms.

Soil adsorbed sulfate content and sulfate adsorption capacity were positively correlated with free iron content but negatively correlated with organic matter content. Organic matter apparently blocks adsorption sites, preventing sulfate adsorption in iron-rich A horizons and Spodosol B2ir horizons. This blockage may account for the accumulation of adsorbed sulfate in B horizons in temperate and tropical soils and the susceptibility of New England Spodosols to leaching by H_2SO_4.

INTRODUCTION

Ecologists and foresters have been concerned about the effects of acid rain on terrestrial ecosystems for over a decade. In particular, suggestions of reduced forest growth in southern Sweden due to accelerated calcium leaching from soils sparked several research efforts to determine the effects of acid rain on soil leaching (Tamm, 1976; Johnson et al., 1972; Cole and Johnson, 1977; Cronan et al., 1978). There is some evidence that atmospheric sulfuric acid inputs are contributing significantly to increased soil leaching rates in the northeastern United States (Johnson et al., 1972;

* Research sponsored by the Office of Health and Environmental Research, U.S. Department of Energy, under contract W-7405-eng-26 with Union Carbide Corporation. Publication No. 1436, Environmental Sciences Division, ORNL.

†Environmental Sciences Division, Oak Ridge National Laboratory, Oak Ridge, Tennessee 37830.

‡U.S. Forest Service, Durham, New Hampshire.

§Tennessee Valley Authority, Oak Ridge, Tennessee.

||U.S. Forest Service, Franklin, North Carolina.

Cronan et al., 1978) and Scandinavia (Malmer, 1976). Tamm (1976) cites evidence suggesting that forest nitrogen cycling processes are disrupted by atmospheric H_2SO_4 additions as well.

Reuss (1976), however, argues that the effects of acid rain on soil nutrient status must be assessed within the perspective of numerous natural acidifying processes within the soil (e.g., carbonic acid production, nitrification, sulfur oxidation, H^+ release from plant roots). In several cases the H^+ production by these mechanisms far exceeds H^+ input via acid rain (Cole and Johnson, 1977; Johnson et al., 1977; Andersson et al., 1979; Sollins et al., 1979).

The anionic components of acid rain, usually sulfate and nitrate, often constitute more significant additions to the ecosystem than H^+. Ecologists have recently begun to turn their attention to the effects of atmospheric sulfur inputs on the cycling of both sulfur (Likens et al., 1977; Shriner and Henderson, 1978) and nitrogen (Turner et al., 1979) in forest ecosystems. The fate of sulfate is of interest not only in itself but also in terms of the effects of acid rain on cation leaching from soils. Unless sulfate is mobile in the soil, inputs of sulfuric acid will have no effect on cation leaching. Due to charge balance considerations, cations cannot leach without associated anions, in this case, sulfate. This fact has been amply demonstrated in both laboratory (Wiklander, 1976) and field (Johnson and Cole, 1977) studies of the effects of sulfate mobility on soil leaching processes.

Input-output budgets for sulfate in various forest ecosystems indicate that some ecosystems accumulate S (Heinrichs and Mayer, 1977; Shriner and Henderson, 1978; Swank and Douglass, 1977), whereas other ecosystems either show a net loss or maintain a balance of S inputs and outputs (Likens et al., 1977; Cole and Johnson, 1977). A likely mechanism of ecosystem sulfur accumulation is sulfate adsorption in subsoil horizons. Sulfate adsorption is known to occur on sesquioxide (and, to a lesser extent, kaolinite) surfaces (Chao et al., 1964; Harward and Reisenauer, 1966). Sulfate adsorption is strongly pH-dependent, however, with little adsorption occurring above pH 6–7 (Harward and Reisenauer, 1966).

Accumulations of adsorbed sulfate in acid subsoils have been noted throughout the world (Williams and Steinbergs, 1964; Ensminger, 1954; Neller, 1959), particularly in sesquioxide-rich tropical subsoils (Hesse, 1957; Hasan et al., 1970). Johnson et al. (1979b) noted that soil sesquioxide content, sulfate content, and sulfate adsorption capacity were much greater in soils from a tropical sulfur-accumulating forest ecosystem than in soils from a temperate non-sulfur-accumulating ecosystem.

Soil sulfate content is a function of adsorption properties and sulfate input integrated over time. Thus, while soil properties (e.g., sesquioxide and clay content) may favor sulfate adsorption, the degree of soil sulfate saturation will influence the net annual accumulation of sulfate at any point in time. In order to properly characterize soil sulfate adsorption capacity and its influence on ecosystem sulfur budgets, we must determine basic soil properties relevant to sulfate adsorption, e.g., sulfate content and the capacity of the soil for further sulfate accumulations.

The purpose of the following study was to characterize soils from a variety of ecosystems in the above manner and to relate these soil characteristics to measured ecosystem sulfate budgets.

STUDY SITES

Soils selected for characterization were obtained from the seven locations listed in Table 1. The sites chosen represent a variety of forest and soil types as well as geologic bases. Additionally, data on soils, vegetation, climate, and sulfur inputs

Table 1. Summary site descriptions for watershed study areas

Location	Elevation range (m)	Precipitation (cm)	Forest type	Dominant species	Basal area (m²/ha)	Parent material	Soil series	Classification
Walker Branch, Tenn.[a]	265–360	151	Mixed hardwood	Quercus, Carya, Pinus	21	Dolomitic limestone	Claiborne	Fine-loamy, siliceous, mesic, typic paleudult
							Fullerton	Clayey, kaolinitic, thermic, typic paleudults
Camp Branch, Tenn.[b]	598–518	144	Mixed hardwood	Quercus, carya	20	Pennsylvanian sandstone	Hartsell	Fine-loamy, siliceous, thermic, typic hapludult
							Philo	Coarse loamy, mixed, mesic, fluvaquentic dystrochrept
Cross Creek, Tenn.[b]	574–495	155	Mixed hardwood	Quercus, Carya	22	Pennsylvanian sandstone	Cotaco	Fine-loamy, mixed mesic, aquic hapludult
							Hartsells	Fine-loamy, siliceous, thermic, typic hapludult
Coweeta, N.C.[c]	720–1740	180	Mixed hardwood	Quercus, Carya, Acer	25	Gneiss, schist, sandstone	Porters	Fine-loamy, mixed mesic, humic hapludult
							Saluda	Loamy, mixed, mesic, shallow, typic hapludult
Hubbard Brook, N.H.[d]	500–800	120	Northern hardwood	Acer, Fagus, Betula	24	Gneiss, schist	Becket	Coarse-loamy, mixed, frigid, typic fragiorthods
Thompson, Wash.[e]	210	136	Douglas fir plantation	Pseudotsuga	37	Glacial outwash	Everett	Loamy-skeletal, mixed, mesic dystric xerochrept
La Selva, Costa Rica[f]	40	430	Tropical rain forest	Pentaclethera, Grias, Palmae		Alluvium	La Selva	Typic hydrandept

[a] Source: Shriner, D. S., and G. S. Henderson. 1978. Sulfur distribution and cycling in a deciduous forest watershed. J. Environ. Qual. 7:392-397.

[b] Source: Kelly, J. M. 1979. Camp Branch and Cross Creek experimental watershed projects; objectives, facilities, and ecological characteristics. TVA/EPA Interagency Energy/Environment R&D Program Report. EPA-600/7-79-053.

[c] Source: Swank, W. T., and J. E. Douglass. 1977. Nutrient budgets for undisturbed and manipulated hardwood forest ecosystems in the mountains of North Carolina. Pp. 343-363 in Correll, D. L. (ed.), Watershed Research in Eastern North America, Vol. 1. Smithsonian Institution, Washington, D.C.

[d] Source: Likens, G. E., F. H. Bormann, R. S. Pierce, J. S. Eaton, and N. M. Johnson. 1977. Biogeochemistry of a Forested Ecosystem. Springer-Verlag, New York.

[e] Source: Cole, D. W., and D. W. Johnson. 1977. Atmospheric sulfate additions and cation leaching in a Douglas-fir ecosystem. Water Resour. Res. 13(2): 313-317.

[f] Source: Johnson, D. W., D. W. Cole, and S. P. Gessel. 1979b. Acid precipitation and soil sulfate adsorption properties in a tropical and temperate forest soil. Biotropica 11:38-42.

and losses were available for each site. In the interest of brevity, pertinent information relating to each site has been summarized in Table 1. Readers are directed to the indicated references for more specific descriptions of individual sites.

METHODS

Soil samples were air-dried and passed through a 2-mm sieve prior to analysis. Free iron and aluminum were determined by the citrate-buffer–dithionite method of Jackson (1973). Total carbon was determined by combustion in a LECO automated furnace. Percent clay data were available for all but the Cross Creek and Camp Branch soils, and determinations of clay in the latter were conducted by the hydrometer method (Day, 1965).

Initial soil sulfate content was determined by sequential leachings with H_2O and NaH_2PO_4. Triplicate 10-g samples from each horizon were placed in Buchner funnels lined with Whatman No. 50 filter paper and leached with 200 ml of distilled H_2O followed by 50 ml of NaH_2PO_4 containing 500 ppm P (Ensminger, 1954). Sulfate removed by water extraction was designated as initial soluble sulfate, and that removed by NaH_2PO_4 was designated as initial insoluble sulfate. Sulfate adsorption and desorption were determined by leaching another set of three samples with 50 ml of Na_2SO_4 containing 75 ppm S followed by H_2O and NaH_2PO_4 extractions. The loss of sulfate from the Na_2SO_4 leachate was termed total sulfate adsorption, and a significant ($\alpha = 0.05$, t test) change in insoluble sulfate (relative to initial insoluble sulfate content) was termed insoluble sulfate adsorption. Sulfate removed from the latter samples by NaH_2PO_4 leaching was termed potential insoluble sulfate content. Figure 1 shows a flow diagram for the extractions.

Sulfate in water and Na_2SO_4 leachings was determined colorimetrically on a Technicon Autoanalyzer. Due to phosphate interferences on the Autoanalyzer, sulfate in NaH_2PO_4 leachates was determined by the barium chloranilate procedure of Bertolacini and Barney (1957).

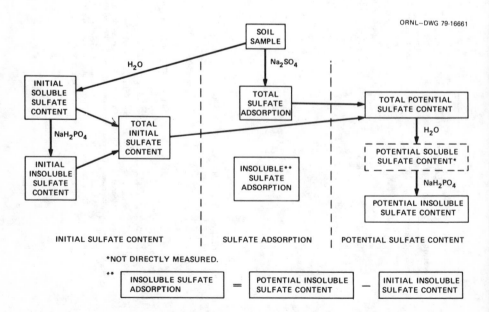

Fig. 1. Flow diagram for soil sulfate extraction and adsorption procedures.

Correlations between sulfate content (soluble and insoluble), sulfate adsorption (total and insoluble), free iron and aluminum, organic matter, and clay content were performed by the CORR procedure of the SAS data analysis system (Barr et al., 1976).

RESULTS

All soils tested except the Becket series contained insoluble sulfate in subsurface horizons (Fig. 2 and 3). The Philo and Hartsells series contained very small amounts of insoluble sulfate relative to other soils.

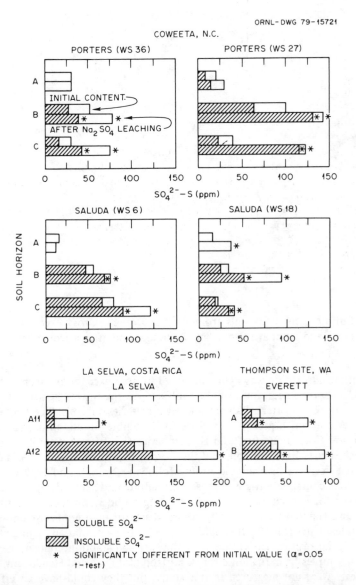

Fig. 2. Sulfate content of soils from Coweeta, N.C.; La Selva, Costa Rica; and Thompson site, Wash., before and after leaching with Na_2SO_4 containing 75 ppm S.

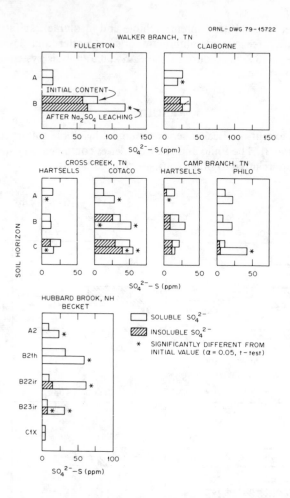

Fig. 3. Sulfate content of soils from Walker Branch, Cross Creek, and Camp Branch, Tenn., and Hubbard Brook, N.H., before and after leaching with Na₂SO₄ containing 75 ppm S.

Total sulfate adsorption was significant in all subsoils except those from the Hartsells and Clairborne series, but insoluble sulfate adsorption was significant only in the Cotaco, Becket, Porters, Saluda, and Everett series subsoils. The Porters and Saluda series had much greater capacities for insoluble sulfate adsorption (15–70 ppm) than the Cotaco, Becket, and Everett series (5–13 ppm). Samples from the Cross Creek, Hartsells C horizon, and the Cotaco B horizon experienced an unexplainable net decrease in insoluble sulfate following Na₂SO₄ and H₂O leachings.

Surface soils contained little or no insoluble sulfate and generally had lower total sulfate adsorption capacities than subsoils did (Fig. 2 and 3). Surface horizons of the Claiborne and Hartsells series actually desorbed significant amounts of sulfate during Na₂SO₄ leachings (Fig. 2). It should be noted, however, that an "aging effect", or increasing adsorption of sulfate with time of soil-solution contact, has been observed in samples from the Fullerton A horizon (Johnson and Henderson, 1979). Aging effects were not measured in this study, but they could have an effect on sulfate retention in the field for any of the soils tested. This is an area deserving of further investigation.

DISCUSSION

Relationships to Ecosystem Sulfur Budgets

The relationship between watershed sulfur budgets and soil sulfate accumulations can be assessed by expressing the latter in units of kilograms per hectare. We did not analyze a sufficient number of samples to get accurate quantitative data for soil sulfate accumulation in each watershed, but the values in Table 2 provide some perspective for selected soil series within the various watersheds. The fact that the soil (Becket) from the one non-sulfur-accumulating watershed (Hubbard Brook) contained no insoluble sulfate is consistent with our hypothesis that soil sulfate adsorption accounts for watershed sulfur retention. It is also noteworthy that the Saluda soil on watershed 6 at Coweeta, which was treated with sulfate-containing fertilizer in 1959 (Swank and Douglass, 1977), contained over three times as much

Table 2. Sulfur budgets and soil sulfate content at various research sites

Site and soil	Annual sulfate sulfur accumulation[a] (kg·ha^{-1}·year^{-1})	Soil sulfate sulfur content[b] (kg/ha)			Depth to parent material (m)
		Soluble	Insoluble	Total	
Watershed Studies					
Walker Branch, Tenn.	6.6[c]				
Fullerton		250	450	700	30+
Claiborne		150	150	300	10+
Coweeta, N.C.					
Saluda-WS 6	8.7[d]	150	500	650	~7
Saluda-WS 18	9.3[d]	100	150	250	~7
Porters-WS 36	6.2[d]	250	250	500	~7
Porters-WS 27	6.6[d]	350	400	750	~7
Hubbard Brook, N.H.					
Becket	−4.9[e]	100	0	100	1–2
Cross Creek, Tenn.	6.5[f]				
Hartsells		150	20	170	<1
Cotaco		250	350	600	1
Camp Branch, Tenn.	8.5[f]				
Hartsells		180	130	310	<1
Philo		150	50	200	1
Lysimeter studies					
Thompson, Wash.	−2.7[g]				
Everett		40	100	140	2
La Selva, Costa Rica	11.0[h]				
La Selva		100	900	1000	2

[a]Input via bulk precipitation minus output in streamwater or soil solution.

[b]Approximate values for top meter of soil.

[c]Source: Shriner, D. S., and G. S. Henderson. 1978. Sulfur distribution and cycling in a deciduous forest watershed. J. Environ. Qual. 7:392–397.

[d]Source: Swank, W. T., and J. E. Douglass. 1977. Nutrient budgets for undisturbed and manipulated hardwood forest ecosystems in the mountains of North Carolina. Pp. 343–363 in Correll, D. L. (ed.), Watershed Research in Eastern North America, Vol. 1. Smithsonian Institution, Washington, D.C.

[e]Source: Likens, G. E., F. H. Bormann, R. S. Pierce, J. S. Eaton, and N. M. Johnson. 1977. Biogeochemistry of a Forested Ecosystem. Springer-Verlag, New York.

[f]Source: Kelly, J. M. 1979. Camp Branch and Cross Creek Experimental Watershed Projects; Objectives, Facilities and Ecological Characteristics, TVA/EPA Interagency Energy/Environment R&D Program Report. EPA600/7-79-053.

[g]Source: Cole, D. W., and D. W. Johnson. 1977. Atmospheric sulfate additions and cation leaching in a Douglas-fir ecosystem. Water Resour. Res. 13(2):313–317.

[h]Source: Johnson, D. W., D. W. Cole, and S. P. Gessel. 1979b. Acid precipitation and soil sulfate adsorption properties in a tropical and temperate forest soil. Biotropica 11:38–42.

insoluble sulfate as the unfertilized Saluda soil on watershed 18. Laboratory adsorption tests indicated that both the Saluda and Porters series soils from Coweeta had large capacities to adsorb additional sulfate into insoluble forms (Fig. 2). These findings suggest that the watersheds at Coweeta can continue to adsorb atmospherically deposited sulfate for longer periods of time than the other sulfur-accumulating watersheds at Walker Branch, Cross Creek, and Camp Branch.

The heterogeneity among soils from a given watershed could have an effect on ecosystem sulfur budgets also; for instance, the Fullerton soil on Walker Branch has accumulated more sulfate than the Claiborne soil (Table 2). Similarly, the Cotaco soil has accumulated more sulfate than the Hartsells soil on the Cross Creek watershed. Thus, it is probable that sulfate accumulations indicated by watershed-level input-output budgets occur primarily within specific soil series (e.g., Fullerton, Cotaco).

Another important consideration in the watershed studies is soil depth. The soil sulfate contents in Table 2 were calculated for a 1-m soil depth, whereas in some cases the soil mantle is many meters thick. At Walker Branch, for example, there is an estimated accumulation of over 2700 kg/ha of sulfate sulfur in the top 30 m in the Fullerton soil (Johnson and Henderson, 1979).

Lysimeter studies of elemental budgets have distinct advantages over watershed studies in that only one soil series and one soil depth are considered at a time. This was the case for the tropical-temperate sulfur comparison reported on previously (Johnson et al., 1979b). Soils from these two sites were reanalyzed for the purposes of this study, since procedures differed somewhat. The La Selva soil (tropical) had a very large accumulation of insoluble sulfate, and the sulfur budget showed a large net annual accumulation (Table 2). No statistically significant additional accumulation of insoluble sulfate was noted following laboratory Na_2SO_4 additions, however. The Everett soil had a lower accumulation of insoluble sulfate, and the sulfur budget indicated a slight net annual loss. Apparently a steady-state condition had been reached in the Everett soil at the current (1974) sulfate input levels ($\sim$ 4 kg of S per hectare per year), which were considerably lower than those at any of the other sites (inputs range from 12 to 18 kg of S per hectare per year). Since sulfate adsorption is concentration-dependent (Harward and Reisenauer, 1966), further accumulations could occur in the Everett soil at higher input levels. According to laboratory adsorption tests the Everett soil has a moderate capacity to accumulate additional insoluble sulfate (Fig. 2). Field studies involving the application of H_2SO_4 and municipal wastewater have also shown that the Everett soil can accumulate additional sulfate (Johnson and Cole, 1977; Johnson et al., 1979a).

Factors Affecting Sulfate Content and Adsorption

Factors potentially influencing retention of sulfate in various soil series include free iron and aluminum oxide content (Chao et al., 1964), clay content (Neller, 1959), and organic matter content (Haque and Walmsley, 1973). Regression analysis of these soil properties indicated that free iron was most closely correlated with initial sulfate content (both soluble and insoluble), sulfate adsorption from Na_2SO_4 solutions (both total and insoluble), and potential insoluble sulfate content (Table 3). Clay content was significantly correlated with both initial and potential insoluble sulfate content but was not significantly correlated with soluble sulfate content or total sulfate adsorption. Free Al content was not significantly correlated with any soil sulfate parameter. It should be noted, however, that the citrate-dithionite method may not remove all free Al oxides in soils (Jackson, 1973).

Table 3. Correlation coefficients (R) between soil sulfate, free iron, free aluminum, carbon, and clay content

| | Soil sulfate | | | | | | |
| | Initial content | | | Adsorption[a] | | Potential content | |
	Soluble	Insoluble	Total	Total	Insoluble	Insoluble	Total
Percent Fe	0.28[b]	0.50[c]	0.55[c]	0.40[d]	0.27[b]	0.41[d]	0.55[c]
Percent Al	0.19	− 0.06	− 0.0001	0.20	0.24	− 0.01	0.11
Percent carbon	0.29[b]	− 0.31[b]	− 0.21	− 0.01	− 0.13	− 0.32[b]	− 0.12
Percent clay	0.05	0.43[c]	0.42[d]	0.21	− 0.16	0.19	0.37[d]

[a]After leaching with 75 ppm S as Na_2SO_4.

[b]90% significance.

[c]99% significance.

[d]95% significance.

Some interesting relationships were found between soil sulfate and percent carbon. The initial content of soluble sulfate was positively correlated with percent carbon, but both the initial and the potential content of insoluble sulfate were negatively correlated with percent carbon (Table 3). Organic matter may be a source of soluble sulfates (e.g., by breakdown of ester sulfates, e.g., Tabatabai and Bremner, 1972), but it may also block sulfate adsorption sites on sesquioxide surfaces (Johnson et al., 1979b; Couto et al., 1979).

Although there were significant correlations between adsorbed sulfate, iron oxide, and clay content, there were notable exceptions in cases where organic matter, sesquioxides, and clay accumulated together. In the Porters and La Selva soils, organic-matter-rich surface horizons contained very little insoluble sulfate despite high sesquioxide and clay contents (Figs. 2 and 5). The Becket soil had a slight accumulation of clay and large accumulations of both sesquioxides and organic matter in B horizons, as is typical of a Spodosol (Fig. 4). Again, there was no insoluble sulfate associated with the sesquioxide and organic matter accumulation (Fig. 3).

In all other soils analyzed in this study, there was a pattern of decreasing percent carbon but increasing insoluble sulfate, sesquioxide, and clay content with depth. Profiles for the Fullerton and Hartsells (Cross Creek) series illustrate this general pattern for high- and low-sesquioxide soils, respectively (Figs. 3 and 4). Excluding the Porters A horizon, La Selva A 11 horizon, and Becket soils, data from regression analysis improved the correlation coefficients of free iron vs insoluble sulfate content (from 0.55 to 0.61), insoluble sulfate adsorption (from 0.27 to 0.38), and potential insoluble sulfate content (from 0.41 to 0.52).

Results of this study support previous speculations concerning the blocking effect of organic matter on soil sulfate adsorption sites (Johnson et al., 1979b; Couto et al., 1979). We are currently testing this hypothesis by measuring soil sulfate adsorption properties with and without organic matter removal. If this hypothesis proves correct, it will help account for the observation that only subsoils with low quantities of organic matter accumulate substantial quantities of sulfate, even where surface soils have appreciable clay and sesquioxide contents. It also has implications in terms of the susceptibility of Spodosols to H_2SO_4 leaching. Although large accumulations of sesquioxides are common in Spodosol subsoils, such accumulations are usually associated with organic matter in the form of the fulvic acids, which transport iron and aluminum to lower horizons (Kononova, 1966).

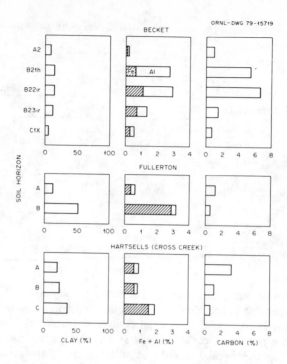

Fig. 4. Percent clay, free iron and aluminum, and carbon in the Becket, Fullerton, and Hartsells soil series from Hubbard Brook, N.H., and Walker Branch and Cross Creek, Tenn., respectively.

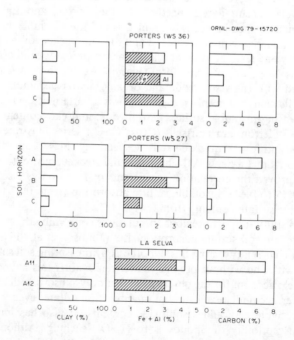

Fig. 5. Percent clay, free iron and aluminum, and carbon in the Porters and La Selva soil series from Coweeta, N.C., and La Selva, Costa Rica, respectively.

Thus, despite their high sesquioxide contents, Spodosols would be very suscepti- ble to leaching by atmospheric H_2SO_4. This does in fact seem to be the case in view of studies by Johnson et al. (1972), Likens et al. (1977), and Cronan et al. (1978) on acid rain effects on nutrient leaching from Spodosols in New England. Further research on this subject is needed.

CONCLUSIONS

Laboratory studies indicated that of the sulfur-accumulating ecosystems ex- amined, soils from La Selva, Walker Branch, Cross Creek, and Camp Branch may be near saturation with respect to sulfate accumulation, whereas soils from Coweeta have considerable additional sulfate adsorption capacity remaining. Potential sulfate content (sulfate content plus laboratory adsorption) was greatest at La Selva and Coweeta, intermediate at Walker Branch, and lowest at Cross Creek and Camp Branch.

Of the two non-sulfur-accumulating ecosystems studied, only soils from the Thompson site contained insoluble sulfate. A steady-state condition had apparently been reached with regard to the relatively low atmospheric sulfate inputs at this site (4 kg·ha^{-1}·year^{-1} vs 12–18 kg·ha^{-1}·year^{-1} at all other sites), but both laboratory and field experiments have demonstrated that further soil sulfate accumulations could occur at greater input levels due to concentration-dependent increases in sulfate ad- sorption.

Despite large free iron and aluminum accumulations in subsurface horizons, soil from Hubbard Brook (a Spodosol) contained no insoluble sulfate, and only minor amounts (5–15 ppm) of sulfate were adsorbed in laboratory tests. The high organic matter accumulations in the Hubbard Brook subsoils and surface soils from other sites appear to block sulfate adsorption sites. Thus, while the well-documented rela- tionship between sulfate adsorption and sesquioxide content does hold true for these soils in general, there are important exceptions in cases where both organic matter and sesquioxides are high. Organic matter blockage of sulfate adsorption sites has important implications with respect to sulfur budgets and the susceptibility of sur- face soils and Spodosols to leaching by atmospheric H_2SO_4.

REFERENCES

Andersson, F., T. Fagerstrom, and S. I. Nilsson. 1979. Forest ecosystem responses to acid deposition—hydrogen ion budget and nitrogen/tree growth model ap- proaches. In Hutchinson, T. C. and M. Havas (eds.), NATO Advanced Research Institute Conference on the Effects of Acidic Precipitation on Terrestrial Eco- systems, May 22–26, 1978, Toronto. Plenum Ecosciences Series, in press.

Barr, A. J., J. H. Goodnight, J. P. Sall, and J. T. Helwig. 1976. A User's Guide to SAS. SAS Institute Inc., Raleigh, North Carolina.

Bertolacini, R. J., and J. E. Barney II. 1957. Colorimetric determination of sulfate with barium chloranilate. Anal. Chem. 29:281–283.

Chao, T. T., M. E. Harward, and S. C. Fang. 1964. Iron or aluminum coating in relation to sulfate adsorption characteristics of soils. Soil Sci. Soc. Am., Proc. 28:632–635.

Cole, D. W., and D. W. Johnson. 1977. Atmospheric sulfate additions and cation leaching in a Douglas-fir ecosystem. Water Resour. Res. 13(2):313–317.

Couto, W., D. J. Lathwell, and D. R. Bouldin. 1979. Sulfate sorption by two oxisols and an alfisol of the tropics. Soil Sci. 127:108–116.

Cronan, C. S., W. A. Reiners, R. C. Reynolds, and G. E. Lang. 1978. Forest floor leaching: contributions from mineral, organic, and carbonic acids in New Hampshire subalpine forests. Science 220:309–311.

Day, P. R. 1965. Particle fractionation and particle-size analysis. Pp. 545–566 in Black, C. A., D. D. Evans, J. L. White, L. E. Ensminger, and F. E. Clark (eds.), Methods of Soil Analysis. Part I. Physical and Mineralogical Properties, Including Statistics of Measurement and Sampling. No. 9 in the Agronomy Series. American Society of Agronomy, Madison, Wisconsin.

Ensminger, L. E. 1954. Some factors affecting the adsorption of sulfate by Alabama soils. Soil Sci. Soc. Am., Proc. 18:259–264.

Haque, I., and D. Walmsley. 1973. Adsorption and desorption of sulphate in some soils of the West Indies. Geoderma 9:269–278.

Harward, M. E., and H. M. Reisenauer. 1966. Movement and reactions of inorganic soil sulfur. Soil Sci. 101:326–335.

Hasan, S. M., R. L. Fox, and C. C. Boyd. 1970. Solubility and availability of sorbed sulfate in Hawaiian soils. Soil Sci. Soc. Am., Proc. 34:897–901.

Heinrichs, H., and R. Mayer. 1977. Distribution and cycling of major and trace elements in two central European forest ecosystems. J. Environ. Qual. 6:402–407.

Hesse, P. R. 1957. Sulfur and nitrogen changes in forest soils of East Africa. Plant Soil 9:86–96.

Jackson, M. L. 1973. Soil Chemical Analysis. Advanced Course. 2nd Ed. Copyright by M. L. Jackson.

Johnson, D. W., and D. W. Cole. 1977. Sulfate mobility in an outwash soil in western Washington. Water, Air, Soil Pollut. 7:489–495.

Johnson, D. W., D. W. Cole, S. P. Gessel, M. J. Singer, and R. V. Minden. 1977. Carbonic acid leaching in a tropical, temperate, subalpine, and northern forest soil. Arct. Alp. Res. 9:329–343.

Johnson, D. W., and G. S. Henderson. 1979. Sulfate adsorption and sulfur fractions in a highly-weathered soil under a mixed deciduous forest. Soil Sci. 128: 34–40.

Johnson, D. W., D. W. Breuer, and D. W. Cole. 1979a. The influence of anion mobility on ionic retention in wastewater-irrigated soils. J. Environ. Qual. 8:246–250.

Johnson, D. W., D. W. Cole, and S. P. Gessel. 1979b. Acid precipitation and soil sulfate adsorption properties in a tropical and temperate forest soil. Biotropica 11:38–42.

Johnson, N. M., R. C. Reynolds, and G. E. Likens. 1972. Atmospheric sulfur: its effect on the chemical weathering of New England. Science 177:514–515.

Kelly, J. M. 1979. Camp Branch and Cross Creek Experimental Watershed Projects; Objectives, Facilities, and Ecological Characteristics. TVA/EPA Interagency Energy/Environment R&D Program Report. EPA-600/7-79-053.

Kononova, M. M. 1966. Soil Organic Matter: Its Nature, Its Role in Soil Formation and in Soil Fertility. Pergamon Press, London.

Likens, G. E., F. H. Bormann, R. S. Pierce, J. S. Eaton, and N. M. Johnson. 1977. Biogeochemistry of a Forested Ecosystem. Springer-Verlag, New York.

Malmer, N. 1976. Acid precipitation: chemical changes in the soil. Ambio 5:231–233.

Neller, J. R. 1959. Extractable sulfate-sulfur in soils of Florida in relation to the amount of clay in the profile. Soil Sci. Soc. Am., Proc. 23:346–358.

Reuss, J. O. 1976. Chemical and biological relationships relevant to the effect of acid rainfall on the soil-plant system. Pp. 791–814 in Dochinger, L. S., and

T. A. Seliga (eds.), Proc., First International Symposium on Acid Precipitation and the Forest Ecosystem. USDA Forest Service Gen. Tech. Rep. NE-23.

Shriner, D. S., and G. S. Henderson. 1978. Sulfur distribution and cycling in a deciduous forest watershed. J. Environ. Qual. 7:392–397.

Sollins, P. E., C. C. Grier, F. M. McCorison, K. Cromack, Jr., A. T. Brown, and R. Fogel. 1979. The internal nutrient cycles of an old-growth Douglas-fir stand in western Oregon. Ecol. Monogr., in press.

Swank, W. T., and J. E. Douglass. 1977. Nutrient budgets for undisturbed and manipulated hardwood forest ecosystems in the mountains of North Carolina. Pp. 343–363 in Correll, D. L. (ed.), Watershed Research in Eastern North America, Vol. 1. Smithsonian Institution, Washington, D.C.

Tabatabai, M. A., and J. M. Bremner. 1972. Distribution of total and available sulfur in selected soils and soil profiles. Agron. J. 64:40–44.

Tamm, C. O. 1976. Acid precipitation: biological effects in soil and on forest vegetation. Ambio 5:235–238.

Turner, J., D. W. Johnson, and M. J. Lambert. 1979. Sulphur cycling in a Douglas-fir forest and its modification by nitrogen applications. Oecol. Plant., in press.

Wiklander, L. 1976. The influence of anions on adsorption and leaching of cations in soils. Grundfoerbaettring 27:125–135.

Williams, C. H., and A. Steinbergs. 1964. The evaluation of plant-available sulfur in soils: 2. The availability of adsorbed and insoluble sulfates. Plant and Soil 21:50–62.

DISCUSSION

Ron McLean, Domtar Research: In a lot of the postulations about the effects of acid precipitation in North America we've extrapolated a lot of the results from the Hubbard Brook studies. It seems to me, looking at your data, that you're suggesting that perhaps Hubbard Brook, as far as soil is concerned, is not a typical situation to look at.

D. W. Johnson: Oh, not at all; I'd say it's a very typical spodosol profile, and, in fact, I'm anxious to do more spodosol profiles, and I fully expect them to behave the same way. It's not typical of an ultasol; I think this brings out the importance of knowing what kind of soil you're dealing with. The ultasol we have at Walker Branch is completely different from the Hubbard Brook soil, and its behavior with regard to leaching is totally different. Hubbard Brook happens to be the only spodosol I've done to date, and I'd like to do more to find out how representative it is. I suspect, though, it is representative.

Bill Graustein, Yale University: How does the capacity of soil to absorb sulfate vary with the pH of the solution that carries sulfate?

D. W. Johnson: Well, the sulfate absorption goes up with the pH, as I'm sure you know. I didn't include pH in this study because pH among the soils didn't vary more than a unit, on the grades between 4 and 5, so that the correlations between that and sulfate absorption were not significant, but I know that it is significant over a wide range of pH's. Also, if you apply your sulfate as an acid rather than as a salt, you get much greater absorption in the acid form than you will in the neutral salt form.

Chris Cronan, Dartmouth: I was reminded of a paper that Rich Barton from UPM published on a test for spodosol character, and as I recall he uses phosphate adsorption in different mineral profiles to test for adsorption characteristics and in the B_2 horizon gets a great increase in phosphate adsorption. Now we already know that phosphate and sulfate have some different affinities for aluminum and iron oxides and hydroxides, but nevertheless he seems to find a significant increase in phosphate adsorption where you have got that alluvial horizon and where you're not seeing sulfate adsorption.

D. W. Johnson: Well, I kind of whizzed through that quickly, but if you notice there was a higher sulfate adsorption associated with the B_2 AR. It was not permanent. It did not adsorb into the insoluble form. After the sodium sulfate treatment I got adsorption in three cases. The trouble is that very little of it was obtained in a permanent form, so I think you have to look at adsorption and desorption in these things, because so what, if it adsorbs and comes back off again. There's another effect I did not include which probably is important and that's aging. If you allow soil to sit in contact for a long time with these solutions, you get further adsorption with time. That's not included in this study, and it could be an important factor too.

Bill McFee, Purdue: I wonder if you'd comment. Is nitrate likely to behave like sulfate, and secondly, at the very end you talked about the need for considering other acid inputs, organic matter, decay, root exudates, and so forth. I wish you'd comment on the availability of good estimates of those.

D. W. Johnson: First of all, I don't think nitrate is going to behave at all like sulfate. It is totally different, nonadsorbing, much more biologically active, and deficient in most of the regions of the world; I expect that nitrate would probably be immobilized biologically but not chemically. And the second part, the estimates of internal hydrogen ion production are few and far between. I only know of about, well, I know of one complete one from Sweden by Andersson, who did an estimate of the hydrogen ion production through various sources including root exudation and carbonic acid and organic acid production, and in his site, which, I don't know if it was subjected to acid rain or not, but the atmospheric input was 10% of that produced within the system, but the estimates are hard to get. I agree with that.

Carl Schofield, Cornell: Did the differences you see in sulfate adsorption capacity in the soil show any relationship to stream discharge sulfate concentration relationships in different areas?

D. W. Johnson: Yes, they do. Grey Henderson did some studies on the sulfate concentration during discharge at Walker Branch and found that the sulfate concentration increased as the hydrograph increased—the stream flow increased—and what we thought happened in this case was that the water table would rise into that A horizon, where we have a substantial pool of soluble sulfate, during these periods of high flow and then pick up that soluble sulfate pool, and then during, as the storm receded, well, it would go back down to the B_2 horizon, and the sulfate there is insoluble, and it would not escape. I'm sure that—I wouldn't be surprised if that's a common effect among these soils.

CHAPTER 50

GEOLOGIC FACTORS CONTROLLING THE SENSITIVITY OF AQUATIC ECOSYSTEMS TO ACIDIC PRECIPITATION

Stephen A. Norton*

ABSTRACT

Minerals, when dissolved, are capable of neutralizing excess acid in precipitation according to the number of positive charges released in excess of negative charges released into solution. However, the kinetics of solution are important in the effectiveness of a particular mineral's capability of neutralizing acid. Thus quartz,

$$SiO_2 + 2H_2O \rightarrow H_4SiO_4 \, ,$$

has no buffering capacity; albite,

$$2NaAlSi_3O_8 + 2H^+ + 9H_2O \rightarrow 2Na^+ + Al_2Si_2O_5 (OH)_4 + 4H_4SiO_4 \, ,$$

and calcite,

$$CaCO_3 + H^+ \rightarrow Ca^{2+} + HCO_3^- \, ,$$

$$HCO_3^- + H^+ \rightarrow H_2CO_3 \, ,$$

have similar ultimate acid-neutralizing capacities. However, $CaCO_3$ dissolves rapidly and consequently is highly effective at neutralizing excess acid.

Rocks have been classified from type 1 (yield no or little buffering capacity) to type 4 ("infinite" buffering capacity). Maps, based on bedrock geology, have been produced which predict vulnerability of aquatic ecosystems to impact from acidic precipitation. Field checks of areas underlain by rock types 1 and 2 verify that acidification of surface waters and impact on biological systems due to acidic precipitation are occurring. The lack of discrimination by these maps for rock types 1 and 2 is due to the generalization of large-scale maps, the local hydrology, and neutralization by soils of precipitation acidity.

INTRODUCTION

Although the phenomenon of acidic precipitation has been developing on a hemispheric scale (Cragin et al., 1975) since the early 1800s, only recently have workers identified those areas in the United States receiving acidic precipitation (Cogbill and Likens, 1974; Likens et al., 1979) and associated heavy metals (Lazarus et al., 1970).

Impact to aquatic ecosystems is largely determined by the chemical characteristics of the bedrock. Limestone terrains yield "infinite" buffering (acid-neutralizing capacity) to acidic precipitation, whereas granites (and related igneous rocks), their

---

*Department of Geological Sciences, University of Maine at Orono, Orono, Maine 04469.

metamorphic equivalents, and noncalcareous sandstones yield minimal buffering. On these bases, Kramer (1976), Galloway and Cowling (1978), and Likens et al. (1979) defined areas of the United States and Canada which were, on the basis of large-scale geologic maps, susceptible to impact from acidic precipitation. These areas were largely concentrated in large Precambrian terrains (the Canadian shield, the Adirondack Massif) and the Precambrian core of the Appalachian mountains.

In the United States, acidified (from atmospheric deposition) surface waters have now been identified from New England and New York to Florida and in the boundary Waters Canoe Area, Minnesota. Acidic precipitation exists (pH less than 5.6) in all states contiguous to the Mississippi River and east to the Atlantic Ocean (NADP, 1979) as well as in areas of California (Liljestrand and Morgan, 1979) and Washington (Barnes et al., 1979).

Because of the widespread aspect of acidic precipitation, it is important to understand the natural characteristics of the landscape which render an area susceptible to impact. Three major factors can be identified as important. They are meteorology, pedology, and geology. This paper focuses on the geologic controls of impact from acidic atmospheric deposition.

Funding from the United States Environmental Protection Agency and the University of Maine at Orono (Agricultural Experiment Station) is gratefully acknowledged. I was assisted in the preparation of this report by Frederick P. Liotta, Susan B. Hall, Lisa A. Thurlow, Roland Dupuis, and Marilyn L. Morrison.

GEOLOGIC CONTROLS

Several hundred common minerals comprise the huge variety of rocks which make up the crust of the earth, the surficial overburden, and soils. Of interest in terms of susceptibility of a landscape or water bodies to acidification from precipitation is the capacity of the minerals, no matter what their configuration, to assimilate protons (H_{aq}^+). Other considerations are solubility of the phases and the kinetics of solution.

Certain minerals are soluble and rapidly dissolved but yield no buffering capacity, e.g.:

$$NaCl_{xl} = Na_{aq}^+ + Cl_{aq}^-, \qquad K_{sp} = 10^{1.58} = [Na^+][Cl^-] . \tag{1}$$

Others are relatively insoluble and yield no buffering capacity, e.g.:

$$\underset{\text{(quartz)}}{SiO_{2_{xl}}} + 2H_2O_{aq} = H_4SiO_{4_{aq}}, \qquad K_{sp} \cong 10^{-4.0} = [H_4SiO_4] . \tag{2}$$

Common rock-forming minerals, when placed in water, yield pH's in excess of 7 (commonly in excess of 8), but the kinetics are sufficiently slow so that the total potential H^+ buffering capacity of these minerals is not realized, e.g.:

$$\underset{\text{(albite)}}{2NaAlSi_3O_{8_{xl}}} + 2H_{aq}^+ + 9H_2O_{aq} = 2Na_{aq}^+ + \underset{\text{(kaolinite)}}{Al_2Si_2O_5(OH)_{4_{xl}}} + 4H_4SiO_{4_{aq}},$$

$$K_{sp} \cong 10^1 = ([Na^+]^2/[H^+]^2)[H_4SiO_4]^4 . \tag{3}$$

Assuming an initial pH of 4.0 and buffering from the CO_2-H_2O system ($HCO_3^- \cong 10^{-5.7}$), the expected pH after complete reaction (3) would be approximately 12. This is literally never achieved.

Additional H^+ sinks available in soils include Al and Fe hydroxides and silicates, e.g.:

$$\underset{\text{(kaolinite)}}{Al_2Si_2O_5(OH)_{4_{xl}}} + 6H_{aq}^+ = 2Al_{aq}^{+3} + 2H_4SiO_{4_{aq}} + H_2O_{aq} , \tag{4}$$

$$Al(OH)_{3xl} + 3H^+_{aq} = Al^{3+}_{aq} + 3H_2O_{aq} , \tag{5}$$

(gibbsite)

$$FeO(OH)_{xl} + 3H^+_{aq} + e^- = Fe^{2+}_{aq} + 2H_2O_{aq} . \tag{6}$$

(goethite)

These minerals are either developed in subsoils in young soils (and thus not necessarily available to overland flow) or dominant in mature old soils, where they may provide extensive buffering capacity for subsurface flow.

The most important mineral for neutralizing acidic waters is calcite (or nearly any other carbonate mineral). The solution of this mineral at low and at intermediate pH is given by:

$$CaCO_3 + 2H^+ = Ca^{2+} + H_2CO_3 \tag{7a}$$

at low pH and

$$CaCO_3 + H^+ = Ca^{2+} + HCO_3^- \tag{7b}$$

at intermediate pH. Further addition of acid to the resulting solution (7b) ($Ca^{2+} + HCO_3^-$) will result in the protonation of the HCO_3^-:

$$HCO_3^- + H^+ \rightarrow H_2CO_{3aq} . \tag{8}$$

Reactions (7) and (8) are rapid, and for each mole of $CaCO_3$ consumed, two moles of H^+ are consumed by pH ~ 5. Reactions (3), (4), (5), and (6) consume H^+ ions but do not contribute HCO_3^- for buffering. However, HCO_3^- is produced by the dissociation of H_2CO_3 as the pH is raised by reactions:

$$H_2CO_3 \rightarrow H^+ + HCO_3^- . \tag{9}$$

This reaction does not produce additional alkalinity because of the equivalents of HCO_3^- and H^+. The amount of HCO_3^- produced in the system will be a function of the total CO_2 in the system. It should be noted that the gain or loss of molecular CO_2 to the water, by itself and with no addition of cations to balance HCO_3^- production, will not change alkalinity although pH does change.

Thus we have a spectrum of response of minerals (and rocks) to the changing acidity of atmospheric precipitation. Accordingly, rocks may be classified according to the buffering capacity which they render to surface waters (see "Methods," below).

Additional geologic and related controls on acidification of aquatic ecosystems include hydrologic characteristics of the terrain (e.g., overland flow vs groundwater flow, soil porosity and permeability, residence time of water in the soil), distribution of precipitation through time, type of precipitation, thickness of soil, types of soils (residual, glacial [till, ice-contact stratified, etc.], aeolian, lacustrine, alluvial, etc.), and age of soil.

Any prediction about the vulnerability of a terrain to acidification from precipitation based solely on bedrock geology must be tempered with consideration of these other factors. For example, Florida is underlain by highly calcareous and phosphatic rocks, suggesting that acidification of lakes and streams is highly unlikely. However, many of the soils (particularly in northern Florida) are very mature, highly leached of $CaCO_3$, and, as a result, acidification of some lakes with minimal deep groundwater influx has occurred (Brezonik, pers. comm.). Conversely, there are areas in Maine underlain by granite and receiving precipitation with an average pH of about 4.3 where lakes have not been acidified because of lime-bearing till and marine clay in the drainage basins of the lakes.

METHODS

An analysis of the vulnerability of aquatic ecosystems to acidification from precipitation must start with the bedrock geology. The scale of variability of rock types is such that one must literally look at the geology on a drainage-basin basis. Igneous rocks commonly have maximum dimensions less than 10 km. Folded or faulted metasedimentary and sedimentary rocks may have essentially one-dimensional map distribution. Flat-lying sedimentary rocks may be widely distributed, but topographic relief commonly intersects many rock types over short distances.

Small amounts of limestone in a drainage basin exert an overwhelming influence on terrains which otherwise would be very vulnerable to acidification. Consequently, in the regional analysis of vulnerability, areas with rocks of varying buffering capacity have been classified according to the less vulnerable (and more influential) rock types. Analysis of geologic terrain has been undertaken on the scale of the most recent state geologic maps (generally 1:250,000 or 1:500,000). Rock formations have been classified according to their potential acid-neutralizing capacity. This judgment was based on map explanations, various stratigraphic lexicons (Keroher et al., 1966, 1970), and the writer's personal knowledge of the geology of areas.

Rock classification was as follows:

Type 1 Low to no buffering capacity
(widespread impact from acidic precipitation expected)
Granite, syenite, or metamorphic equivalent
Granitic gneisses
Quartz sandstones or metamorphic equivalent

Type 2 Medium to low buffering capacity
(impact from acidic precipitation restricted to first- and second-order streams and small lakes; complete loss of alkalinity unlikely in large lakes)
Sandstones, shales, conglomerates, or their metamorphic equivalent (no free carbonate phases present)
High-grade metamorphic felsic to intermediate volcanic rocks
Intermediate igneous rocks
Calc-silicate gneisses with no free carbonate phases

Type 3 High to medium buffering capacity
(impact from acidic precipitation improbable except for overland runoff effects in areas of frozen ground)
Slightly calcareous rocks
Low-grade intermediate to mafic volcanic rocks
Ultramafic rocks
Glassy volcanic rocks

Type 4 "Infinite" buffering capacity
(no impact to aquatic ecosystems)
Highly fossiliferous sediments or metamorphic equivalents
Limestones or dolostones

RESULTS

The development of vulnerability maps for three types of terrains is illustrated below.

Figure 1a represents, unmodified, a tracing of all boundaries of geologic formations shown on part of the Tennessee state geologic map. This area is characterized

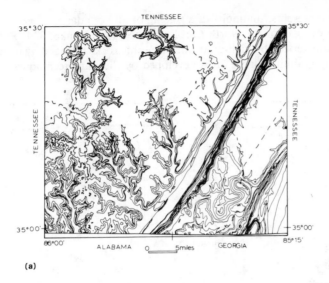

(a)

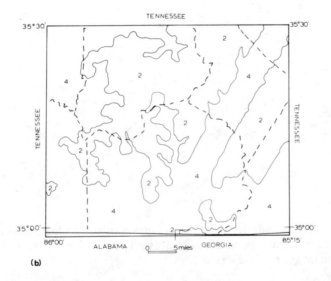

(b)

Fig. 1. Geologic boundaries shown on a portion of the geologic map of Tennessee at the western edge of the Valley and Ridge Province. Dashed lines are county boundaries. (a) Unmodified map; (b) with boundaries between formations of comparable buffering capacity deleted and other boundaries smoothed. Small areas of rock with buffering capacity less than the surrounding rocks have been deleted.

by valley and ridge topography (east) and flat-lying rocks with dendritic drainage (west). Figure 1b differs from 1a in that boundaries between contiguous formations with comparable acid-neutralizing capacities are not differentiated and contacts have been smoothed, favoring rocks with higher buffering capacity, and small areas of linear outcroppings of low-buffer-capacity rocks have been deleted.

Figure 2*a* represents, unmodified, a tracing of all boundaries of geologic formations shown on part of the North Carolina state geologic map. This Precambrian
terrain is characterized by high-grade metamorphic rocks intruded by many igneous
rocks, largely granitic. Figure 2*b* is developed by the same techniques as Fig. 1*b*.

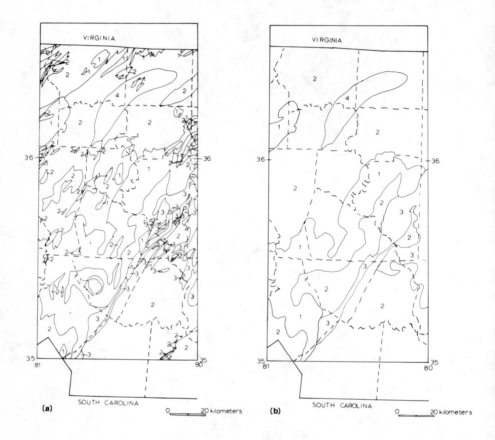

Fig. 2. Geologic boundaries shown on a portion of the geologic map of North
Carolina in the Precambrian core terrain. Dashed lines are county boundaries. (*a*)
Unmodified map; (*b*) modified as in 1*b*.

Figure 3 represents a similar development for a portion of the state of Maine in an
area of folded, metamorphosed, and intruded rocks.

The state geologic maps already represent a smoothing of geologic boundaries.
Most state maps utilized in this synthesis have been drawn at a scale of 1:250,000 or
1:500,000 from maps at a scale of 1:24,000 or 1:62,500, with the loss of much small
detail. At a scale of 1:250,000 the largest geologic unit that is commonly deleted is
approximately 0.3 km in diameter or width. Our final pictorial analysis depicts only
rocks which have an outcrop width greater than 0.7 km.

"Zero-dimensional" rocks (generally intrusions or erosional remnants of flat-
lying strata) will not normally affect the chemistry of a large number of streams or
lakes. However, one-dimensional rock map units (generally the intersection of
moderately to steeply dipping strata or dikes with the land surface) commonly affect
large numbers of primary and secondary streams or lakes which lie in topographic

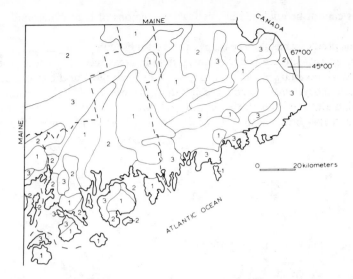

Fig. 3. Geologic boundaries shown on a portion of the geologic map of Maine in the Paleozoic eugeosyncline. Dashed lines are county boundaries. Modified as in Fig. 1b.

lows occupied by certain strata. Thus, thin limestones in the Valley and Ridge topographic province (Fig. 1) may not be shown on a scale of 1:500,000 (and thus not on these maps) but may dominate the aereal water chemistry and vulnerability to acidic precipitation.

Because soils and vegetation data are being analyzed for vulnerability and these data are commonly based on a county division, I undertook an analysis by county of susceptibility to acidification of surface waters. Each county for states in the study area was analyzed by planimetry for percentage of area underlain by rock classes 1 and 2. These data have been compiled on a 1:2,500,000 map of the United States. This type of analysis is shown in Fig. 4. Although individual drainage systems or

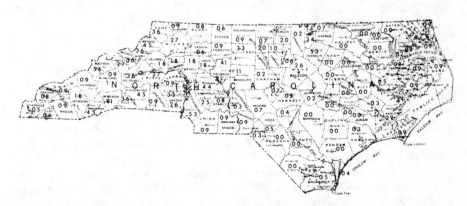

Fig. 4. Vulnerability map for North Carolina, by county. The first number of the binary indicates the decade percentage of rock type 1 (0, 0-9%; 1, 10-19%; 2, 20-29%; etc.) The second number indicates the decade percentage of rock type 2. See Fig. 2b for comparison.

lakes cannot be analyzed in this fashion, regions of high vulnerability can be identified.

The predictiveness of the maps can be established by comparing the geology-based maps against surface-water chemical data in areas receiving acidic precipitation [an approach taken by Hendrey et al. (1979)] or by looking at the biological response of aquatic ecosystems to acidic precipitation. Schofield (1979) has examined a number of high-altitude lakes in the Adirondack Mountains of New York in terms of lake pH and fish populations. He finds a close correlation between the two. Figure 5

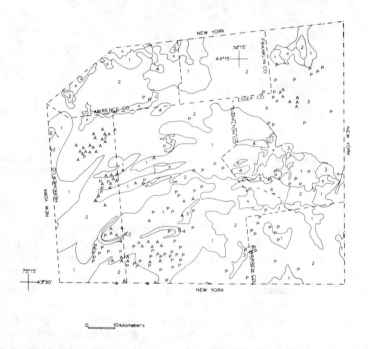

Fig. 5. Geologic map of a portion of the Adirondack Mountains area in New York, modified as for Fig. 1*b*. P, fish present in lake; A, fish absent in lake. After Schofield, 1979.

depicts those lakes which have lost their fish populations and those areas underlain by rocks of type 1 (low or no buffering capacity) and type 2. A good geographic correlation exists, but the maps do not discriminate well between rock types 1 and 2. Either the loss of detail in going from maps with scales of 1:24,000 to generalized maps with a scale of 1:250,000 causes the loss of discrimination, or other factors (soils, lake hydrology, etc.) are important. What is clear is that areas underlain by rock types 1 and 2 (Figs. 1–5) are subject to impact to aquatic ecosystems from acidic precipitation.

Similar correlations exist between acidification of surface stream waters and bedrock geology (Hendrey et al., 1979) in a number of states and lakes in areas including New England (Davis et al., 1979). This suggests that bedrock geology exerts the most important influence on the extent of acidification of aquatic ecosystems in response to atmospheric loading of acid. Soils and vegetative types are of secondary importance. However, locally, soils may overwhelm bedrock influences (McFee, 1980).

REFERENCES

Barnes, R. S., D. E. Spyridakis, and P. B. Birch. 1979. Atmospheric input of trace metals to lake sediments in western Washington State (abstract). Am. Chem. Soc. Annual Meeting, Environ. Div.

Cogbill, C. V., and G. E. Likens. 1974. Precipitation in the northeastern United States. Water Resour. Res. 10:1133–1137.

Cragin, J. H., M. M. Herron, and C. C. Langway, Jr. 1975. The chemistry of 700 years of precipitation at Dye 3, Greenland. CRREL Research Report No. 341.

Davis, R. B., S. A. Norton, and D. F. Brakke. 1979. Heavy metal deposition in bottom sediments and acidification of New England lakes from atmospheric inputs, and effect on lake biota (abstract). Conf. on Great Lakes Res.

Galloway, J. N., and E. B. Cowling. 1978. The effects of precipitation on aquatic and terrestrial ecosystems: a proposed precipitation chemistry network.

Hendrey, G. R., J. N. Galloway, S. A. Norton, and C. L. Schofield. 1979. Geological and hydrochemical sensitivity of the eastern United States to acid precipitation. E.P.A. report, Contract no. EY-76-C-02-0016.

Keroher, G. C., et al. 1966. Lexicon of geologic names of the United States for 1936–1960. Parts 1, 2, and 3. U.S. Geol. Survey Bull. 1200.

Keroher, G. C., et al. 1970. Lexicon of geologic names of the United States for 1961–1967. U.S. Geol. Survey Bull. 1350.

Kramer, J. R. 1976. Geochemical and lithological factors in acid precipitation. In Proceedings of the First International Symposium on acid precipitation and the forest ecosystem. U.S. Dept. Agric. Forest Service Gen. Tech. Rept. NE-23, pp. 611–618.

Lazarus, A. C., Elizabeth Lorange, and J. P. Lodge, Jr. 1970. Lead and other metal ions in United States precipitation. Environ. Sci. Technol. 4:55–58.

Likens, G. E., R. F. Wright, J. N. Galloway, and T. J. Butler. 1979. Acid rain. Sci. Am. 241:43–51.

Liljestrand, H. M., and J. J. Morgan. 1979. Acidic precipitation in southern California (abstract). Am. Chem. Soc. Annual Meeting, Environ. Div.

McFee, W. W. 1980. This volume.

National Atmospheric Deposition Program. 1979. NADP first data report: July 1978 through February 1979. Natural Resource Ecology Laboratory, Colorado State University.

Schofield, C. L. 1979. Acidification impacts on fish populations (abstract). Am. Chem. Soc. Annual Meeting, Environ. Div.

DISCUSSION

Bill Graustein, Yale University: Your points about this method as a regional survey are very well taken, but I would like to make a comment about its applicability to specific areas, particularly in the Northeast. It's really the bedrock that is exposed to the surface, and it's usually mingled with debris, many of which are glacially transported. Even if you do represent the local rock type, the depth and permeability of the debris on top of the bedrock determine to a large extent how long the water is in contact with the rock before it ends up in the body of water of interest, usually a lake or river. The longer the contact time, the greater the chance for weathering to occur, and given enough time the feldspars will react enough to neutralize waters, so the bedrock type is an excellent indicator to the danger area, but it's by no means the kiss of death to a particular site.

S. A. Norton: I have two responses to that. I thoroughly agree that the soils may be very important, and you'll hear later in the day how important they may be. Be that as it may, your point about the glacially derived material is important. Glacially derived material, in general, is not transported very far except in the case of stratified sands and gravels and marine clays. Nonetheless, using these maps and using this approach, we still have been able to predict with a high degree of accuracy that we're going to run into water quality of a certain type and thus have aquatic ecosystems that are vulnerable in certain areas. In certain areas, though, the soils in fact will overwhelm the bedrock. The water never sees the bedrock.

Leonard Newman, Brookhaven: I wonder if you're not presenting an over-alarming position by not being able to differentiate between classifications 1 and 2 as areas that are vulnerable and then therefore saying that they're both vulnerable; then essentially 50% of your categories become vulnerable. If you had divided your categories into two categories and couldn't differentiate between 1 and 2 and you found them both vulnerable, you say everything is a vulnerable area. Shouldn't you proceed with a little bit more caution and say, I can't differentiate between 1 and 2, so maybe there's a factor involved in there in terms of the vulnerability that I'm not describing and that it's limiting the vulnerability?

S. A. Norton: I think your point is well taken, the fact that we, at least in the Adirondacks, were not able to discriminate between areas 1 and 2. We don't have enough water quality data yet to in fact see if things do sort out according to 1 and 2. There obviously are other factors that are smoothing things out. One, of course, as Bill Graustein mentioned, is soils, another is the local hydrology—there are any number of factors—and given more time and maybe even some more money, we might be able to come up with something that has a little better resolution. The break between my rock classifications 2 and 3 is based on the presence or absence of free carbonate in the rock, and between 1 and 2 it was a somewhat biased judgment about how reactive certain rock types are, and it perhaps might have to be honed a bit, and it might be more desirable to just lump 1 and 2 and say, those are the areas where we expect to see impact, and in fact we've seen it.

Stephen Wilson, New York State ERDA: This question has to do mostly with the data. Can I assume that you're going to document pretty carefully the data sets as you reduce the maps into smaller scales? Let me be specific as to what I mean. It would seem to be useful possibly to us in New York State to have some of those maps which you created on the way to your statewide map before you began to rub out lines, so that when we want to get more detail we'll go to a larger map scale, then go back into that lower level of generalization here.

S. A. Norton: First of all, our end product was a map of the United States with a scale of 1:2.5 million. Counties end up being the sizes of postage stamps, and all that I can represent on that scale is that binary number which you saw, which is a very gross generalization. We will also have hopefully in the future a state-by-state atlas which will give at the same scale or slightly reduced scale of the state geologic map an interpretation of the acid-consuming capabilities of the various types of bedrock.

Stephen Wilson: Excellent. The second part of my question has to do with data handling. Have you looked at any grid system for a bit-by-bit compilation? It seems to me that we have meteorological data and other data that are available in some cases in grid systems, and you may want to think about some work there.

S. A. Norton: We have not done this; it is something that might be done. George Hendry has a comment relative to that.

George Hendry, Brookhaven: Not so much a comment, it's a plea. Galloway, Schofield, Norton, and Hendry are interested in expanding our data base, so we'd be very interested in identifying in each state east of the Mississippi River any type of automated or unautomated stacks of old records that contain water quality information or other information relevant to the type of projects he's described.

CHAPTER 51

IMPLICATIONS OF REGIONAL-SCALE
LAKE ACIDIFICATION

D. W. Schindler*

ABSTRACT

Large areas of terrain on the Canadian Precambrian Shield contain lakes currently being impacted by acid deposition. Although data from which to analyze trends
are currently scanty, enough evidence is available to suggest that serious impacts to
fish populations of the region are possible with little or no additional acidification.
The economics of the loss of sport fisheries or of mitigation of the impact cannot be
ignored.

INTRODUCTION

Due to the combination of prevailing wind directions and acid-sensitive terrain,
most of the effects of acid precipitation on North American fresh waters will be seen
north of the U.S.-Canadian border. An area of 1,430,000 km², or about 30% of the
Canadian Precambrian Shield, is currently receiving precipitation of pH 4.6 or
lower. Such values, representing roughly 10 times the natural inputs of hydrogen
ion, have caused aquatic problems which are well documented in Scandinavia
(Leivestad et al., 1976; Wright and Snekvik, 1977; Dickson, 1975; Oden, 1976). The
heavily affected area runs to the northeastern coast of Labrador, including some of
the most remote arctic sites on the continent. Roughly 30% of the Precambrian
Shield is covered by water, so that the magnitude of the potential problem is truly
enormous.

I have selected three sites from the acid-sensitive area for further discussion, in
order to illustrate the effects which are being seen under precipitation regimes of different average pH values. The first of these, southeastern Ontario, currently has
precipitation of pH 4.0 to 4.2 (Dillon et al., 1978). It is among the most heavily
acidified areas on the continent. Precipitation and lake chemistry have been summarized by the Ontario Ministry of Environment (1978).

The second area, in Nova Scotia, was one of the first regions of the continent to
be studied from the standpoint of lake acidification, thanks to Eville Gorham's
work in the mid 1950s. The area currently has precipitation which averages about
pH 4.6.

*Department of Fisheries and Oceans, Freshwater Institute, 501 University Crescent, Winnipeg,
Manitoba R3T 2N6.

The third area which I will talk about is the Experimental Lakes Area in northwestern Ontario, just to the east of Lake of the Woods. Four surveys of 100 or more lakes in the area were made between 1967 and the present. While older surveys were not designed with lake acidification in mind, conductivity, pH, and alkalinity were measured, allowing some assessment of deterioration due to acid precipitation. Precipitation in the area currently averages pH 4.9 to 5.0.

The assessments which I can give are miserably scanty. While we now have relatively good programs on this continent for measuring deposition and trajectories and a good start on mapping of sensitive areas is being made by Dr. Norton and his colleagues at Maine, we still have no idea of how rapidly waters are being acidified under different precipitation regimes. Some of the older limnological stations in acid-sensitive regions have historical data bases that should provide a reasonable baseline for current assessments, but no one has extracted the relevant data files so far. A regional assessment of the effects of acidification on fresh water is currently our greatest shortcoming if we wish to predict the consequences of acid precipitation.

RESULTS

The waters of the southeastern part of Ontario are certainly among the world's more heavily damaged by acid. The area appears to receive acid precipitation both from the industrial United States and from the world's largest single SO_2 source at Sudbury, Ontario. Hundreds of lakes near Sudbury are now devoid of fishes, and several thousand more appear to be doomed in the next few decades (Beamish and Harvey, 1972; Beamish, 1976; Harvey, 1975; Dillon et al., 1978). pH values have decreased by an average of 0.09 unit per year in the area (Ontario Ministry of Environment, 1978).

In the Muskoka-Haliburton area, 100 to 200 km southeast of Sudbury, similar problems are developing. Background data are scarce, but those that exist show rapid deterioration. Dillon et al. (1978) found that many lakes in the area for which there were older data had lost a high proportion of their alkalinity in the past five to ten years. Once this buffering nears exhaustion, rapid decreases in pH, to values lethal to many fishes, will take place.

Large areas of eastern Ontario and western Quebec are receiving precipitation as acid as Haliburton-Muskoka. A serious problem is undoubtedly developing, but at present we do not have background data which we can use to assess the rate at which damage is occurring.

In Nova Scotia, W. D. Watt and his colleagues at the Canadian Department of Fisheries and Oceans have discovered that the pH of lakes has decreased significantly between Gorham's measurements in 1955 and their own in 1977. For lakes above pH 5, the change has averaged about 0.5 unit in that period. Many of the lakes surveyed were at or below pH 6 to begin with. Most sport fishes are killed at pH 5.0 to 5.5, so that even slight acidification of these waters has lethal consequences.

The same investigators have found that many eastern Canadian rivers, including the Tusket, the Jordan, the Clyde, the Broad, the Sackville, the Barrington, and the Sable, have become too acid to support runs of Atlantic salmon. Spring pH values are 4.5 to 4.8. Other rivers show signs of decreasing fish stocks. Habitat restoration by liming of the rivers is planned but will be extremely costly—from $10 to $30 per recovered fish in the early stages of the program!

At ELA, where precipitation is currently only 5 times more acid than the assumed baseline of pH 5.6, there appears to be little, if any, change in the acidity of lakes.

The data are of interest because of the proximity of the area to the much-disputed Boundary Waters canoe area of northeastern Minnesota.

Interpretation of changes in pH is complicated by the fact that many watersheds in the ELA were denuded by a windstorm in 1973 and burned by a large forest fire in 1974. In general, pH values in lakes of these modified watersheds have increased, which we would expect from measurements of runoff chemistry (Schindler et al., unpublished data).

In unaffected basins, there is a tendency for more lakes to decrease than to increase in pH (Fig. 1), although the change is not statistically significant.

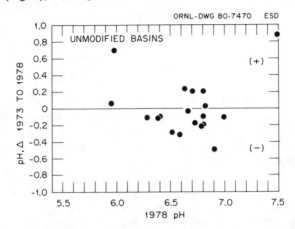

Fig. 1. The change in pH (ΔpH) of lakes in unmodified watersheds at ELA between 1973 and 1978. 1973 data are from Beamish et al. (1976).

A longer time period, 1967 to 1978, can be examined if we analyze conductivity. pH measurements were not made in 1967. However, hydrogen has a very high specific conductivity: therefore, accurate measurements should be a sensitive means of detecting lake acidification. We find that if there is a detectable trend, it is for conductivity to decrease slightly, suggesting that there is little, if any, tendency for accelerating acidification in northwestern Ontario.

I would like to emphasize once more that the ELA lakes analyzed have conductivities in the 10–40 μmhos range, with alkalinities of 0–200 μeq/liter, so that they should be extremely sensitive to changes in H^+ input.

A simple experiment demonstrates how sensitive these lakes can be. In 1979, we decided to simulate changing the pH of precipitation falling on a very sensitive ELA lake by one unit, from pH 5.0 to pH 4.0. To simplify the study and calculation, we added the acid equivalent of a month's precipitation once each month. The terrestrial watershed was ignored, which is equivalent to assuming that the watershed's buffering capacity was unaltered by acidification. Terrestrial systems appear to be much more resistant to acidification than aquatic ones (M. Alexander, unpublished data), so that the assumption is probably reasonable.

The pH of lake 114 decreased almost exactly as we had predicted from laboratory titrations (Fig. 2). There was a slow increase during the next few weeks, which seems likely to be due to buffering from sediments, because runoff and seepage should have been negligible, during the driest summer on record.

The overall effect of a half year's precipitation has been to drive the pH down by 0.8 pH unit. Lake 114 is not uniquely sensitive in the Precambrian Shield. I think that from its rapid response, we can assume that many small Shield lakes have

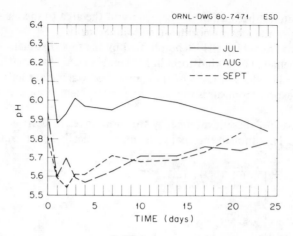

Fig. 2. The change in pH of lake 114 after additions of enough H₂SO₄ each month to be equivalent to changing the pH of precipitation falling on the lake to 4.0. The pH value predicted by Gran titration of the lake water on day 0 (before acidification) was within 0.02 pH unit of the measured pH value on day 1 (first day after acidification).

already been severely acidified, because they lie in regions which have had very acid precipitation for years. We have no background data on such small, shallow lakes. While they do not contain sport fishes, they have enormous populations of cyprinids which support a lucrative bait-fishing industry (Beamish et al., 1976), as well as furnishing food for loons and other fish-eating birds.

While most lakes containing sport fishes are larger and better buffered than lake 114, we cannot dismiss the possibility that acidification will affect some of them very rapidly. Lake trout embryos exposed in a lake at pH 5.84 have a high incidence of mortalities and deformities (Kennedy, 1980). This fact is in disturbing agreement with statistics from the Ontario Ministry of Natural Resources, which reveal that only 9% of Ontario's lake trout lakes have pH values less than 6.0 (Martin and Olver, 1976). Almost half—47%—had pH's of 6.0 to 6.5. It is possible that acidifying these lakes by only a few tenths of a unit will severely affect lake trout populations.

CONCLUSIONS

The economics of the sport fishery in Canada cannot be ignored. A preliminary calculation (D. Cauvin, pers. comm.) has placed the value of the inland sport fishery in the Precambrian Shield at over $600 million per year. If anadromous salmon and char fisheries were included, this value would increase substantially. United States studies have already estimated high costs of acid precipitation or the gases which cause it, for architectural damage and effects on human health. Studies are under way to estimate the damage to crops, forests, and soils. For the continent as a whole, the total is certain to be tens of billions of dollars per year—a fact which should discourage even those who insist that all values must be expressed in dollars and cents.

The regions which produce the pollutants are usually not the ones to suffer the consequences. They are practicing a subtle chemical warfare on vulnerable regions—equivalent to spraying them with low concentrations of defoliants and to slowly increasing the content of toxins in the waters which are used for drinking, recreation, and the growing of food. If these chemicals had recognizable fearsome names and were ejected from gun barrels instead of smokestacks, we would certainly have a tense military situation in the world, but we have been conditioned to disregard these toxins, and their ejection upon unsuspecting regions is done legally. Not only resource managers but national governments and international relations are faced with an enormous challenge in solving this problem.

REFERENCES

Beamish, R. J. 1976. Acidification of lakes in Canada by acid precipitation and the resulting effects on fishes. Pp. 479–498 in Proc. of the First Intl. Symp. on Acid Prec. and the Forest Ecosystem.

Beamish, R. J., and H. H. Harvey. 1972. Acidification of the La Cloche Mountain lakes, Ontario, and resulting fish mortalities. J. Fish. Res. Board Can. 29:1131–1143.

Beamish, R. J., L. M. Blouw, and G. A. McFarlane. 1976. A fish and chemical study of 109 lakes in the Experimental Lakes Area (ELA), northwestern Ontario, with appended reports on lake whitefish ageing errors and the northwestern Ontario baitfish industry. Can. Fish. Mar. Serv. Tech. Rep. 607.

Dickson, W. 1975. The acidification of Swedish lakes. Institute of Freshwater Research, Drottningholm, Sweden. Rep. No. 54.

Dillon, P. J., D. S. Jeffries, W. Snyder, R. Reid, N. D. Yan, D. Evans, J. Moss, and W. A. Scheider. 1978. Acidic precipitation in south-central Ontario: recent observations. J. Fish. Res. Board Can. 35:809–815.

Harvey, H. H. 1975. Fish populations in a large group of acid-stressed lakes. Verh. Int. Ver. Limnol. 19:2406–2417.

Kennedy, L. A. 1980. Teratogenesis in lake trout (Salvelinus namaycush), in an experimentally acidified lake. Can. J. Fish. Aquat. Sci., under review.

Leivestad, H., G. Hendrey, I. P. Muniz, and E. Snekvik. 1976. Effects of acid precipitation on freshwater organisms. In F. H. Braekke (ed.), Impact of acid precipitation on forest and freshwater ecosystems in Norway. SNSF Project, March 1976.

Martin, N. V., and C. H. Olver. 1976. The distribution and characteristics of Ontario lake trout lakes. Ont. Min. Environ., Fish Wildlife Res. Br., Res. Rep. 97.

Oden, S. 1976. The acidity problem—an outline of concepts. Water, Air, Soil Pollut. 6:137–166.

Ontario Ministry of Environment. 1978.

Wright, R. F., and E. Snekvik. 1977. Acid precipitation: chemistry and fish populations in 700 lakes in southernmost Norway. SNSF TN 37/77.

DISCUSSION

R. A. N. McLean, Domtar Inc.: I think your liming figures were only based on one year. The cost of fish was only based on recovering the fisheries for one year?

D. W. Schindler: That's right.

R. A. N. McLean: So in actual fact the economics is somewhat faster than it would appear, slight if you're actually going to recover the lake by liming.

D. W. Schindler: Well, he's expecting that you'd have to add this to these streams each year.

R. A. N. McLean: I would hope that we'd be doing something about abatement measures in the meantime.

D. W. Schindler: I'd like to mention also that when you look at what salmon sells for in the store, those aren't bad figures; you can still make money on them if you're ranching salmon.

Gunnar Abrahamson, from Norway: Did you suggest that the pH in lakes outside the Sudbury area had jumped from about 6 to 4.8 or something like that?

D. W. Schindler: Some of them have dropped from 6 to at least the low 5's. Peter Dillon and his colleagues have a paper published in the *Journal of the Fisheries Research Board of Canada* in 1978 on some lakes in the Haliburton-Muskoka area. One of them is a lake that I worked when I first came to Canada 12 years ago, and they went back and surveyed it. This is Clear Lake, just south of the Algonquin Park boundary. Using the same methods and so forth, the alkalinity dropped 70% in the ten years between their survey and mine. There's another small lake with an initial conductivity of about 15–20 and in an area where the precipitation ranges from annual means of 4.0 to 4.2 during that period.

J. H. B. Garner, EPA, North Carolina: I might say that the figures you have stated agree very closely with those from the New York State Department of Conservation. They have been working at liming some of the lakes, and they figure it costs them about $50 an acre, the same as the Swedish people.

CHAPTER 52

A REGIONAL ECOLOGICAL ASSESSMENT APPROACH TO ATMOSPHERIC DEPOSITION: EFFECTS ON SOIL SYSTEMS*

Jeffrey M. Klopatek†

W. Frank Harris†

Richard J. Olson†

ABSTRACT

A regional ecological overview of the potential effects of acid precipitation on soils is presented. Computer maps of soil pH, CEC, base saturation, and base content in the eastern United States are displayed using county-level data. These maps are then overlain with a computer map of the hydrogen ion loading, and resultant maps of acid-sensitive and acid-insensitive soils are presented. Of 1572 counties in the eastern United States, only 117 are classified as being acid-sensitive. A number of qualifications concerning the data and the implications for management are discussed.

INTRODUCTION

The need for a regional ecological approach in environmentally related problems, planning, and decision making should be axiomatic. Such an approach views the ecological attributes of regional systems not simply as additions of ecosystems but as attributes specific to the scale of regions. When human intervention into ecological processes affected small localized areas, nature often compensated for man's ignorance of the complex interactions of the physical and biological environment. Today, however, the resiliency of environmental systems can become stretched beyond any possible recuperative capacity due to disruptive actions and pollutant residuals resulting from cultural and technological systems. Thus, these widespread effects dictate the need for evaluation on a macroscale level, not only because of their large spatial nature (e.g., location of coal-fired power plants in the Ohio River valley, development of western United States energy reserves, or conversion of vast natural land areas to subsidized monocultures) but also because of the multiplicative or cumulative effects from aggregated activities.

The phenomenon of acid precipitation is a prime example of a regional ecological problem. At one time, air pollution problems were attributable to large, uncontrolled point sources. However, the problems today are associated with regional-scale

*Research sponsored by the Office of Health and Environmental Research, U.S. Department of Energy under contract W-7405-eng-26 with Union Carbide Corporation. Publication No. 1541, Environmental Sciences Division, Oak Ridge National Laboratory.

†Environmental Sciences Division, Oak Ridge National Laboratory, Oak Ridge, TN 37830

increases in concentration of pollutants from a variety of multiple sources. This is particularly true in the eastern United States, where the density of fossil-fuel combustion plants, frequency of air stagnation events, elevated levels of oxidants over widespread areas, and increases in the acidity of precipitation in recent decades combine to expose large areas of valuable agricultural and forest lands to chronic air pollution stress (Shriner et al., 1977; McLaughlin et al., 1978). This paper examines the potential effect of acid precipitation in the eastern United States from a regional perspective and displays, by computer graphic techniques, those geographic areas which may be most sensitive to acid precipitation and may exhibit the greatest perturbations.

There exist three fundamental steps in any regional ecological assessment: (1) characterize the region's environment, (2) determine where the stress areas of the region exist, and (3) evaluate the possible impacts from the perturbation being assessed. The characterization of the region includes two constraints: the problem addressed determines the regional boundaries, and the data must be of uniform resolution. It may appear that these constraints are inconsequential; however, if they are not dealt with objectively at the start, the assessment is doomed to failure, at least from a regional aspect. Location of the stress points can be accomplished by using extant data and/or by employing simulation models. Three techniques can be utilized to evaluate impacts: (1) visual interpretation, (2) statistical and computer analysis techniques, and (3) predictive models and systems analysis.

BACKGROUND

Geoecology Data Base

Regional studies require data, analysis, and display resources which can describe the spatial and functional characteristics of regions. The Geoecology Data Base project at Oak Ridge National Laboratory (ORNL) has developed an extensive county-level data base of environmental data with associated analysis and display capabilities (Olson et al., 1979). Data on various aspects of the environment can be easily associated through the use of common spatial cells (counties). In addition, the cells serve as building blocks which can be aggregated into "regions" whose sizes and shapes may vary depending on the needs of the study.

Over 1200 variables on file for county-subcounty units provide data on terrain, water resources, forestry, vegetation, wildlife, agriculture, land use, climate, air quality, population, and energy. Variables are selected from existing data sources or are digitized from maps, e.g., Kuchler's (1964) *Potential Natural Vegetation of the U.S.* The geographic coverage of the standardized files is the conterminous United States (3071 county units). The temporal aspect of each data set reflects the most current conditions. In the case of rapidly changing resources, data sets contain observations from a series of years. The reference year for most files lies between 1976 and 1978. The regional systems are characterized by cartographic display of environmental parameters within the Geoecology Data Base. However, the maps are often created by computing new variables from the combination and mathematical manipulation of variables within the data base.

Air Pollutant Impacts Assessment

A central component of the developing National Energy Plan is the expanded use of coal. Despite significant improvement in air pollution abatement technologies, one of the primary impacts from burning more coal appears to be the increased levels of sulfur dioxide (SO_2) and its transformation products (Shriner et al., 1977).

As part of a regional assessment to estimate the potential impacts of increased levels of SO_2 on vegetation, we employed maps generated from long-range trajectory models based on coal-use scenarios for 1985 (Davis, 1978). Soybean cropland (Fig. 1a) and softwood forests (predominantly pine species) were chosen because of their known sensitivity to SO_2 (regional characterization). Estimates of SO_2 levels were transcribed from a rectangular grid to county centroid coordinates and mapped with SYMAP (Fig. 1b). Since it was impossible to ascertain what effect average ambient SO_2 levels would have on macroscale systems, the map of predicted 1985 levels was overlain with maps of the vegetation most likely to be impacted (location of stress points). The darkest regions on the resulting maps indicate the coincident areas with the highest crop productivity (Fig. 1c) or highest softwood productivity (Fig. 1d) and the highest levels of atmospheric SO_2. While the role of emissions from coal-fired electric plants alone in the magnitude and occurrence of these SO_2 levels may be debated, there is a high probability that the effluents from steam electric plants will interact with those from other industrial and urban sources. Thus, while the SO_2 levels depicted here may not cause acute vegetation injury, they may, when combined with ambient levels of SO_2 and other air pollutants, contribute to chronic injury. Furthermore, levels represent annual averages for county cells, so that there still may be periodic episodes sufficient to result in acute injuries to the vegetation. The darkest areas on the maps, therefore, indicate areas of justifiable concern (evaluation of impacts). For example, if just those counties which are shaded darkest on the SO_2-soybean map (Fig. 1c) were to suffer a 10% decrease in yield, the result, other factors remaining constant, would be a loss of 17 million bushels, or over \$100 million per year. Although the results are very general at this point, they indicate areas of concern for potential impact of SO_2 on two vegetation systems. This coupling of the Geoecology Data Base to the synagraphic mapping systems illustrates not only the applicability of the data base but the versatility of the combined systems for regional analysis.

Acid Precipitation Impacts

As has been extensively discussed in other symposium papers, acid precipitation can have direct and indirect effects on environmental systems, both terrestrial and aquatic. Direct effects on vegetation involve the response of the vegetation to direct exposure to the acidified precipitation, either alone or in combination with other pollutants. Direct aquatic effects follow from the acidification of watershed soils and receiving freshwater bodies. The potential for such direct effects can be surveyed by comparing hydrogen ion loading vs dominant land use categories or classifications (e.g., cropland, forests, lakes).

Indirect effects of acid precipitation involve soil processes. For terrestrial systems, soil acidification can result in loss of exchangeable cations, with subsequent potential effects on productivity. For aquatic systems, soils (and geologic parent material) can ameliorate the rainfall acidity, thus mitigating against further acid input to water bodies. At the other extreme, extremely acid soils (often in combination with base-poor geologic parent material) have a lower buffering capacity and thus do little to affect the activity of percolating water. Percolating acid water can solubilize aluminum ions, resulting in toxic effects on biota. Comparison of hydrogen ion loading with various combinations of soil characteristics should yield some insights into extent of areas sensitive to acidification or those having little capacity to buffer excess acidity.

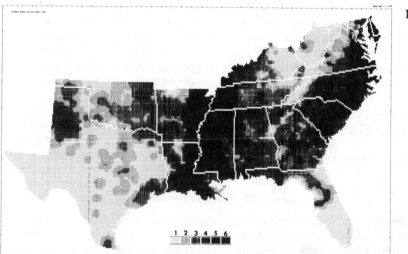

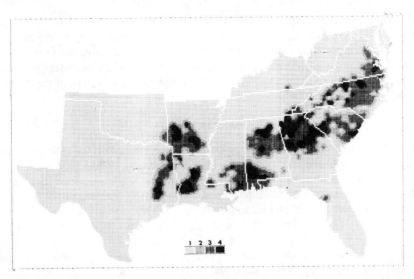

1(d)

Fig. 1. Potential impacts of increased levels of SO_2 on vegetation. (*a*) Projected 1985 regional concentrations of SO_2 resulting from electrical generating facilities. Ambient concentrations not included. From Davis (1978). (*b*) Soybean yield from 1969 Census of Agriculture expressed as total yield divided by county area. Interval limits in bushels per acre: (1) <0.001, (2) 0.001–0.01, (3) 0.01–0.1, (4) 0.1–1.0, (5) 1.0–10.0, (6) >10.0. From Davis (1978). (*c*) Potential SO_2 impact on soybeans. Interval limits: (1) no impact due to low soybean or SO_2 levels, (2) low impact due to medium soybean and SO_2 levels, (3) medium impact due to medium (high) soybean and high (medium) SO_2 levels, (4) high impact due to high soybean and SO_2 levels. From Davis (1978). (*d*) Potential SO_2 impacts on softwood forests, with softwood productivity determined as cubic feet of growing stock divided by county acreage. Interval limits: (1) no impact due to low softwood or SO_2 levels, (2) low impact due to medium softwood and SO_2 levels, (3) medium impact due to medium (high) softwood and high (medium) SO_2 levels, (4) high impact due to high softwood and SO_2 levels. From Davis (1978).

Soil Acidification

Soil acidification is a natural process occurring continuously in most humid regions. Soils of low pH, depleted nutrients, and a high degree of weathering stand as evidence of this pedogenic process. There are several sources of hydrogen ions: (1) the reaction of CO_2 and water, (2) mineral acid from nitrification, (3) organic acids as a product of organic matter decomposition, (4) oxidation of soil parent materials (e.g., pyrite), and (5) acids deposited in precipitation or from other external sources (e.g., fertilizers). Under natural conditions, hydrogen ion production within the soil profile greatly exceeds the increments of hydrogen ions in precipitation (Cole and Johnson, 1977; Reuss, 1976).

Soil acidification processes are complex, and there are many soil factors which modify these processes. Discussions by Wiklander (1974) and McFee et al. (1976) adequately review the chemistry of acidification. For a general case, the soil cation exchange capacity (CEC) provides a measure of the soil's capacity to buffer pH. The degree to which this capacity is saturated with cations (base saturation, BS) provides a measure of the degree to which pedogenic processes have advanced. Soil pH gives a measure of soil acidity and roughly approximates the "replacing efficiency" (after

Wiklander, 1974) of additional H⁺. Classifying soils on the basis of these three parameters serves to identify those geographic areas sensitive to accelerated soil acidification, those areas insensitive to acidification with lowered buffering capacity, and those areas of intermediate response. There are, of course, localized exceptions, but the analysis should yield interesting insights into the extent of the acid rain problem with respect to soils.

METHODS

The soils data from the Geoecology Data Base were based on the soils map of the United States published in the National Atlas (USGS, 1970). This map, which depicts the association of phases of great soil groups, was transcribed to a map of counties in the eastern United States. Thus, the result was a record of the percent of a county dominated by a particular association of great soil groups.

The three parameters—pH, CEC, and base saturation—plus soil organic matter and clay content were keyed into the soils data base for each of the soils groups. Values were obtained for the A horizons of typic soils whenever possible. Generally, the soils data represent averages for the upper 20–25 cm. If a typic soil was not described with these parameters, an atypic but widely distributed soil was used, e.g., alfic. Soil descriptions were obtained from USDA-SCS (1975), Buol et al. (1973), Hole (1975), Hoyle (1973), and USDA-SCS (1960). The resultant maps for soil pH, CEC, and base saturation are shown in Figs. 2, 3, and 4, respectively. In order to observe the total base contents of the soils (20–25 cm), the base saturation was multiplied by the CEC. This resultant map is shown in Fig. 5. This value (milliequivalents per 100 g) is more representative of the total potential base reserves than base saturation or CEC by itself.

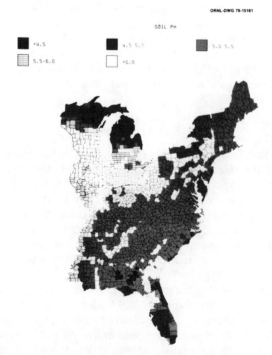

Fig. 2. Map of soil pH of counties in the eastern United States. Map is based on pH values ascribed to typic great soil groups.

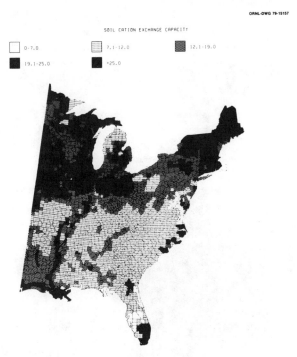

Fig. 3. Map of soil cation exchange capacity of counties in the eastern United States. Map is based on soil cation exchange capacity values ascribed to typic great soil groups.

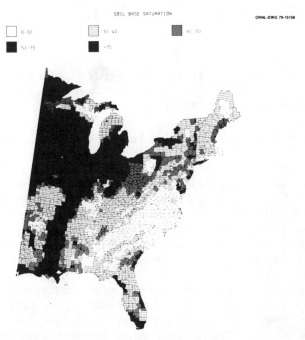

Fig. 4. Map of soil base saturation of counties in the eastern United States. Map is based on soil base saturation values ascribed to typic great soil groups.

ORNL-DWG 79-15160

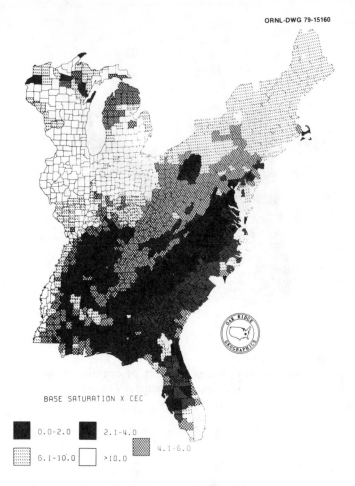

Fig. 5. Map of base contents of soils in the eastern United States. Base content is determined by multiplying percent base saturation times cation exchange capacity (milliequivalents per 100 g of soil).

Soil data were averaged on a county basis depending on the proportion of the county occupied by a great soil group. Such an averaging procedure tends to obscure minor soil types or a typic variation. At the spatial scale of this study, consideration of detailed variation was not of interest. Clearly, local variation must be explicitly included when analyzing specific sites. Rocky outcrops and stony land were not included in the ratings. Therefore, the resultant county data and regional averages are likely overestimates of some terrestrial systems' capacity to buffer the acid precipitation before entry into the surrounding water bodies, e.g., the Adirondacks. Nonetheless, the patterns illustrated by the maps are useful approximations of zones to examine in more detail.

Rainfall pH data were obtained from the Cogbill and Likens (1974) map of acid precipitation of the eastern United States for 1972–1973. These were converted to county values, multiplied by the annual average precipitation (in centimeters), and adjusted to reflect the hydrogen ion concentration per square meter. Thus, the resultant map (Fig. 6) represents the annual average hydrogen ion loading rates (equivalents per square meter) for the eastern United States. These loadings,

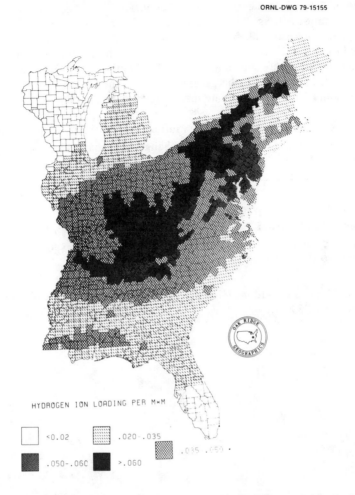

Fig. 6. Hydrogen ion loading map of the eastern United States. In equivalents per square meter. Hydrogen ion data are from Cogbill and Likens (1974) and are multiplied by county annual average precipitation.

however, neglect the cation chemistry of the precipitation, which will buffer the acid precipitation, as well as ammonia (NH_4^+), which may be transformed to nitrate (NO_3^-) in the soil (Malmer, 1976; Wiklander, 1974). In addition, the precipitation acidity data used are based on annual averages for one year. Since there are large year-to-year variations in the acidity of precipitation, there will also be changes in the potential effects on the soils.

Data storage, retrieval, and manipulation were accomplished on IBM 360/91 and 3330 computers at ORNL, extensively using the Statistical Analysis System (Helwig and Council, 1979). All soils, vegetation, and climatic data are part of the Geoecology Data Base (Olson and Strand, 1978; Olson et al., 1979).

RESULTS

Using soil parameters generally available from survey documents, soils of the United States east of the Mississippi River were classified according to the scheme outlined in Table 1. Two extremes of soil conditions are of concern with respect to acid precipitation effects. Soils with a pH above 5 but with low CEC (total buffer capacity) are those soils likely to exhibit accelerated acidification as a result of accelerated inputs of H$^+$ from the atmosphere. Areas broadly mapped as typified by these soil conditions are shown in Fig. 7. Should atmospheric H$^+$ input continue at current levels or increase further, these areas have the highest risk of soil acidification and could require appropriate management (e.g., liming).

Table 1. Classification of parameters influencing sensitivity to soil acidification
Soil acidification is defined as a large change in exchangeable cations per small addition of H$^+$ ion. The classification follows that of Wiklander (1974, 1979)

<table>
<tr><td colspan="2" align="center">More sensitive</td></tr>
<tr><td>I. Low CEC (<12 meq/100 g)
Med.-High BS (30–50%) pH > 5
Noncalcareous
sandy soils</td><td>II. Med.-High CEC (>12 meq/100 g)
Med. BS (30–40%) pH > 5
Noncalcareous clays,
cultivated soils</td></tr>
<tr><td colspan="2" align="center">Less sensitive</td></tr>
<tr><td>III. Med.-High CEC (>12 meq/100 g)
Low BS (< 30%)
pH < 5
Acid soils</td><td>IV. High CEC (>20 meq/100 g)
High BS (> 50%)
pH > 6
Calcareous soils</td></tr>
</table>

ACID SENSITIVE SOILS
CEC<12   BS>30   PH>5.0

ORNL-DWG 79-15159

<0.02   .020-.035   .035-.050

.050-.060   >.060

Fig. 7. Map of counties with soils judged to be sensitive to acid precipitation in the eastern United States. The average soil characteristics of these counties are low CEC (less than 12 meq per 100 g), medium to high base saturation (30–50%), and pH above 5. Sensitive counties are overlain with hydrogen ion loading rates as depicted in Fig. 6.

The second soils of concern are those that because of their already acidic nature (low BS and pH below 5) are resistant to further acidification (Fig. 8). While these soils may be insensitive to further pH change, these same soils would not be expected to buffer pH of percolating water. Therefore, regions containing these soils may pose the risk of chronic or pulse acid runoff to streams, thus creating conditions potentially resulting in direct acid effects on aquatic biota. Combined with conditions of base-poor geologic parent material (generally a condition for development of soils with low buffering capacity), the risk of direct aquatic effects is heightened.

Generally, cultivated soils (already receiving salt additions as lime and mineral fertilizers) and clayey soils are well buffered or sufficiently weathered that a small addition of H^+ does not displace a large amount of exchangeable cations. These soils

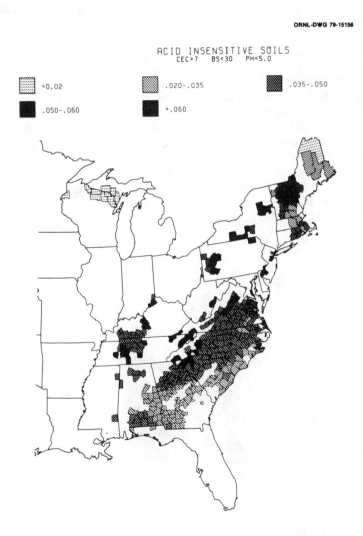

ORNL-DWG 79-15156

ACID INSENSITIVE SOILS
CEC>7   BS<30   PH<5.0

<0.02   .020-.035   .035-.050

.050-.060   >.060

Fig. 8. Map of counties with soils judged to be acid-insensitive (potential for acid runoff) in the eastern United States. The average soil characteristics of these counties are medium to high CEC (more than 12 meq per 100 g), low base saturation (less than 30%), and pH below 5. Insensitive counties are overlain with hydrogen ion loading rates as depicted in Fig. 6.

comprise our class II (Table 1). Soils high in CEC and with high BS and pH above 6 are sufficiently base-rich that acidification from atmospheric sources is unlikely. Areas containing such soils are rare in the humid zone east of the Mississippi.

Only 117 of 1572 total counties in the eastern United States are classified as potentially sensitive to acidification based on a CEC of less than 12 meq per 100 g. Our initial classification using a CEC of less than 7 meq per 100 g classified only one county in the eastern United States in the acid-sensitive soils. The criteria were relaxed to CEC of 12 meq per 100 g based on the actual distribution of calculated county values rather than experimental evidence that these soils indeed are sensitive.

While more conservative than some general soils maps, the areas identified as sensitive to acidification reflect averages over county units. Not all soils in a unit are necessarily sensitive, while locally important areas that are sensitive may have gone undetected at this scale of resolution. Therefore, a considerable amount of analysis remains to examine soil associations within a county and corresponding land uses, in order to finally assess the actual soils which are sensitive.

Soils classified as insensitive to acidification based on the criteria in Table 1 are probably more representative of county-scale conditions than the more atypical sensitive soils. Their more general distribution within a survey unit exacerbates, to some extent, the subsequent potential effects to receiving water bodies.

As summarized by Tamm and Cowling (1976), acid precipitation can have a number of detrimental effects on vegetation, with the end product being a reduction in growth. However, the effects of acid precipitation are dependent on a large number of factors, and for the majority of agricultural crops, many of these have yet to be quantified. Therefore, we decided to simply overlay the hydrogen ion loading map over those counties which are covered by 20% or more of cropland (a minimum of 20 to 40,000 ha per county). The resultant map is shown in Fig. 9. What

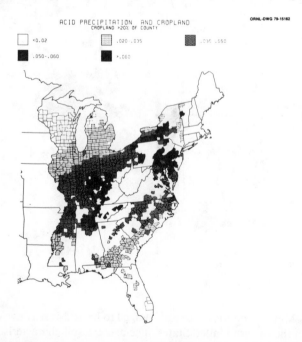

Fig. 9. Acid precipitation overlay of cropland in the eastern United States. Counties depicted have more than 20% of their surface area covered by cropland. Counties shown are overlain with hydrogen ion loading rates as depicted in Fig. 6.

this map indicates is not where the impacts from acid precipitation *will* occur but rather the areas where the impacts are most possible, i.e., where the combination of use of land for crops and hydrogen ion loading is such that a significant effect might occur with a sensitive crop species.

DISCUSSION

At this stage in the analysis, these results should be qualified in a number of important ways. First, our hydrogen ion loading data do not consider the base content of rainfall. These bases vary geographically in both composition and amount and reduce the flux of "excess acidity" to the landscape. Precipitation acidity does not include dry deposition of acid aerosols. This variable source of acidity can account for a significant fraction of the acidity delivered to the landscape. As data accrue from programs such as NADP or MAP3S (e.g., J. Miller, this volume), base content of rainfall and dry deposition should be included.

Accumulation of acidity in snowpack is another factor certainly influencing timing and amount of acidity which impacts streams. The susceptibility of soil parent material weathering further moderates acidity.

Our hydrogen ion loading values do not include consideration of the sequence of reactions occurring at the vegetation surface, stem flow, and litter. Lindberg et al. (1979) found that rainfall acidity was significantly buffered as water moved through the canopy to beneath the surface litter.

Finally, land management practices, particularly fertilization, already add an excess of base to agricultural soils. The addition of hydrogen ions in precipitation is a small contribution under all but the most extreme conditions.

While these several qualifications of the existing data base and its interpretation make firm conclusions elusive at this time, several important purposes are served. First, the potential extent of acid rain effects in the eastern United States is large, extending well beyond the northeastern corridor of states. Thus, the national level of effort toward this problem is warranted. Second, the landscape is a heterogeneous mix of factors. We are just beginning to understand how these are distributed and how the combinations of landscape factors influence response to acidic deposition. Cataloging data in their geographic juxtaposition serves to illustrate the complexity and extent of potential problems. Regional ecological assessment, therefore, is an important synthesis tool. However, it is not intended to replace site-specific studies. Rather, it points out locations where there is a high probability of impact and areas where site-specific studies should take place.

REFERENCES

Buol, S. W., F. D. Hole, and R. J. McCracken. 1973. Soil Genesis and Classification. Iowa State University Press, Ames, Iowa.

Cogbill, C. V., and G. E. Likens. 1974. Acid precipitation in the northeastern United States. Water Resour. Res. 10:1133–1137.

Cole, D. W., and D. W. Johnson. 1977. Atmospheric sulfate additions and cation leaching in a Douglas fir ecosystem. Water Resour. Res. 13:313–317.

Davis, R. M. (project coordinator). 1978. A preliminary assessment of coal utilization in the South. ORNL/TM-6122. Oak Ridge National Laboratory, Oak Ridge, Tennessee.

Helwig, J. T. and K. A. Council (eds.). 1979. SAS User's Guide, 1979 Edition. SAS Institute, Inc. Raleigh, North Carolina.

Hole, F. D. 1975. Soils of Wisconsin. University of Wisconsin Press, Madison, Wisconsin.

Hoyle, M. C. 1973. Nature and properties of some forest soils in the White Mountains of New Hampshire. USDA Forest Service Research Paper NE-260. Upper Darby, Pennsylvania.

Küchler, A. W. 1964. Manual to accompany the map—potential natural vegetation of the conterminous United States. American Geographical Society Spec. Pub. No. 36.

Lindberg, S. E., C. Harriss, R. R. Turner, D. S. Shriner, and D. D. Huff. 1979. Mechanisms and rates of atmospheric deposition of selected trace elements and sulfate to a deciduous forest watershed. ORNL/TM-6674. Oak Ridge National Laboratory, Oak Ridge, Tennessee.

Malmer, N. 1976. Acid precipitation: chemical changes in the soil. Ambio 5:231–233.

McFee, W. W., J. M. Kelly, and R. H. Beck. 1976. Acid precipitation effects on soil pH and base saturation of exchange sites. Pp. 725–735 in Dochinger, L. S., and T. A. Seliga (eds.), Proc., First International Symposium on Acid Precipitation and the Forest Ecosystem. USDA Forest Service General Technical Report NE-23. Upper Darby, Pennsylvania.

McLaughlin, S. B., D. C. West, H. H. Shugart, and D. S. Shriner. 1978. Air pollution effects on forest growth and succession: applications of a mathematical model. Pp. 1–16, paper 78-24.5 in H. B. H. Cooper, Jr. (ed.), Proc., 71st Annual Meeting of the Air Pollution Control Association. Air Pollution Control Association, Pittsburgh, Pennsylvania.

Miller, J. 1980. National atmospheric deposition program: analysis of data from the first year. This volume.

Olson, R. J., and R. H. Strand. 1978. Management of diverse environmental data with SAS. Pp. 200–206 in Proc., Third Annual Conference, SAS Users Group, International. SAS Institute, Inc., Raleigh, North Carolina.

Olson, R. J., J. M. Klopatek, and C. J. Emerson. 1979. Regional environmental studies: application of the Geoecology Data Base. Second Annual International Users' Conference on Computer Mapping, Hardware, Software and Data Bases. July 15–20, 1979, Harvard University, Cambridge, Massachusetts.

Reuss, J. O. 1976. Chemical and biological relationships relevant to the effect of acid rainfall on the soil-plant system. Pp. 791–813 in Dochinger, L. S., and T. A. Seliga (eds.), Proc., First International Symposium on Acid Precipitation and the Forest Ecosystem. USDA Forest Service General Technical Report NE-23. Upper Darby, Pennsylvania.

Shriner, D. S., S. B. McLaughlin, and C. F. Baes. 1977. Character and transformation of pollutants from major fossil fuel energy sources. ORNL/TM-5919. Oak Ridge National Laboratory, Oak Ridge, Tennessee.

Tamm, C. O., and E. B. Cowling. 1976. Acid precipitation and forest vegetation. Pp. 845–855 in Dochinger, L. S., and T. A. Seliga (eds.), Proc., First International Symposium on Acid Precipitation and the Forest Ecosystem. USDA Forest Service General Technical Report NE-23. Upper Darby, Pennsylvania.

U.S. Soil Conservation Service (USDA-SCS). 1960. Soil classification—a comprehensive system: 7th approximation. U.S. Dept. Agriculture, Washington, D.C.

U.S. Soil Conservation Service (USDA-SCS). 1975. Soil taxonomy—a basic system for making and interpreting soil surveys. U.S. Dept. Agriculture. Agriculture Handbook No. 436. Washington, D.C.

U.S. Geological Survey (USGS). 1970. Distribution of principal kinds of soils: orders, suborders and great groups. Pp. 85–88 in The National Atlas of the United States of America. U.S. Geological Survey, Washington, D.C.

Wiklander, Lambert. 1974. The acidification of soil by acid precipitation. Grund-foerbaettring 26:155–164.

Wiklander, L. 1980. The sensitivity of soils to acid precipitation. In Hutchinson, T. C. and M. Havas (eds.), Effects of Acid Precipitation on Terrestrial Ecosystems. Plenum Press, New York. Pp. 553–567.

DISCUSSION

Moderator Galloway: How severe are the differences between what you are saying about sensitivity of soils and what Bill McFee was saying about sensitivity of soils?

J. M. Klopateck: I don't think there is too much. We talked about this last night somewhat, and where our sensitive and insensitive soils—lump them together—they come up underneath his sensitive to moderately sensitive soils. If we could project our maps over his state maps I think that they should correspond fairly closely.

Bill McFee, Purdue: There are significant differences in our data base. The data base that he has for cation exchange capacity is surface soil cation exchange capacity. As an example, when you work in the Northeast, where you have high organic matter in the surface, but it is frequently shallow, you get a high exchange capacity in your data base, whereas I averaged the cation exchange capacity in the top 25 cm, and it drops mine considerably, because, typically, those soils have organic-rich surface layers underlain by an A_2 horizon which is essentially all sand and low in cation exchange capacity. That makes a difference that we'll have a hard time ironing out. There is a difference in what we were trying to do also. I was trying to give an evaluation of sensitivity as it might relate to an aquatic ecosystem as well as potential for changes in the soil that might be important to the terrestrial system, whereas he pointed out very carefully that he was talking about the sensitivity to acidification. Of course where the soil is already acid, as acid as the rainfall, as it is in many of my "sensitive" regions it is not going to be further acidified by acid rain.

LIST OF PARTICIPANTS

AUTHORS, SPEAKERS, PROGRAM COMMITTEE

Dr. G. Abrahamsen
Director, Forest Research
Norwegian Forest Research
  Institute
NORSK Institutt for
  Skogforskning
Postbox 61-1432 Ås-NLH,Norway

Dr. Donald F. Adams
Route 2, Box 596
Pullman, Washington 99163

Dr. Viney Pal Aneja
Environmental Science Center
Northrop Services, Inc.
Research Triangle Park,
  North Carolina 27709

Dr. Alan Bandy
Department of Chemistry
Drexel University
32nd and Chestnut Streets
Philadelphia, Pennsylvania 19104

Dr. Cyrill Brosset
Swedish Water and Air
  Pollution Research Laboratory
Sten Sturegaten 42
Box 5307
S-402 24 Goteborg
Sweden

Dr. A. C. Chamberlain
Atomic Energy Research
  Establishment
Harwell, Berks
England

Dr. Ellis B. Cowling
Associate Dean for Research
School of Forest Resources
Box 5488
North Carolina State University
Raleigh, North Carolina 27650

Dr. Chris S. Cronan
Department of Biological
  Sciences
Dartmouth College
Hanover, New Hampshire 03755

Dr. M. T. Dana
Atmospheric Sciences Section
Pacific Northwest Laboratory
Richland, Washington 99352

Dr. W. E. Dalbey
Biology Division
Oak Ridge National Laboratory
P.O. Box Y
Oak Ridge, Tennessee 37830

Dr. J. G. Droppo
Atmospheric Sciences Section
Pacific Northwest Laboratory
Richland, Washington 99352

Dr. Lance S. Evans
Brookhaven National
  Laboratory
Building 318
Upton, Long Island,
New York 11973

Dr. James N. Galloway
Department of Environmental
  Sciences
University of Virginia
Charlottesville, Virginia 22903

Dr. Donald F. Gatz
Illinois State Water Survey
P.O. Box 232
Urbana, Illinois 61801

Dr. Charles H. Goodman
Southern Company Services, Inc.
800 Shades Creek Parkway
P.O. Box 2625
Birmingham, Alabama 35202

Dr. Jack D. Hackney
Environmental Health Service
Rancho Los Amigos Hospital
Medical Sciences Building,
  Room 51
7601 East Imperial Highway
Downey, California 90242

Dr. Ronald J. Hall
Section of Ecology and
  Systematics
Langmuir Laboratory
Cornell University
Ithaca, New York 14583

Dr. Gray S. Henderson
1-30-Agriculture Building
University of Missouri
Columbia, Missouri 65211

Dr. George Hendry
Brookhaven National Laboratory
Upton, Long Island,
  New York 11973

Dr. Bruce B. Hicks
Atmospheric Physics Section
Radiological and Environmental
  Research
Argonne National Laboratory
Argonne, Illinois 60439

Dr. T. W. Horst
Atmospheric Sciences Section
Pacific Northwest Laboratory
Richland, Washington 99352

Dr. Rayford P. Hosker
National Oceanic and
  Atmospheric Administration
461 Illinois Avenue
Oak Ridge, Tennessee 37830

Dr. Dale W. Johnson
Oak Ridge National Laboratory
Building 1505
P.O. Box X
Oak Ridge, Tennessee 37830

Dr. Ulysses S. Jones
International Rice Research
 Institute
P.O. Box 933
Manila, Philippines

Dr. J. Michael Kelly
Tennessee Valley Authority
Oak Ridge National Laboratory
Building 1505
P.O. Box X
Oak Ridge, Tennessee 37830

Dr. J. M. Klopatek
Oak Ridge National Laboratory
Building 1505
P.O. Box X
Oak Ridge, Tennessee 37830

Dr. Sagar V. Krupa
Department of Plant Pathology
Stakman Hall
University of Minnesota
St. Paul, Minnesota 55108

Dr. William K. Lauenroth
Natural Resource Ecology
 Laboratory
Colorado State University
Fort Collins, Colorado 80521

Dr. Morton Lippmann
New York University Medical
 Center
Institute of Environmental
 Medicine
550 First Avenue
New York, New York 10016

Dr. Steven E. Lindberg
Oak Ridge National Laboratory
Building 1505
P.O. Box X
Oak Ridge, Tennessee 37830

Dr. Sati Mazumdar
School of Public Health
University of Pittsburgh
130 De Soto Street
Pittsburgh, Pennsylvania 15213

Dr. William W. McFee
Agronomy Department
Purdue University
West Lafayette, Indiana 47907

Dr. Peter H. McMurry
125 Mechanical Engineering
University of Minnesota
Minneapolis, Minnesota 55455

Dr. James F. Meagher
Air Quality Research Section
Tennessee Valley Authority
E & D Building
Muscle Shoals, Alabama 35660

Dr. John M. Miller
Air Resources Laboratory
National Oceanic and
 Atmospheric Administration
8060 13th Street
Silver Spring, Maryland 20910

Dr. Thomas H. Moss
U.S. House of Representatives
2342 Rayburn Building
Washington, D.C. 20525

Dr. John S. Nader
2336 New Bern Avenue
Raleigh, North Carolina 27610

Dr. Leonard Newman, Head
Environmental Chemistry
 Division
Department of Energy and
 Environment
Brookhaven National Laboratory
Upton, Long Island,
 New York 11973

Dr. J C Noggle
Tennessee Valley Authority
Muscle Shoals, Alabama 35660

Dr. Steven Norton
Department of Geological
 Sciences
Boardman Hall
University of Maine
Orono, Maine 04473

Mr. Geoffrey Parker
Department of Environmental
 Sciences
Clark Hall
University of Virginia
Charlottesville, Virginia 22903

Dr. Dennis C. Parzyck
Oak Ridge National Laboratory
Building 4500S
P.O. Box X
Oak Ridge, Tennessee 37830

Dr. David E. Patterson
CAPITA
Washington University
Box 1185
St. Louis, Missouri 63130

Dr. Chester R. Richmond
Oak Ridge National Laboratory
Building 4500N
P.O. Box X
Oak Ridge, Tennessee 37830

Dr. David W. Schindler
Department of the Environment,
 Fisheries, and Marine Sciences
Freshwater Institute
501 University Crescent
Winnepeg, Manitoba
Canada R3T 2N6

Dr. Carl L. Schofield
Department of Natural
 Resources
Cornell University
Ithica, New York 14850

Dr. G. A. Sehmel
Atmospheric Sciences Section
Pacific Northwest Laboratories
Richland, Washington 99352

Dr. David S. Shriner
Oak Ridge National Laboratory
Building 1505
P.O. Box X
Oak Ridge, Tennessee 37830

Dr. Frank E. Speizer
180 Longwood
Harvard Medical School
Boston, Massachusetts 02115

Dr. John M. Skelly
Department of Plant Pathology
 and Physiology
Virginia Polytechnic Institute
 and State University
Blacksburg, Virginia 24061

Dr. Wayne T. Swank
U.S. Forest Service
Coweeta Hydrologic Laboratory
P.O. Box 601
Franklin, North Carolina 28734

Mr. David W. Tundermann
Office of Policy and Planning
PM 221
Environmental Protection
 Agency
401 M Street, SW
Washington, D.C. 20460

Dr. John Turner
Wood Technology and Forest
 Research Division
Forestry Commission of
 New South Wales
P.O. Box 100
Beecroft, New South Wales
 2119
Australia

Dr. Philip J. Walsh
Oak Ridge National Laboratory
Building 4500S
P.O. Box X
Oak Ridge, Tennessee 37830

OTHER PARTICIPANTS

Anders W. Andren
Water Chemistry Program
University of Wisconsin
Madison, Wisconsin 53706

Pal S. Arya
Department of Geosciences
North Carolina State
 University
Raleigh, North Carolina 27650

Stanley I. Auerbach
Environmental Sciences
 Division
Building 1505
P.O. Box X
Oak Ridge, Tennessee 37830

John Bachmann
Environmental Protection
 Agency
MD-12
Research Triangle Park,
 North Carolina 27711

David S. Ballantine
Office of Health and
 Environmental Research
MS E-201, Germantown
Department of Energy
Washington, D.C. 20545

Jill S. Baron
Office of Science & Technology
492 Air & Water
National Park Service
1100 L Street, NW
Washington, D.C. 20240

Charles J. Barton
Science Applications, Inc.
800 Turnpike
Oak Ridge, Tennessee 37830

David Balsillie
Ontario Ministry of
 Environment
Sudbury, Ontario
Canada

Jim Behrmann
California Air Resources Board
P.O. Box 2815
Sacramento, California 95812

Alan R. Biggs
211 Buckhout Lab
Pennsylvania State University
University Park,
 Pennsylvania 16802

Gail E. Bingham
Lawrence Livermore Laboratory
L-524
P.O. Box 5507
Livermore, California 94550

Charles A. Bleckmann
Research and Develoment
 Department
CONOCO, Inc.
Box 1267
Ponca City, Oklahoma 74601

Donald C. Bogen
Environmental Measurements
 Laboratory
376 Hudson Street
Department of Energy
New York, New York 10014

Patrick L. Brezonik
Department of Environmental
 Engineering Sciences
University of Florida
Gainesville, Florida 32611

Eugenia E. Calle
Health and Safety Research
 Division
Oak Ridge National Laboratory
P.O. Box X
Oak Ridge, Tennessee 37830

Norbert C. Chen
Oak Ridge National Laboratory
Building 9204-1
P.O. Box Y
Oak Ridge, Tennessee 37830

R. O. Chester
Health and Safety Research
 Division
Oak Ridge National Laboratory
P.O. Box X
Oak Ridge, Tennessee 37830

Boris I. Chevone
Department of Plant Pathology
University of Minnesota
St. Paul, Minnesota 55108

Chris Chouteau
Pacific Gas and Electric
3400 Crow Canyon Road
San Ramon, California 94583

Jan M. Coe
Oak Ridge National Laboratory
Building 1505
P.O. Box X
Oak Ridge, Tennessee 37830

Cynthia J. Cohen
Crop Science Department
Oregon State University
Corvallis, Oregon 97330

Charles E. Comiskey
Science Applications, Inc.
800 Oak Ridge Turnpike
Box 843
Oak Ridge, Tennessee 37830

Emily D. Copenhaver
Health and Safety Research
 Division
Oak Ridge National Laboratory
Building 4500S
P.O. Box X
Oak Ridge, Tennessee 37830

Kenneth E. Cowser
Oak Ridge National Laboratory
Building 1505
P.O. Box X
Oak Ridge, Tennessee 37830

Karen R. Darnall
Sandia Labs
4449 Kellia Lane, NE
Albuquerque, New Mexico 87111

Cliff I. Davidson
Department of Civil
 Engineering
Carnegie-Mellon University
Pittsburgh, Pennsylvania 15217

Donald D. Davis
Pennsylvania State University
211 Buckhout Lab
University Park,
 Pennsylvania 16802

Christine Eason
Graduate Program in Ecology
University of Tennessee
102 Tilson Drive
Knoxville, Tennessee 37920

Robert W. Elias
Virginia Polytechnic Institute
 and State University
Biology Department
Blacksburg, Virginia 24061

Susan L. Falvo
Edison Electric Institute
1111 Nineteenth Street, NW
Washington, D.C. 20036

Edward G. Farnworth
Institute of Ecology
University of Georgia
Athens, Georgia 30602

William A. Feder
Suburban Experiment Station
University of Massachusetts
240 Beaver Street
Waltham, Massachusetts 02154

Birney R. Fish
McDowell Cancer Network
400 Colonia Trace
Frankfort, Kentucky 40601

Robert M. Friedman
Office of Technology
 Assessment
U.S. Congress
Washington, D.C. 20510

Marian Gabriel
M. H. Gabriel and Associates
Argonne National Laboratory
680 W. 97th Street
Lemont, Illinois 60439

David Gardner
Virginia Polytechnic Institute
 and State University
RFD 2, Box 477
Newport, Virginia 24128

J.H.B. Garner
ECAO, MD-52
Environmental Protection
 Agency
Research Triangle Park,
 North Carolina 27711

James H. Gibson
Natural Resource Ecology
 Laboratory
Colorado State University
Fort Collins, Colorado 80523

Robert A. Goldstein
Electric Power Research
 Institute
P.O. Box 10412
Palo Alto, California 94303

William C. Graustein
Department of Geology &
 Geophysics
Yale University
2161 Yale Station
New Haven, Connecticut 06520

Jane E. Hagner
Institute of Ecology
Athens, Georgia 30602

Charles S. Hakkarinen
Electric Power Research
 Institute
P.O. Box 10412
Palo Alto, California 94303

W. Franklin Harris
Environmental Sciences
 Division
Oak Ridge National Laboratory
Building 1505
P.O. Box X
Oak Ridge, Tennessee 37830

Allen S. Heagle
Plant Pathology Department
USDA
North Carolina State
 University
Raleigh, North Carolina 27607

David J. Helvey
Timber and Watershed Lab
Box 445
Parsons, West Virginia 26287

Charles Ross Hinkle
Science Applications, Inc.
800 Oak Ridge Turnpike
Oak Ridge, Tennessee 37830

F. Owen Hoffman
Oak Ridge National Laboratory
Building 4500S
P.O. Box X
Oak Ridge, Tennessee 37830

Boyd A. Hutchison
ATDL/NOAA
P.O. Box E
Oak Ridge, Tennessee 37830

Patricia M. Irving
RER/203
Argonne National Laboratory
9200 S. Cass Avenue
Argonne, Illinois 60439

Danny R. Jackson
505 King Avenue
Columbia, Ohio 43201

J. William Johnston
Environmental Sciences
 Division
Oak Ridge National Laboratory
Building 1505
P.O. Box X
Oak Ridge, Tennessee 37830

Jon D. Jones
Department of Forestry
Southern Illinois University
Carbondale, Illinois 62901

Stephen V. Kaye
Health and Safety Research
 Division
Oak Ridge National Laboratory
Building 4500S
P.O. Box X
Oak Ridge, Tennessee 37830

Lance Whitaker Kress
RER/203
Argonne National Laboratory
9200 S. Cass Avenue
Argonne, Illinois 60439

Malcolm Ko
Atmospheric and Environmental
 Research, Inc.
827 Massachusetts Avenue
Cambridge, Massachusetts
 02139

Joseph Laznow
United Engineers and
  Constructors, Inc.
100 Summer Street
Boston, Massachusetts 02110

Edward H. Lee
Plant Stress Laboratory,
  USDA/SEA
Building 001, Room 225
Beltsville, Maryland 20705

Allan H. Legge
Kananaskis Center for
  Environmental Research
University of Calgary
Calgary, Alberta, Canada
T2N 1N4

W. E. Lotz
Electric Power Research
  Institute
1800 Massachusetts Avenue,
  NW
Suite 700
Washington, D.C. 20036

Orie L. Loucks
The Institute of Ecology
Holcomb Research Building
Butler University
Indianapolis, Indiana 46208

Philip D. Lowry
Environmental Sciences
  Division
Oak Ridge National Laboratory
Building 1505
P.O. Box X
Oak Ridge, Tennessee 37830

Maris Lusis
Air Resources Branch
Ministry of the Environment
880 Bay Street, 4th Floor
Toronto, Ontario M5S 1Z8
Canada

R. J. Luxmoore
Environmental Sciences
  Division
Oak Ridge National Laboratory
Building 1505
P.O. Box X
Oak Ridge, Tennessee 37830

Ron McConathy
Oak Ridge National Laboratory
P.O. Box X
Oak Ridge, Tennessee 37830

David J. McKee
Environmental Protection
  Agency
Research Triangle Park,
  North Carolina 27711

Samuel B. McLaughlin
Environmental Sciences
  Division
Oak Ridge National Laboratory
Building 1505
P.O. Box X
Oak Ridge, Tennessee 37830

Ronald A. N. McLean
Domtar Inc.
P.O. Box 7210, Station "A"
Montreal H3C 3M1
Quebec, Canada

J. Dain Maddox
USFS/Monongahela National
  Forest
Sycamore Street
Elkins, West Virginia 26241

David E. Maschwitz
Minnesota Pollution Control
  Agency
1935 West Co. Road, B-2
Roseville, Minnesota 55113

Ray Mathews
National Park Service
Uplands Field Research
  Laboratory
Twin Creeks Area
Gatlinburg, Tennessee 37738

James R. Newman
Environmental Science and
  Engineering
P.O. Box 13454
University Station
Gainsville, Florida 32604

Charles W. Miller
Health and Safety Research
  Division
Oak Ridge National Laboratory
Building 4500S
P.O. Box X
Oak Ridge, Tennessee 37830

Rufus Morison
ORD RD682
Environmental Protection
  Agency
401 M Street, SW
Washington, D.C. 20460

B. D. Murphy
Computer Sciences Division
Oak Ridge National Laboratory
P.O. Box X
Oak Ridge, Tennessee 37830

Charles E. Murphy
E. I. du Pont de Nemours
  and Co., Inc.
Savannah River Laboratory
Aiken, South Carolina 29801

Dave Odor
Public Service Indiana
1000 E. Main Street
Plainfield, Indiana 46168

Don W. Ott
Tennessee Valley Authority
1110 Chestnut Street Tower II
Chattanooga, Tennessee 37401

Danny L. Rambo
Northrop Services, Inc.
200 SW 35th
Corvallis, Oregon 97330

Robert M. Reed
Environmental Sciences
  Division
Building 1505
P.O. Box X
Oak Ridge, Tennessee 37830

David E. Reichle
Environmental Sciences
  Division
Oak Ridge National Laboratory
Building 1505
P.O. Box X
Oak Ridge, Tennessee 37830

Gilles Robitaille
Laurentian Forest Research
  Centre
P.O. Box 3800
Ste-Foy, Quebec G1V 4C7
Canada

Paul S. Rohwer
Health and Safety Research
  Division
Oak Ridge National Laboratory
Building 4500S
P.O. Box X
Oak Ridge, Tennessee 37830

Don A. Rolt
British Embassy
3100 Massachusetts Avenue, NW
Washington, D.C. 20016

Karen L. Rourke
Massachusetts Department of
Environmental Quality
Engineering
600 Washington Street, Rm. 320
Boston, Massachusetts 02111

Richard E. Saylor
Oak Ridge National Laboratory
Building 9204-1, MS-3
P.O. Box Y
Oak Ridge, Tennessee 37830

R. Kent Schreiber
U.S. Fish and Wildlife Service
NPPT
2929 Plymouth Road
Room 206
Ann Arbor, Michigan 48105

Roy A. Schroeder
U.S. Geological Survey
P.O. Box 744
Albany, New York 12201

Carole R. Shriner
Science Applications, Inc.
800 Oak Ridge Turnpike
P.O. Box 843
Oak Ridge, Tennessee 37830

Harald Siem
Norwegian Institute for
  Air Research
c/o I Nielsen
P.O. Box 130
N-2001 Lillestrom
Norway

Steven A. Silbaugh
Lovelace Research Institute
P.O. Box 5890
Albuquerque, New Mexico
87115

Wayne H. Smith
University of Florida
3038 McCarty Hall
Gainesville, Florida 32611

Lars F. Soholt
Argonne National Laboratory
9700 S. Cass Avenue
Argonne, Illinois 60437

Attila A. Sooky
Graduate School of Public
  Health
University of Pittsburgh
Pittsburgh, Pennsylvania 15261

R. D. Samuel Stevens
United Technology and
Science, Inc.
859 Magnetic Drive
Downview, Ontario M3J 2C4
Canada

Glen Suter
Environmental Sciences
Division
Oak Ridge National Laboratory
Building 1505
P.O. Box X
Oak Ridge, Tennessee 37830

Fred G. Taylor
Environmental Sciences
Division
Building 1505, Rm. 366
P.O. Box X
Oak Ridge, Tennessee 37830

William E. Thompson
Research Triangle Institute
50 Park
Research Triangle Park,
  North Carolina 27709

J. B. Truett
The MITRE Corporation
W 252
1820 Dolly Madison Boulevard
McLean, Virginia 22102

Ralph R. Turner
Environmental Sciences
Division
Oak Ridge National Laboratory
Building 1505
P.O. Box X
Oak Ridge, Tennessee 37830

David M. Umbach
The Pennsylvania State
University
208 Mueller Lab
University Park, Pennsylvania
  16802

James E. Vancil
Science Applications, Inc.
318 East Drive
Oak Ridge, Tennessee 37830

S. Dirk Van Hoesen
Oak Ridge National Laboratory
P.O. Box X
Oak Ridge, Tennessee 37830

John Viren
MS E-201 (Germantown)
Department of Energy
Washington, D.C. 20545

George T. Weaver
Department of Forestry
Southern Illinois University
Carbondale, Illinois 62901

David E. Weber
Environmental Protection
Agency
401 M Street, SW, RD-682
Washington, D.C. 20460

Richard J. Wedlund
Minnesota Pollution Control
Agency
2343 Cleveland Street, NE
Minneapolis, Minnesota 55418

Stephen O. Wilson
New York State Energy
  Research and Development
  Authority
Agency Building #2
Empire State Plaza
Albany, New York 12223

Robert O. Woods
Sandia Labs
ORG 4533
Albuquerque, New Mexico
  87187

James B. Worth
Research Triangle Institute
P.O. Box 12194
Research Triangle Park,
  North Carolina 27709

Gerald L. Zachariah
Agricultural Engineering
Department
9 Rogers Hall
University of Florida
Gainesville, Florida 32611

Bernard D. Zak
Environmental Research
Division-4533
Sandia Laboratories
Albuquerque, New Mexico
87187

Larry Zaragoza
Environmental Protection
Agency
Mail Drop 12
Research Triangle Park,
  North Carolina 27711

INDEX

acid deposition 533
acid precipitation 15,48,245,299,378
    384,453,469,521,539
acid sensitive soils
    See soils
acidified lakes 345,369,453
    clarity of 369
    transparency of 369
adsorption, inorganic sulfate 507
Aerochem Metric collector 470
aerodynamic roughness 228
aerosols 136,210
    ammonium 137,145
    charger 203
    coagulation 157
    condensation 55,56,154
    crustal 255
    deposition velocity 224,225,229
    dynamics 138
    emission 205
    formation 159
    growth 164
    hygroscopic 238
    nitrate 137
    residence time 163
    resuspension 205
    sampling 247
    sea salt 57
    secondary 154
    scavenging 264,268,272
    size distribution 71,125,154,272
    stratospheric 52
    sulfuric acid 145-152
    suspension ratio 279
    visibility effects 163-176
    water soluble 145
agroecosystems 377
air monitoring 99
air pollution control 13
Air Quality Control Regions (AQCR)
    7,123,163
air stagnation events 540
    Also see stagnation
airway resistance 89
aitken nuclei 136
alfalfa 289
algae 458

alternative standards 8,9
    full scrubbing 8
    partial scrubbing 8
    regional 8
    sliding scale 8
    variable 8
aluminum 345
    discharge 439
    in plant tissue 377
    leaching 335
    mobilization 337,345,443
    toxicity of 345,357
    transport 335
ambient standards 9
ammonia neutralization 78
antagonism (Al^{+++} to H$^+$ toxicity)
    353
aquatic ecosystem
    vulnerability maps 524
artificial surfaces 237,238,378
asthmatics 81,82,86
Atlantic salmon 536
atmospheric stability 225
atomic absorption spectrophotometry
    (flameless) 247
attack rate 113

Bacillariophyceae 361
barley 379
base content 539
base saturation 406,539
bedrock 432,521
    geology 521
    weathering 432
beneficial effects (SO$_2$) 291
benzo (a) pyrene 85,86,91
Best Available Control Technology
    (BACT) 17,18
Betula pubescens 410
bicarbonate
    buffering 345
    alkalinity 353
biogenic sulfur 47-53
    climatic factors 51
    emission rates 50-53
    global flux 56

marine flux   51
marsh emissions   53,54
seasonal effects   39
soil emissions   36-44
soil order effects   39,40
sources   35,48
temprature effects   39
terrestrial flux   40,41,51
tidal effects   53-55
birch   410
boundary layer   189
Brassica rapa   379
bronchitis   90
experimental   69,72
bronchial clearance half-time (TB$_{1/2}$)   88
broncho constriction   69,70
brook trout   346,445
Brownian diffusion   228
budget studies
crops   186
ecosystem   507
global sulfur   51
receptor uptake   210
buffering capacity   522

cabbage   289
calcite   523
calcium mobilization   443
Canadian Precambrian Shield   453,533
canopy
absorption   384
element cycling   380
emission   205
filtering ability   388
forest   191,377-390
leaching   379-390
loblolly pine   199,201
resistance   191
response to rain   380-382
resuspension   205
surface area   230
wash-off   388-389
wet   192
carbon disulfide   39-41,44,48,49,51,52,
56,61
production of SO$_2$ 61
reaction with OH   60,61
carbon flow   417
carbonyl sulfide   39,41,44,48-53,56,58,
59,61,210
generation of   60
ozone influence   58
production of SO$_2$ 60
reaction with OH   58-61
carcinogenesis   86
cardiopulmonary function   89
catalyst depletion   133
cation leaching   378,431,433,508
in soils   378,397

Catostomus commersoni   355
CEC   495,539
Chamaenerion augustifolium   410
char   536
CHESS study   23,112
chickweed wintergreen   410
Chlorella   458
chlorophyll   419
chlorophyta   361
chronic respiratory disease   86
cigarette smoking   109
Clean Air Act   6,14,22
of 1964   14
1967 ammendment   14
1970 ammendment   14
1977 ammendment   6,14
Section 111   6
clearance
bronchial   87
mucociliary   87
clover   394
clinical studies   77,86
coal
Electric Utilities Model   18
high sulfur   8
market   7
prices   7
coal combustion
electricity generation   127
emissions   125,128,310,378
plume studies   132,210,310
power plants   309,539
cocarcinogen   85
conifer needles   407
nutrient content   407
consumers   421
grasshoppers   421
corn   289,379
cost-benefit analysis   6,20,21,23,26
cost-effectiveness analysis   6
cotton   289,293-295,394
coughing   87,90
Covesius plumbeus   355
Chrysophyta   362
crayfish   458
crops
argonomic   394
economic   396
horticultural   394
nutrition   289
sulfur balance   186
wheat   189
Cryptophyta   361
cucumber   379
Cucumis sativus   379
cultivated soils   377
cumulative impacts   418
cuticular absorption   184
Cyanophyta   362
cyclone collectors   126
cyprinids   536

decomposition 457
  microarthropods 423
  microbial in lakes 367
deposition
  bare soil 192
  dry 470
  forests 190,377-390
  snow 191
  spatial variations 377,469
  temporal variations 377,469
  wet
  See wet deposition
deposition velocity 186,199
  diurnal variation 191
  gases 226-228
  model 228
  particles 224,225,229
  particulate sulfur 213,214
  prediction 228
  SO_2 199,210
  soil 193
  snow 192
  waves, effects 238
  water surfaces 239
dew 191
Dicranum spp 400
"Die Away" method 190,193
dimethyl sulfide 39,49-52
diurnal variations 191,201,202,204
  ammonia 149
dry deposition 309,377,410
  Also see deposition, deposition
  velocity
Duke Forest 201

economic comparisons 20
economic growth 5
ecosystem sulfur budgets 507
eddy correlation 241
eddy flux 212
eddy diffusivities 200,212,228
  momentum 215
Edison Electric Institute 18
education level 100
effective dose 86
electricity
  demand growth 7
  generation 13,17
  industry 13
  supply optimization model 18
electrostatic precipitator 125,137
embryo 459
  deformity 459
  mortality 459
emerging benthos 458
emission factors 125,126
  biogenic sulfur 50-53
  coal 128
  industrial 126
  oil 128
  residential 126

sulfate 129
emission flux reactor 48,49
  chamber 38
emissions ceiling 9
emphysema 86
energy policy 5
environmental
  policy 5
  protection 5
  regulation 23
  risk 13
EPA
  emissions report 123
epidemiologic studies 109
epilimnion 455
epilithiphyton 458
EPRI 13
euphotic zone 369
evapotranspiration 403
Executive Order 12044 6

factor analysis 247
fertilization 291,425,497
filter pack 132
fish
  embryological effects 453
  mortality 345,346
  physiological effects 453
  populations 345
  stress symptoms 350
  survival 345
fisheries 443
flue gas desulfurization (FGD) 18
fly ash 126,134
foliar pathogen 324
forced expiratory flow (FEF) 81
forced expiratory volumes (FEV) 80,100
forced vital capacity (FVC) 80,100,113
forests
  canopy 402
  coniferous 191
  cutting 433
  ecosystems 339,397
  evapotranspiration 191
  loblolly pine 199,201
  management 329
  nutrition 401
  oak-hickory 378-390
  pine 212
  rain interception 402
  soil 335
  watersheds 431
Fourier analysis 110
friction velocity 204
fulvic acids 515
fungicide sprays 389
  containing sulfur 389

GAMETAG 56
gas chromatography 38,48-50

gas cooking   105
gas-particle conversion   135,166,379
    homogeneous   136,154
Geoecology Data Base   540
geologic factors   521
gill
    filaments   351
    lamellae   351
    structure   349
glacial till   502
gradient method   190,240,241,387
Gran titration   456
granitic bedrock   449,521
grassland ecosystems   417
gravitational settling   224,228,275,276
Glycine max   293,379
    See soybean
Gossypium hirsutum   289,293-295,394

HASL collector   378
haze   55,58,172
health effects   23,24,78,99
    acute exposures   69,70,71
    confounding factors   109
    chronic exposures   72
    cigarette smoking   109
    excess mortality   110
    toxicity   71
heart disease   111
heterogeneous reactions   134,154
    mechanisms   137
homogeneous reactions   131,136,154
    mechanisms   137
Hordeum vulgare   379
House Commerce Committee   7
human populations   86
humic substances   368
hydrogen ion loading   539
hydrogen sulfide   39,41,44,49-52,56,210,
    291
hypolimnion   457
    oxygen deficit in   457
    sulfate reducers in   457

infection   73
inflation   13
inflationary impact   6
Integrated Lake Watershed Acidification
    Investigation (ILWAI)   258
invertebrate   447
    drift   445
    functional groups   447
iodine   224
    scavenging   269
ion separation   350
    during thaws   350
irritant aerosol   71

Kingston-Harriman, Tennessee   99-107

laboratory animals   69-73
lake chub   355
Lake Michigan   238
lake trout   355,458,536
lakes
    acidification   453,533
    alkalinity   453,535
    bicarbonate buffering   345
    conductivity   535
    experimental acidification   453
    transparency   458
land use   495
leaf
    abscission   306
    expansion   306
legume   300
lime   385,497
    treatment with   400,534
limestone   521
lodgepole pine   409
London fog   186
lung
    diseases   93
    fibrosis   86
Lycopersicon esculentum   379
lysimeter studies   397,507

MAP3S   173,551
Maianthemum bifolium   410
marine atmosphere   56,57
    remote   59,60
maritime sulfur inputs   321
mass respirable particles (MRP)   100,101
may lilly   410
metals
    See trace metals
methyl mercaptan   39,49
methyl sulfide   210
METROMEX   245-259
microcosms   339
midmaximal expiratory flow (MMEF)
    110
modeling
    air parcel trajectory   165
    Chamberlain   276,277
    Csanady   280
    deposition   275-282
    diagnostic transport   163-177
    Gaussian diffusion   275-282
    Horst   280
    linear programming   7
    Monte Carlo trajectory   164-177
    photochemistry   138
    plume   275-282
    regional scale   164
    source depletion   276,277,280
    surface depletion   277
morbidity   111
mortality   111
moss   400
Mougeotia   458

mucus discharge 87
municipal incinerator 254
Mysis relicta 458

National Ambient Air Quality Standards
(NAAQS) 14-17,22
National Atmospheric Deposition
Program (NADP) 25,469,551
National Energy Plan 19,540
National Rural Electric Cooperative
Association 18
nephelometer 241
New Source Performance Standard
(NSPS) 7,14,15,17
New England 166
nitrate 137
nitric oxide 137,227
scavenging 269
nitrification 436
nitrogen
deficiency 323
deposition of 398
discharge 433
fixation 420
mineralization 412
requirement 289
nitrogen: sulfur relationships 321,425
non-attainment area 15
non-compliance penalties 15
Norway spruce 409
nutrient availability 398
in lakes 357
in soils 398
nutrient cycling 417
nutrient discharge 431
nutrient leaching 339
NWS sites 164

oasis effect 189
occupation 100
oil combustion 138
emissions 128
sources 126
oil prices 7
oligotrophic lakes 357,453
Orconectes virilis 458
osmoregulatory stress 346
effect of calcium on 346
gill permeability 346
sodium imbalance 346
oxidation rates 132,133
liquid phase 269
modeling 138
oil plume 133
photochemical 134
ozone 210
bulge 136
flux 212
SO_2 synergism 78,79
uptake 202

Pacific Ocean 56
parent material
See soil
particles
See aerosols
peach 380
Perca flavescens 355
Phaseolus vulgaris 301,379
phlegm production 102-104
phosphorus 289,368
aluminum complexing of 368
requirement 289
photosynthesis 419
physiological stress 345
Also see fish
phytoplankton 357,458
biomass 357
chlorophyll a 357
density of 368
nannoplankton 357
net plankton 357
productivity 363
ultraplankton 357
pinto bean
growth rates 302
seed yield 301
See Phaseolus vulgaris
plant
growth 377
nutrients 497
productivity 399
plume *Also see* power plant, coal, oil,
smelter
BNL studies 137
depletion 224
fringe reactivity 136
Gaussian diffusion 275-282
Labadie studies 189
studies 132-139
podsolized soils 337
Spodosols 338
pedogenesis 336
policy alternatives 8
pollutant mixtures 77
pollution tax 23
Polytrichum commune 400
population structure 345
Portage, Wisconsin 99-107
potassium requirement 289
power plant
coal 7,125,128,210
oil 126,128,138
plume studies 132-139
precipitation 245,263,377
Also see wet deposition
canopy response 380-382
rate 265
relationships with deposition 254
sampling 246,378,386
sequential sampling 386
spatial variability 250
sulfur in 377

volume-concentration relationships
380
precipitation scavenging 264-272
aerosols 264-268
below cloud 267
coefficients 267
gases 269,272
in cloud 259,266
mechanisms 257
nucleation 258,267
ratios 270
SO$_2$ 258
washout 268
Prevention of Significant Deterioration
(PSD) 15,16,18-20
Class I area 15
Class II area 15
Class III area 15
primary producers 418
in lakes 357
net primary production 418
profile studies 212,213
particle 215,216
PNL 213
silicon 216
Prunus persica 380
pulmonary
function 69,113
irritation 70
lesions 72
resistance 69,70
Pyrrhophyta 361

radioactive fallout 237
rainfall
See precipitation
RAPS 258
receptor uptake budgets 210
refugia 354
regional ecological assessment 539
regional patterns 507
regional transport 313
air mass trajectory analysis 313
climatological dispersion model 310
regulation, benefits of 6
Regulatory Analysis Review Group 7
reproductive inhibition 345
Also see fish
resistances
aerodynamic 200,239
atmospheric 212
canopy 191
cuticular 190
diurnal 201,202,204
stomatal 190,202
surface 186,202,212,239
respiratory epithelium 346
aluminum toxicity to 346
respiratory mechanics 88,89
respiratory symptoms 80,100
respiratory tract

infectivity 86
risk assessment 20
rose-bay willow herb 410

Savelinus namaycush 355,458,536
satellite imagery 172
sandstone 522
salt marsh 48-53
salmon 536
scavenging
See precipitation scavenging
Scots pine 409
scrubbing of emissions 6
costs 7
desulfurization 126
dry 9
wet 9
secondary pollutants 15
seed yield 306
flowering 306
pod development 306
pod set 306
pollination 306
sesquioxide surfaces 508
Six City Study 99-107,114
smelter 129
copper 254
plumes 135
smoking habits 100
Smokey Mountains 156
smog chamber 155-158
snow deposition 192
snow melt 335,345,403,450
episodic acidification 345
snapbean 379
softwood forests 541
soil 397,499
aluminum losses 496
aluminum oxide 514
associations 498
base saturation 496
base saturation 496
biogenic productivity potential 37
buffer capacity 496
carbonate 495
chemical properties of 397
classification 499
depth 514
emission of gases 36-44
flooding of 495
iron oxide 514
lime potential 496
microbiological processes 496
nutrient leaching 496,507
nutrient status 377,508
organic matter 418,499,507,508,514
parent material 499
pH 539
sulfate adsorption 406,507
systems 539
texture 386

uptake of gases 193
weathered 500
soil acidification 543
influence of sulfur on 377
soil sensitivity 495
definition of 497
solution kinetics 522
soybean 293,379,541
growth rates 305
seed yield 303
specific activity 293
Spodosol 507
sport fisheries 533
stack height credits 15
stagnation, air 474
Standard Metropolitan Statistical Area
(SMSA's) 111,112
standard setting 7
State Implementation Plans 15,22
stationary sources 123
Steubenville, Ohio 99-107
stirred chamber 193,210
St. Louis, Missouri 99-107
precipitation network 246
urban plume 156
stomata 189
resistance 190,202,210
response studies 210
stream acidification 443
stream chemistry 431
streamflow 432
base flow 432
direct channel input 432
surface storm flow 432
subsurface storm flow 432
strong acid 137,353
concentration of 137,353
snowpack storage of 137,353
sulfate adsorption 507
accumulation in acid subsoils 508
pH dependent 508
sulfate-reducing bacteria 457
sulfur
accumulation 507
balance, crops 186
deficiency 291,321
incremental emissions cost 5
incremental emissions damage 5
isotopes 133,186,187,293,294
nutrition, forests 321
requirement 289,321
toxicity 321
uptake efficiency 290
sulfuric acid 55,60,78,125,135,145
phase diagram 146,147,151
sulfuric acid aerosol 78,145-152
fate 146,151
reaction with ammonia 145-152
reaction with water vapor 145-152
SURE 36,164-176
surface flux 209,212,213
surface waters, acidified 522

surrogate surfaces
See artificial surfaces

throughfall
asymptotic concentration 382-384,
386
deposition rates 380,384-390
emission source proximity effect
384-390
enrichment in 378-384,410
flux 379-384
leaching 384-390
net 378-384
relationship to SO_2 levels 388
seasonal trends 379-381
sources of elements 382,384-390
species effects 384
volume-concentration relationship
380
titration curve 345
tomato 379
Topeka, Kansas 99-107
toxic metals 345
trace metals
arsenic 135
cadmium 254
catalysts 269
copper 135
crustal elements 255
dry deposition 247
iron 217
lead 136,254
manganese 388
scavenging mechanisms 257
uptake of 257
vanadium 126
wet deposition 245-259
zinc 255,388
tracheal mucus transport rate (TMTR)
87
Trientalis europa 410
Trifolium repens 394
Triticum aestivum 394
trophogenic layer 369
turnip 379

urban plume 310
ozone episodes 310
urea fertilization 436
utermöhl technique 359
Utility Air Regulatory Group (UARG)
13,18
utility boilers 6

vegetation 313
biomonitoring 316
canopy 432
effects of atmospheric sulfur 397

injury 313
photosynthesis 313
stress 315
transpiration 313
visibility
contours 164
impairment 15
protection of 15
regulations 10
volcanic emissions 59

water chemistry 455
water quality 431
dissolved minerals 521
temporal fluctuations 345
water surface 237
deposition to 237
exchange with atmosphere 238
particle flux 241
watershed 336,359,377,453,534
Camp Branch 378
Cross Creek 378
sulfur budgets 513
Virginia 378
Walker Branch 378
Watertown 99-107

wet deposition 245,263,377,397,470
Also see precipitation
crustal elements 255
factor analysis 247,248,249
incinerator influence 254
location effect 385,386
mechanisms 257
rain volume relationships 254
smelter influence 254
soil elements 247
spatial variability 250
trace metals 254,255
urban influence 255,257
wheat 289,394
wheezing, rate 90,103
Whitby Aerosol Analyzer 158
white sucker 355,458
wind tunnel 210,228

x-ray fluorescense (XRF) 214,216,247

yellow perch 355

Zea mays 289,379
zooplankton 458